Machine Shop Essentials: Questions and Answers

Second Edition

Frank M. Marlow, P.E.

Illustrations by
Pamela Tallman

Metal Arts Press
Huntington Beach

Metal Arts Press
www.MetalArtsPress.com
8461 Valencia Drive
Huntington Beach, CA 92647-6033
USA

Machine Shop Essentials: Questions & Answers
Second Edition
ISBN-13 978-0-9759963-3-1

Printed and bound in the United States of America.

Cover art by Jeremy Long

10 9 8 7 6 5 4 3 2 1

In Memoriam

Connie Louise Randal

1948–2004

Without whose help this book
would not have been possible

Books by Frank Marlow

Machine Shop Essentials: Questions & Answers

Welding Fabrication & Repair: Questions & Answers

Welding Essentials: Questions & Answers

Contents

Acknowledgements

The author acknowledges the assistance of the following individuals, companies and organizations that made contributions to this book:

Clausing Industries, Inc.
Cooper Industries, Inc.
Robert J. DeVoe
Emhart Inc.
Neil Gitter
Hardinge Inc.
International Fastener Institute
Jacobs Chuck Manufacturing Company
William Johnson
Kent Industrial (USA) Inc.
KEO Cutters
Guy Lautard
Craig Libuse
Louis Levin & Son, Inc.
L.S. Starrett Company
Morse Cutting Tools
Eddie Torres
Myford Ltd.
Phase II Machine & Tool, Inc.
Connie Randal
David Randal
Phil Samuels
Sherline Products Inc.
Bernie Wasinger

Introduction

Machine Shop Essentials covers the use of manually-controlled metal lathes, milling machines and drill presses to make one-of-a kind parts, prototypes and industrial models and to modify and repair existing equipment.

Although NC machines dominate today's production environment, manually controlled machine tools are indispensable to R&D labs, tool and die makers, industrial model makers, scientific instrument makers, prototype designers, auto racers, custom motorcycle and car builders and gunsmiths.

When just a few parts or modifications are needed, manual machines can do the job faster than an expert can write the NC code. Also, manual machines are the best way to learn machine tool basics before going on to NC machines.

Every day valuable engineering developments are made by machinists who, using their ability to make and modify devices, turn a matchbook sketch into a practical and elegant working model. This back-to-basics book will help you accomplish this too.

Machine Shop Essentials includes:

- A simple question-and-answer format.
- Material divided into small, easy-to-understand blocks.
- Over 500 clear, concise drawings.
- Step-by-step instructions for common operations on the lathe, drill press and milling machine, including typical problems and their solutions.
- Short cuts, specialized tools and expert tips on indispensable shop-made tools that will quickly expand the user's capabilities.
- The use of machine tool accessories to simplify making complex parts.
- Screw threads and non-threaded fasteners.
- How to incorporate purchased components such as bearings, gears, snap rings and roll pins to quickly make sophisticated and durable devices.
- How to heat-treat steel in the machine shop.
- A review of basic precision measuring and marking tools and methods.
- General safety issues and special precautions for each tool.
- Guidelines for avoiding metal fatigue failures.
- Cutting, drilling and shaping plastic, rubber and glass are also covered.
- Detailed instructions for frozen or broken tap and fastener removal and damaged thread repair methods.

Machine Shop Essentials is written for a machine shop equipped with a drill press, lathe, milling machine, vertical band saw, bench grinder and disk grinder. Several hand-held power tools are also needed. They include a ⅜-inch variable-speed reversible electric drill, a die grinder, (Dremel®-type grinder, pencil-style electric- or air-driven die grinder) and a Sawzall®-type reciprocating saw.

This book does not cover commercial production where time, tooling and material costs are paramount. These considerations are not important when making a prototype, model or replacement part. For this reason the book covers carbide insert cutters and HSS toolbits and does not go into cutting tools with exotic, wear-resistant coatings. These tools, though great for production, are rarely justified in the prototype shop.

Two men made major contributions to this Second Edition and deserve special recognition. David Randal, a machinist, mold maker and welder with nearly 40 years experience, answered hundreds of questions. He often provided shop demonstrations to illustrate his points. Robert J. DeVoe, a retired professor of production management and a world-class machinist and tool collector, spent many days with me reviewing the book line-by-line, pointing out the additions and refinements needed. His extensive shop and its machine tools were the basis for many illustrations. These men also brought their collective—and often undocumented—hands-on experience and practical methods that can only be obtained by working with other master machinists. I am thankful for their generous help. Naturally, I remain responsible for all remaining errors.

HUNTINGTON BEACH, CALIFORNIA — FRANK MARLOW
MARCH, 2008

Chapter 1

Measurement Tools, Layout & Job Planning

Where observation is concerned,
chance favors the prepared mind.
—Louis Pasteur

Introduction

Fewer than twenty-five different tools are required to perform basic machine shop layout and measurement. Many of the hundreds of other tools are merely special-purpose versions of these that speed, simplify or improve the accuracy of a particular task. We will examine these basic twenty-five, see what they do and how to use them. Each machinist will add more tools to these based on his specific needs.

We will also cover the use of digital slide calipers, which for many jobs are easier and faster than traditional vernier calipers and their cousins, inside and outside calipers. Toolmakers' buttons and gage blocks, common to high-precision work, are also presented.

We will then look at layout methods, and finally, we will present work planning rules. Although these rules are not universally valid, they can often prevent problems.

Section I – Basic Measurement Tools

Measuring & Marking Tools for Layout

What are the *essential* measuring and marking tools and how are they used?

- *Layout fluid* puts a deep-blue background on the workpiece so scribed layout lines and punch marks stand out sharp and clear. It is applied to clean, dry metal from a brush or spray can and dries quickly. Remove it by wiping it off with denatured alcohol or acetone on a rag. Tip: Using a blue felt-tip marker which is transparent instead of layout fluid is often more convenient when a small area or single point must be scribed. Using layout fluid minimizes the depth needed to make scribe lines visible, which is very desirable because scribe marks are a source of fatigue failure on parts subject to cyclic loads and vibration.

- *Scribers,* Figure 1–1, apply scratch marks to work indicating the position of holes, openings and cut lines. Good scribers have hardened and finely tapered points so they can get close to the rule or straightedge, minimizing errors. The last 0.030 inches (0.8 mm) of the scriber should be sharpened to a 60° point by spinning it rapidly in a lathe and tapering it with a flat oilstone. Premium scribers have carbide points. To protect scriber points from becoming blunted, reverse them in their holder or store them inside a drilled-out dowel.

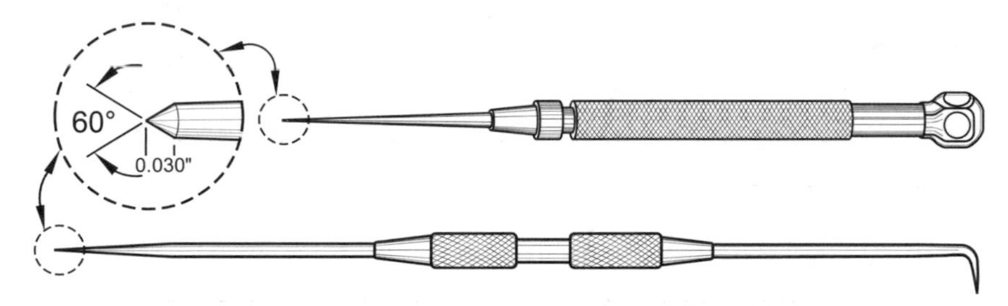

Figure 1–1. Starrett scribers: No. 70 is available with a steel or carbide point which stores reversed in its handle (top) and No. 67 with a steel point (bottom).

- *Straight edges* are useful for scribing straight lines and testing surfaces for flatness. They can also be positioned and clamped in place (Vise-Grip® welding clamps work well for this) then used as a guide for cutting with a utility knife. This method is useful in precision cutting of cardboard, gasket materials and foils. Clamped straight edges also work well for scoring deep lines on acrylic sheet goods so they can be snapped along the score line, somewhat like glass. Because rulers could be damaged when used as cutting guides, straight edges are preferred for cutting. Use an oilstone to remove small nicks. Straight edges may be reground if damaged. See Figure 1–2.

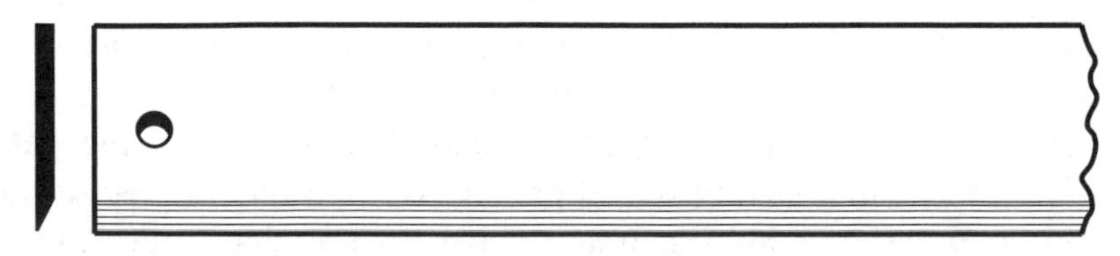

Figure 1–2. Stainless steel straight edge.

- *Steel rules,* Figure 1–3, have many uses and are a shop essential. They can be used to measure distances directly on work, to set dividers or surface gages or as straight edges. There are dozens of sizes and styles from 1 to 144 inches and from 150 to 1000 mm. Precision rules are steel or stainless steel, not aluminum. Their edges are ground, not punched. Because they are machine-divided, not photo-engraved or silk-screened, the tips of dividers can be felt dropping into their scribed division lines. Tip: When using a rule for accurate measurements, the eye can read better by measuring between two lines rather than from the end of the rule to a line.

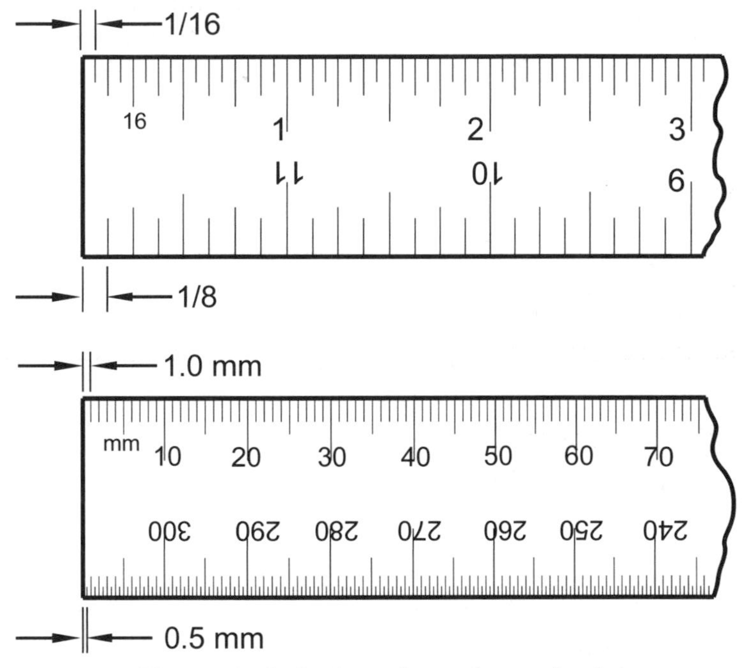

Figure 1–3. Inch and metric steel rules.

- *Prick punches and center punches,* Figure 1–4, are needed for fine layout work. Use the prick punch to locate the exact point on the work and strike it with a hammer. If the point is located at the junction of two scribe marks, the prick punch can be used to *feel* their crossing point. Now that the point has been marked, use the center punch to enlarge and deepen it. The larger punch mark gives the drill a better start. Both punches must be reground when their points dull. Their points can be sharpened by holding the punches tangent to the grinding wheel and rotating them to apply a new cone-shaped point. Sharpening them this way puts the sharpening scratches from the grinding wheel parallel to the punching action. Holding the punch *parallel* with the axis of the motor shaft puts the grinding scratches at right angles to punching action and reduces punch depth.

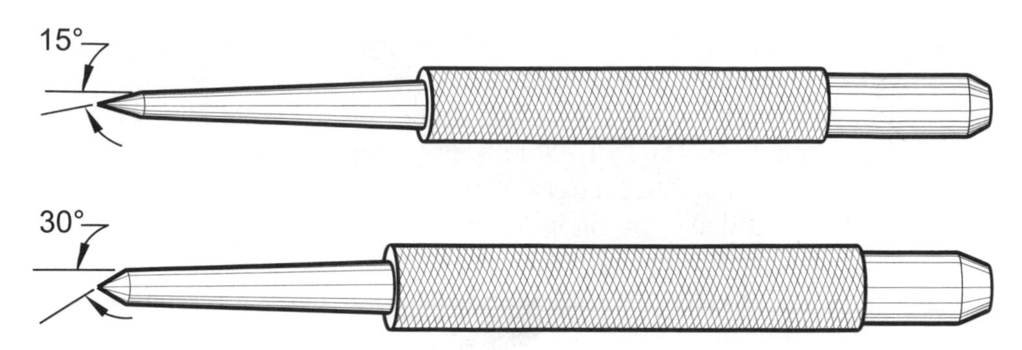

Figure 1–4. Prick punch (top) and center punch (bottom).

- *Machinists' combination squares*, Figure 1–5, are versatile tools. They measure right angles, have a protractor head for other angles and a center-finder. They can be used to scribe a line either perpendicular or parallel with the edge of the work. Common sizes have rulers from 4 to 24 inches in length, but most machinists will need both the 4-inch and the 12-inch models. Many models have an integral level and scriber, handy features. Square heads are offered in both cast iron, or the better and more expensive, forged and ground steel. Figure 1–6 shows how a center-finding head works.

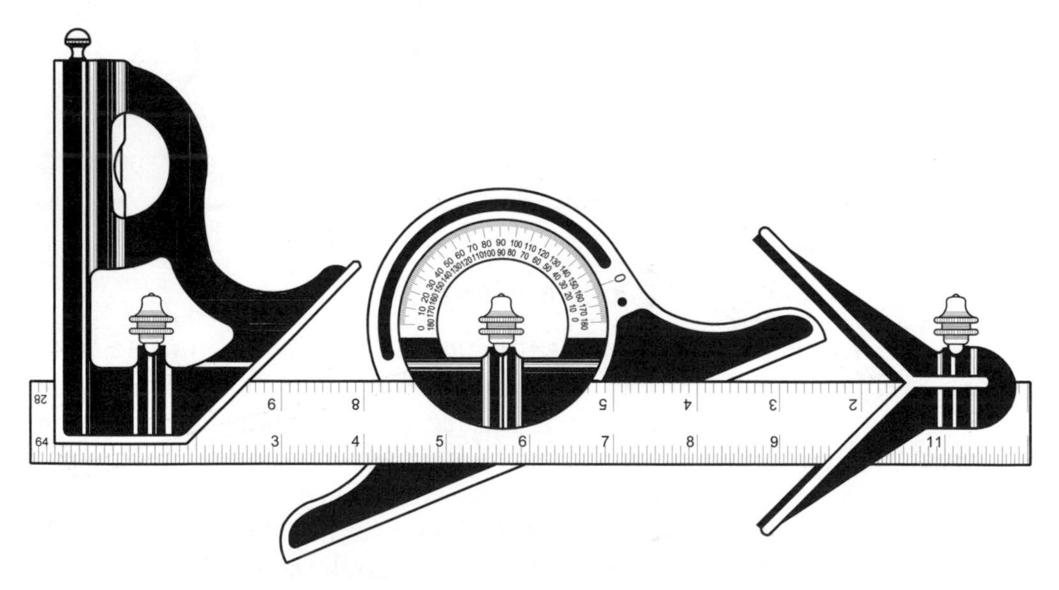

Figure 1–5. Combination square set with right-angle head, angle gage and center-finding head. There is also a scriber and level on the right-angle head.

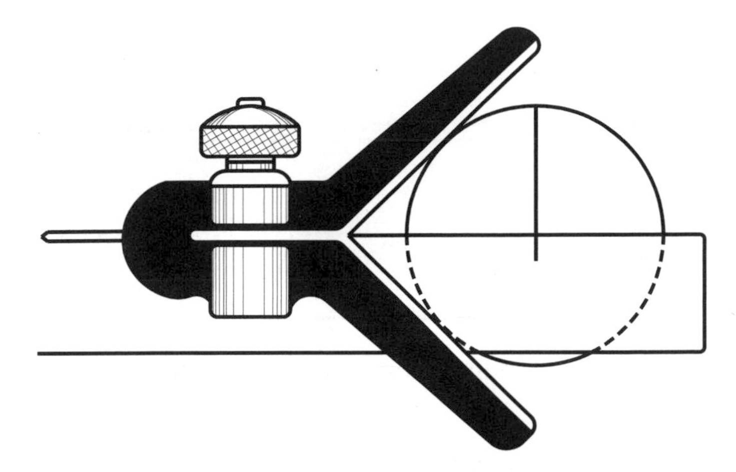

Figure 1–6. To find the center of a cylinder using the center-finding head of a combination square, scribe two perpendicular lines.

- *Spring dividers* scribe circles, arcs and mark off equal increments along a line, Figure 1–7. They are commonly supplied in lengths from 4 to 12 inches. Sharpened divider points are needed to locate scriber marks and ruler divisions by feel. Trammels are needed for larger circles and arcs.

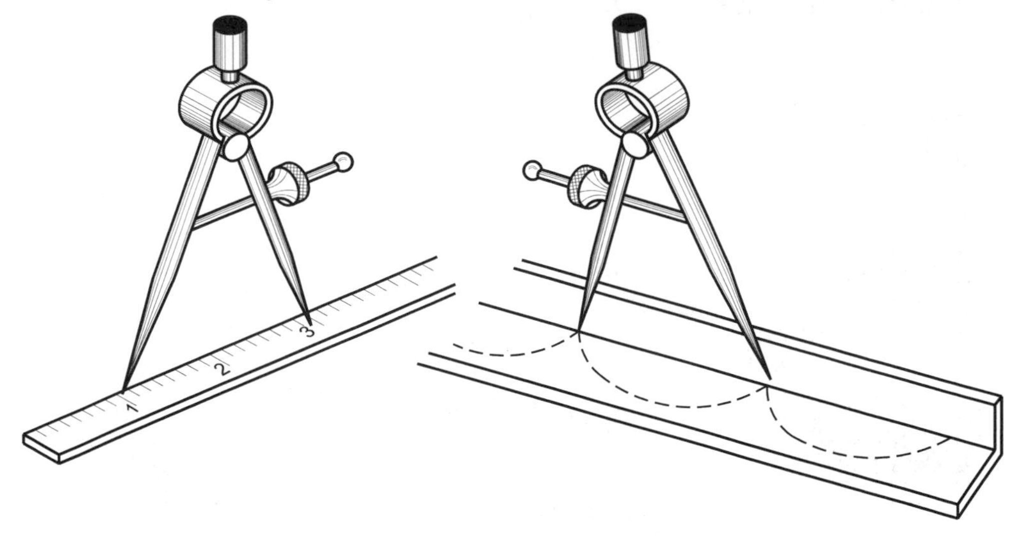

Figure 1–7. Spring dividers: setting a radius dimension along a scale (left) and marking off equal intervals along a line (right).

- *Hermaphrodite calipers,* Figure 1–8, are used to locate and test centers of cylinders for laying off distances from the edges of flat stock and for measuring back along a cylinder's side from its face. They are especially useful for scribing lines parallel to an inside or outside curve, something squares cannot do.

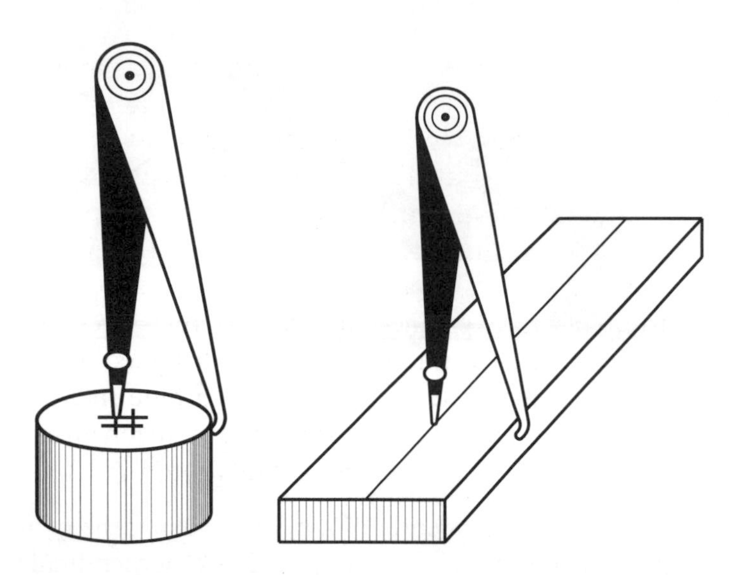

Figure 1–8. Hermaphrodite calipers finding a center on round stock and scribing lines parallel with the edge of a workpiece.

- *Magnifiers* are necessary for precision layout work and are useful in examining cutting tools and small parts. The OptiVISOR, a binocular headband magnifier, is comfortable to wear, works well over regular or safety glasses, and can be pushed up and out of the way when not needed. It offers magnifications from 1½- to 3½-power, Figure 1–9 (left). The lenses are glass, not plastic. This headband magnifier is the first step down from a binocular microscope. If you prefer, your optician can order eye glasses in these low magnifications. When more magnification is required, a 15-power pocket magnifier does the job, Figure 1–9 (right). Any machinist over age forty will probably need these tools.

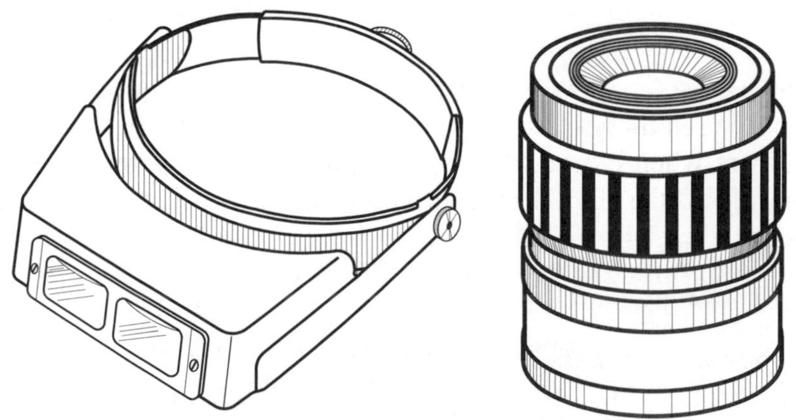

Figure 1–9. OptiVISOR magnifier (left) and 5-power hand-held magnifier (right).

- *Micrometers* are the machinists' oldest precision measurement tools. All are capable of measuring to 0.001 inches and the high-precision ones to 0.00005 inches. Most shop work does not require such precision. Many machinists prefer to use digital slide or dial calipers because of their direct digital read out and their 6-inch measurement range, rather than the 1-inch range of the typical micrometer. However, machinists revert to micrometers for fine work, measuring hole gages and checking caliper accuracy. Starrett offers over one-hundred different sizes and models of outside micrometers. Many micrometers are designed for one specific application. Quality micrometers have lapped carbide measuring faces, a locking mechanism to hold a dimension and a mechanism to provide constant spindle pressure when tightening. See Figure 1–10.

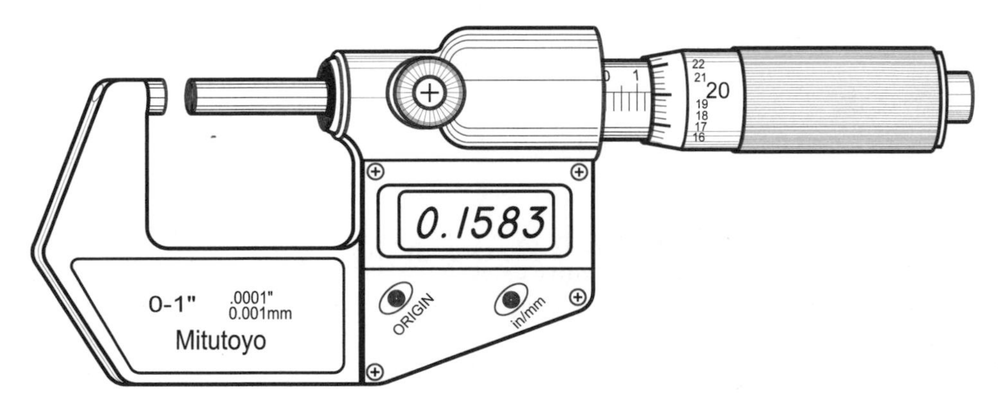

Figure 1–10. Digital micrometer.

- *Digital slide calipers*, Figure 1–11, are the principle measuring tools in many shops today, replacing traditional inside and outside calipers. Additionally, they make depth measurements and measure accurate distances from an edge for making a line parallel to an edge. They may be set to zero or set to store a reading at any point. Digital slide calipers switch between inches and millimeters at the push of a button. By zeroing the calipers when closed, then opening them to the target dimension and zeroing again, the digital readout shows the exact distance remaining to the target dimension. This function speeds work and helps eliminate errors, particularly on lathes and milling machines.

  Most digital slide calipers are stainless steel and the best ones have carbide faces for better wear resistance. They are precision measuring tools typically accurate to ±0.001 inches (±0.03 mm) over 6 inches and have a resolution of 0.0005 inches (0.01 mm).

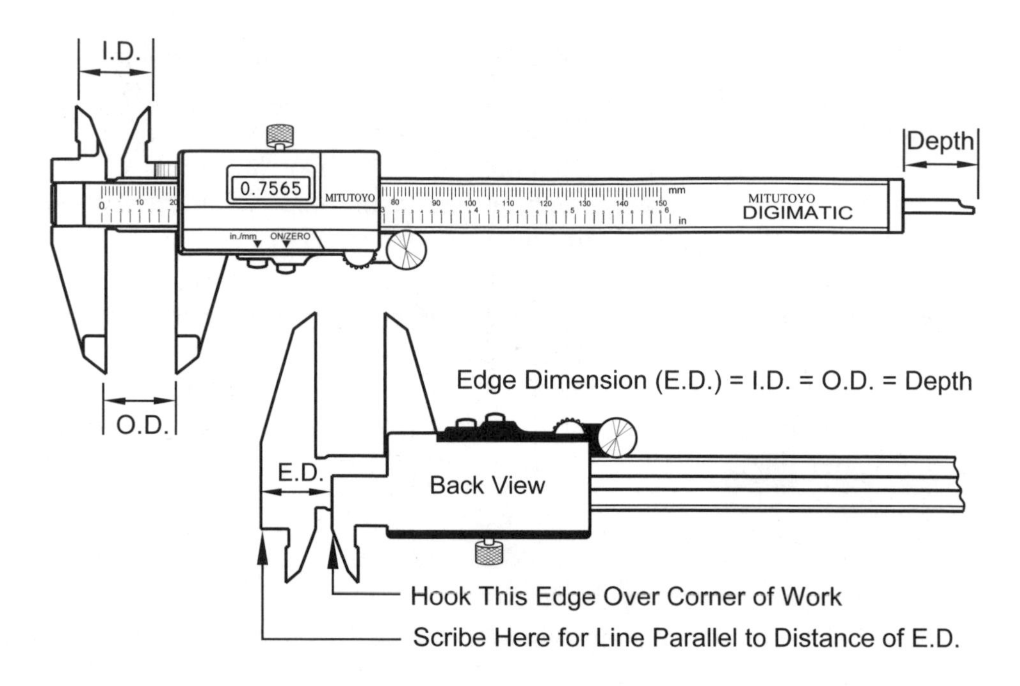

Figure 1–11. Digital slide caliper.

Tip: While not considered the best machine shop practice, many machinists open the caliper's jaws to the wanted distance and use the pointed tips of the caliper jaws to transfer this distance with a tiny scratch onto the workpiece. In other words, the calipers are used the way a pair of dividers would be used after being set to an opening on a rule. The cost of digital calipers has fallen considerably from when they were first introduced 15 years ago, so many machinists consider this practice cost-effective based on the time saved. Common digital caliper sizes are 6, 9, 12 and 24 inches. Even larger caliper models are available, too.

- *Surface gages* and *height gages* both place a layout mark at a precise height on the workpiece or measure the height of a feature on the work. They are particularly convenient for scribing long, parallel lines. To use and set a height gage on work that is not free-standing, the work and ruler must be clamped to a 90° angle iron or plate, Figure 1–12 (left). Surface gages and height gages are usually used on surface plates. The curved end of the surface gage scriber is used to reach the lower parts of the workpiece where the pointed scriber makes too acute an angle with the work, Figure 1–12 (right). To store and protect the scriber when not in use, tuck the curved tip of the scriber into the groove under the ball. This will also prevent user injuries, Figure 1–12 (inset).

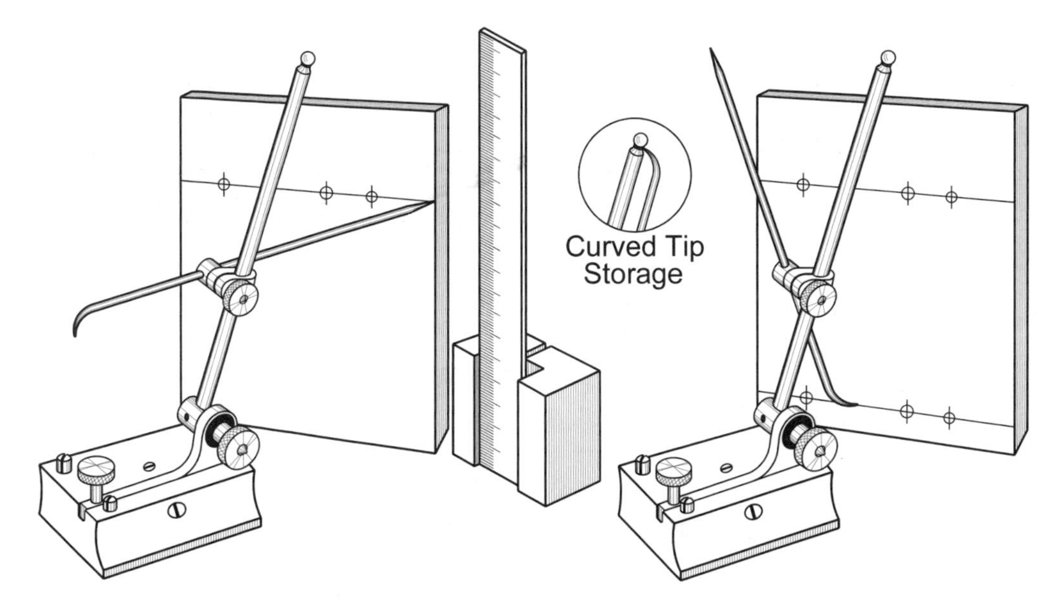

Figure 1–12. Surface gages: scribing a parallel line on work (left) and using the curved end of the scriber to scribe the lower end of the work (right). Inset shows storage of the curved tip against ball end.

- *Height gages*, Figure 1–13, are more convenient than surface gages because they indicate the height dimension on a mechanical dial or digital display. They offer the advantages of a digital caliper in that they can be zeroed at any point to indicate the distance from a target dimension. Better height gages have carbide scribers and hardened bases.

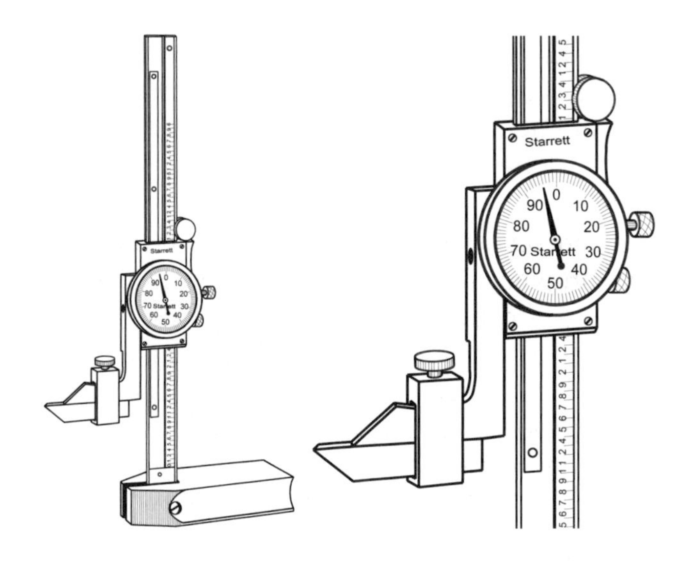

Figure 1–13. Height gage.

- *Surface plates*, Figure 1–14 (top), are rigid blocks of granite or cast iron that provide a very flat, smooth surface. For example, on 3 × 4-foot plate, no point deviates by more than 0.0005 inches from the rest of the plate. Surface plates are available as small as 9 × 12 inches and as large as 6 × 12 feet in several grades of accuracy. Granite is lapped flat, but cast iron must be hand scraped, a tedious and expensive process, so granite is replacing cast iron in many shops. When the surface of a granite plate is damaged or dinged, material removed leaves a crater and does not much reduce the accuracy of the plate. However, dings in a cast iron plate leave a crater surrounded by a raised edge leaving permanent damage. When using a surface or height gage for layout and inspection work, surface plates provide a reference plane for vertical measurements. Care must be taken to avoid scratching or nicking the plate's surface. Do not perform center punching on surface plates and cover them when not in use. To maintain their accuracy, all surface plates should be cleaned periodically to remove films, dirt and grit. Lacking a surface plate, a piece of ⅜- or ½-inch plate glass, supported at three points by wood blocks to prevent rocking, can be used.

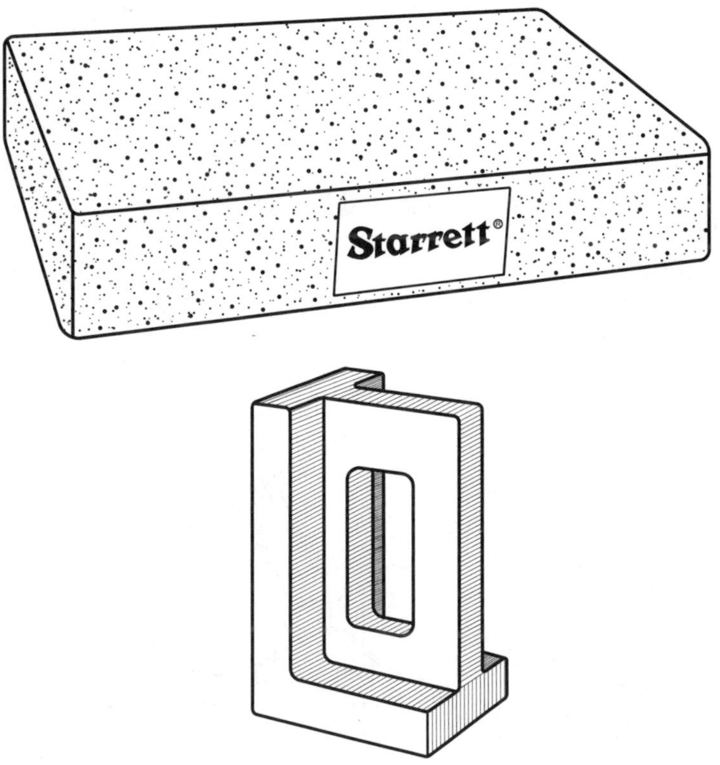

Figure 1–14. Granite surface plate (top) and universal right angle iron (bottom).

- *Universal right angle irons,* also called *angle plates,* Figure 1–14 (bottom), are used to hold flat workpieces vertically for layout on a surface plate, to hold rulers vertically for setting a surface gage and to provide a rigid vertical support for milling machine work when clamped to the milling table. Angle plates are also used on lathe faceplates to position work. Usually they are cast iron and come in a wide variety of sizes from a few inches to several feet on a side. For more precise work granite angle plates are also available.
- *V-blocks* are essential for laying out and machining cylindrical work because they insure the workpiece axis is horizontal and parallel to the surface plate, drill press or milling table. They also provide rigid support against machining forces. Figure 1–15 (top left) shows the most common V-block design, which includes a top strap. This V-block design can be used on its side or in a milling vise. Figure 1–15 (top right) shows a magnetic V-block whose magnetic force can be turned on and off making part removal easy. Figure 1–15 (bottom) shows a large V-block being used to hold a workpiece. Sometimes two V-blocks, machined as a pair, one at each end of the work, are used to hold long workpieces.

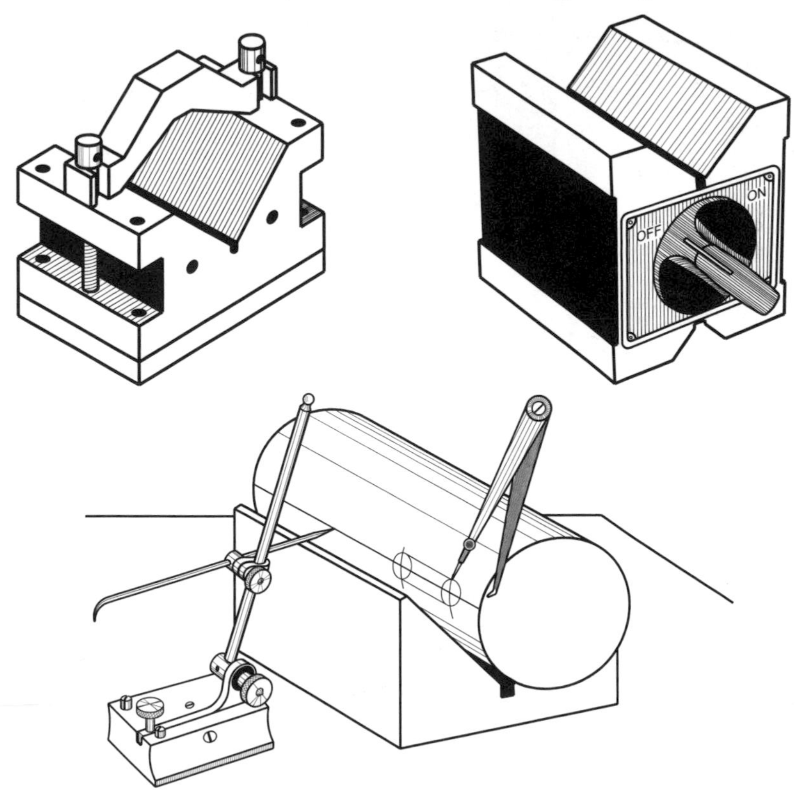

Figure 1–15. V-blocks.

Other Precision Measuring Tools

- *Test indicators,* also called *universal dial test indicators*, come in two styles: plunger and lever, Figure 1–16. Both styles can be used to center work in a lathe or milling machine and to check runout, but the lever-style works best for comparing dimensions. Plunger-style test indicators are larger and easier to read than lever-style test indicators. Usually both types have about 2½ revolutions of travel, but there are extended-range plunger styles with measurement ranges of several inches. Many different dials are available, including metric. For most shop work, indicators with 0.001- or 0.0005-inch dial graduations are suitable; for finer work, particularly inspection, indicators with 0.0001-inch markings are needed. An indicator with graduations too fine for the task makes the job harder without improving the results.

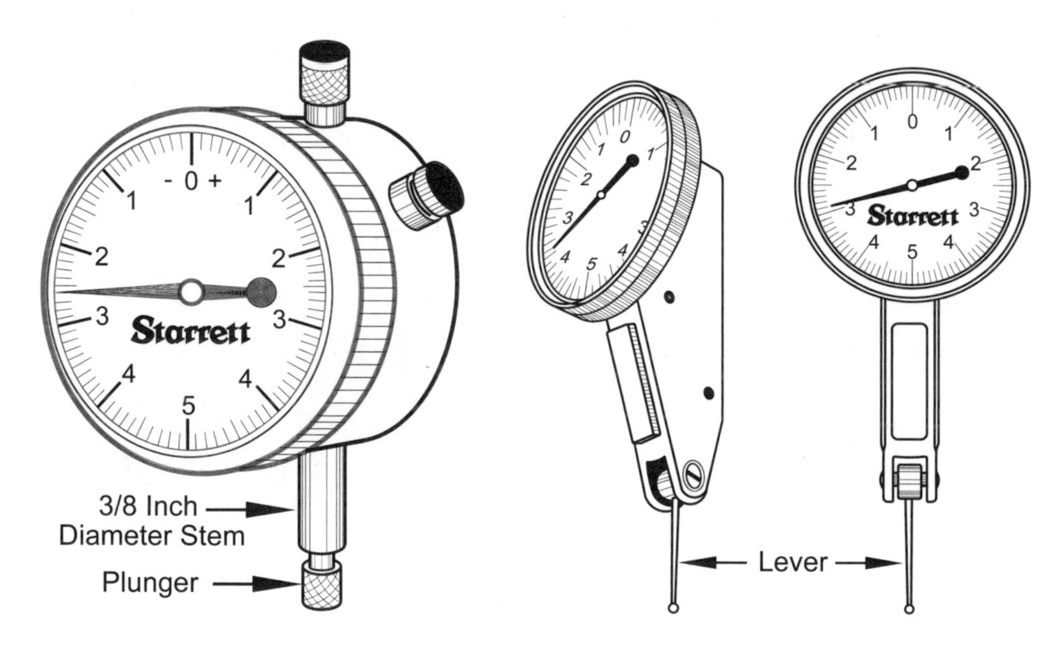

Figure 1–16. Plunger-style dial test indicator with a 2¼-inch diameter dial and 3/8-inch stem (left) is a common size. Lever-style universal dial test indicators (center and right) usually have smaller 1-inch diameter dials.

To achieve maximum accuracy when making measurements with the lever-style indicator, position the indicator so its leg is *relatively* parallel with the work surface being measured, Figure 1–17. This will reduce *cosine error*, which can account for a 6% error in an indicator measuring 0.010 inches with its leg at a 20° angle to the work. This precaution is not necessary when not making absolute measurements such as when centering work.

Do not center punch work in contact with an indicator, do not actuate indicators rapidly with your finger and do not oil their mechanisms as the indicator is likely to be damaged.

There are many test indicator holders, clamps, brackets and fittings for supporting and positioning indicators. Some have magnetic bases, which are very convenient on lathes and mills. Others are for mounting indicators on height or surface gages or the spindles of milling machines. Look at the Starrett catalog or an industrial tool catalog to see the many variations.

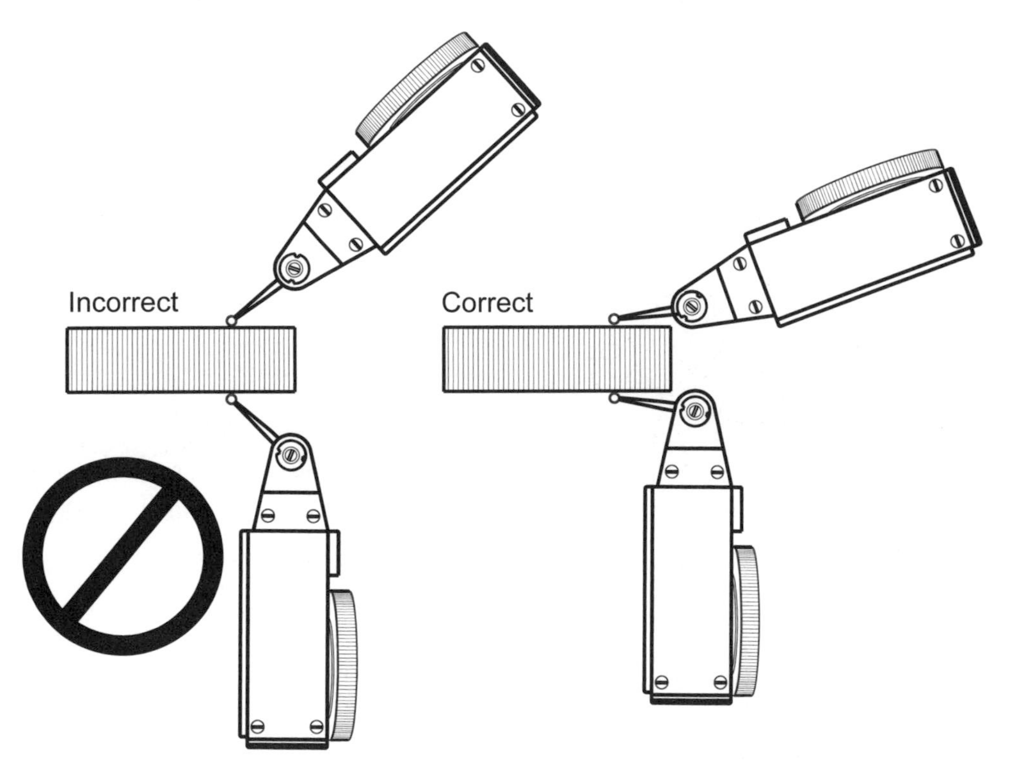

Figure 1–17. Positioning the test indicator lever parallel to the work increases measurement accuracy.

- *Hole measuring devices* are not always needed since dial or digital calipers often provide the necessary accuracy. However, for small holes, those under about 0.5 inch, and when a dimensional tolerance of ± 0.002 inches or better is needed, tools specifically designed for measuring holes are required. There are two relatively inexpensive tools for this purpose. *Small hole gages,* Figure 1–18, for holes from 0.125 to 0.500-inches (3.2- to 12.7-mm) diameter and *telescoping gages*, Figure 1–19, for holes from 0.500- to 6-inches (13- to 150-mm) diameter. Both tools capture the diameter and make it available for measurement by a micrometer or caliper. By checking in several places of the hole, out-of-round and tapered conditions can be checked. Neither device provides a direct readout, but more expensive devices do and these should be used when hole diameter measurements are frequently made.

Accurate measurements using a small hole gage are obtained by slightly "rocking" the gage in the hole to be measured. This will guarantee contact at the true diameter. The final size is obtained by measuring over the ball with a micrometer.

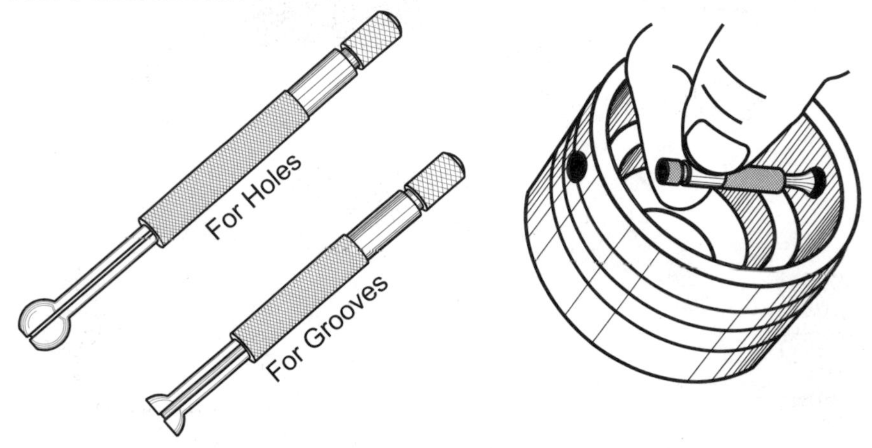

Figure 1–18. Small hole gages: Spherical head for holes (left), hemisphere for slots and grooves (center) and gage checking hole size in a part (right).

Like small hole gages, telescoping gages must be slightly "rocked" back and forth when making a measurement to insure that the tool is on the true diameter before it is locked and withdrawn. The final size is obtained by measuring over the gage contacts with calipers or a micrometer.

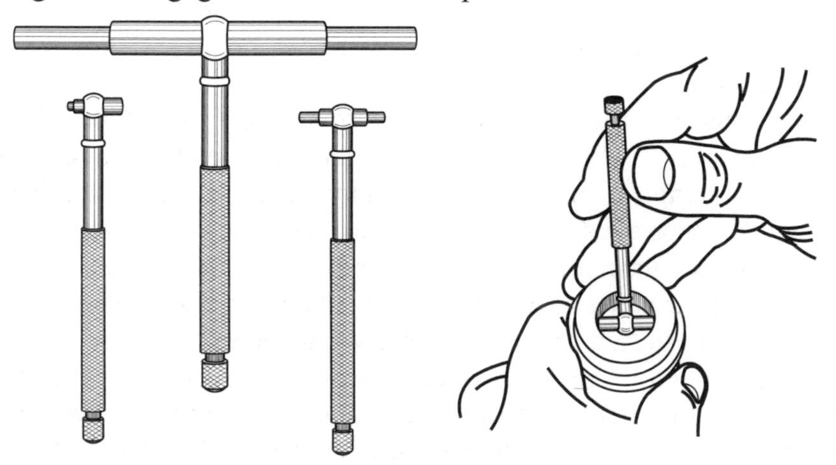

Figure 1–19. Centering-type telescopic hole gage in use.

- *Radius gages* are used to check or lay out concave or convex radii or fillets. They are made of hardened steel or stainless steel sheet, have smooth edges and are marked with their radius. They are available individually or in sets. Sets are available in fractional and decimal inches and in metric. Each gage has five radii for use in different situations, Figure 1–20.

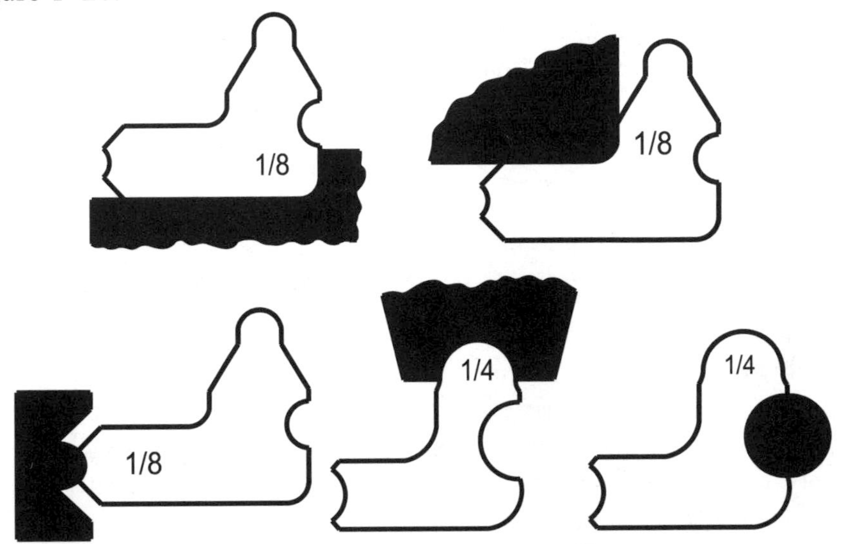

Figure 1–20. Five ways to use a radius gage.

Additional High-precision Measuring Tools

Note: These tools may or may not be needed depending on the precision of the work. Shops requiring positional accuracy better than ±0.005 inches will benefit from these tools. Remember that the "rule of ten" calls for measurement accuracy ten times better than the smallest units of measure you are concerned with.

What are *toolmakers' button sets*?

Button sets consist of four steel cylinders either 0.300-, 0.400- or 0.500-inches diameter, Figure 1–21. One cylinder is longer than the others to distinguish it on the workpiece and to make it easier to move if it is close to another button. They are hardened, ground and lapped to size and squareness. When the buttons are clamped in place with their screws, their cylindrical surfaces will be perpendicular to the workpiece surface. The over-sized center hole allows the buttons to be shifted around the screw into an exact position. Use them when one or more holes must be bored accurately in relation to one another or with regard to a reference point and when a jig borer is not available. *Buttoning* can improve hole position tolerance from ±0.005 to 0.0005 inches, but very careful work can bring tolerance down to ±0.0002 inches. They are considered obsolete and no longer offered by Starrett, but still essential for the precise spacing of holes for gear axles drilled on a lathe faceplate when better, alternative drilling equipment is unavailable.

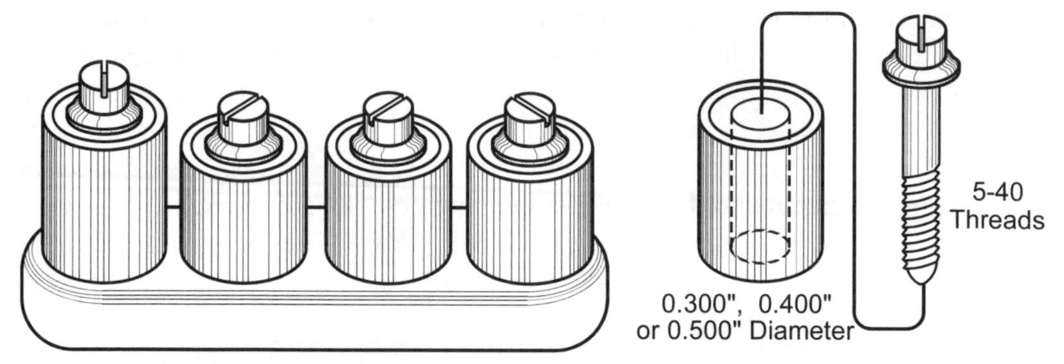

Figure 1–21. Toolmakers' button set.

How are these buttons used?

Here is how to use them:

1. Machine at least two square reference edges on the workpiece.
2. Wipe down the work surface with solvent to remove dirt and oil.
3. Apply the layout lines and hole center punch marks. High precision for hole positions is not needed.
4. Drill and tap button screw holes using a No. 38 drill and a 5-40 tap.
5. Attach buttons to the work. Tighten the screws so the buttons are firmly in place, but can be moved with a light tap from a soft-faced hammer.
6. Place the workpiece on a surface plate and against a right-angle plate, similar to the work secured in Figure 1–23 and clamp it to the right-angle plate.
7. Adjust the vertical height of the button using a height gage or gage blocks. Remember to compensate for the radius of the button.
8. Tap the button into position with a soft-faced hammer.
9. Rotate the workpiece 90° so it rests on its *other* reference surface.
10. Adjust the vertical height of the button using a height gage or gage blocks.
11. The button should now be in its correct position, so tighten the hold down-screw and recheck the button's position.
12. Repeat steps 7–11 for the other buttons.
13. Place the workpiece on a lathe faceplate and center it with a test indicator. Or, secure it to the worktable of a milling machine and center it using a test indicator held in the spindle.
14. Remove the button.
15. Repeat steps 13–15 for the other buttons.
16. Use the largest suitable center drill, then drill, bore or trepan each hole.

This hole spacing could also be located with gage blocks, adjustable parallels between buttons or with a micrometer over the outside of the buttons.

What are *gage blocks* and what are they used for?
They are rectangular or square steel, chrome-plated steel, ceramic or chromium carbide blocks of precise thicknesses, Figure 1–22. The top and bottom of the blocks has been lapped and polished to achieve an optically flat surface, typically within 2 to 8 millionths of an inch (50 to 200 millionths of a mm) of the size marked on their edge. Gage blocks are temperature sensitive and only equal their marked length at 68°F (20°C). They are available in both inch and metric sizes. Usually they are sold in sets so they can be stacked together to make up a desired length. U.S. Government specifications cover the dimensions of the blocks in sets as well as their accuracy tolerance. Using various combinations from the basic set of 81 blocks, you can make up stacks from 0.1001 inches to over 10 inches in length in 0.0001-inch increments. A combination of atmospheric pressure and intermolecular forces between the steel surfaces makes the blocks cling to each other. This behavior makes them easy to handle as they form stacks or sticks of blocks, which are easy to move. Block faces are so close to each other that no compensation need be made for the space between them, so any number of blocks can be *wrung* together. The smallest standard set of gage blocks contains 81 blocks.

They have many uses:
- To calibrate micrometers, calipers and other gages.
- To set surface gages and height gages to a precise height.
- To make set-ups on machine tools.
- To check the accuracy of finished parts.

In general, gage blocks do not make direct measurements. The length of a stack of blocks is transferred to the work or compared with an existing dimension.

What precautions must be followed in using and handling gage blocks?
- When using gage blocks, do not allow the blocks to strike each other, the work, tools or the layout table.
- Keep your fingers off the measuring surfaces of blocks as oil from your hands will eventually stain the blocks.
- Before using blocks, meticulously clean the working surface of oil, dirt and any foreign substances.

- After their use, all the blocks must be re-lubricated with acid-free oil before storage.
- Do not leave gage blocks wrung together for extended periods.
- Remember that gage blocks are temperature sensitive and that handling them affects their length.

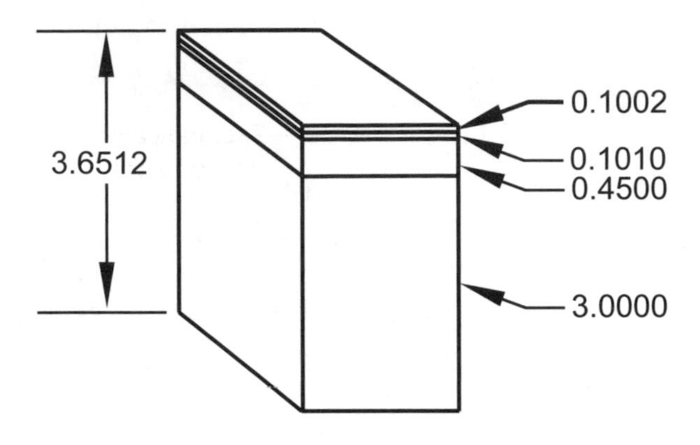

Figure 1–22. Stacked gage blocks.

How are gage blocks used to set surface and height gages?

Here is the procedure:

1. Determine the *minimum* number of blocks required to make up the desired length.
2. Wipe each gage block with a clean, soft cloth to remove preservative oil.
3. Wipe off each measuring surface again with the clean palm of your hand to remove dust.
4. Lay each block over the other, one at a time and push and twist them together simultaneously. This is called *wringing* the blocks. Block pairs that fail to adhere to each other are dirty. Re-clean them and try again.
5. Attach a test indicator either to a surface gage or to a height gage, Figure 1–23.
6. Wipe off the surface plate and place the stacked gage blocks on it.
7. Touch the test indicator leg to the top of the gage block stack and adjust the surface gage or height gage so the test indicator deflects at least 1/3 of a revolution. This insures the gage leg is firmly in contact with the gage blocks. If you are using a height gage, ignore its readout; it is not used.
8. Carefully adjust the surface or height gage so the test indicator reads zero. At this zero point the test indicator leg is at *exactly* the same height as the gage block stack. Do not change the height of the surface or height gage when moving it to the work.

9. Use the test indicator and surface gage to transfer this height to the toolmakers' button. When the test indicator reads zero, the point on the work is at the same height at the block stack. When using a test indicator on a toolmakers' button, be sure to move the surface gage back and forth over the button to find its topmost point. This is the point whose dimension we want to set. See Figure 1–23.

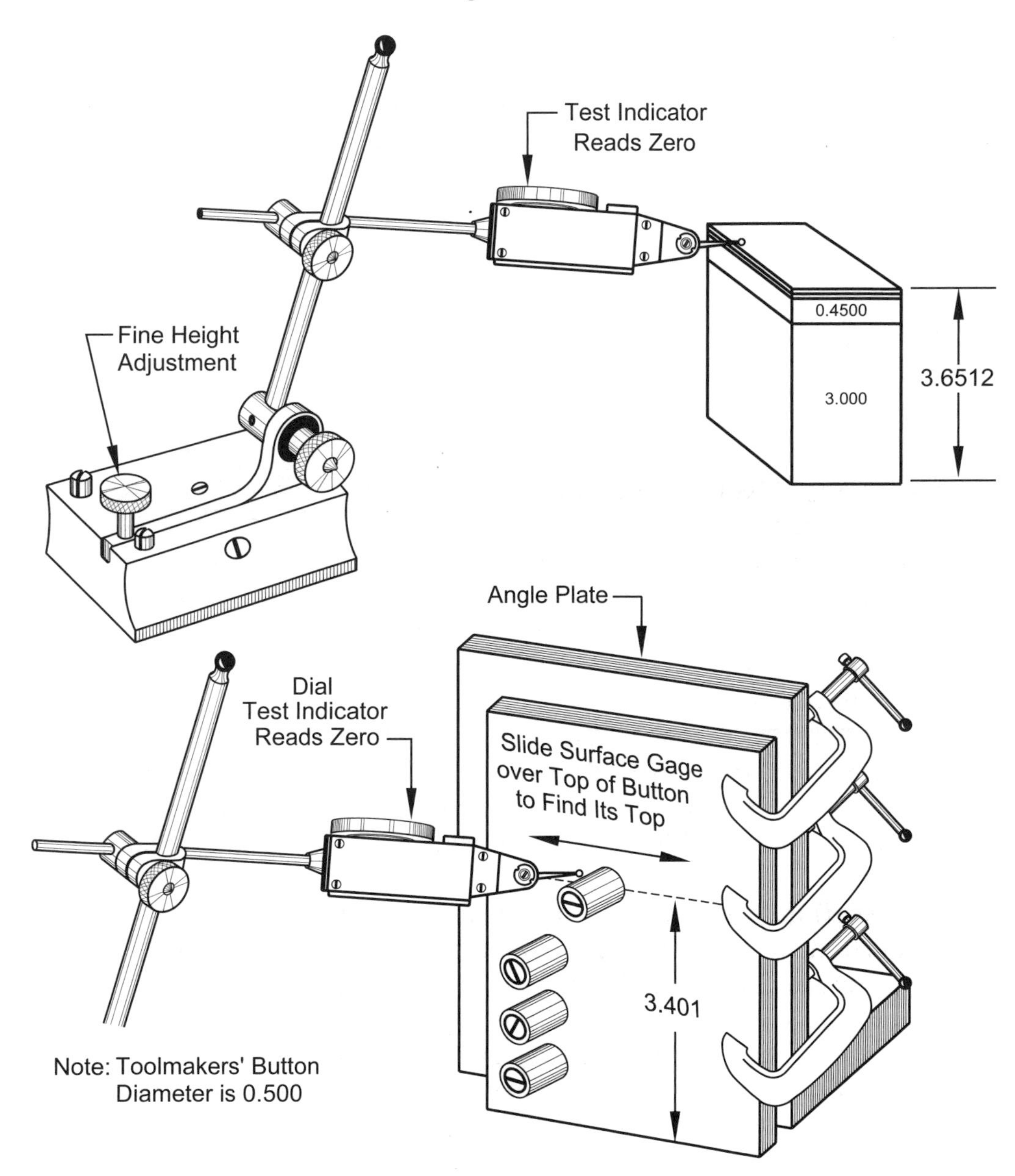

Figure 1–23. Using gage blocks to set a height gage and to set the height of a toolmakers' button.

What is a master precision square, also called an engineers' square?
A master precision square, unlike a combination square, has no graduations, Figure 1–24 (bottom). Its two components are riveted together, hardened, ground and lapped for straightness and parallelism. They have an error of less than 0.0001 inch in 6 inches and so are more accurate than machinists' combination squares. They come in a wide range of sizes, but the most popular size has a 4½-inch blade. There are 1½-inch miniature models that are very handy in tight quarters. A miniature die makers' square is shown in Figure 1–24 (top).

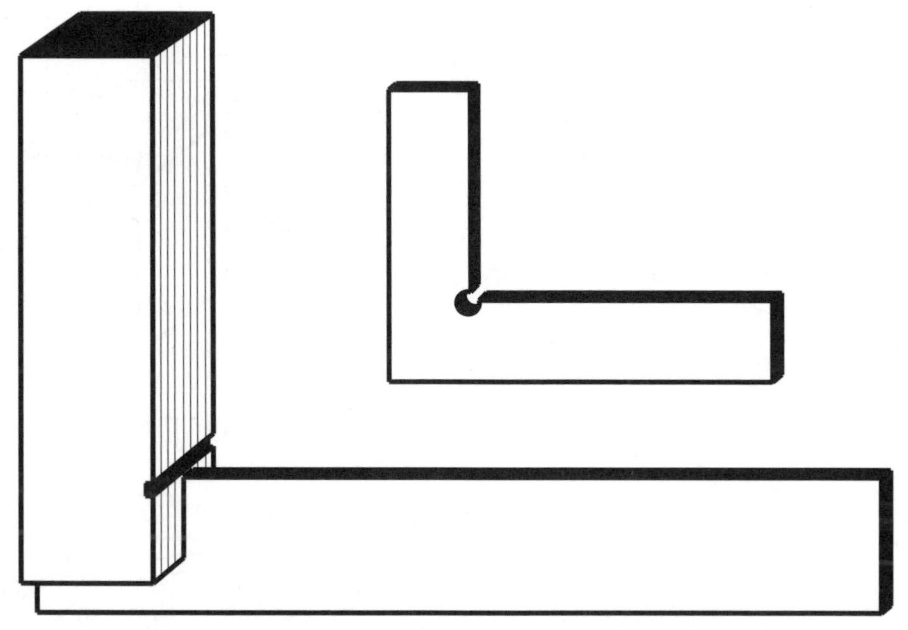

Figure 1–24. Engineers' square with miniature die makers' square.

Section II – Layout Work

General Layout Guidelines

What are the steps to lay out work?

1. Cut off stock with enough extra material to square it if needed.
2. Remove all burrs, clean the workpiece and apply layout fluid.
3. Establish reference lines on the work. On flat work, this is usually done by machining two perpendicular edges and making measurements from them. Sometimes flat stock comes from the supplier with two square edges and these may be used. On three-dimensional work, three reference surfaces are needed. All critical measurements must be made from these reference

lines or surfaces or the inaccuracies of each measurement will build up. For cylindrical parts, a V-block and surface gage are essential.

4. Use the machinists' combination square to transfer measurements from the working drawing onto the work with a scriber. If higher accuracy is needed, or the work is not flat, work on a surface plate and use a height or surface gage to locate dimensions. For extremely high accuracy, toolmakers' buttons and gage blocks are needed. Remember to protect the surface plate from the sharp edges of castings. Space rough work, like castings, off the surface plate with parallels or other same-size metal spacers.
5. With the principle features of the part transferred to the workpiece, identify the hole locations and mark them first with a prick and then with center punch marks. It is a good idea to use a divider to scribe the outlines of the holes to check on the overall accuracy of the layout and to prompt you on the hole size needed when machining them. Complicated parts deserve written notes as to hole sizes, tapped-hole thread sizes, radii and other details. A Sharpie® marker works well for this. If the casting has cored holes that must be machined to final dimension, glue a piece of thin brass over the cored hole and lay out the center of the hole on the brass. Scribe the hole outline, too.

Section III – Job Planning

General Guidelines

What are some common guidelines for planning machining work?

- Measure twice, check dimensions twice, mark once and cut once. A simple measuring mistake can be very expensive if a casting must be discarded or a workpiece with a lot of machining time is ruined.
- Let work cool after rough machining or heavy drilling operations, otherwise measurements will be "high." When the work has cooled, measure and finish machining to final dimensions.
- Always perform operations that may ruin the workpiece (as would a broken tap threading a small-diameter hole) as soon as possible in the machining cycle to minimize lost machining time should the part be damaged. Some work will allow early threading, some will not.
- Always study the work print carefully to decide the sequence of your operations *before* starting to cut metal.

What are the general rules for round work?

1. Rough-turn the largest diameters first; this minimizes bending forces on the work and chances of damage.
2. Rough-turn the diameters to within 1/32-inch (0.79-mm) diameter.
3. Turn steps and shoulders to within 1/32 inch.
4. Make sure to measure all lengths from the same end.
5. Complete any grooving or knurling now.

What general rules apply for flat work?

1. Cut off raw stock oversized.
2. Machine work to proper thickness.
3. Lay out outside dimensions of the part. Consider using light prick punch marks to outline cut lines.
4. Remove large irregular areas with a band saw or use chain drilling to remove the excess stock and file to finished dimensions.
5. Complete dctailed machining.
6. Lay out holes and center punch.
7. Drill or bore holes, complete spot facing.
8. Tap and ream holes. Always *try* to do this as soon as the situation permits because if the part is ruined by a snapped tap or drill, the loss of machining time is minimized.

Chapter 2

Basic Hand Tools

The chief cause of problems is solutions.
—Richard Sennett

Introduction

Having just the right tool for the job makes all the difference in getting the job done quickly and correctly. No place is this more important than in the machine shop. This chapter looks at fourteen different categories of basic machine shop hand tools and how they are used. For every category of tool there are dozens of variations in size, shape and design, so it is easy to see why most machinists have *hundreds* of tools in their rollaway chests. There are another dozen types of hand tools, such as reamers, taps and dies, which are equally important and will be covered in later chapters. Although socket, open-end and box-end wrenches are basics and often used in the machine shop, their use is familiar to most and will not be covered here.

We will also examine two hand-held power tools, the reciprocating saw and the portable band saw. They are important for cutting metal stock into smaller sizes prior to machining.

From quality tools to budget imports there is a bigger selection of tools available today than ever before. Actually seeing and handling the quality and budget alternatives before making a choice between them is always a good idea. Sometimes there is a big difference between them and sometimes not, so be aware of this before making a purchase. Many budget tools look good, but that is the *only* good thing about them. Paying list price for a quality tool is often the better value. Industrial tool catalogs, aside from being fun wish books, are a good way to learn about the myriad of tools now available which may someday help you solve a problem.

Bench Vise & Jaw Covers

What are the uses of a bench vise?

Bench vises:

- Hold work securely when filing, sawing or drilling.

- Support work across its open jaws when driving out a pin with a drive punch.
- Bend small metal bars, rounds and sheet goods by inserting them up to the bend point and striking the work with a ball peen hammer.
- Assemble force-fit parts by squeezing them together.
- Hold bench blocks and miniature anvils at a convenient level.
- Position work for soldering or brazing.
- Provide a small anvil behind the back jaw for light hammering.

See Figure 2–1.

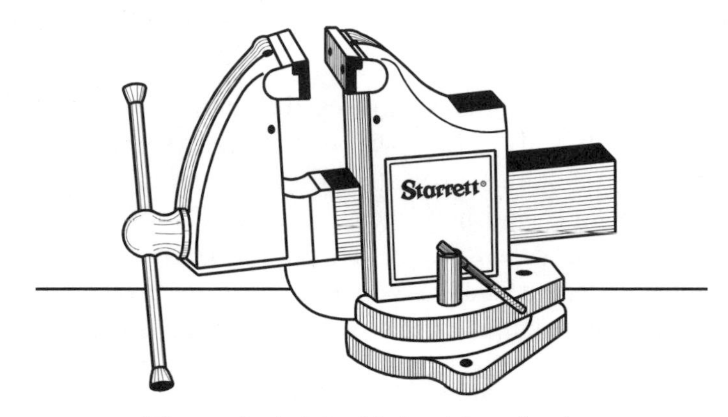

Figure 2–1. Machinists' bench vise

Why use protective *vise jaws* and how are they made?

They prevent damage to the workpiece from the hardened and diamond-serrated vise jaws. There are many different materials used to make them. Rubber, lead, Teflon® and polyethylene make very soft vise jaws, aluminum and brass, slightly harder ones. Fiberboard, Masonite® and wood blocks also work well. Some protective jaws lay over the vise jaws, Figure 2–2, some have tabs bent to clasp the vise, Figure 2–3, and others are held against the steel vise jaws with magnets.

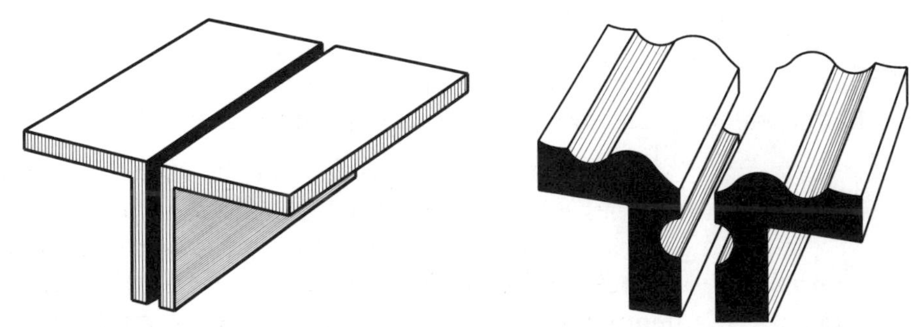

Figure 2–2. Protective vise jaws: brass angles that sit on vise jaws (left) and molded rubber jaws for holding gun barrels (right).

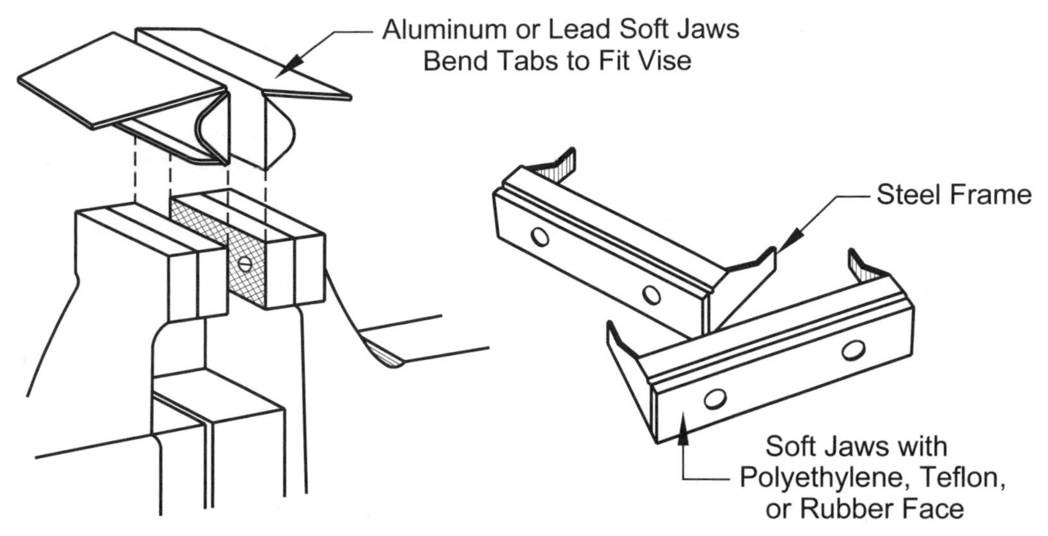

Figure 2–3. Protective vise jaws held in place by metal tabs.

Sometimes the workpiece must be protected from the *inside*. To do this, place a wooden dowel *inside* the thin-walled tubing to prevent the vise from crushing the tubing, Figure 2–4. The dowel must fit snugly inside the tubing.

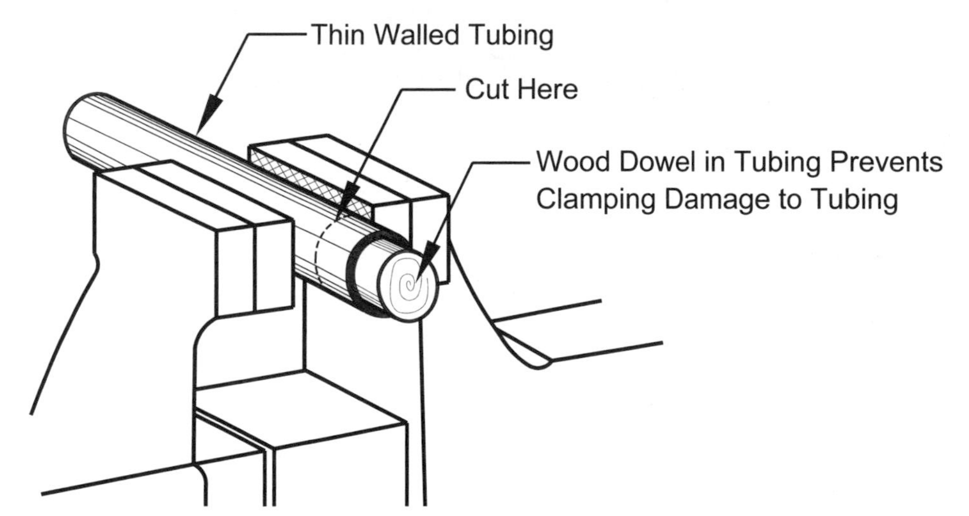

Figure 2–4. Using a dowel to safely clamp thin-walled tubing.

To hold a specific part, sometimes the best approach is to replace the original vise jaws with shop-made ones with cylindrical or V-shaped grooves, Figure 2–5. Two screws hold each jaw in place, making it easy to install new ones. Often custom jaws do not need to be hardened and serrated, so they can be made of brass or aluminum for easy fabrication. Besides gripping a part firmly without marking it, custom jaws will not come loose from the vise as other designs will.

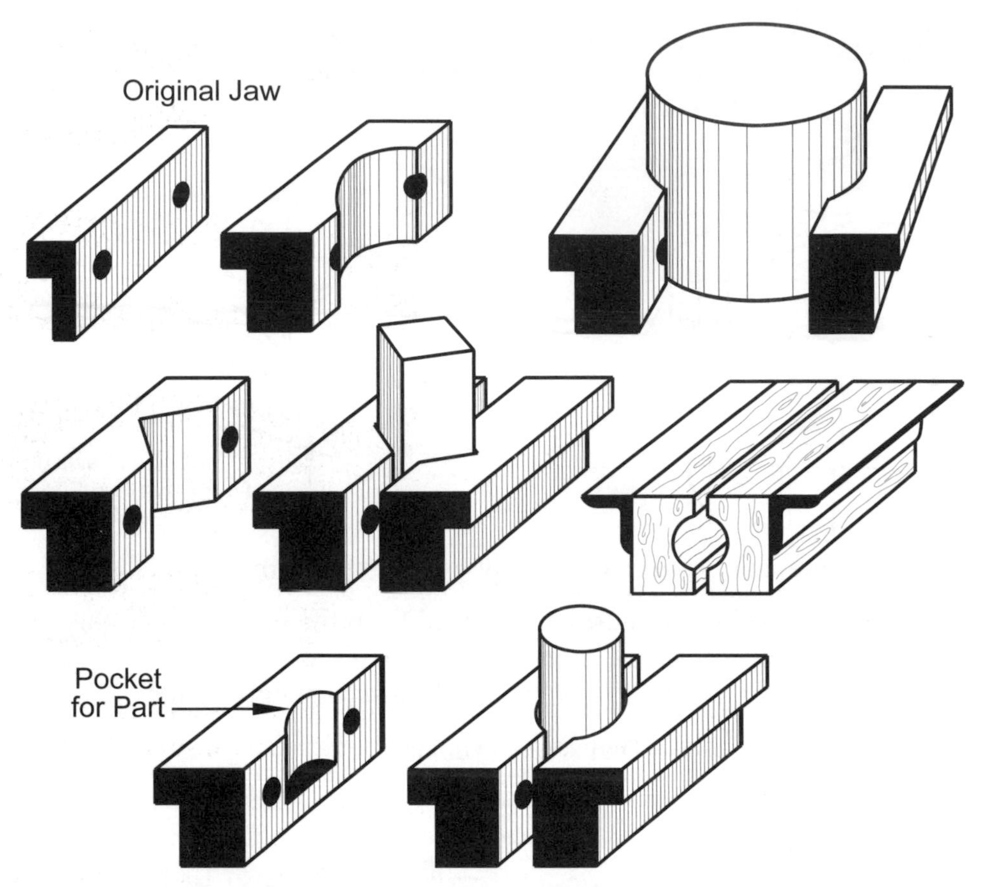

Figure 2–5. Custom vise jaws shaped to hold particular parts. The wood jaws (center row, right) have angle iron sections attached to their sides so they can sit in a vise without falling through the jaws.

Screwdrivers

What are the four most common *screwdriver designs*?

From the top of Figure 2–6, they are the:

- Stubby or close-quarters screwdriver with a standard blade.
- Phillips screwdriver.
- Standard screwdriver.
- Electricians' screwdriver.

Both the Phillips and standard screwdrivers pictured are heavy-duty industrial models with hexagonal wrenching flats for extra torque. Phillips screwdrivers are available in six sizes: #00, #0, #1, #2, #3, #4 with #00 the smallest. The most common size is #2. The #00 and #0 sizes are quite small and are used on instruments, clocks and electronic equipment such as CD players. The electricians' screwdriver not only has an extra-long blade, it has an insulated shaft to prevent accidental contact with live circuits.

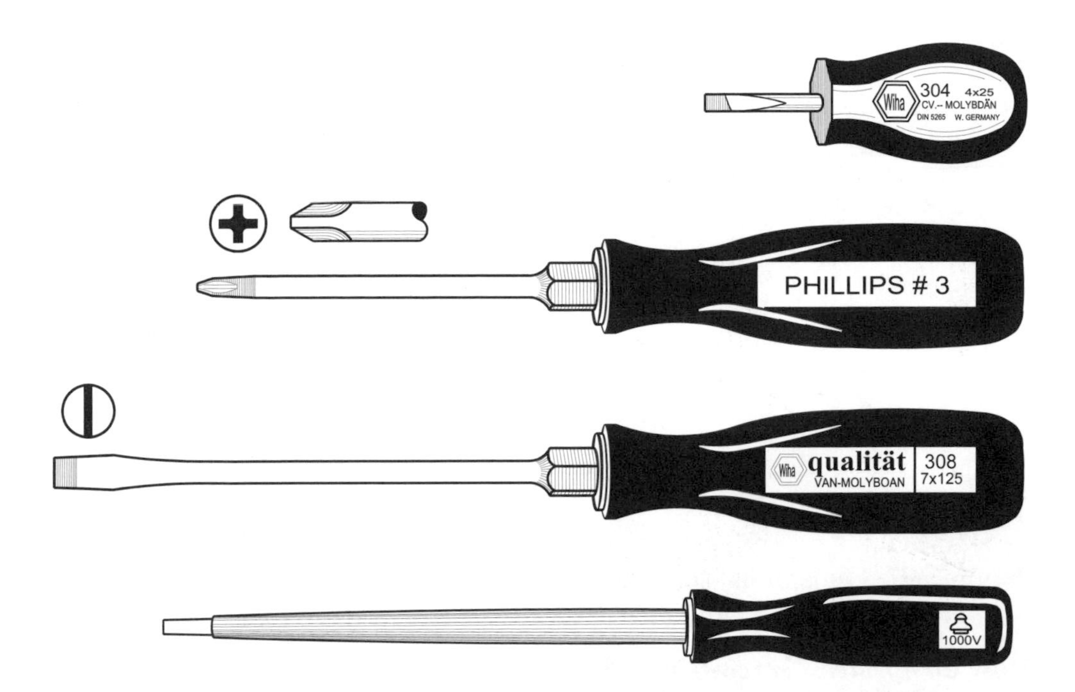

Figure 2–6. Screwdrivers (top to bottom): close-quarters standard blade, Phillips head, standard blade and electricians' with fully insulated shaft.

To avoid damaging the screw, match the screwdriver blade to the screw head. Too small a screwdriver and the screw slot will become nicked, too large and the material surrounding the screw head will be gouged and the blade will fail to seat. A good selection of screwdrivers is needed to make the proper match.

Do not use screwdrivers as pry bars, paint can openers, chisels or wedges. Redress the tips on a bench grinder if they become broken or worn.

What are *gunsmiths' screwdriver sets* and what is their advantage?
Gunsmiths' screwdriver sets consist of twenty screwdrivers held in a stand and are also available in kits with two handles and several dozen different interchangeable blades. A typical set has screwdrivers in three to six blade thicknesses in *each* of nine standard blade widths. For example, a screwdriver with a 0.360 wide blade has six blade thicknesses: 0.020, 0.025, 0.030, 0.035, 0.040 and 0.050 inches. This insures that a screwdriver with the proper blade width and slot thickness is available for most screw heads. See Figure 2–7. The screwdriver bits, or tips, fit into the screwdriver handles and are held in place by a magnet. The bits are inexpensive. Most kits contain standard, Phillips, Torx®, square and hex bits.

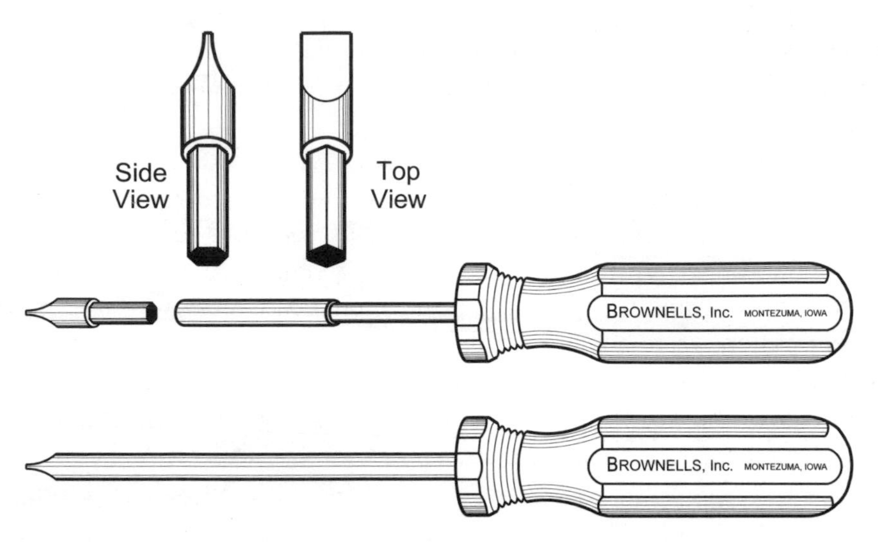

Figure 2–7. Gunsmiths' screwdrivers: interchangeable blades and handle (top and middle) and individual screwdriver with a fixed blade (bottom).

How do *jewelers' screwdrivers* differ from standard screwdrivers?
Jewelers' screwdrivers, Figure 2–8, have much smaller blades, typically between 0.025- and 0.100-inches wide. The blades fit into their handles and so are easily replaced. The blades can be reversed into the screwdriver handles so they can safely be carried in a shirt pocket. Standard blade screwdrivers are shown, but they are also available in Phillips sizes.

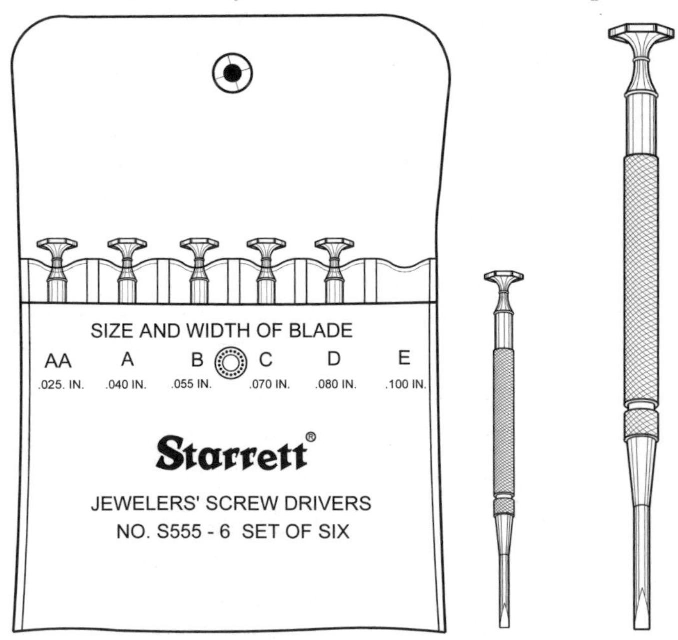

Figure 2–8. Jewelers' screwdrivers.

Allen & Torx® Wrenches

What are *Allen* and *Torx (*pronounced *TOR'-Ex) wrenches* used for and what are their advantages?

Allen wrenches, also called *hex socket head keys*, *Allen set screw wrenches* and in Europe *Unbrako* or *Inbus keys,* fit the matching depressions in socket head set screws, cap head screws and cap head bolts. The advantage of hex-head fasteners over the standard slotted-head design is that hex heads are less likely to strip out under high torque and their heads can be made in smaller diameters. Allen wrenches are available in both inch and metric sizes and are measured by the distance across their flats. This distance is usually about half the diameter of the fastener. These wrenches come in three common forms: L-shaped wrenches, T-shaped wrenches and screwdrivers with fixed-hex ends, Figure 2–9. There are also hex-shaped bits for use in power tools, socket wrenches and replaceable bit screwdrivers (like gunsmiths' screwdrivers). Several tool companies offer Allen wrenches with ball-ends, which allow the wrench to work in close quarters and as much as 25° off axis. When the ends of Allen wrenches become worn, they can be ground off exposing a fresh hex. When grinding a new end, dip it in water to avoid taking the temper out of the key by overheating.

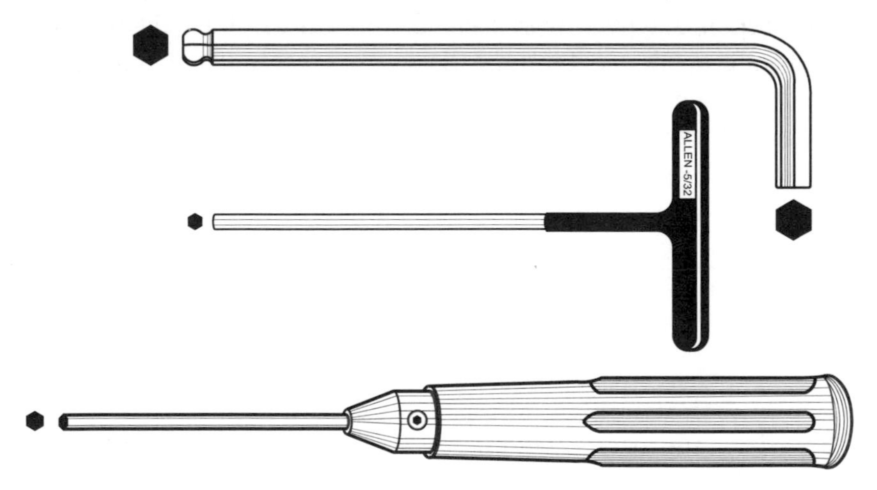

Figure 2–9. Allen hex socket head wrenches: L-shaped wrench (top), T-shaped wrench (middle) and screwdriver (bottom). Note the ball tip on the L-shaped wrench for off-center, close-quarters use.

Torx wrenches, Figure 2–10, fit the matching depressions in machine screws, self-tapping screws, cap screws and bolts. The Torx shape is that of a six-pointed star with rounded tips. Torx head fasteners are less likely to "clutch out" or strip out under high torque than Allen heads and are well suited for production work with power screwdrivers. They increase in size going from

T5 to T50. Torx-head fasteners are available in both inch and metric threads, but the same T5 to T50 sizes are used. Torx replaceable bits and screwdrivers are also available with a male-shaped Torx head instead of a female recess.

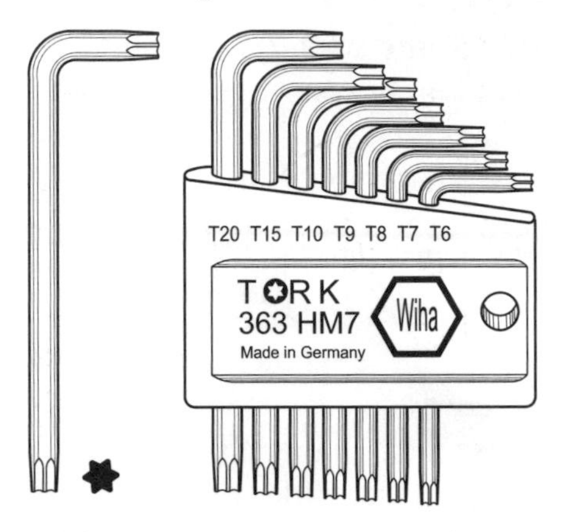

Figure 2–10. Torx wrenches.

Automatic Center Punch

What does the *automatic center punch* do and what are its advantages?
The automatic center punch, Figure 2–11, automatically strikes a blow on its punch-like tip when downward pressure is applied to its cap. No hammer is needed. Their one-handed operation allows the machinist to see the scribe marks, impossible when holding both a hammer and a punch. Turning the adjustable knurled cap regulates the force of the blow. Automatic center punches are typically about 5-inches long and have screw-on replaceable tips. Use them for the rougher classes of work. Fine work requires first a prick punch, then a center punch to achieve on-the-mark accuracy.

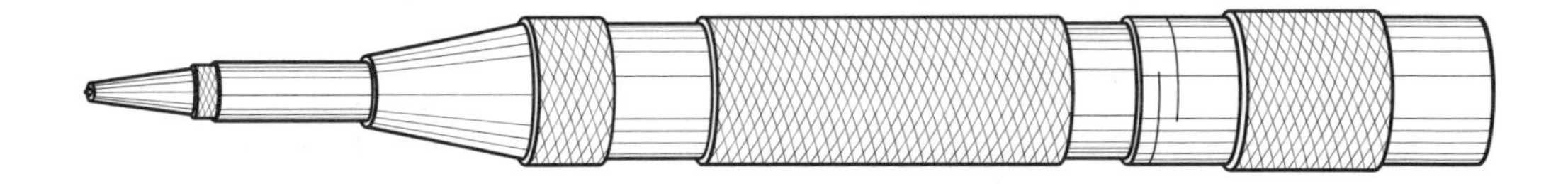

Figure 2–11. Automatic center punch.

Bench Blocks

What are *bench blocks* and how are they used?
Bench blocks are miniature anvils of hardened and ground steel, Figure 2–12. They range from 1¼- to 5-inches diameter and are used as a work support surface when driving out pins, filing and drilling. Because of their smooth top surface, work placed on them does not pick up surface marks. Some designs have V-grooves for holding round work, while others have holes, slots and

slits. Some designs are hexagonal to permit mounting them in a vise. Bench blocks are for fine work, not heavy hammering, and are useful for instrument, clock, watch and firearms work.

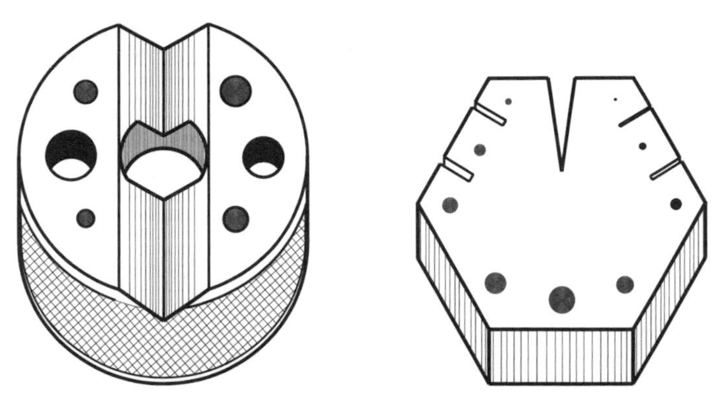

Figure 2–12. Two bench block designs.

Pliers

What *pliers designs* are most common in the machine shop?

From the top of Figure 2–13, they are:

- *Tongue-and-groove pliers,* sometimes called *Channellocks* after one of their manufacturers, are the most popular pliers designs. They have dislodged the traditional slip-joint pliers from their leading position because of their two advantages: first, the tongue-and-groove design is very strong, so once its jaw distance is set, it will not change, and second, the jaws are set at an angle to the handles and are more comfortable for many jobs. Channellock® Inc. makes 17 varieties of this tool in 8 lengths.
- *Slip-joint pliers* are a classic design and are still popular.
- *Needle-nose pliers* can get into tight spots and make sharp bends and loops in wire. Although usually used in electrical and electronics work, needle-nose pliers are great for bending and adjusting springs and sheet metal parts. They are available in several lengths and needle tapers.
- *Diagonal cutters* snip copper, brass, iron and steel wire. They will be damaged if used on piano wire, which must be cut with special purpose nippers. Diagonal cutters are also available in small sizes for electronics work.
- *Multipurpose pliers* were originally designed for crimping AMP-style solderless electrical connectors. But because they can also strip insulation from wire, trim five common sizes of brass and soft steel screws to length without destroying their threads, cut wire, crimp solderless connectors and crimp smaller clamps for steel and stainless cable, they have become popular in their own right. They are available from several manufacturers.

- *Vise-grip® pliers*, shown in two models, are excellent for holding small parts for drilling, grinding, wire brushing and sanding. They hold pipe, rounds and square stock and are useful in clamping one part to another for welding, soldering, drilling or patterning. They can even remove some "one-way" screws by clamping onto their edges. Their disadvantage is that their jaws leave marks on the work.
- *Soft-jaw pliers* can safely be used on easily marked or scratched workpieces. However, they cannot exert a lot of torque before their nylon or rubber pads are damaged. Soft-jaw pliers are mostly used on camera lenses and other optical equipment.

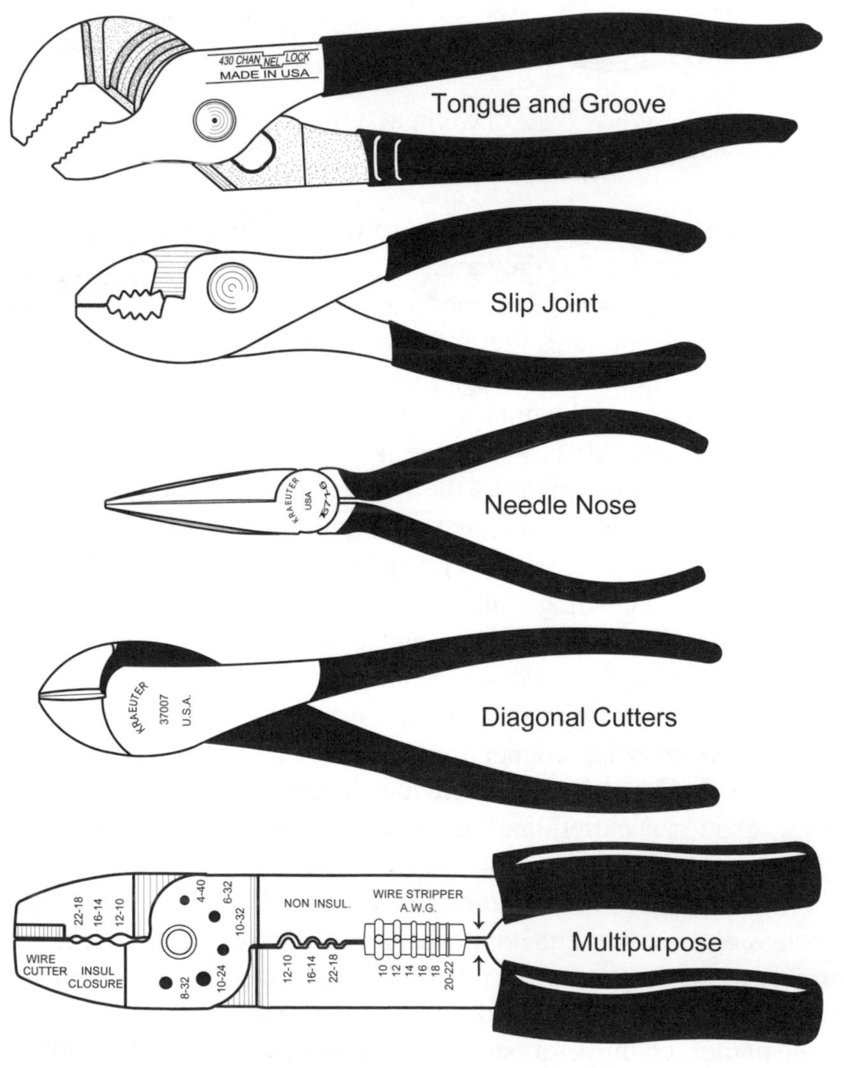

Figure 2–13. Pliers.

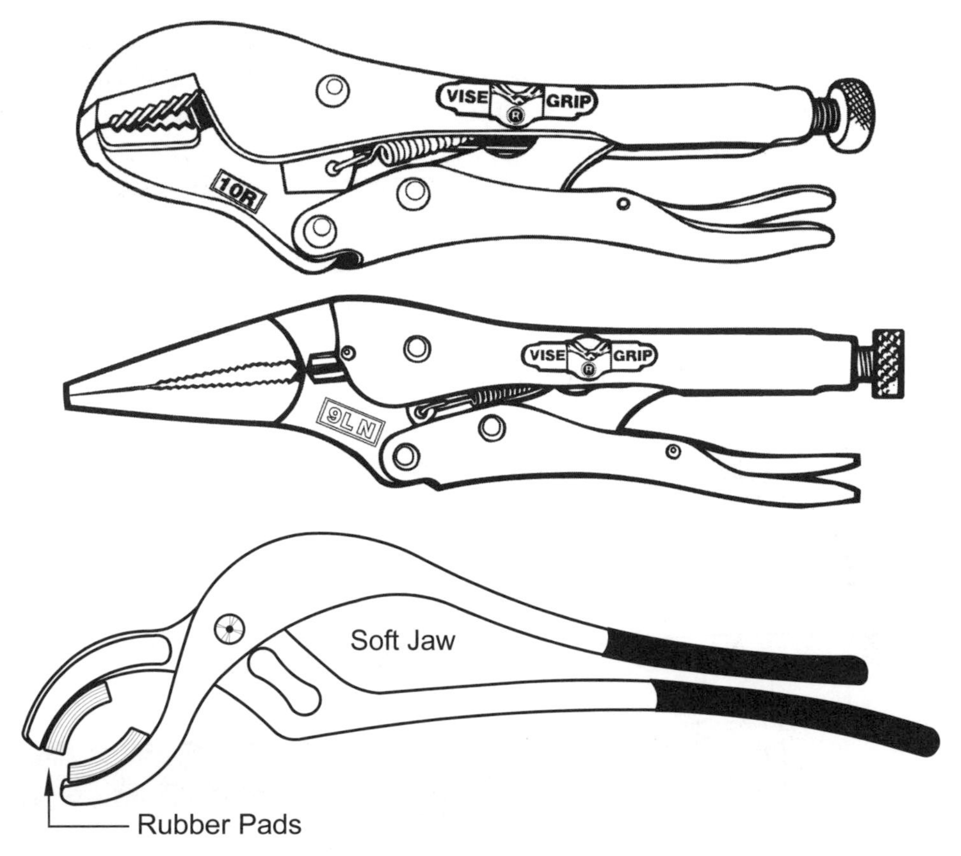

Figure 2–13. Pliers (continued).

Never use pliers in place of a wrench. Do not attempt to cut hardened steel or piano wire with pliers, use a cut-off wheel.

Hammers

What are the four most common *hammers* found in the machine shop and what are their uses?

See Figure 2–14.

- *Ball-peen hammers* have forged steel heads. Their flat face is for striking work or tools, like punches or chisels. Their rounded end is for peening over (flattening out or spreading out) rivets and other materials. A wide range of sizes is available from 2-ounce heads for instrument work to 32-ounce heads for heavy-equipment work.
- *Lead or brass hammers* do not mar the workpiece surface, yet can apply heavy blows to the work. They are often used to seat work in lathe chucks and milling machine vises.
- *Soft-face hammers* have rawhide, plastic or rubber faces for more delicate work surfaces.

- *Dead-blow hammers* contain lead or steel shot and do not bounce back after striking the work. They are available in steel, rubber, plastic or non-sparking faces.

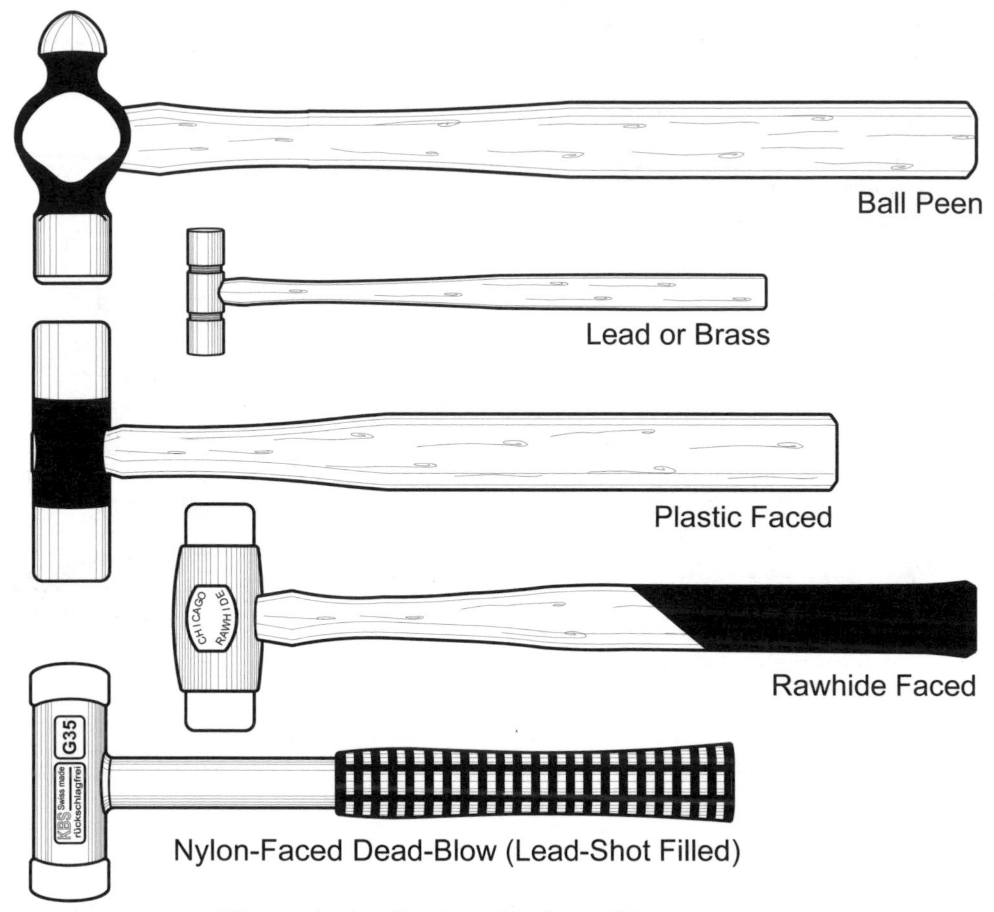

Figure 2–14. Basic selection of hammers.

Chisels

What are the *basic chisels* needed in a machine shop?

Four basic chisel designs, Figure 2–15, are needed. They are the:

- *Cold chisel* for removing rivets and general cutting and chipping. The cold chisel is so named because they are used on *cold,* not hot work.
- *Cape chisel* for cutting square-bottomed grooves like those needed on motor shafts to key a pulley to a shaft.
- *Round-nose chisel* for cutting round-bottom grooves and to cut radii.
- *Diamond-point chisel* for squaring up corners and cutting V-grooves. Before machine tools were widely available, machinists depended on chisels and files for much of their metal shaping.

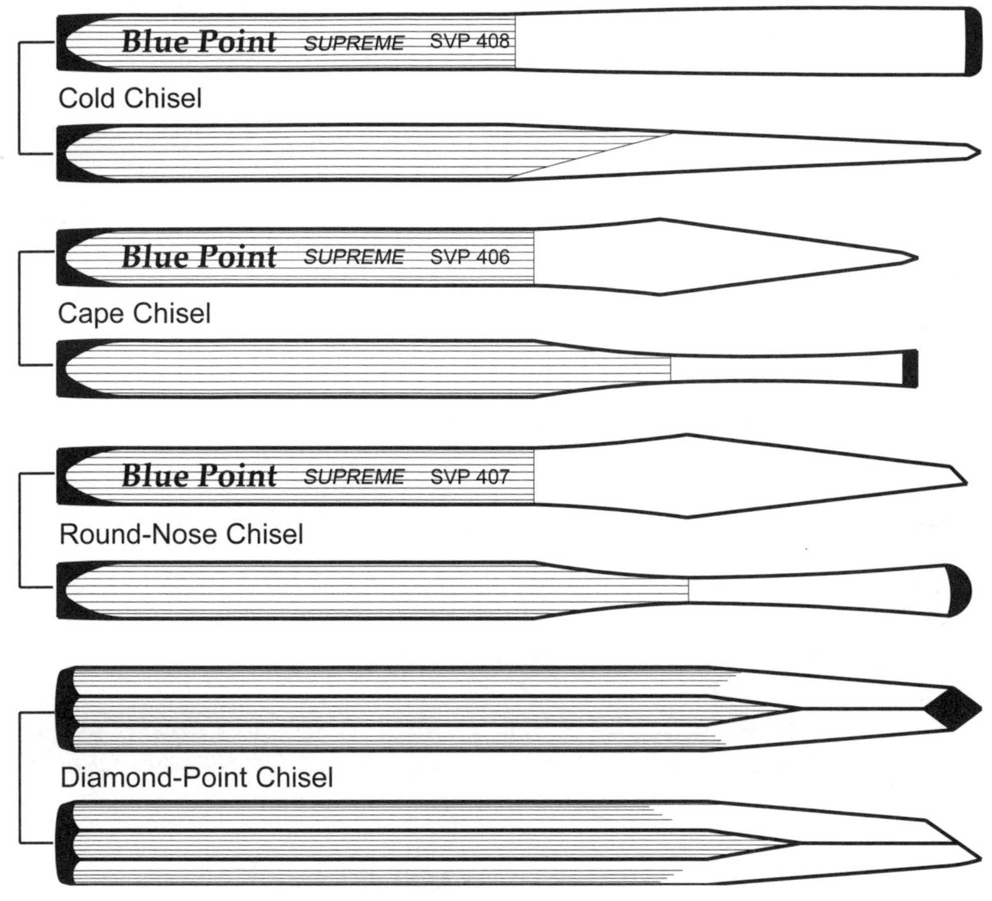

Figure 2–15. Chisels.

Drive Punches

What are *drive punches* used for and what are their common designs?

Drive punches are struck with a ball-peen or brass hammer to drive pins out of their holes. The one-piece design, Figure 2–16 (top), is the most common. The longer, two-piece design, Figure 2–16 (bottom), provides extra reach for deep holes. Both designs are easily reground on top to remove the "mushroom" and on the punch-end when it is no longer flat. There are other designs made of brass or nylon, which are used in gunsmithing to adjust drift sights.

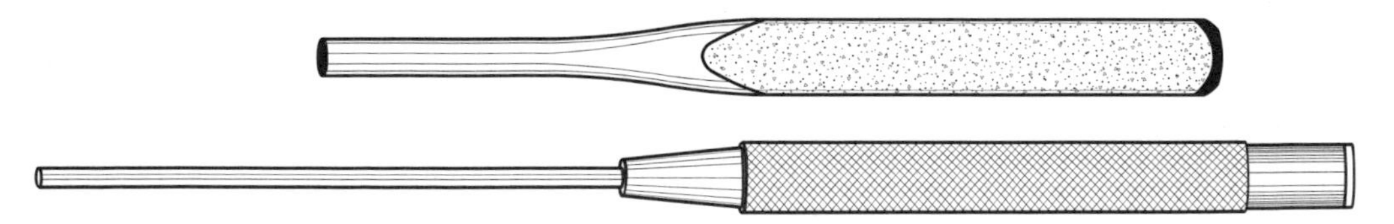

Figure 2–16. Two drive punch designs.

Clamps

What are the two basic style clamps seen in the machine shop?

They are the C-clamp and the machinists' clamp, Figure 2–17. The C-clamp is applied with its single screw. The machinists' clamp is applied by snugging up the jaws so they are parallel using the center clamp screw, then pinching the jaws together by tightening up the other clamp screw. Clamps in these two designs are available in a wide range of sizes. They are essential when holding a metal pattern against a workpiece or drilling two pieces of stock at the same time. Sometimes brass, plastic or cardboard placed between the faces of the clamp and the work are used to prevent marring the work.

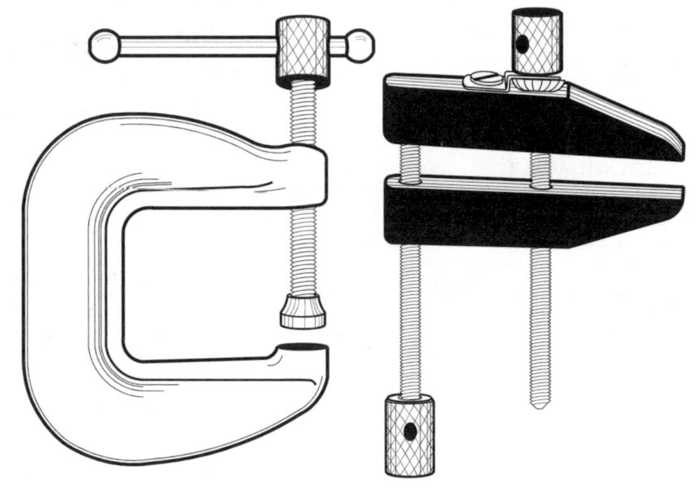

Figure 2–17. C-clamp (left) and machinists' clamp (right).

Knives

What two knife designs are useful in the machine shop?

The Xacto®-style hobby knife and the retractable Stanley®-style utility knife are useful and inexpensive tools. Both feature easily replaceable blades in a variety of shapes. See Figure 2–18.

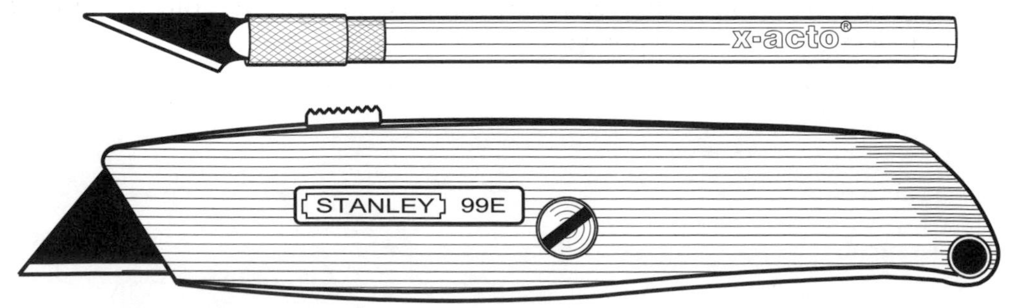

Figure 2–18. Xacto®-style knife (top) and utility knife (bottom).

Scrapers & Deburring Tools

What is a *metal shop scraper*, how is it made, and what is it used for?
Figure 2–19 shows two commercial metal shop scrapers. Scrapers are used to remove burrs and lightly chamfer the edges of drilled and machined holes.

How are *deburring tools* used?
These are commercially made tools with a swivel blade to follow around circles and curves to remove sharp edges and leave a chamfer, Figure 2–19 (bottom).

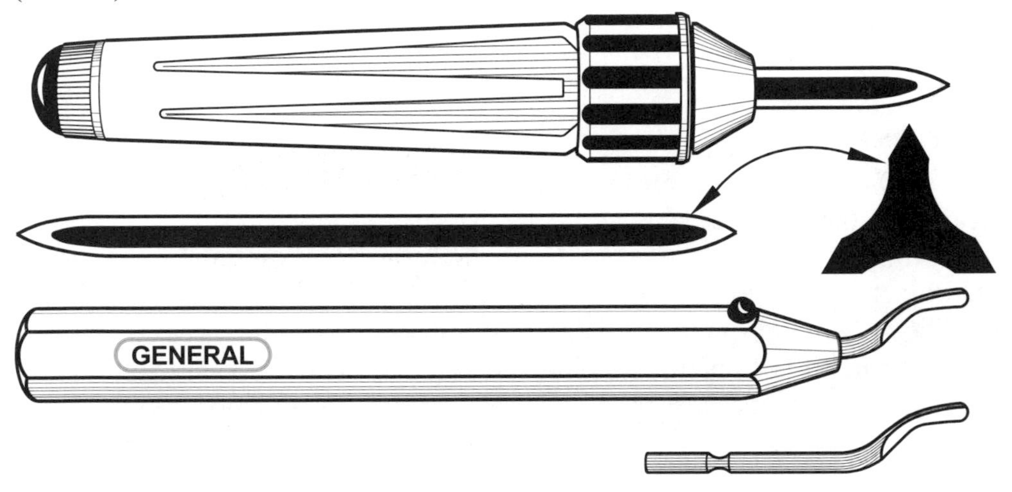

Figure 2–19. Scraper holder and blade (top) and deburring tool (bottom).

Trammel Points

What are *trammel points* used for?
When trammel points, Figure 2–20, are fastened to a metal bar or wood beam, they form an oversized compass and divider. They are useful for scribing circles and marking off equal intervals beyond the capacity of ordinary dividers. The knurled screw on the left-hand leg allows fine adjustment between points without repositioning the heads. Other trammel designs have caliper points for comparing and measuring distances or hold a marking pencil in place of one of the points.

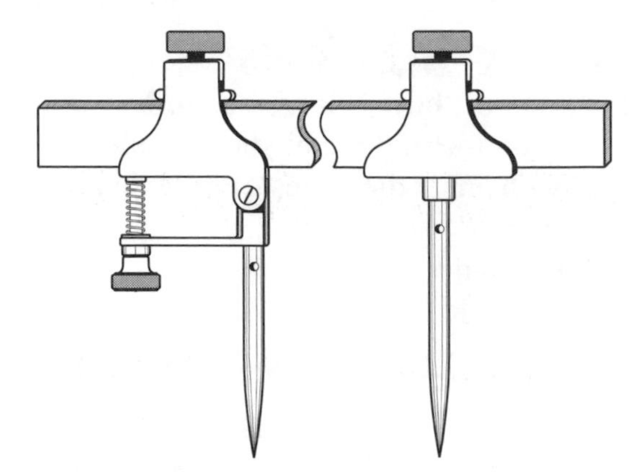

Figure 2–20. Trammel points for scribing larger circles.

Hemostats

How can *hemostats* be useful in the shop?

Because of their ratchet locking mechanism, hemostats can pick up and hold parts firmly with very little effort. Sometimes they are used to pick up hot or cold objects, or to dip or swab an object with fluid while keeping your fingers dry. They handle objects too small for pliers and too big for tweezers such as wire, nuts and other small parts. They are made of stainless steel, so they will not rust. Some hemostats have ends with mating teeth that work well for gripping cloth, rubber, cotton balls and other soft materials. Sometimes hemostats are converted into miniature tongs by grinding off their ratchet mechanisms. See Figure 2–21.

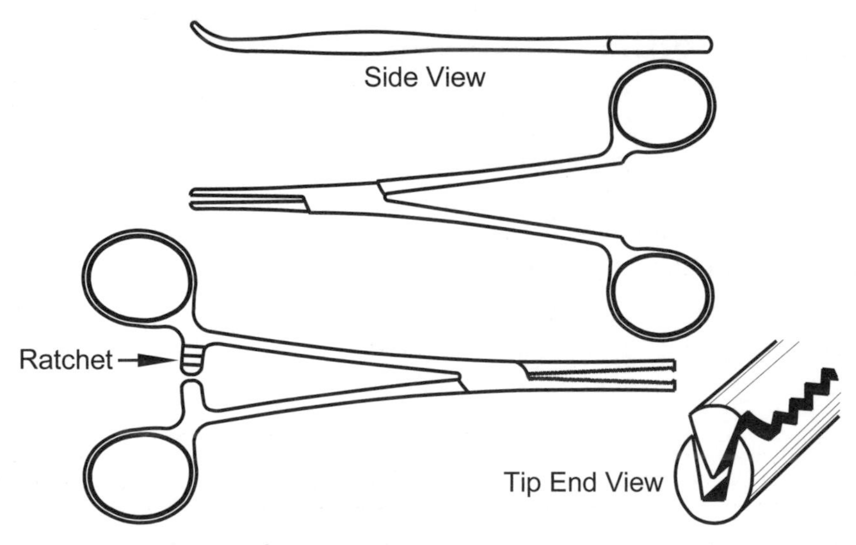

Figure 2–21. Two hemostat designs: without ratchet (top) and with ratchet (bottom).

Chapter 3

Filing & Sawing

The ability to focus attention on important things
is a defining characteristic of intelligence.
—Robert J. Shiller

Section I – Filing

Introduction

Crude files were used since 1100 years BC in the form of stones with ridges running at right angles across them. Like modern files, they were used to sharpen tool edges. The first attempt to make files by machine began in 1490, but the first successful production began in France about 1750. Hammers, cold chisels and punches are blacksmiths' tools adopted by machinists, but files are the machinists' first and most basic tool. Before the development of machine tools, files were the *only* way to cut and shape metals precisely. With skill and patience, early machinists made clocks and locks with files, usually out of brass. Because today there are faster methods of metal removal, files are mainly used to sharpen, smooth edges, remove burrs or make small adjustments, but they remain an important and handy tool. In some third-world countries today, copies of modern firearms are still made using files. Files have the advantages of being inexpensive, portable, able to get into tight spots, and require no electric power.

File Nomenclature

What are the parts of a file?

See Figure 3–1.

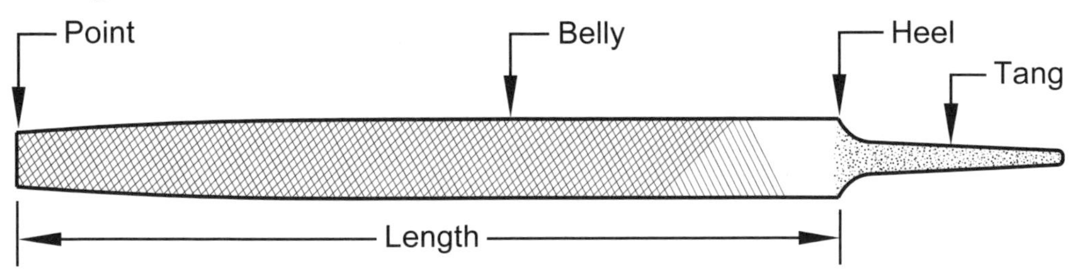

Figure 3–1. Parts of a file.

File Characteristics

What characteristics distinguish one file from another?

There are many, including:

- *Length* for most files lies between 4 and 14 inches. The most common lengths are 4, 6, 8 and 12 inches. File length is measured from the heel to the point and does not include the handle length, except for jewelers' files where the handle length is included.
- *Cross sectional shape* often makes a file more suitable for a specific task. For example, flat files should be used for general-purpose work, a square file for enlarging rectangular holes, and a round file for enlarging round holes. A half-round file can be used for dual purposes: the flat surface for filing flat surfaces, and the half-round one for grooves. Figure 3–2 shows some common file cross sections.

Cross Section	Name	Shape
	Flat	Rectangular
	Hand	Rectangular
	Pillar	Rectangular
	Warding	Thin
	Square	Square
	Three-Square	Triangular
	Round	Circular
	Half-Round	Third-Circular
	Knife	Knife-Shaped

Figure 3–2. File cross sections.

- Figure 3–3 shows four file shapes especially useful for making or enlarging grooves.

Figure 3–3. Files for making grooves.

- Type of teeth:
  - *Single-cut files* have one row of parallel teeth diagonally along their face. With relatively light pressure, they produce a smooth finish on hard materials, see Figure 3–4 (top).
  - *Double-cut files* have two intersecting rows of teeth which produce fast stock removal and good clearing of chips, but leave a coarse surface. Heavy cutting pressure is often used. See Figure 3–4 (bottom).

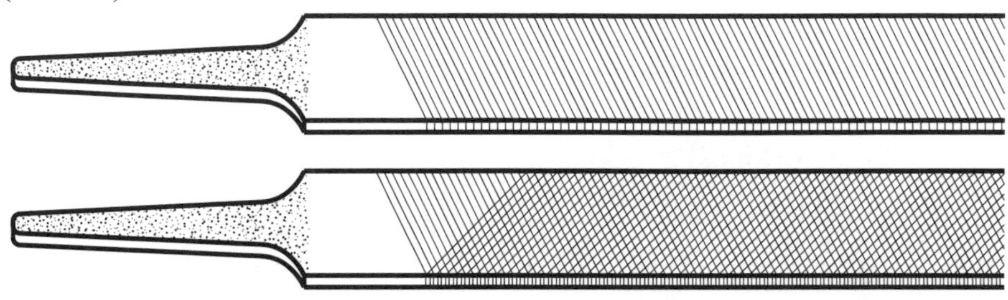

Figure 3–4. Single-cut (top) and double-cut file teeth (bottom).

  - *Rasp-tooth files,* Figure 3–5 (top), have a series of individual rounded teeth raised from the surface of the file blank by a sharp narrow, punch-like cutting tool. They are used for soft materials like wood, leather and lead, where fast material removal is needed. Heavy cutting pressure may be used.
  - *Curved-tooth files,* Figure 3–5 (bottom), work well on softer materials like lead, aluminum, brass, copper, plastics, wood, die cast zinc, auto body fillers and hard rubber. This tooth design tends to resist pinning, filling the teeth with filed particles.

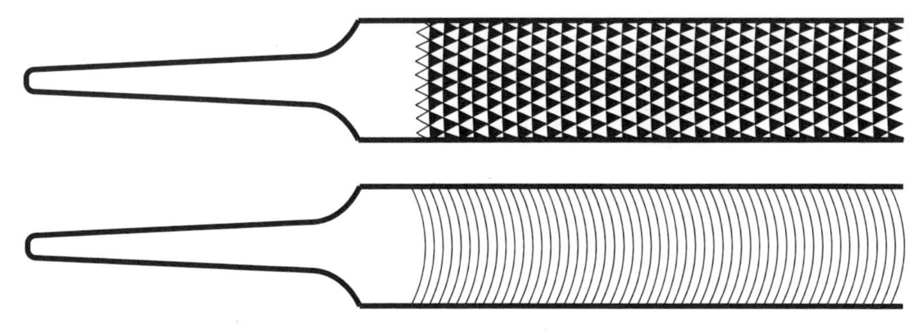

Figure 3–5. Files with special purpose teeth: rasp-tooth file (top) and curved-tooth file (bottom).

- *Coarseness* is the number of teeth per inch of file length, Figure 3–6. Both single- and double-cut files are manufactured in various degrees of coarseness, Table 3–1. No particular number of teeth represents a degree of coarseness because the number of teeth for a given coarseness is proportional to the length of the file. Varying degrees of coarseness are comparable only when files of the same length and shape are considered.

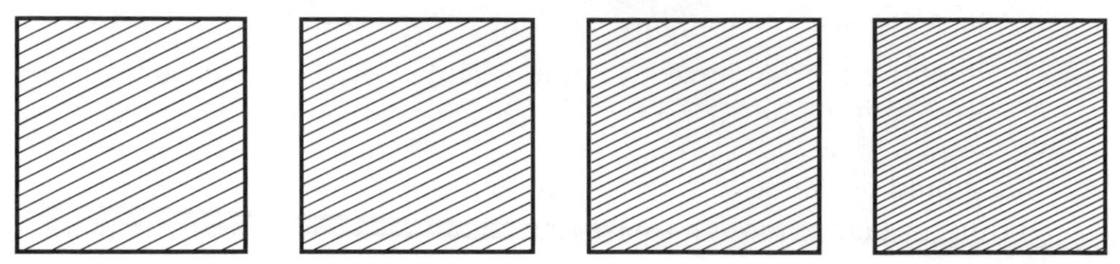

Figure 3–6. File coarseness: coarse to smooth (left to right).

Single Cut	Double Cut
Rough	—
Coarse	Coarse
Bastard	Bastard
Second Cut	Second Cut
Smooth	Smooth
—	Dead or Super Smooth

Table 3–1. File coarseness.

- *Tooth angle* on most files runs between 65° and 85° to the file's length. Most files have a 65° angle.
- *Edge design,* Figure 3–7, with and without a safe edge. Safe edges allow the face of the file to cut while the edge does not. It is a good idea to run a stone along the safe edges of files the first time they are used to insure that the edges *are* smooth.

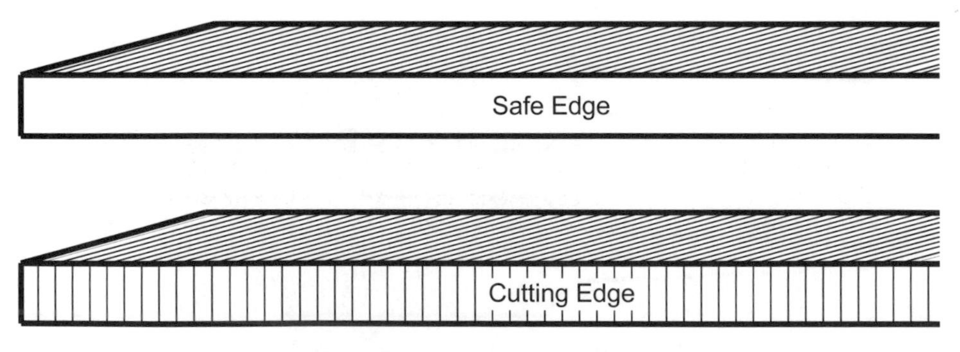

Figure 3–7. Files with and without a safe edge.

Special File Designs

What are two special file designs the machinist will use?

- *Long-angled lathe files* have teeth set at 45° to the file's length instead of the usual 65° and are designed for lathe filing of most metals and for bench filing of aluminum and copper alloys. The single-cut, bastard tooth is self-clearing when used at right angles to work. It cuts clean without chatter and does not drag or tear. Both edges are safe to protect the shoulders of work, which are not to be filed. This file has a rectangular cross section and tapers slightly in width toward the point. In the lathe the file should not be held rigid or stationary, but stroked constantly. A slight gliding or lateral motion assists the file to clear itself and eliminate ridges and grooves. Use a long steady stroke across the work and move laterally about half the width of the file. Lathe filing is most often used to remove sharp edges from shoulders and for sizing shafts when only a little material must be removed. When lathe filing must be performed with a conventional file (65° teeth), hold the file at a 20° angle clockwise to the lathe axis, Figure 3–8. This will place the file teeth at 45° to the work as a lathe file does. Dipping the file in cutting oil or filling its grooves with blackboard chalk when lathe filing improves surface finish. Run the lathe the same speed as for turning the same metal. Uneven pressure on the file leads to an out-of-round condition on the work. Excessive lathe filing or abrasive cloth use usually leads to an out-of-round situation on the work.

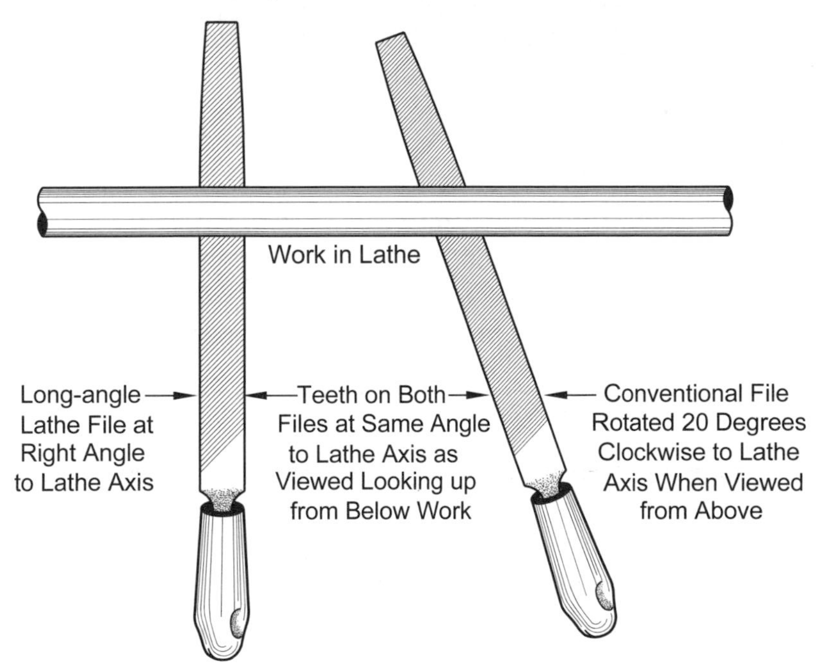

Figure 3–8. Rotating a conventional file by 20° to the lathe axis puts its teeth at the same angle to a long-angle lathe file. This is a view from *below* the work looking up to show the lathe teeth angle in relation to the work axis.

- *Needle* or *Swiss-pattern files,* Figure 3–9, are designed for jewelers, clockmakers, watchmakers, instrument makers and diemakers. They are available in 2-, 3-, 4-, 5½- and 6-inch versions. These files are ideal for making small adjustments on delicate mechanisms. They have integral round, knurled handles or have vinyl-dipped handles like those in the figure.

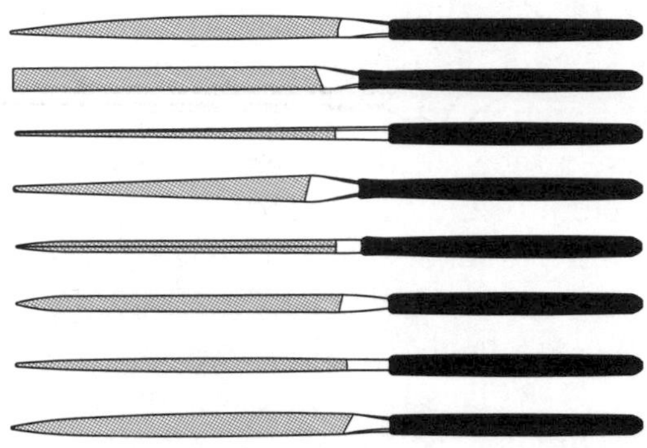

Figure 3–9. Needle files.

Similar size and shape needle files are available with a synthetic diamond coating in place of metal file teeth. The nickel plating holds tiny diamonds on the file. They are excellent for hardened steel, carbide, ceramics and glass. They are for making small adjustments, not for removing lots of metal. Needle files without handles are best held in pin vises. This provides excellent control and is easier on your fingers.

Filing Techniques

What are the common filing methods and how are they performed?

- For *two-handed operation*, the handle should be grasped in the right hand and the point of the file in the left, Figure 3–10. The left-hand thumb reduces file bending under pressure. All filing operations require pressure on the forward cutting stroke and relaxed pressure on the reverse stroke to avoid damaging the file teeth.

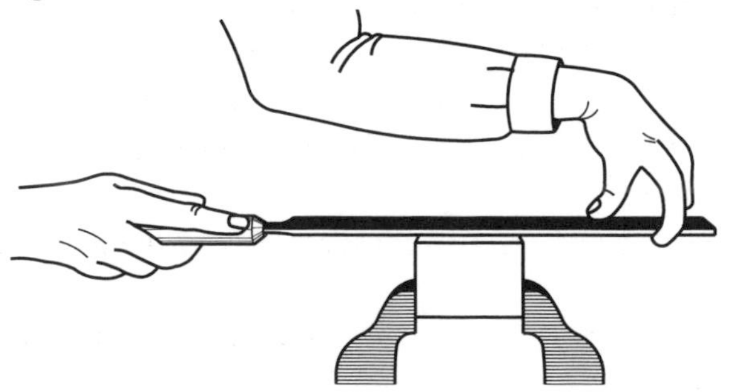

Figure 3–10. The grip for a right-handed person.

- *Flat filing*, Figure 3–11, which removes material rapidly, uses the left-hand thumb to apply forward pressure directly over the work.

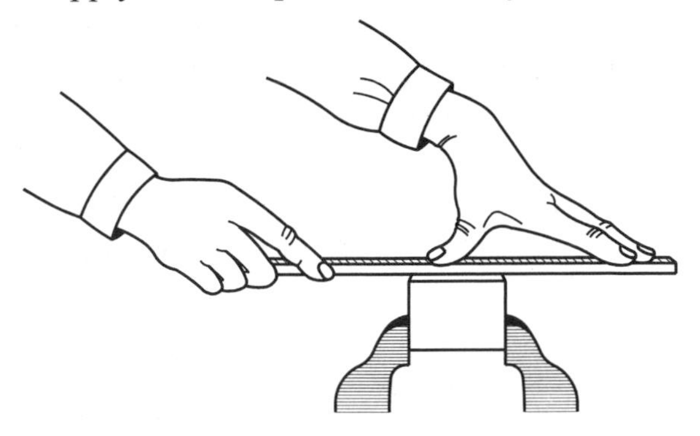

Figure 3–11. Flat filing method for a right-handed person. A left-handed person reverses hands from this drawing.

- *Draw filing*, Figure 3–12, produces a smooth, flat surface and removes the marks and scratches from cross filing. Begin with a coarse file and move to progressively smoother ones. Apply pressure to the file only when pulling it toward yourself. Very flat and smooth surfaces can be produced.

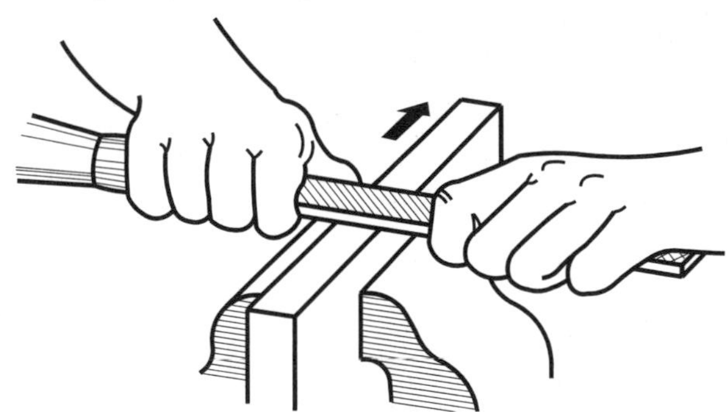

Figure 3–12. Draw filing is a two-handed operation using a single-cut file.

Good Practice

What are some important rules of file use?

- Apply pressure only on the *forward* cutting stroke.
- Always use the widest file that fits the work since a wide blade is easier to keep flat and square. A long file is also easier to hold and keep straight even on short strokes.
- Store files in wall racks, not in drawers where they can bang against each other.
- Using a bench grinder, a file can have a safe edge added or its shape changed to fit the job.

- Using a steel file on hardened steel will destroy the file. Consider using a grinder or disc grinder.
- Too much pressure on a file will crush its teeth, permanently damaging it.
- Some machinists chalk or oil their files to reduce *pinning*, the accumulation of small metal particles in the file teeth.
- Keep the file teeth clean to produce a smooth surface finish. Use a file card, Figure 3–13, or wire brush to remove trapped metal particles. Sometimes a pointed tool must be used to dislodge stubborn file cuttings.

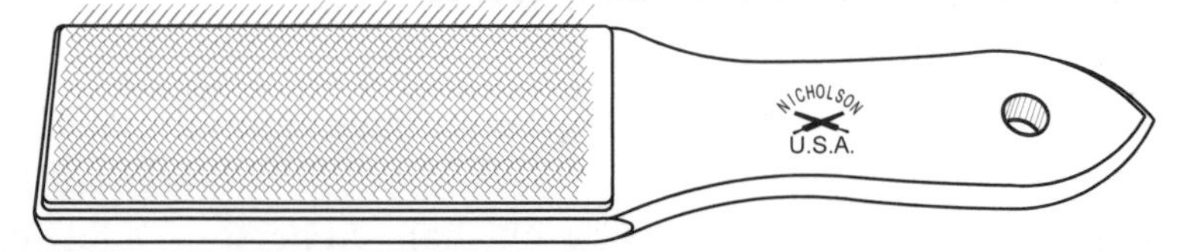

Figure 3–13. File card for removing particles trapped in file teeth.

Securing a File Handle

How should a file handle be mounted on a file?

Hold the file in one hand and push the handle onto the file tang with the other. Then invert the file so its point is up and tap the bottom end of the handle on a hard surface several times to drive the tang further into the handle. *Do not hammer on the handle or point of the file to seat the handle; the file may shatter.*

File Safety

What are the safety precautions for files?

- Never use a file without a handle because there is a chance the tang or handle-end will pierce your palm.
- Do not use files for prying. They are brittle and will snap.
- Do not hammer on files; they will shatter.
- Clean files with a file card, or stiff brush only, not with your hand, or by striking the file against the bench.

Section II – Sawing

Introduction

Saws are ancient tools, but the development of steel, its hardening techniques, and finally steam and waterpower, brought out their full potential. Saws, like most machine tools, remove metal by converting it to chips using sharpened metal teeth. Sawing is really a milling operation. Although saws lack the precision of other machine tools, they sever materials quickly with a narrow kerf. Very often saws convert raw stock into a semi-finished workpiece of the approximate size and shape needed, much like a raw iron casting. Saws can remove large volumes of unwanted stock in just a few cuts. Critical areas on the part are then machined to final size in lathes and milling machines. There are many different saw designs and we shall examine them. Sawing machines differ in their initial cost, portability, accuracy and tool life. Each design offers some advantages.

Saw Applications

What are the saw applications in the machine shop?

There are three types of saw applications:

- *Cutting off* or *severing* raw stock into sizes suitable for machining. Limited accuracy is needed here since stock is cut slightly oversized to allow for machining.
- *Contour cutting* or *cutting out* a part to shape, usually before more machining in mills and lathes, removes large chunks of metal in solid pieces rather than in chips. This saves electrical energy, reduces horsepower requirements, extends tooling life and lowers machining time.
- *Slitting* operations for making collets. These are usually performed in milling machines and are covered in *Chapter 7 – Milling Machines*. The accuracy of slitting operations is higher than severing and cutting.

Hacksaws

What are the main hacksaw designs?

- *Hacksaws*, Figure 3–14, cut most unhardened materials. They are most often used to cut off stock prior to machining. They cut thick stock slowly and require a lot of machinist muscle.

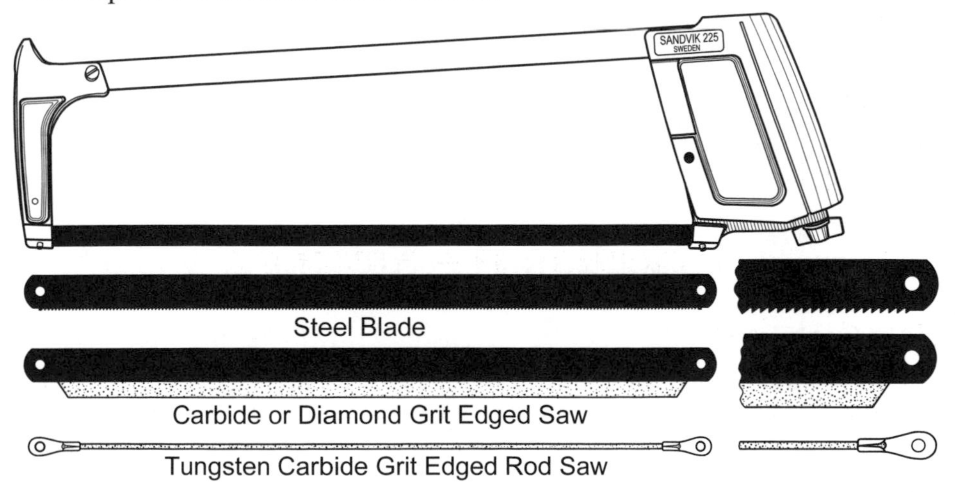

Figure 3–14. A hacksaw with several different blades.

- *Jewelers' saws,* Figure 3–15, are small versions of conventional hacksaws. They cut intricate shapes in thin stock, most often thin metal sheet goods. Blades for them are between 5- and 6-inches long and are available in 22 different teeth counts. Conventional saw tooth designs cut only in one direction, but designs based on a spiral-toothed wire cut in all directions. Watchmakers' supply houses offer a better selection of blades than most industrial tool suppliers.

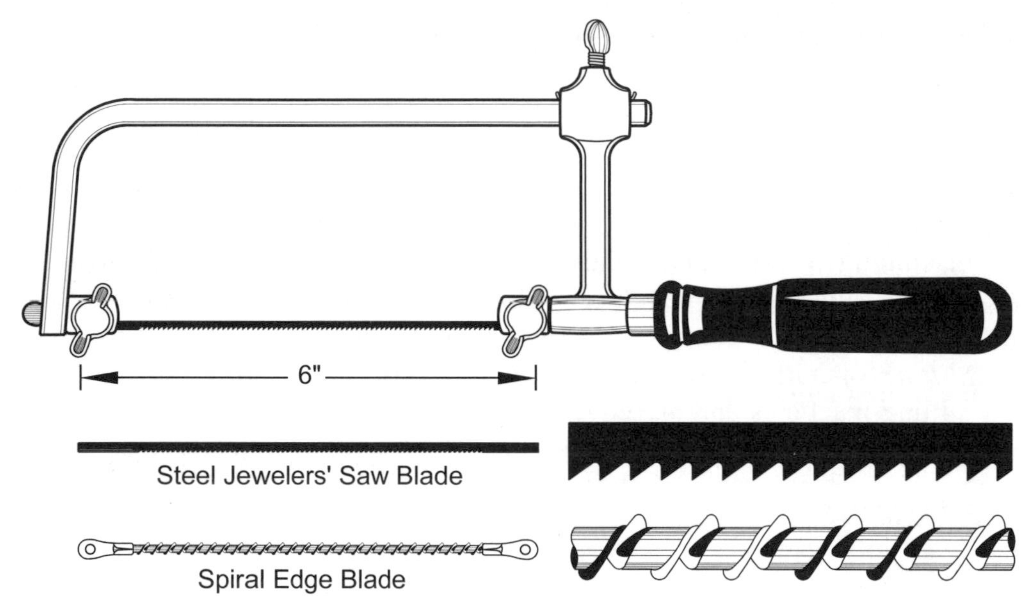

Figure 3–15. Jewelers' saw.

- *Power hacksaws* sever all types of unhardened metal stock including rounds, flats, structural shapes and tubing. They are not precision tools and are used mainly for cutoff operations on raw stock. They are inexpensive compared with bandsaws. Most models flood the cutting area with coolant, so the blades last a long time.

How are hacksaw blades constructed?

Blades for hand hacksaws are 10- or 12-inches long, either ½- or ⅝-inches wide, and about 0.030-inches thick. Blade length is measured between the two blade mounting holes. In order of increasing cost, there are three choices for toothed blades:

- *Carbon steel* is an economy grade blade for mild steel, copper, brass and aluminum.
- *Molybdenum HSS steel* blades are hardened from its teeth to the top of its back. They work well on hard materials held securely, but their blades are brittle and likely to snap if the saw twists in the cut.

- *Bi-metal blades* are the next step up in cost and cutting life. These blades feature hardened teeth coupled with a softer, more flexible back, which is less likely to snap if twisted. These saws last 20 to 30% longer than carbon steel blades and are the most economical on a per cut basis.

There are two more choices for blades and they use carbide grit:

- *Carbide grit blades* cut hardened steel, stranded cable, glass and tile. Their lower edge is grit-coated, Figure 3–14.
- *Carbide grit rod saws* cut similar materials as carbide grit blades by cutting in any direction. They may be threaded through an access hole and remove a section of material without a cut from the outer edge.

What is the rule of thumb for selecting the number of teeth on a hacksaw?

Keep at least two teeth on the cutting edge to prevent the blade from snagging on the work. Blades for hand hacksaws are available in 14, 18, 24 and 32 teeth per inch.

What is the proper procedure for using a hand hacksaw?

1. Check that the saw has the proper pitch for the work.
2. Make sure the blade teeth point away from the handle and the blade is pulled up tight.
3. Secure the work in a vise if possible. Keep the work within ¼ inch from the vise jaws to keep the work rigid.
4. To insure an accurate starting point for the cut, file a small notch in which to start the blade.
5. Hold the hacksaw as in Figure 3–16.
6. Apply pressure to the saw only on the forward stroke. Backstrokes do not cut and dull the teeth.

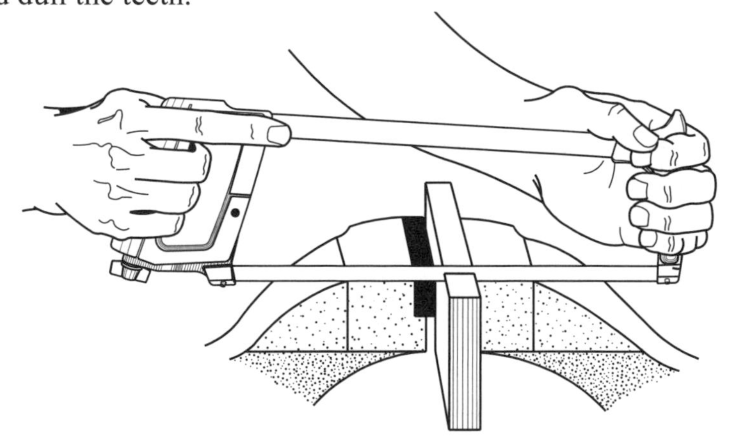

Figure 3–16. How to hold a hacksaw.

When cutting thin sheet metal, clamp the metal to a piece of wood and make the cut through both. This will prevent bending the metal when cutting. For very thin sheet goods, sandwich the work between *two* pieces of wood.

What is the best way to cut a new slot in a bolt, screw or cap screw when the original slot is damaged?
Mount *two* hacksaw blades side by side in the saw frame to make a slot of adequate width in a single cut.

If a blade breaks while hacksawing, what procedure should be followed?
Replace the blade, then if possible, turn the work one-half revolution and begin cutting from the other end of the work. This eliminates starting in the narrower groove made by the old blade which will damage the new blade's teeth. Wear causes the older saw's teeth to be narrower than the teeth on a new blade of the same type.

Bandsaws

What is the most common bandsaw design used in machine shops and how is it constructed?
The *vertical metal-cutting bandsaw* or *contour saw*, Figure 3–17, is the most common design. Although it can perform cutoff operations on raw metal stock, its real value is contour sawing.

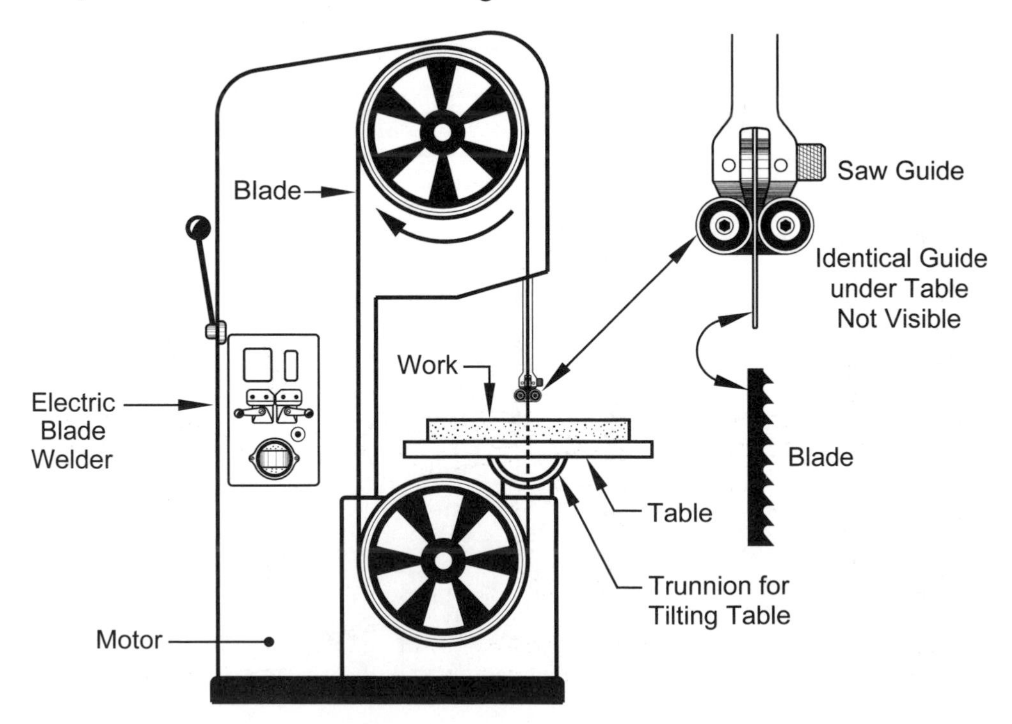

Figure 3–17. Vertical metal-cutting bandsaw.

Here are some of its characteristics:

- It has a continuous, endless blade, called a *band*, which is tensioned and turned on two or three wheels.
- Bands are made by electrically welding a strip of steel-toothed blade material end to end. Industrial-grade bandsaws have an electric welder built into them for blade welding.
- The band can be cut, fed through a hole in the workpiece and rewelded to remove internal sections without an access cut, Figure 3–18.
- The band, which turns continuously in one direction, has teeth on one side only. The workpiece is fed into a portion of the band where the teeth mill away the workpiece material.
- Bands remain sharper longer than a hacksaw blade of the same design because wear is distributed evenly over many more teeth.
- The length of cut is unlimited and the cut can be made at any angle or direction.
- Because cutting action is downward, blade forces pull the workpiece against the table and usually eliminate the need for clamping.
- Contour sawing is efficient as the kerf is narrow and little material is turned to chips and wasted.
- Bandsaws for contour cutting typically have motors from 1½ to 5 horsepower.
- Heavy industrial bandsaws usually have a mist or flood lubrication system. Lacking a lubrication system, solid lubricant is applied directly to the blade.

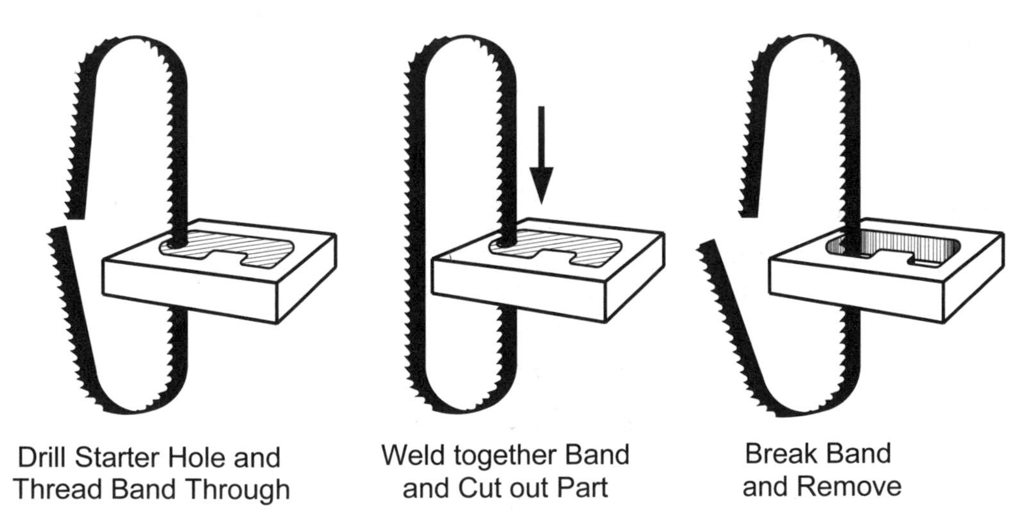

Figure 3–18. Cutting out an internal section.

How do metal-cutting bandsaws differ from those used for wood?
Metal-cutting bandsaws generally have:

- More and smaller teeth than wood-cutting blades.
- More chip clearance as with skip-tooth blades.
- Blade speeds in the 70 to 250 fpm range, whereas wood-cutting bandsaws have blade speeds as high as 5000 fpm. A metal-cutting blade in a wood-cutting bandsaw may not cut properly due to lack of chip clearance.

With what accuracy can parts be cut on a bandsaw?
An experienced machinist can cut parts ±0.010 inches. Some automated machines, either digital or tracing, can make cuts within ±0.002 inches.

What are the main factors in choosing a bandsaw blade and setting up the saw?

- *Blade pitch* is the number of teeth per inch. Pitches from 2 to 32 teeth/inch are available. Based on the thickness of the workpiece, choose a blade with enough teeth so at least three teeth are in contact with the work at all times. For metal cutting, it is ideal to have 6 to 12 teeth in contact with the work. More teeth will provide a smoother finish.
- *Blade width* is measured from the tooth tip to the back or other side of the blade. A wide blade is stronger than a narrow one, but it cannot cut as small a radius, Table 3–2. Wide blades are ideal for straight cuts. Choose the widest blade consistent with the smallest required radius.

Blade Width (inches)	Smallest Radius (inches)
1/16	1/16
3/32	1/8
1/8	7/32
3/16	3/8
1/4	5/8
5/16	7/8
3/8	1¼
1/2	3

Table 3–2. Minimum cut radius for a given width bandsaw blade.

- *Set* is the clearance teeth provide for the blade back. The more blade teeth depart from their centerline, the wider the set.
- *Tooth form* is the shape of the tooth.
  - *Standard-tooth blades* have well-rounded gullets (the dips between the blade teeth) and work well for ferrous metals, hard bronze and brass.

  - *Skip-tooth blades* have more gullet and chip clearance than a standard tooth blade and work well for aluminum, magnesium, brass and wood.
  - *Hook-tooth blades* work well for plastic, hardwood and nonferrous metal.
  - *Variable-pitch blades* offer reduced noise and vibration and more even wear on their teeth.
- *Blade material* offers three main choices:
  - *Bi-metal* is an all-purpose blade. It cuts a wide range of materials and shapes, but has the highest initial cost and the longest life, up to ten times longer than carbon steel. These blades are the most cost-effective for cutting steel pipe and tubing with welded seams. Weld metal becomes hardened with the absorption of carbon and rapid cooling.
  - *Hard-back premium carbon steel* for light- to medium-duty sawing. Its cost is between bi-metal and flex-back carbon steel.
  - *Flex-back carbon steel* is the lowest initial cost blade, for light-duty cutting. The flexible back provides greater fatigue resistance at higher speeds than hard-back blades.
- *Saw blade speed* is determined by the work material. Most industrial band saws have a blade selector which will indicate the best blade and cutting speed for a given thickness and material. *Machinery's Handbook* also has blade speed settings.

What is the procedure for using a metal-cutting bandsaw?

1. Layout the shape to be cut, then prick punch it. This will insure the cut lines are visible even if the scribe lines are not.
2. Select the saw blade design to match the work (see section above) and set the blade speed.
3. Turn on the lubrication system, if present.
4. Cut along the layout lines.
5. If the workpiece is small, use a work holder, really a big clamp, which will keep your fingers away from the blade while giving you complete control of the work.

A vertical bandsaw is a great convenience and timesaver in the prototype and model shop, but it is not an essential tool. Milling machines, drill presses, hacksaws and files can get the job done in the absence of a bandsaw, just not as fast.

Other Saw Designs

What other type saws are used in the machine shop?

- *Sawzall®-type reciprocating saws* are principally used for cutting off raw metal stock—plates, rounds, bars, pipe, tubing and blocks—into smaller pieces prior to being machined. They cut metal, plastics, wood, and with the right abrasive blade, ceramics. These saws save the machinist from using a hacksaw. However, reciprocating saws cannot hold to tolerances less than 1/16 of an inch and are not for precision work. Because of the single projecting blade, cuts into the center of large, thick pieces of metal are possible, particularly with Cobalt HSS blades. See Figure 3–19. Many different blade designs are available. In general, purchasing industrial-grade blades will be most cost-effective.

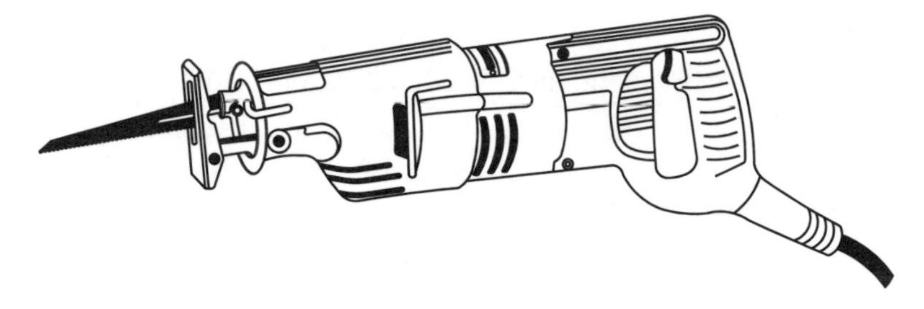

Figure 3–19. Sawzall®-type reciprocating saw.

- *Portable bandsaws*, Figure 3–20, like the reciprocating saw above, cut raw metal stock into smaller pieces. Although its geometry prevents it from cutting into the center of large plates, because of its continuous cutting action, it cuts faster than the reciprocating saw. They are mainly used on bar stock, shapes, pipe and tubing.

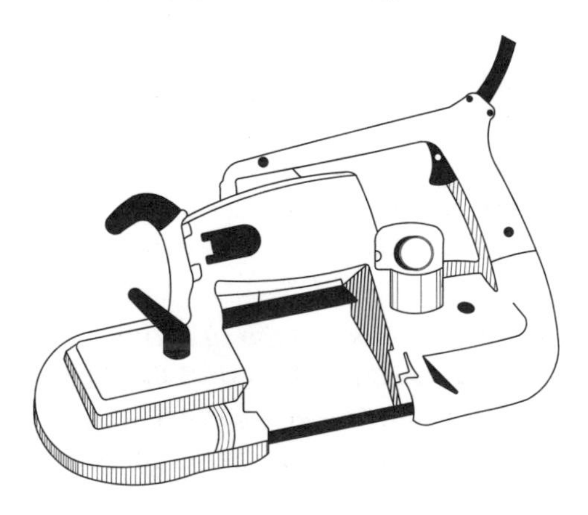

Figure 3–20. Portable bandsaw.

Chapter 4

Grinding, Reaming, Broaching & Lapping

In the republic of mediocrity, genius is dangerous.
—Robert G. Ingersoll

Introduction

Although grinding, reaming, broaching and lapping remove a relatively small metal volume compared with turning and milling, they are important processes because they are often the last machining operation and determine final size and finish.

The basic versions of these processes are easy to learn and master. Steps to perform them are detailed here. Only the specialized grinding operations, center-less, cylindrical, internal and surface grinding, that use dedicated machines, are beyond the range of most machine shops. Although this type grinding may be sent out to a grinding shop, it is often possible to purchase components that already have the ground surfaces needed and incorporate them into the project. This avoids the delay and inconvenience of sending out work. Examples of already-ground components are ready-to-machine ground flat stock, bearings, drill rod, drill blanks, reamer blanks and dowel pins.

Section I – Grinding

Grinding Mechanics

How does the *grinding process* remove material?

When work is brought into contact with a rotating grinding wheel, each abrasive grain on the wheel's surface acts as a cutting tool and removes a tiny metal chip. When a grain becomes dull, the extra force between the wheel and the workpiece causes the dulled grain to break away due to a soft bond or friable grains. These fractures expose new sharp edges. Grinding wheels are the only self-sharpening cutting tools.

Grinding Processes

What are the main *grinding processes* and their applications?

- *Offhand grinding* applies size and shape to workpieces. The grinding wheel may be fixed and work hand-held or vice versa.
- *Cutoff grinding* works well for severing and slotting hard materials, particularly HSS stock, drill rod, dowel pins and reamer blanks. Damaged end mills are often cut off with abrasive wheels in preparation for regrinding.
- *Tool grinding* involves the sharpening of drills, taps, milling machine and lathe cutters.
- *Surface grinding* applies a flat surface and brings the workpiece to the desired thickness and surface smoothness. It is the most common grinding operation and comprises over 75% of production grinding.
- *Cylindrical grinding between centers* brings parts to exact diameter, concentricity and surface smoothness. Diameter tolerances in the tenths of thousandths of an inch are common. Precise tapers may also be applied. Although purpose-built machines perform cylindrical grinding in production, this process may be done in lathes with a grinding attachment (See *Chapter 7 – Turning Operations*).
- *Centerless grinding* produces the same results as cylindrical grinding, but is done when the work cannot be held between centers, usually because the work is too thin to resist bending. Also used for high production applications where it eliminates the need for centers.
- *Internal grinding* brings a bore to size and surface finish. Tapers can also be applied this way. A chuck holds the workpiece so it can be rotated as its bore is ground. This operation may also be performed on a lathe with a grinding attachment.
- *Form grinding* uses a shaped grinding wheel whose contour is transferred to the workpiece. Precision threads are often applied by form grinding.
- *Snagging* removes relatively large amounts of metal from billets, castings and welds where tight tolerances or surface finish requirements are not important. This process is used in steel mills and foundries.

Table 4–1 shows where these processes are usually performed. Cylindrical, centerless, internal and form grinding usually require dedicated, expensive machines and skilled, full-time operators. This chapter focuses on the abrasive processes common in the general machine shop, the last three items in Table 4–1.

Process	Where the Process Is Usually Used		
	Production	Tool Room	General Machine Shop
Cylindrical grinding between centers	●	●	Note
Centerless grinding	●	●	
Internal grinding	●	●	Note
Form grinding	●		
Snagging	●		
Surface grinding	●	●	
Cutoff grinding	●	●	●
Offhand grinding	●	●	●
Tool grinding	●	●	●

Note: Limited grinding, such as truing lathe centers, can be done in a lathe equipped with a tool post grinder.

Table 4–1. Where grinding processes are performed.

Bench & Pedestal Grinders

What are *bench* or *pedestal grinders* used for?

These grinders, Figure 4–1, perform offhand grinding to:

- Sharpen pointed tools like dividers, center punches, chisels and scribers.
- Form and dress HSS lathe cutters.
- Grind new blades on screwdrivers.
- Remove burrs from drill shanks and the ends of shortened fasteners.
- Apply a straight, angled or rounded edge to metal stock.
- Remove mushrooms from chisel heads.

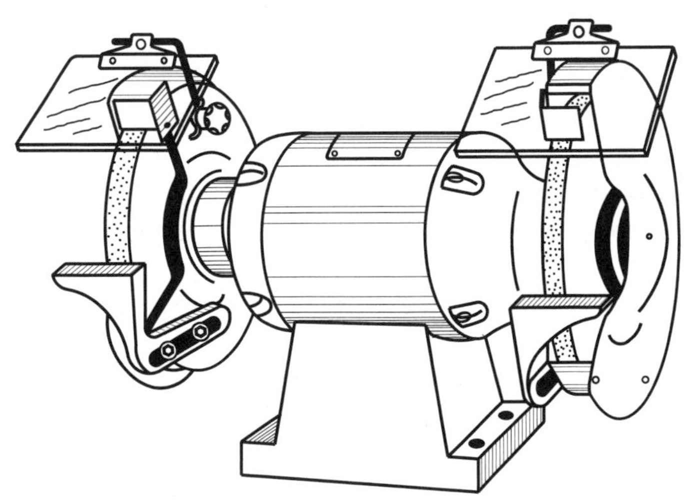

Figure 4–1. Bench grinder.

Typical grinders usually have 6-, 8- or 10-inch diameter wheels and ⅓-, ¾- or 1½-horsepower motors. Usually one abrasive wheel is medium (60 grit) and the other is coarse (36 grit).

Why are there two speeds of bench and pedestal grinders?
Most bench and pedestal grinders rotate at 1700 or 3400 rpm. Two speeds are offered so the wheel speed can be matched to the abrasive. A higher wheel speed works better with softer wheels; a slower wheel speed works better with harder wheels. Unfortunately, new 3400-rpm grinders are often supplied with hard wheels and so have a tendency to burn HSS cutting tools. If such an unfavorable combination of abrasive wheel and motor speed must be used, dip the cutter in water frequently when grinding to cool it.

What is the difference between *truing* and *dressing* an abrasive wheel, and how are these operations performed?
Truing removes enough wheel material from its periphery to bring the wheel back into roundness. That is, the wheel edge is again a perfect circle and its center concentric with the grinder spindle. Truing is performed with a wheel dressing tool. The steps are:

1. Start the grinder and allow it to come up to full speed.
2. *Wearing a face mask and eye protection,* position the handle of the wheel dresser parallel with the grinding wheel and bring the steel wheels on the wheel dresser into contact with the edge of the rotating abrasive wheel, Figure 4–2.
3. Move the wheel dresser side to side along the face of the wheel until the wheel surface is clean and approximately square with the sides. Considerable pressure may be required against the dresser on larger diameter wheels. Never apply the wheel dresser to the sides of the grinding wheel.

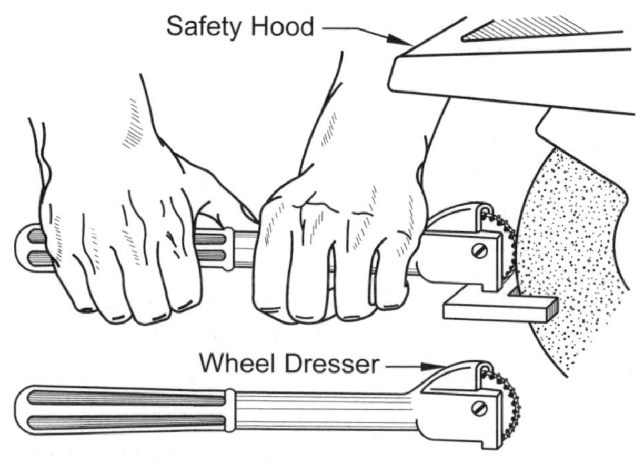

Figure 4–2. Truing an abrasive wheel with a wheel dresser.

Dressing an abrasive wheel with an *abrasive stick*, also called a *wheel dressing stick*, exposes fresh cutting surface and flattens the wheel face, Figure 4–3. Dressing removes much less material than truing. The sticks are either boron carbide or silicon carbide. Dressing can also be done with a diamond dresser such as those used to shape surface grinder abrasive wheels.

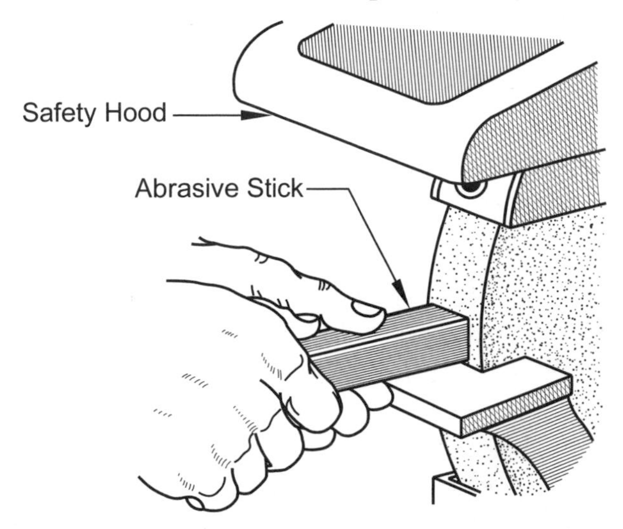

Figure 4–3. Dressing an abrasive wheel with a silicon carbide special purpose abrasive stick. In Britain these are called *Devil's stones*.

Grinding Wheels

What materials are used to make grinding wheels?

The most common abrasives in grinding wheels are:

- *Aluminum oxide.* This is the most common abrasive, usually gray in color, but can also be white or pink. It is used for grinding steels.
- *Silicon carbide.* The second most common abrasive, usually green in color. It is used for grinding nearly all other materials except steel, including cast iron, bronze, copper, aluminum, stone, rubber and cemented carbides.
- *Synthetic diamond.* These grinding wheels are frequently used to sharpen carbide cutters. However, diamond cutoff wheels are common too. They work well on most high-tensile strength materials.

What type grinding wheels are used in bench or pedestal grinders?

Most often they are vitrified bonded aluminum oxide which works fine for steels, but carbide tools require green silicon carbide or diamond wheels.

How are the characteristics of grinding wheels identified by their manufacturers?

Wheel manufacturers use an ANSI specification in which a string of numbers and letters indicates the wheel's characteristics:

- *Abrasive type,* aluminum oxide (A), silicon carbide (C) or diamond (D).
- *Abrasive grain size* or *grit*, see Table 4–2.

Grain Size	Grit Number
Coarse	8–24
Medium	30–60
Fine	80–180
Very Fine	220–600

Table 4–2. Grain size and grit numbers.

- *Grade* or *hardness* of the grinding wheel refers to the tenacity with which the particles of grit are held in the wheel, not to the hardness of the abrasive particles themselves.
- *Structure* indicates the wheel density, the spacing between the grains and the bonding material.
- *Bonding method* indicates what materials hold the abrasive in place. Many wheels are vitrified, others held by shellac, rubber, reinforced rubber or epoxy resins.

Machinery's Handbook has the details of this wheel classification system.

Is it true that the harder the workpiece material, the harder the grinding wheel should be?

No, it is just the opposite. Hard workpiece materials use soft wheels so that new, sharp abrasive grains rapidly replace the ones dulled in cutting.

Grinder Safety

What are the most important steps for grinder safety?

- Before installing new grinding wheels, they should be *ring tested* to insure they are free from cracks. Suspend wheels from a finger or pin through their center holes and strike the wheel at the 2, 4, 8 and 10 o'clock positions with a wooden mallet or plastic hammer. A crack-free wheel will produce a clear, resonant tone when struck; a damaged wheel will produce a dull thud. The bond inside these wheels is ceramic and they can suffer cracks like a dish.
- Never exceed the maximum operating speed or the wheel may explode. This never-exceed speed is marked on the wheel or, if it is small, on its packaging.
- When mounting a new wheel, insure the wheel-mounting flanges are clean, that the wheel fits over the spindle easily, and the nut holding the

wheel on the spindle is snug, but not tight. *The left-hand wheel is held on the spindle by a left-handed nut,* so they will self-tighten.
- Allow a newly mounted wheel to run for a minute or two before using it.
- Always stand to one side of a grinding wheel when starting it.
- *Never* grind on the sides of a wheel. Grinding on the face of the wheel applies forces opposing the centrifugal forces trying to pull the wheel apart, but side grinding applies forces which *add* to the forces trying to pull the wheel apart and may cause wheel failure.
- Do not abruptly force work into the face of the wheel. Bring the work gently to the wheel and gradually increase the force against the wheel.
- Keep rags and clothing away from the wheel as they can draw the rag and you into the wheel.
- Always wear eye protection, position safety shields properly and set the tool support to within 1/16 of an inch to the face of the wheel.

Disc Grinders

What are *disc grinders* used for and how does their use differ from bench or pedestal grinders?

Disc grinders, also called *disc sanders* and *metal finishing machines*, Figure 4–4, are used to:
- Grind almost all materials including metals, plastics, wood, rubber and fiber.
- Remove metal, paint, plating or dirt from a large area.
- Flatten, square up or bevel a workpiece edge or surface.
- Round the corners on flat stock.
- Deburr drilled holes and machined parts.
- Chamfer threads that have been shortened.
- Grind soft materials, like aluminum and brass, that tend to quickly load up grinding wheels and make them stop cutting.
- Grind a small flat on a round part so it can be center punched and drilled.

A 12- or 20-inch grinder with a 1½-horsepower motor can remove large volumes of metal rapidly and is effective in squaring or rounding the ends of steel bars as large as 2-inches square. While many operations can be performed on either bench grinders or disc grinders, bench grinders usually shape points and ends of smaller workpieces, while disc grinders smooth large areas and thick edges. Also, bench grinders work better and last longer when grinding hardened materials than disc grinders whose abrasive discs are thin compared with grinding wheels. Disc grinder tables often have a groove for a miter gage to hold the work at a precise angle to the disc.

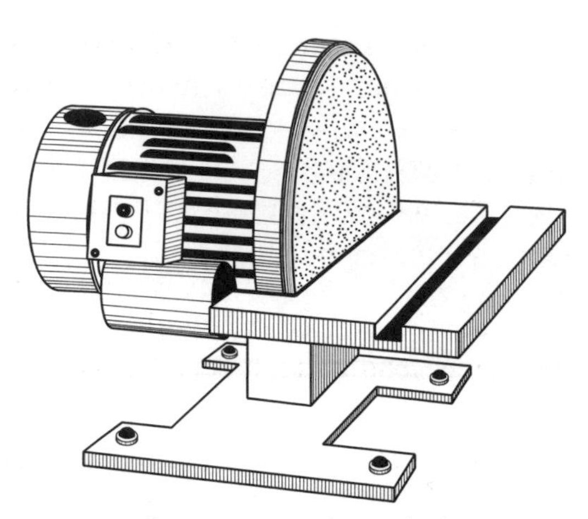

Figure 4–4. Disc grinder.

What are the standard abrasive disc materials?

Aluminum oxide abrasive resin bonded to cloth with a pressure sensitive adhesive (PSA) backing is the most popular abrasive disc design. Older style abrasive discs without the PSA backing require adhesive application to both the disc sander wheel and the abrasive disc. Discs are available from 50 to 120 grit.

Are there tasks for which a bench or pedestal grinder works better than a disc grinder?

Yes, the square edge of grinding wheels allows cutting shapes like fish mouths in pipe and sharp corners in lathe cutters, which cannot be done with disc grinders.

Disc Grinder Safety

- Eye protection is required.
- If the workpiece material produces hazardous dust when ground, a protective mask is required.
- Keep rags, hair and clothing away from the disc. They may be drawn into the space between the table and the disc.
- Do not wear gloves.
- Adjust the gap between the abrasive disc wheel and the worktable to under 1/16-inch wide. *Never adjust the table position with the wheel turning.*

Hand-held Grinders

What are *hand-held grinders* used for?

Hand-held grinders, Figure 4–5, are for free-hand grinding, where the work is fixed or clamped and the grinder is hand-held, just the opposite of free-hand grinding with a bench or pedestal grinder. There are both electric- and air-driven grinders in many sizes and designs. Grinders usually turn between 10,000 and 65,000 rpm. The largest grinders smooth the edges of castings or grind out welds, the medium-sized ones remove metal or smooth parts in gunsmithing, and the smallest ones make adjustments in dental work, manufacture and repair jewelry and precision instruments. There are very inexpensive hobby-type grinders that work well for small jobs, but they overheat under continuous duty and will eventually burn up. Air-driven grinders are suitable for continuous, heavy work with few problems. Even repeatedly stalling them under load will not cause damage.

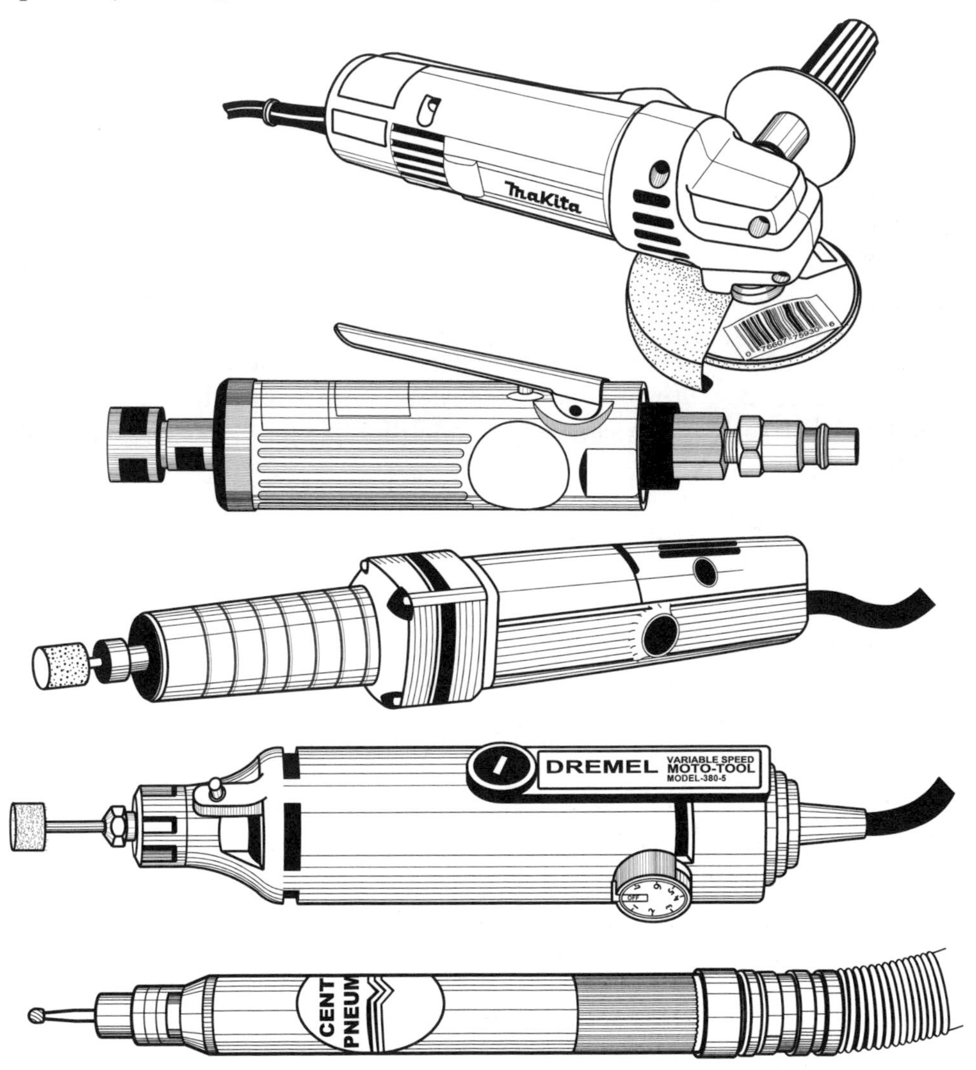

Figure 4–5. From the top, 4-inch right-angle grinder, air-powered die grinder, large electric die grinder, Dremel® grinder and pencil-size air-powered die grinder.

Why are some air-powered grinders *front-exhaust* and other *rear-exhaust?*

In a confined space, corner or hole, the air stream of a front-exhaust grinder blows the cuttings back into the operator's face. A rear-exhaust grinder works better here, but a front-exhaust is desirable out in the open where its exhaust stream blows cuttings away from the operator and the work. Most air-driven grinders are rear-exhaust.

Abrasive Cutoff Saws and Wheels

What are the advantages of using abrasive cutoff wheels?

Abrasive wheels are:

- Available in a wide range of sizes from the industry standard of 14-inch diameter wheels used in cutoff saws, Figure 4–6, to medium-size wheels for bench and pedestal grinders, down to ½-inch diameter wheels for electric- and air-powered hand-held grinders, Figure 4–7.
- Able to cut hardened alloys, shapes, angles, pipe and tubing rapidly.
- Thin, and therefore little material is lost in their kerfs.
- Used for cutting new slots in damaged screw heads, shortening threads on screws and bolts (even hardened ones like socket head cap screws), shortening drills, cutting off nails and removing frozen nuts.
- Used to cut round and square steel tubing that has internal seam welds that are hard on bandsaw blades.

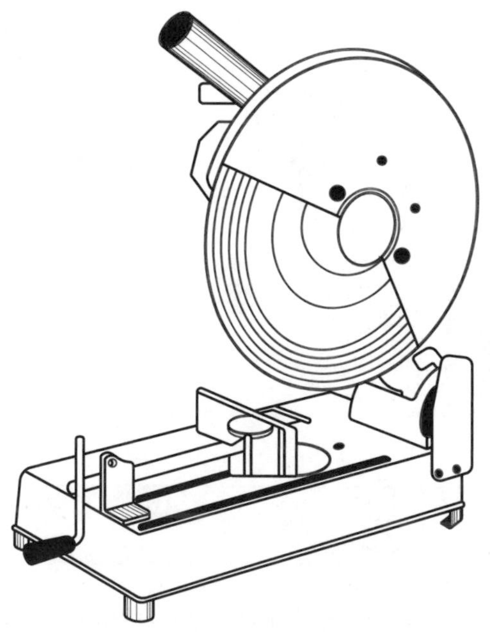

Figure 4–6. Abrasive cutoff saw with 14-inch abrasive wheel that is suitable for cutting steel pipe, rod, angles and tubing.

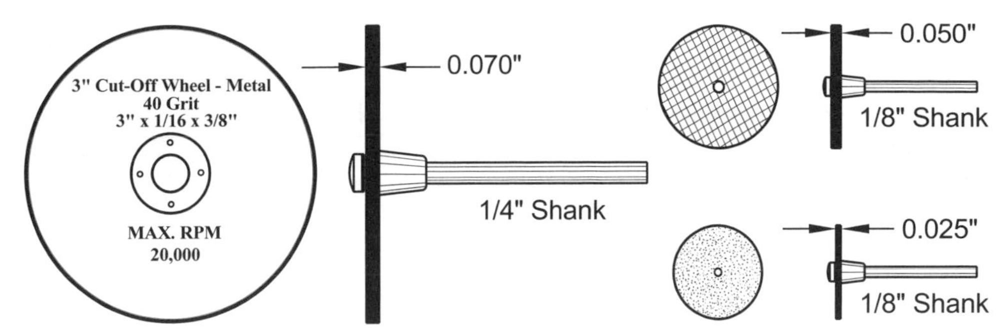

Figure 4–7. Small abrasive cutoff wheels: 3-inch wheel on ¼-inch shank (left) and ⅛-inch shank for small hand-held grinders (right).

Other Grinder Functions

What other wheels can be held by bench, pedestal or hand-held grinders?

- *Wire wheels* for removing dirt, rust and paint.
- *Flap wheels* for smoothing welds, removing rust and working on large areas.
- *Cloth* and *felt buffing wheels* for polishing, usually with a mild abrasive which is applied to the wheel itself.
- *Scotch-Brite*™ *Multi-Finishing wheels* for light polishing and deburring.

These wheels come in sizes to fit most grinders. See Figure 4–8.

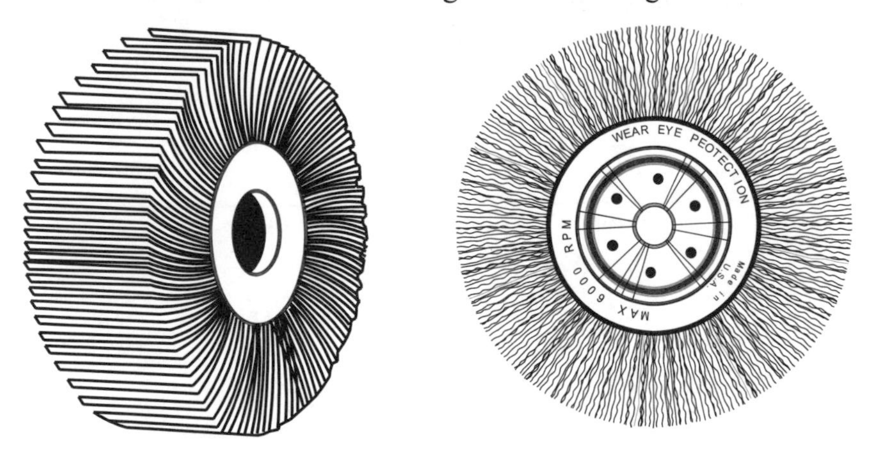

Figure 4–8. Flap wheel (left) and wire wheel (right).

Hand-held grinders also work well for powering other cutting tools mounted on ¼- or ⅛-inch diameter shanks:

- *Shaped abrasive stones* mounted on shanks are available in aluminum oxide and diamond, and in many different shapes like discs, cylinders and cones. Both vitrified and resin bonded stones are available.

- *HSS* or *carbide burrs*, which are really small rotary files, come in many shapes and sizes and are useful for removing small amounts of metal freehand. While not really abrasives, burrs are mentioned here because of their use in high-speed, hand-held grinders and their ability to grind HSS.

See Figure 4–9.

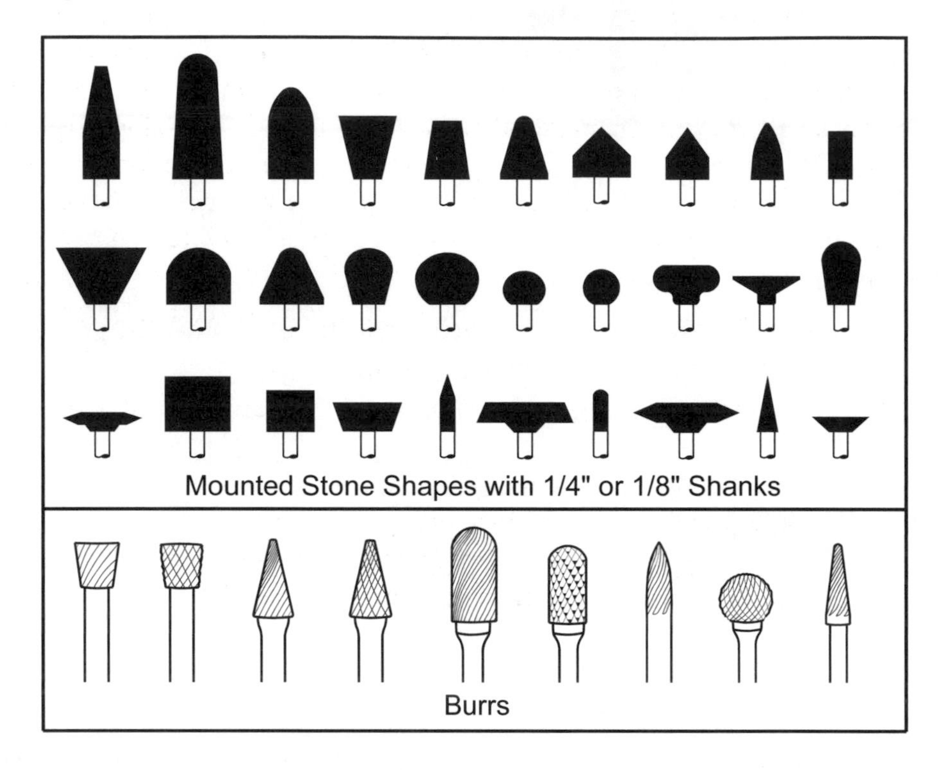

Figure 4–9. Mounted grinding stones and burrs work well in hand-held grinders.

Section II – Reaming

Definition & Purpose

What is *reaming*, why is it important, and how is it done?

Reaming is a cutting operation that slightly enlarges an existing hole to an exact size and smoothes its walls in the process. Although larger holes can be bored or ground to exact size, smaller holes must be reamed. Reaming is also the fastest way to smooth the walls of drilled holes, which are often too rough for many applications. Reaming can be performed by manually turning a reamer in the work hole, but is best done in a drill press, milling machine or in a lathe where the reamer is held concentric with the hole and turned by machine.

Mechanics

How does the reaming process work?

Most metal removal occurs on the 45° chamfered front ends of a chucking reamer's flutes, Figure 4–10. Although the sides of the flutes remove a small amount of metal, they mainly center the reamer, provide chip clearance room and smooth hole walls. Best reaming action and accuracy occurs when the hole is undersized so the reamer removes the proper amount of material. If the hole is too small, too much material must be removed, and chips have no place to go. When this happens they interfere with reaming action and leave rough holes. If the hole is too large, too little material will need removal, and the reamer burnishes rather than scrapes away material, leaving an undersized hole. A reamed hole will be about 0.0005 inches smaller than the reamer itself because the hole material springs back into place when the reamer is removed.

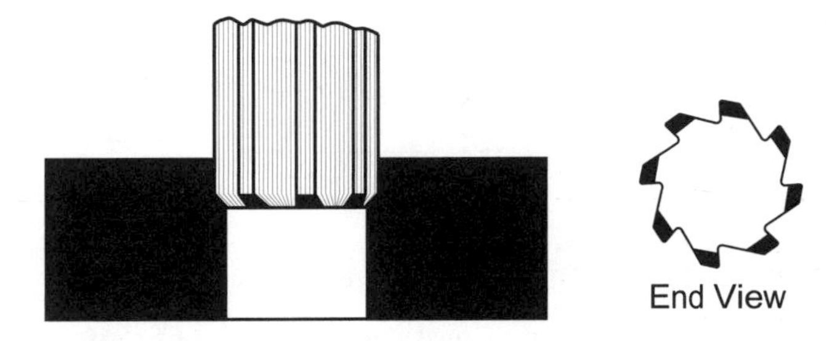

Figure 4–10. On a chucking reamer the chamfered end removes the most metal.

What are the two most common reamer designs and what are the differences between them?

- *Chucking reamers*, which are used under machine power, have either a straight shank for a drill chuck or a Morse taper, and are the most common designs. In general, smaller diameter chucking reamers have straight shanks, and larger ones need Morse tapers because of the larger torque required. See Figure 4–11.

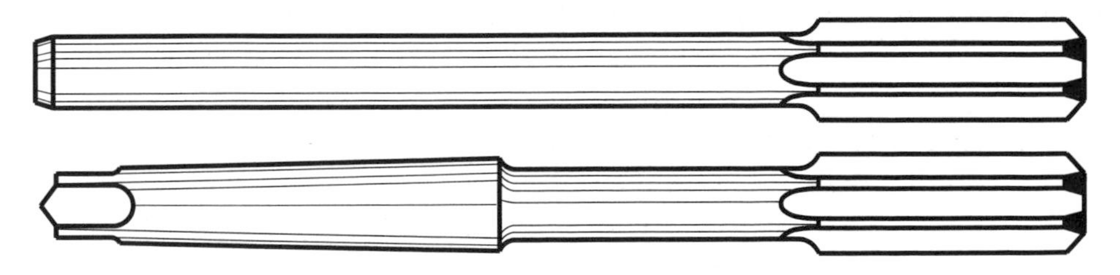

Figure 4–11. Straight-shank chucking reamer and Morse taper-shank chucking reamer.

- *Hand reamers* have a square cut on the shank for turning with a tap wrench, Figure 4–12. Hand reamers may also be clamped in a vise by their square end and the work turned onto them. There is another important difference between the two reamers. Hand reamers have a slight starting taper, typically the length of one diameter, to make it easier to get the reamer started. Chucking reamers do not have this starting taper. Hand reamers remove much less stock than chucking reamers, usually between 0.001 and 0.010 inches. Usually hand reamers are used to "clean up" an existing previously reamed hole or when machine reaming is not possible due to workpiece size or location.

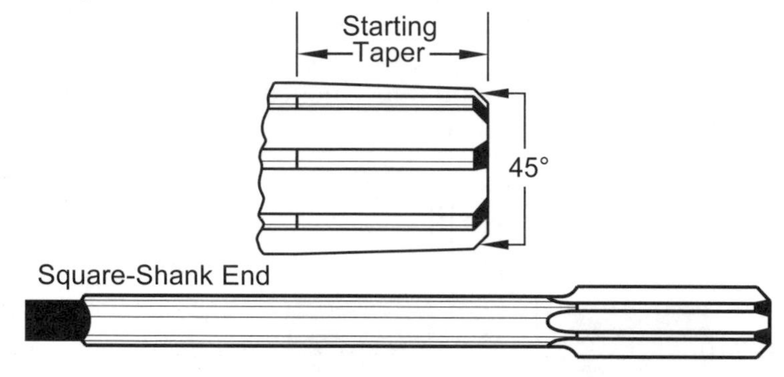

Figure 4–12. Hand reamer.

Why are chucking reamers also available with spiral flutes?
Spiral flutes provide better cutting action in interrupted holes (or keyways), that is, holes that have another hole intersecting them. Spiral flutes bridge the interference of crossing holes and make for smoother, chatter-free cutting.

Right-hand spiral, right-hand cutting reamers tend to lift chips out of the hole and are the most common spiral reamer design. Do not use this design on copper and aluminum as these metals tend to pull the reamer into the hole. Use a left-hand spiral, right-hand cutting reamer on copper, aluminum and other soft, gummy metals. However, these reamers will require more pressure for cutting action. Also, do not use these reamers in blind holes since they push chips *ahead* of themselves and will bind in the hole. See Figure 4–13.

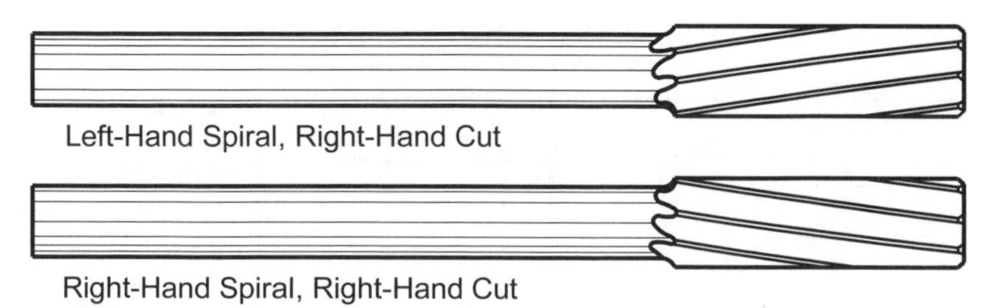

Figure 4–13. Spiral chucking reamers, both designs cut on right-hand rotation.

What materials are used to make reamers?

- HSS is used for most hand and chucking reamers.
- Cobalt HSS is attractive because it can be run 25% faster than HSS, has better abrasion resistance and longer tool life. It is often used on high-nickel alloys, stainless steel, Inconel® and titanium.
- Carbide flutes and solid carbide reamers are used for cast iron, nonferrous alloys, plastics and hardened steels. They have a higher abrasion resistance and retain a sharper, harder edge at higher temperatures than HSS. A rigid machine is needed when using carbide reamers.
- TIN and TiAN coated HSS reamers are used in production.

What sizes of reamers are available?

Reamers are available in:

- Numbered, lettered, fractional inch and metric drill sizes.
- Decimal-inch sizes, for example, one major tool supplier stocks straight shank HSS reamers from 0.0325- to 1.000-inches diameter in 0.001- or 0.0005-inch increments.
- Undersize and oversize reamers, by − 0.001 and +0.001 inch respectively, for standard fractional inch sizes are widely available, too. These reamers bring holes to size for a slip or driven fit for standard diameter shafts, dowel pins and drill rod.

How much undersize should a hole be for proper reaming action?

This depends on the diameter of the hole and the workpiece material. Table 4–3 offers a starting point for chucking reamers.

Reamer Diameter (inches)	Recommended Stock Removal (inches)
< 1/16	0.003–0.005
1/16–1/8	0.004–0.008
1/8–1/4	0.006–0.012
1/4–3/8	0.008–0.014
3/8–1/2	0.010–0.015
1/2–3/4	0.012–0.018
3/4–1	0.014–0.021
1–1½	0.018–0.035

Table 4–3. Recommended stock removal for chucking reamers.

Chucking Reamer Procedure

What are the steps to using a *chucking reamer*?

1. Leave enough material for proper reaming action according to Table 4–3.
2. Check the reamer diameter before starting on the actual workpiece. If there is any doubt about its size, test the reamer in a piece of scrap of similar material. New reamers may have small burrs causing them to cut oversize on the first few holes.
3. Mount the workpiece and reamer securely. If the chucking reamer has a straight shank, insert it only about three diameters into the drill chuck. This allows the reamer to "float" and find its own center. This increases hole-center accuracy. Inserting the reamer shank deep into the chuck increases set-up rigidity and increases any alignment errors from the tailstock, chuck adapter and chuck. Check that the reamer runs true and is not bent. Small reamers may be straightened by bending them to run true. A floating reamer holder is an excellent way to hold reamers.
4. Only usc sharp reamers.
5. Set the cutting speed for a HSS reamer to ½ to ⅔ of that for a similar sized drill. *On a critical job, determine optimum speed on scrap first. Too high a reaming speed may cause chatter and spoil the job.*
6. Apply plenty of cutting fluid.
7. Use as high a feed as possible for a good finish and accurate hole size. *Too low a feed causes glazing, rather than reaming action.*
8. Never turn the reamer backwards or it will be damaged.
9. Withdraw the reamer from the work before stopping the machine.

Getting excellent results from a chucking reamer depends on technique and requires a few test runs first.

What can be done to save a part if the reamer chatters before entering its hole?

There are a couple choices:

- Reface the end of the part to remove the chattering marks and provide a fresh start for the reamer.
- Use a countersink to remove the chattering marks and re-ream the hole.

Drills for Hand-Reaming Fractional-Inch Holes

What are some good drill size choices for producing the most common fractional-inch size hand-reamed holes?

Table 4–4 shows some twist drill sizes which leave enough stock for reaming to final fractional-inch diameters with a hand reamer.

Finished Reamed Size (Inches)	Drill Size	Drill Diameter (Inches)
1/4	'D' Letter drill	0.2460
5/16	7.80 mm	0.3071
3/8	9.40 mm	0.3701
7/16	11.0 mm	0.4331
1/2	12.5 mm	0.4921

Table 4–4. Reamed hole diameters and the starting drill sizes.

Problems and Precautions

What is the most common reaming problem?

The most common problem is chatter. Minimize chatter by:

- Reducing speed.
- Increasing or decreasing feed.
- Increasing tool holder rigidity, usually by reducing the length of the unsupported shank.
- Applying a chamfer to the hole before reaming.
- Using a piloted reamer.

Hand Reaming Procedure

What are the steps to using a *hand reamer*?

1. Verify that the hole is about 0.005-inch smaller than the final reamed hole.
2. Insert the reamer into the hole and fasten the tap wrench.
3. Turn the reamer clockwise applying just a little pressure to allow the reamer to enter the hole and adjust the reamer so it aligns with the hole. Use a square to check this.
4. Apply cutting fluid over the reamer cutting teeth.
5. Turn the reamer clockwise steadily applying pressure to make the reamer cut. Each turn should advance the reamer about ¼ of the reamer diameter.
6. When the reaming is complete do not stop turning it, either push it out through the other end of the hole or pull it out backwards while continuing to turn it. *Never turn the reamer backwards to remove it.*

Adjustable Blade Reamers

What do adjustable reamers look like and how do they work?

A ¼-inch adjustable reamer consists of:

- A threaded steel mandrel with four grooves at 90° increments and parallel to the axis of the reamer, Figure 4–14 (top). The grooves are cut deeper at the point of the reamer than at the shank. There is also a larger center section of the mandrel that serves to hold the blades vertical and four turning flats on the shank.

- The reamer has four carbon steel or HSS blades that fit into the mandrel grooves. Because the angle of the mandrel grooves matches the angle of the blades, the cutting edges of the blades remain *parallel* wherever they are positioned along the grooves, Figure 4–14 (middle). Blades may be sharpened or replaced as they dull.
- Nuts at each end of the blades capture the tips of the blades and keep the blades in the mandrel grooves. Each nut has a pair of turning flats.
- When the blades are positioned closest to the tip, the right-hand end of the reamer in Figure 4–14 (middle), the blades cut the reamer's smallest diameter. Turning the nuts to move the blades toward the shank, cuts the largest diameter.
- Turn these reamers with a straight-handle tap wrench, not machine power, and use the hand-reaming procedure above.
- Adjustable blade reamers are available from ¼- to 3-inches diameter in 20 sizes. Each size overlaps the size ranges of the adjoining sizes. Although the small reamers have only four blades, larger ones have up to eight.

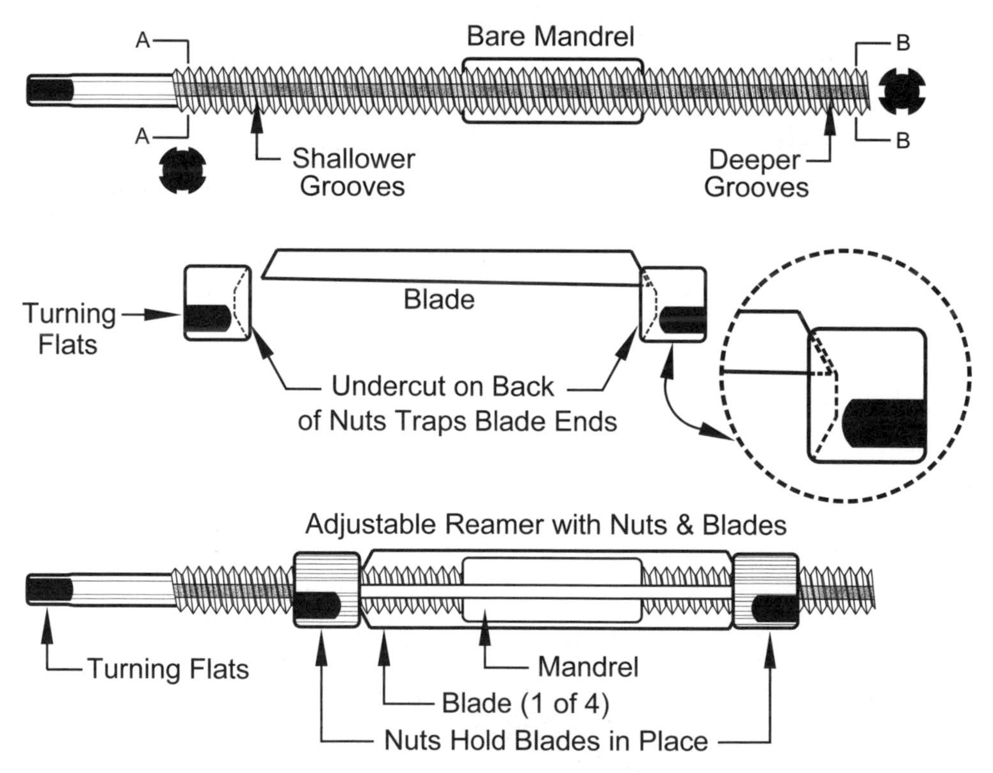

Figure 4–14. Adjustable reamer.

It may take several test cycles to get the diameter set to where you want it. This makes these reamers a second choice whenever a chucking or hand reamer of the exact size is available. Also, while the expensive versions of these reamers work well because their blades fit their grooves precisely and adjustment is predictable, cheaper versions of this reamer allow the blades to shift. This movement can be troublesome and cause parts to be ruined.

Adjustable Chucking Reamers

How do adjustable chucking reamers differ from other adjustable reamers?

Their screw-end adjustment is *only* for expanding them back to their original size after sharpening, not for adjusting them to a range of sizes, Figure 4–15.

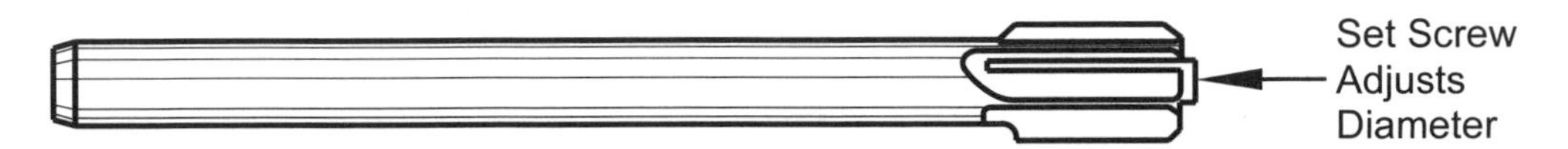

Figure 4–15. Adjustable *after sharpening* chucking reamers.

Other Reamer Designs

Beside hand and chucking reamers, what other reamer designs are commonly found in the machine shop?

- *Pipe threading reamers*, see *Chapter 5 – Threads & Threading.*
- *Morse taper reamers*, see *Chapter 7 – Turning Operations.*
- *Taper pin reamers*, see *Chapter 9 – Fastening Methods.*
- *Repairmens' reamers*, Figure 4–16, are used for slightly enlarging an existing hole. They are tapered HSS and usually hand-powered. They work especially well in sheet metal and soft materials.
- *Hand countersink* and *burr-removal reamers*, Figure 4–17.

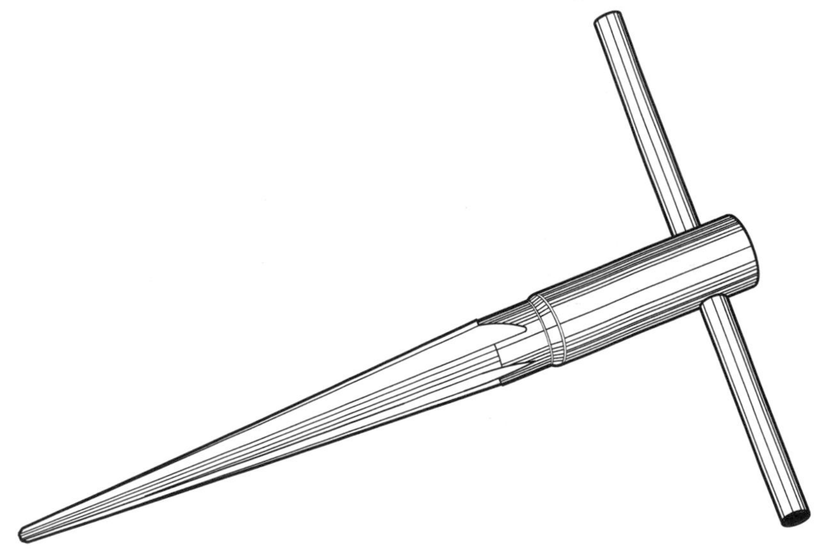

Figure 4–16. Repairmens' tapered hand reamer.

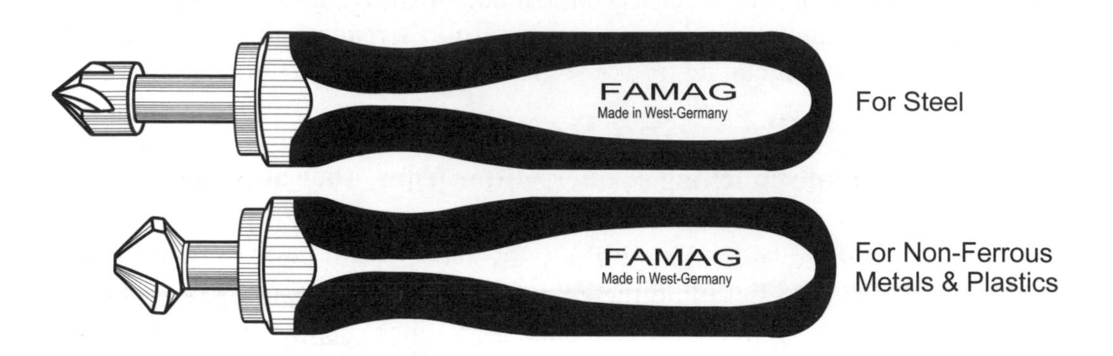

Figure 4–17. Hand reamers for countersinking and burr removal.

Section III– Broaching

Definition

What is *broaching*?

Broaching is a machining process that uses a tapered, multitoothed cutter. Most broaches are HSS. Broaching forces the cutter through an opening in a part or along its exterior. This action:

- Changes the shape of the hole into a square, hexagon, octagon or triangle. This is common practice in fabricating chucks to hold non-cylindrical parts. Figure 7–65 shows two collets with broached openings. Virtually any shape opening is possible.
- Adds slots or grooves in the walls of a hole. Two examples are cutting keyways in shafts and rifling firearms barrels. Turning the broach while it moves through the barrel forms the helical barrel grooves.

In the typical machine shop, broaching is most often used for cutting keyways. But, in a production setting, many specialized parts are also made by broaching, particularly in the automotive industry. Broaching takes little time and holds dimensions well. See Figure 4–18.

Broaching Keyways

What is needed to broach keyways?

- A *broach* of the proper width. Because each successive tooth projects slightly farther out than the previous tooth, the broach makes a series of small, but continuous cuts, moving through the work.

- A *bushing* that fits into the center hole in the gear or pulley holds the broach in a vertical position and guides it as it cuts the keyway. The bushing must fit the center hole and broach snugly. Bushings are available with and without collars, sometimes called *flanges*. Flanged bushings are easiest to use because they cannot drop through the center hole in the work, but uncollared bushings are used when the shape of the work prevents using collared ones.
- *Shims* inserted behind the broach in the bushing slot on successive passes move the broach teeth into the keyway slot so it continues to cut after the initial pass. A typical broach cuts 0.050-inch depth during an 8-inch stroke. The number of shims depends on the size of the broach and the depth of the desired groove.

Broaching Procedure for Keyways

What is the procedure for broaching a keyway?

1. Select the broach and choose the bushing matching the workpiece hole diameter and hole length. The bushing must be as long as the length of the hole in the work to cut a full-length groove.
2. Insert the bushing into the work, insert the broach into the bushing and place the work in the arbor press so the ram of the arbor press bears on the top of the broach. Insure that the broach is perpendicular to the work and the arbor ram face.
3. Force the broach through the work with the arbor press. Use smooth, even pressure on the broach. Do not stop in the middle of a stroke. As the broach completes its cutting stroke and exits the bottom of the work, catch it so it does not fall. *Never pull the broach backwards up through the keyway. This dulls the teeth.*
4. Clear chips from the broach.
5. Insert the broach into the top of the bushing and place an L-shaped shim behind the broach in the keyway. The shim moves the broach deeper into the keyway to make the next cut. Add cutting fluid to the broach.
6. Repeat the downward cutting stroke, catching the broach as it emerges from the work. Do not allow it to fall.
7. Remove the broach, shims and bushing.
8. Apply cutting fluid to the broach teeth, back and sides.
9. Repeat steps 6 through 8 until the keyway is the required depth.

Broaches for cutting extra-long keyways advance their teeth 0.030 to 0.035 inches along an 8-inch length, instead of the more usual 0.050-inch advance. This reduces the force on the broach when a long keyway engages many teeth at once. Hundreds of pounds of force are applied to the broach to make it cut.

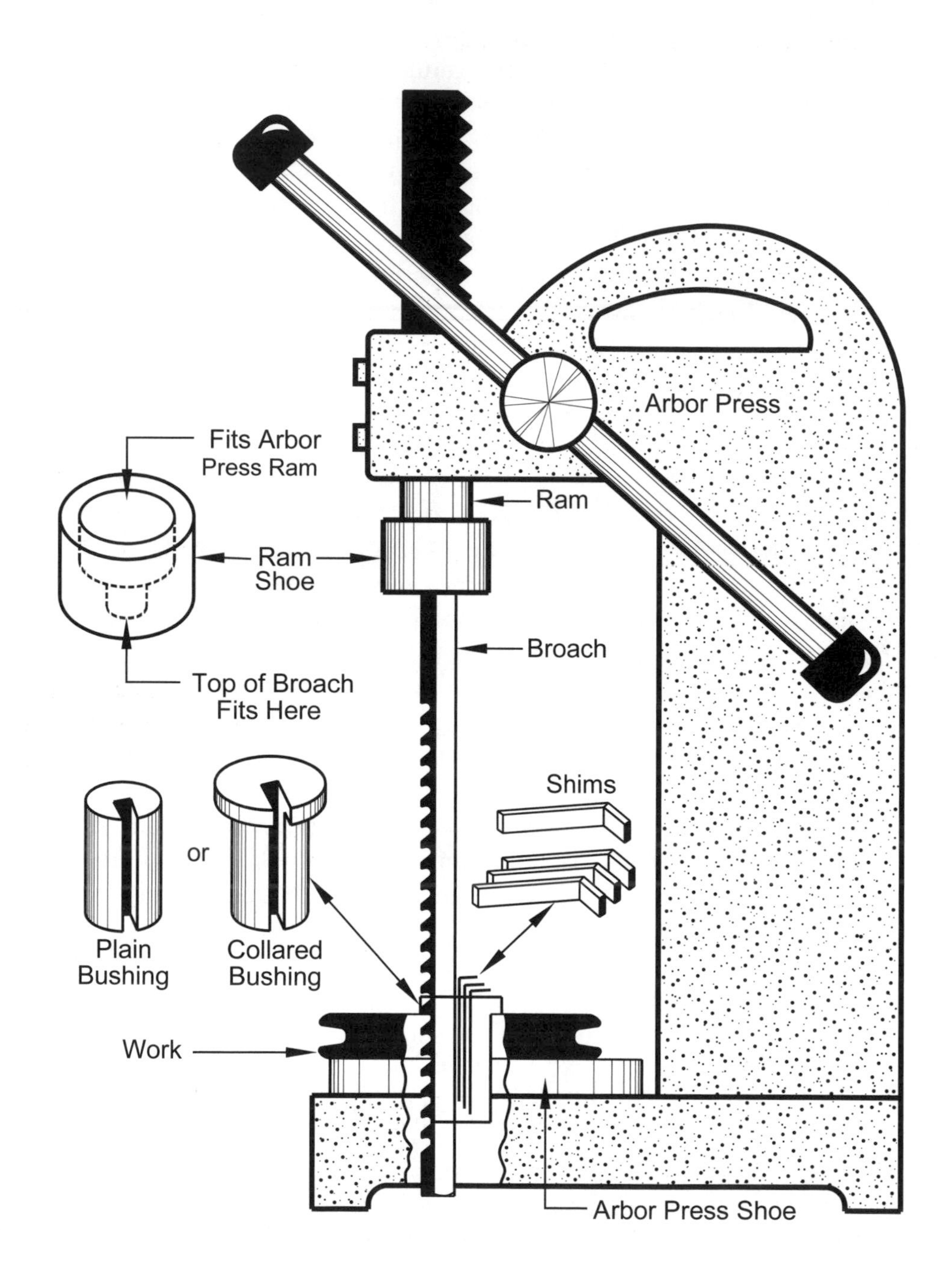

Figure 4–18. A broach, collared bushing, a plain bushing and shims.

What cutting lubricants should be used when broaching?
See Table 4–5.

Workpiece Material	Lubricants
Aluminum	Kerosene or aluminum tapping fluid.
Brass	Dry, but lube back of broach in bushing.
Bronze	Oil.
Cast Iron	Dry, but lube back of broach in bushing.
Steel	Cutting oil.

Table 4–5. Recommended cutting lubricants for broaching.

Trouble Shooting

What are common broaching problems and their solutions?

- There is a maximum and minimum length cut for each broach. Two or more workpieces may be stacked to establish the minimum length of cut. Extra-long keyways require extra-long broaches, shims and bushings.
- Proper alignment of the arbor press ram, broach and bushing are critical to avoiding drifting, deflection and breakage of the broach.
- Never attempt to broach any material harder than Rockwell R_c35.
- Good practice is to use a ram adapter that positions the broach against the face of the arbor press ram and reduces the chances of deflection. On larger or extra-long broaches, use a hydraulic press to get the force needed.
- Lubricant must be applied to all four sides of the broach as it enters the bushing, not just the cutting teeth.

Section IV – Lapping

Lapping Basics

What is *lapping* and what does it accomplish?

It is a machining process that uses loose abrasive materials usually in oil, grease or water either manually or with a machine to:

- Remove machining tool marks.
- Polish and refine the surface to make it extremely smooth.
- Make small adjustments to a part's final dimensions. Tolerances between 0.0005 and 0.00002 inches (0.013 and 0.00005 mm) can be achieved.
- Shape the part, rather than just smooth its surface. Examples are:
    - Dead flat surfaces as in optical flats, mirrors or reference plates.
    - Accurate cylindrical surfaces.
- Correct out-of-shape conditions like:

  - Out-of-round or tapered.
  - Bell-mouthed or barrel-shaped.
  - Off-axis.
- Produce two mating surfaces. Examples are:
  - Piston and cylinder pairs.
  - Opposing flat surfaces in a sliding valve.

Most lapped work is either flat or cylindrical, but tapers and other shapes are possible, too. Lapping is characterized as a two-directional process: flat lapping is performed by moving the work in a figure-eight pattern, and cylindrical lapping turns the lap *and* moves it back and forth along the work axis.

What makes lapping different from other machining processes?

- It removes material slowly and makes it easier to hit a target dimension.
- Lapping is done at a slow-speed, so it generates little heat in the work, and little thermal distortion results.
- Mechanical force on the workpiece is low, typically just a few psi.

How much workpiece material can lapping remove?

While considerable material may be removed by lapping, as in forming shaped optical surfaces, as a practical matter, metal parts are usually machined to within 0.0005 and 0.005 inches (0.013 and 0.13 mm). The rest is removed by lapping.

How does the lapping process work?

The lap is “charged” with abrasive either by rolling the lap in the abrasive or applying a mixture of the abrasive and its carrier liquid onto the lap. The abrasive embeds in the softer lap. When the lap and workpiece are rubbed together, a small amount of workpiece material is scratched off by the abrasive. This removes high spots and smoothes the surface. Were the lap material harder than the work, the abrasive would embed in the work and remove material from the lap. By using progressively finer abrasives, making increasingly finer scratches, a very smooth polished surface of the desired shape may be produced.

What materials can be lapped?

Although all materials can be lapped, harder materials—metals, ceramics and glass—lap best and are the most lapped materials. The harder plastics can be successfully lapped and the bases of many microprocessor chips are lapped to increase their flatness for better heat transfer.

In what machines is lapping performed?
In addition to dedicated lapping machines, internal and external cylindrical lapping may be performed in any machine with a rotating spindle—drill presses, lathes or milling machines. Flat lapping is done on dedicated lapping machines or by hand. The *edge* of a granite surface plate can be used to lap small parts, but larger parts should use a cast iron or granite surface plate.

Lap Designs

How are laps made and of what materials are they made?
There are four basic types of laps:

- *Flat lapping plates* for roughing are usually cast iron plates with a series of small grooves at right angles about ½-inch apart in concentric circles or in a helix, Figure 4–19. These grooves improve the abrasive distribution. Flat laps for finishing are smooth cast iron plates. However, ½- or ⅜-inch thick plate glass may also be used for less critical jobs.

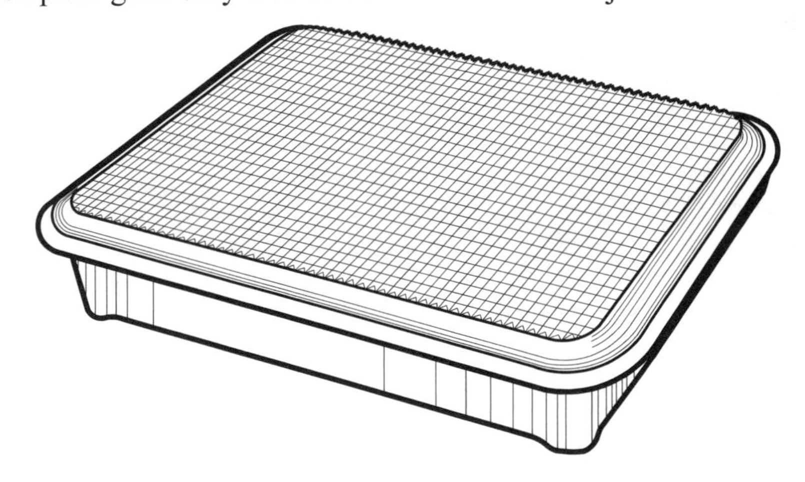

Figure 4–19. Cast iron lapping plate.

- *Internal laps* have an outer lapping surface and an inner tapered mandrel or expansion plug. Helical slots are usually cut in the lap so it can expand and contract. Longitudinal slots are fine for many jobs and are used on many shop-made laps for simplicity. Cast iron laps are best for precision work, but brass, copper, aluminum or lead are also used. Many laps have radial grooves for abrasive distribution.
- *Needle eye laps* are used for small holes between 0.026 and 0.25-inches diameter. The adjustment tool fits into the grooves on the side of the lap and when tapped or squeezed, expands the lap. These laps are usually carbon steel.
- *External laps* may have either helical or longitudinal slits.

See Figure 4–20.

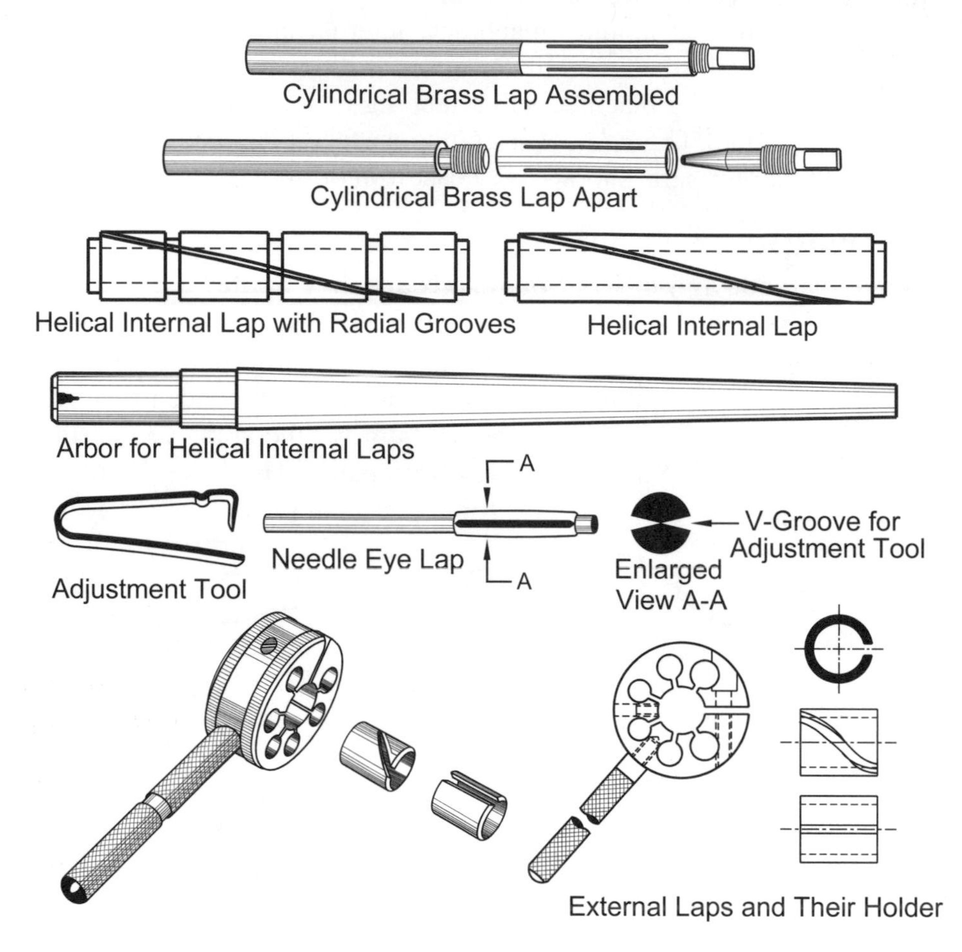

Figure 4–20. Lap designs.

Shop-made Laps

What do shop-made laps look like?

See Figure 4–21. They are simple and easy to make.

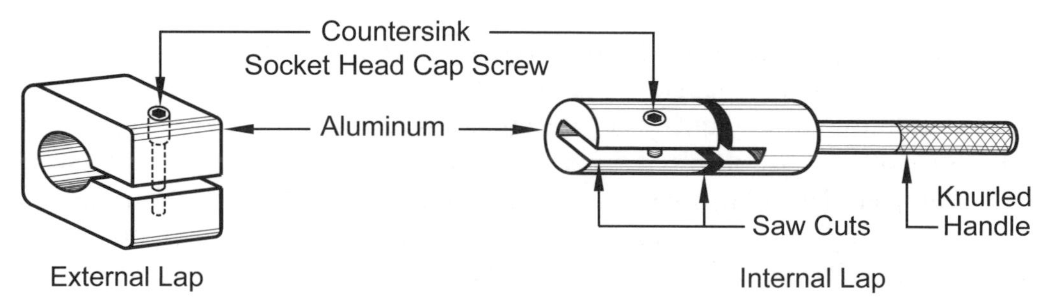

Figure 4–21. Shop-made laps.

Lapping Abrasives

What abrasives are used for lapping?

- *Aluminum oxide* (Al_2O_3), also called *Alundum*® and *fused alumina.*
- *Silicon carbide* (SiC) is available in two forms, green and black. Green is just slightly softer than diamond. The black form, also known as *carborundum*, is slightly softer than green silicon carbide.
- *Boron carbide.*
- *Diamond* is the hardest abrasive.

Lapping abrasives are suspended in oil, grease or water.

Lapping Procedure

How is flat lapping done?

The precise procedure is based on experience, experiment and the condition of the workpiece. This is the general approach:

1. Spread a thin layer of abrasive over the lapping plate and press it into the lap with a steel roller. Remove loose abrasive and inspect the lap for a uniform grey appearance. Repeat if necessary. Do not apply or leave excessive abrasive on the plate as it will tend to "roll" rather than cut.
2. Work is placed on the lapping plate and moved in a figure-eight pattern for ½ to 2 minutes. Never add loose abrasive directly to the lap. Charge it as in Step 1. Examine the work.
3. Lapping may continue using this abrasive or the plate may be cleaned and a finer abrasive applied.
4. Lap with progressively finer abrasive until the desired surface is achieved. Do not apply excessive pressure to the work as the abrasive will only cut so fast.

Careful cleaning of the work and lapping plate is needed when changing to finer abrasives because residual abrasive particles will scratch the work and prevent achieving a smoother surface. Ultrasonic cleaning is often used to get all the abrasive off the part.

How is *external cylindrical* lapping done?

Here is the general approach:

1. Secure the work in a 3-jaw chuck and adjust the lap to a running fit over the work.
2. Charge the lap with abrasive. There are three methods:
   - Remove the work from the machine and roll it on a hard steel plate covered with abrasive as with a flat lap.

- Sprinkle abrasive into the lap and use a steel pin to distribute and imbed it.
- Abrasive may be brushed onto the work itself. This is the poorest method, but adequate for rougher work.

3. Replace the work in the lathe, fit the lap over the work and set the machine for 150 to 200 rpm for a 1-inch diameter workpiece. Use higher speeds for smaller diameter work.
4. Hold the lap to keep it from turning with the work and turn on the machine.
5. Move the lap back and forth over the work.
6. After a minute, remove the work, clean it and gage it to see if additional lapping (and lap charging) is required.

External laps should be no more than ⅓ of the work length. A shorter lap provides better sensitivity so the operator can run the lap on areas with the greatest resistance.

How is *internal cylindrical* lapping done?
Here is the general approach:

1. Internal laps must first be adjusted for a running fit to the work.
2. Roll the lap on a steel surface on which the abrasive has been evenly applied. Press hard enough to embed the abrasive in the lap. Remove the excess or loose abrasive.
3. Mount a lathe dog on the lap.
4. Slip the lap over the work and mount the work between centers with the lathe dog engaged to a turning plate.
5. Hold the lap to keep it from turning and switch on the lathe, 150 to 200 rpm for 1-inch diameter work.
6. Move the lap back and forth over the work.
7. After a minute, remove the work, clean it and gage it to see if additional lapping (and lap charging) is required.

Lap length should be about 130% of the work length. Stroke length should be enough so the center of the lap moves about 5% beyond the center of the workpiece.

Be sure to thoroughly clean the lap and the work of one grade of abrasive when going to the next finer one. It is often necessary to change to a new lap to avoid the problem of the previous grade of abrasive remaining imbedded in the lap itself.

Chapter 5

Drills & Drilling Operations

Procrastination is the thief of time.
—Edward Young

Introduction

Drilling holes is the most common of all machining processes and the tool most commonly used to make them is the twist drill. Twist drills remove metal from holes and reduce it to chips in a fast, simple and economical process. Nearly 75% of the metal removed by machining is drilled out. Drilling is often the starting point for other operations such as reaming, counterboring, countersinking and tapping. Though usually used in a drill press, also called a *drilling machine*, twist drills can be used in lathes and milling machines. There are many twist drill designs. Some optimize drilling speed, diameter control and wall smoothness, while others perform deep-hole drilling or drill very hard materials. This chapter covers the designs most useful to non-production, small- and medium sized shops.

The most popular drill press design is the *sensitive drill press*. We'll cover two sizes of this machine using current commercial models. Also, we'll examine how drill press size is measured, drill press components and controls, twist drill sizing systems, the mechanics of drilling, securing work and drill speeds and feeds. Procedures for transferring hole position, correcting off-position drills, extending drills for a longer reach, the use of bushings and the making of drilling jigs are included. In addition, this chapter looks at other, non-twist drill designs, such as flycutters and punches. Although drilling metal is our primary interest, the drilling of wood, plastic, glass and rubber is also covered. Two other hole-related procedures, reaming and tapping, are covered in *Chapter 4 – Grinding, Reaming, Broaching & Lapping,* and in *Chapter 6 – Threads & Threading*, respectively.

Section I – Drill Presses

Specifying Drill Press Size

What are six common ways of specifying the capacity of a drill press?

- *Maximum throat depth,* the distance between the spindle center and the drill column. This measurement indicates how far the drill press can place a hole from the edge of a workpiece. This depth is also the radius of the largest circle whose center can be drilled.
- *Maximum hole diameter,* the largest hole which can be made in steel and in cast iron using twist drills based on available speed and motor torque.
- *Spindle travel,* the maximum tool stroke that can be taken without repositioning the table. It indicates the deepest hole easily drilled.
- *Maximum vertical clearance,* the distance between the end of the chuck and the worktable.
- *Range of spindle speeds.*
- *Motor horsepower.*

All six parameters are needed to characterize a drilling machine.

Small & Medium Drilling Machines

- *Cameron Series 164 miniature drill press,* Figure 5–1, has a single-speed motor and delivers three spindle speeds of 9500, 17,000 and 30,000 rpm via stepped pulleys. It has a rack and pinion feed and drills holes from 0.002- to 0.150-inches diameter to the center of a 5-inch circle. Spindle runout at the chuck mounting taper is less than 0.0001 inches. Spindles, pulleys and motors are dynamically balanced for smooth operation. A clear plastic cover allows viewing of the belt and pulley positions. The 4 × 4-inch worktable is slotted and has a ⅝-inch center hole. Some of the available options are: variable speed controls, foot operated speed controls, two-speed motors to provide 6 spindle speeds of 7500, 9500, 14,000, 17,000, 24,000 and 30,000 rpm and Jacobs key-type chucks or Albrecht precision keyless chucks. Dial indicators to determine hole depth are available, too. Motors are 240/120 VAC at 2.5/5 A. These drill presses are manufactured by Cameron Micro Drill Presses, a Division of Treat Enterprises in Sonora, California 95370.

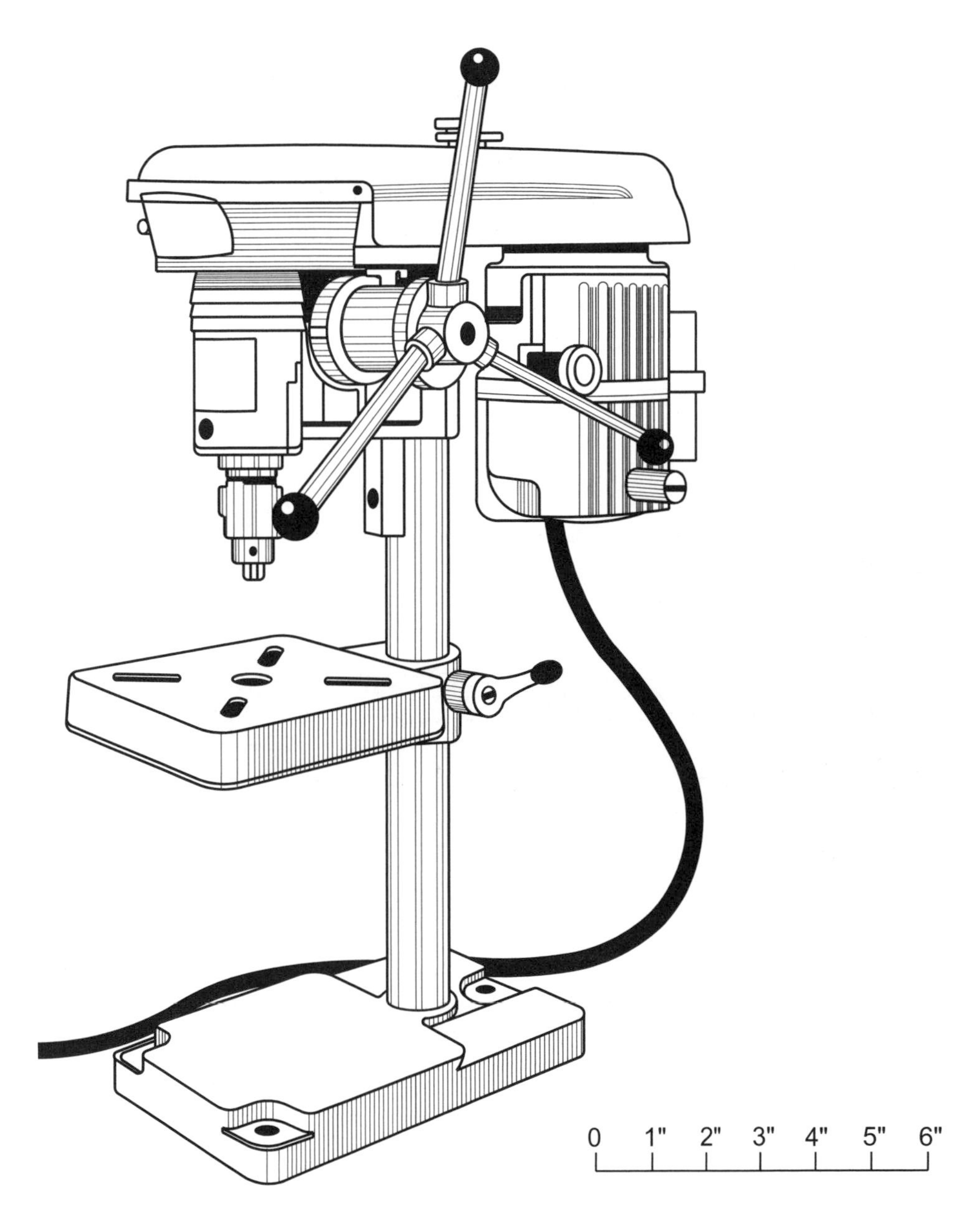

Figure 5–1. Cameron Series 164 miniature drill press.

- *Clausing 15-inch floor drill press,* Figure 5–2, is a medium-heavy industrial-grade machine suitable for production work, a general machine shop or an industrial prototype lab. It has a maximum throat depth of 15 inches and can drill a ⅝-inch hole in mild steel or cast iron. The maximum spindle-to-table distance is 41 inches and maximum spindle-to-base distance is 47 inches. Its tilting table is 14 × 10 inches and its base is 13 × 10 inches. The 3-inch diameter ground steel vertical column is 3/16-

inches thick for rigidity. The 2-inch diameter quill is ground steel. All models include a micrometer depth stop and a push button safety switch. There are also two choices of infinitely variable speed units (330/2460 rpm and 500/4000 rpm) and two five-speed step-pulley models (260/3600 rpm and 400/5400 rpm). In addition, there is a choice between a ½-inch capacity key chuck and a #2 Morse taper spindle. An optional air-hydraulic feed is available for production work. Both single-phase 115/230 volt, 60 Hz motors and three-phase 208/220/440 volt, 60 Hz ball-bearing, ¾-horsepower motors are available. Overall dimensions are 72.6 (H) × 15.5 (W) × 28 (D) inches. This drill press weights 300 pounds. Clausing Industrial, Inc. makes it in Kalamazoo, Michigan 49007.

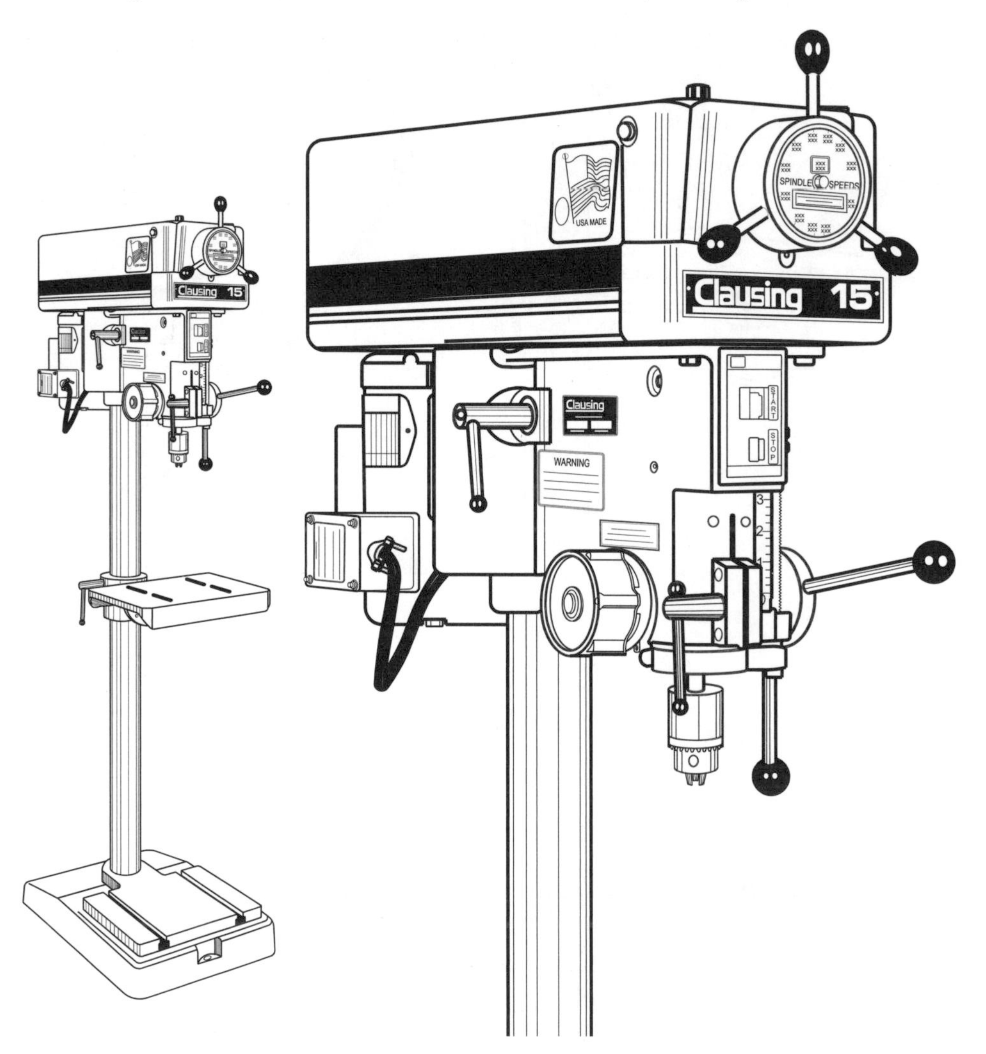

Figure 5–2. Clausing 15-inch floor drill press.

Drill Press Components & Controls

What are the major components of a drill press?

There are four major components of a drill press: the *base,* the *column*, the *head* and the *quill assembly.* The base and column are usually of cast iron or steel, and they support all the other components. The head, typically a large casting, contains the motor, drive belts, speed changing mechanism and the quill assembly. The head fits on the top of the column, and the quill assembly inside the head holds the spindle and the chuck.

What are the typical controls and adjustments on a drill press?

- *Power switch.*
- *Downfeed handle,* controls chuck position and can exert considerable force on the drill to maintain cutting action.
- *Quill lock,* secures the chuck at any point in its travel range. It is used when the drill press performs light buffing, grinding or milling.
- *Depth stop*, Figure 5–3, controls hole depth by limiting quill travel with a mechanical stop. Adjusting the stop nut position along a threaded rod controls depth, and the lock nut secures the stop nut from vibration.

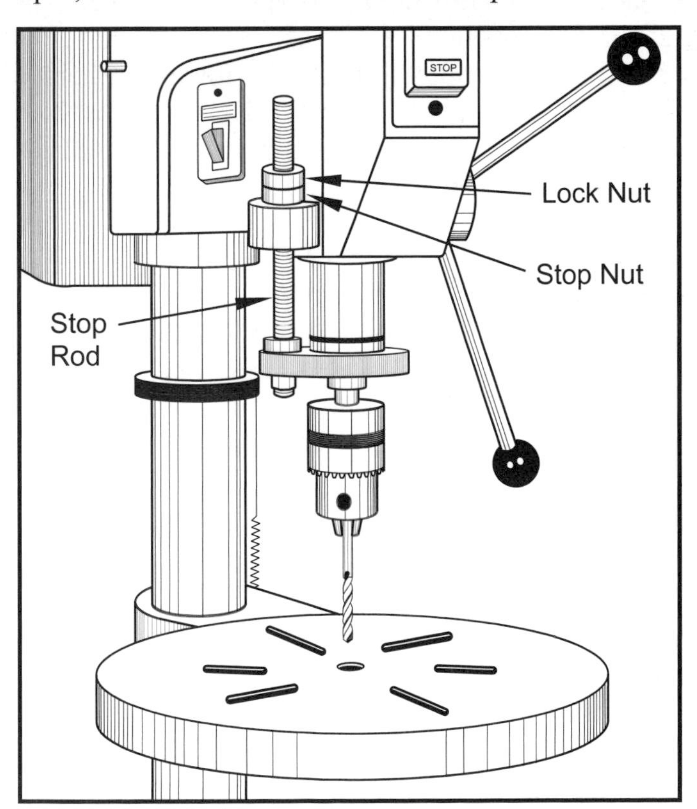

Figure 5–3. Parts of a drill press depth stop.

- *Locks for table vertical position* and *table tilt adjustment.* Drilling tables that tilt simplify drilling angled holes.
- *Speed controls* on miniature machines control motor voltage to vary spindle speed. Full-size drill presses vary spindle speed with either stepped pulleys or a variable-speed belt drive.

Drill presses are sometimes referred to as being *sensitive*. What does this mean?

Hand-fed drill presses with a counter balance spring that pulls upward on the quill assembly are called *sensitive.* This removes the weight of the quill from the axial drilling force and allows the operator to better *feel* the drilling pressure to apply the right pressure for efficient cutting. This sensitive feel also reduces drill breakage.

Why is it bad practice to add a *drilling table*, also called a *compound slide table*, to a drill press to perform milling?

The drill press quill, bearings, spindle and tapers are designed for axial loads, not the sideways forces of milling. Morse tapers on these machines will loosen with sideways force leading to an unrestrained spinning chuck. Drilling tables and compound slides should be used to position work accurately for drilling, but not for milling. However, there are *mill/drills,* which *are* designed to handle both axial and lateral forces. These are usually economy machines, a cross between a drill press and a milling machine, and usually lack the features and accuracy of a full-sized milling machine.

Drill Press Operations – Overview

What are common drill press operations?

- *Drilling,* the making of holes by removing a solid mass of metal with a cutting tool, is the most common drill press operation and is most often performed with a twist drill.
- *Center drilling* uses short, rigid drills to make a cone-shaped starting hole for twist drills or lathe centers. Using a center drill insures more accurate hole placement on the spindle axis than starting a twist drill on a flat work surface or center punch mark. Because twist drills flex and wobble when starting a hole, there is no certainty they will begin drilling on the spindle axis.
- *Counterboring* produces a larger, square-bottomed hole in the upper portion of an existing hole, and provides space and seating for a bolt or cap screw head below the surface of a workpiece. The cylindrical guide, or integral pilot, on the end of the counterbore insures the enlarged diameter is concentric with the original hole, Figures 5–4 and 5–5.

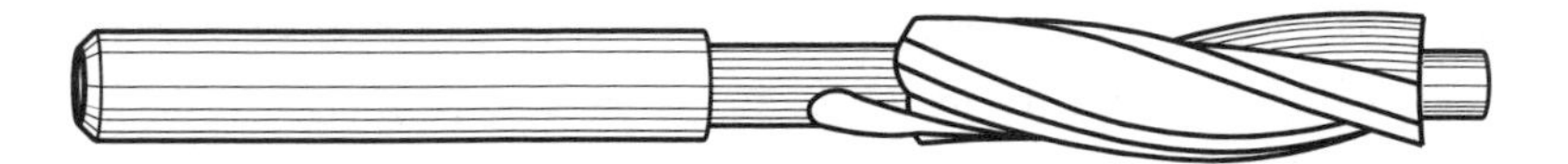

Figure 5–4. Counterbore with integral pilot.

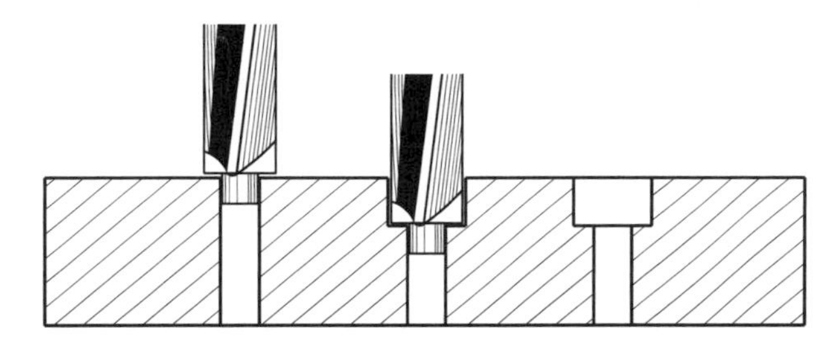

Figure 5–5. Counterboring: pilot end enters existing hole (left) and aligns cutter, counterbore increases the diameter of the existing hole (center) and completed counterbore (right).

- *Spotfacing,* Figures 5–6 and 5–7, mills a flat area around an existing hole to make a flat seating surface for a bolt or washer and is usually necessary on castings and parts with sloping surfaces. Failure to spotface an uneven workpiece puts excessive stress on the bolt.

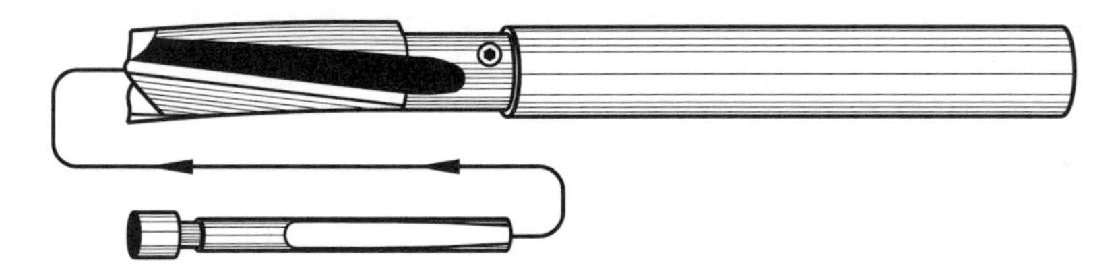

Figure 5–6. Spotfacing cutter (top) with its interchangeable pilot (bottom), which is held in place by a set screw.

Remember to use a counterbore or milling cutter slightly *larger* than the washer, cap screw or bolt head diameter called for on the print.

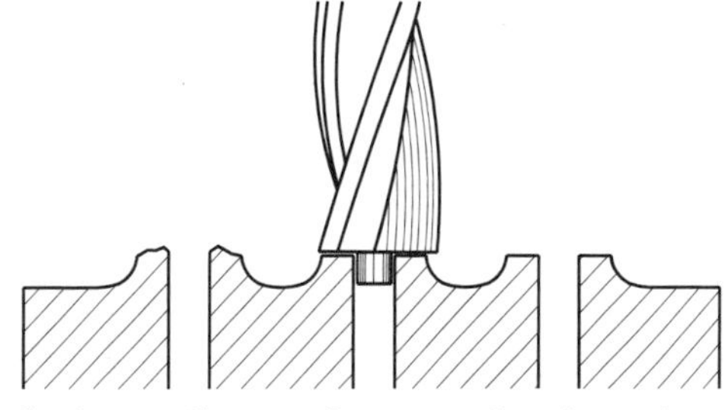

Figure 5–7. Spotfacing a flat surface can be done in a drill press with a spotfacing bit. The left hole shows the workpiece as cast, the middle hole with spotfacing in progress and the right hole with spotfacing completed.

On flat work, use a spotfacing bit or counterbore in a drill press, but on sloping surfaces, because of the interrupted cut, use an end mill in a milling machine, Figure 5–8. Spotfacing cutters differ from counterbores because they cut larger diameter holes and sustain greater cutting forces, but they work the same way. Large spotfacing cutters have separate centers allowing the same cutter to work in different hole diameters. On sloped surfaces, spotfacing is done *before* drilling. This provides a flat surface on which to start the drill and makes it easier to follow-in straight.

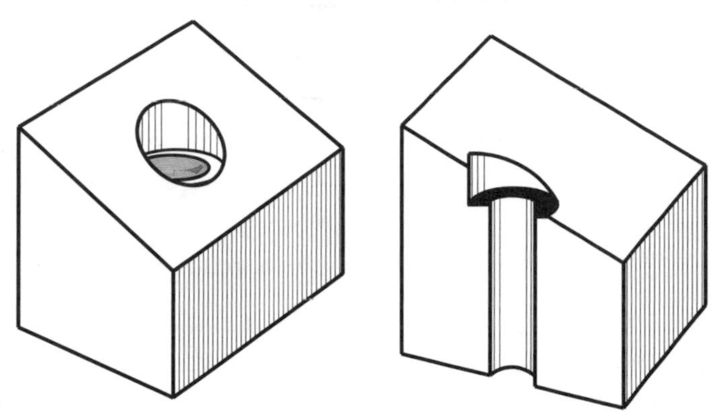

Figure 5–8. Because of the interrupted cut, spotfacing on a sloping or angled surface is best done with an end mill in a milling machine, but in many cases may be done in a drill press using a long pilot and slow feed.

- *Countersinking* uses a cone-shaped cutting tool to chamfer or bevel the edges of an existing hole, usually so flat-head screws can be seated below the workpiece surface, Figures 5–9 and 5–10. Countersinks also deburr holes. These tools are available in HSS, solid carbide or with carbide inserts in a variety of point angles. Countersinking can be done in a drill press, lathe, milling machine or a hand-held drill.

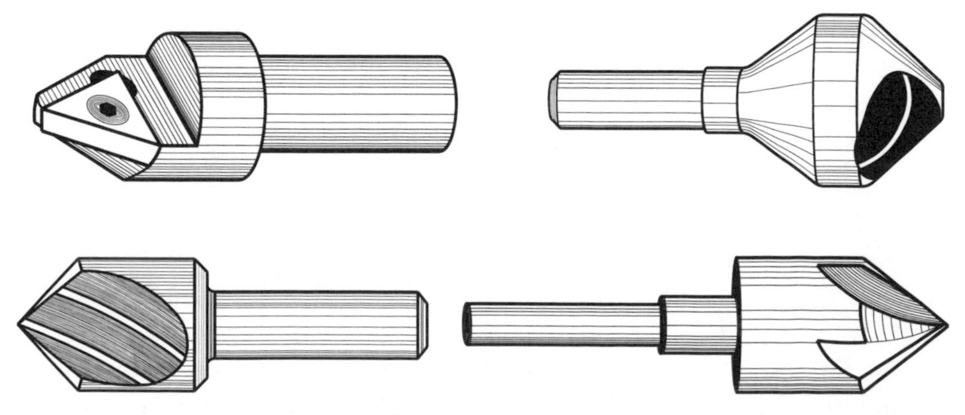

Figure 5–9. Four countersink designs.

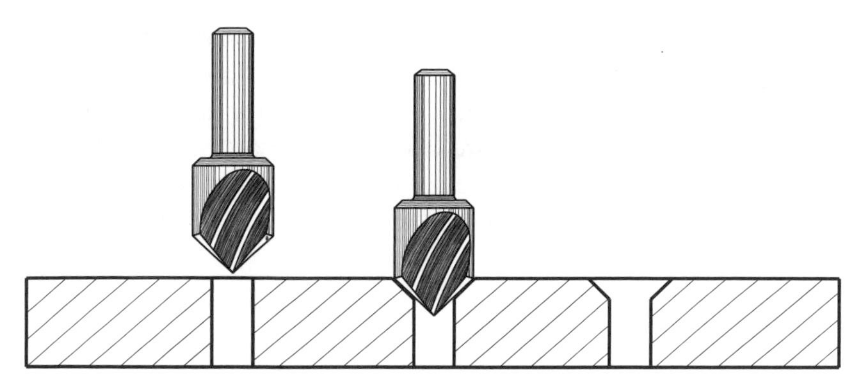

Figure 5–10. Countersinking: existing hole (left), countersinking operation (center) and completed countersink (right).

- *Flycutting* is a trepanning operation which makes a hole by cutting a narrow, annular (donut-shaped) band of material from the outer edges of the hole. Most of the material removed is a solid circle. Flycutting uses a single-point lathe tool, Figure 5–11, to perform trepanning and is used on thin material, usually under 0.062-inches (2-mm) thick. A flycutter can produce larger holes than possible with a twist drill for a given size drill press. This is because it reduces only a thin band of material to chips, not the entire hole volume. Flycutter hole diameter is easy to adjust. Sometimes flycutting is not used to cut a hole, but to cut out a circular workpiece from stock. Trepanning in the lathe is discussed further in *Chapter 7 – Turning Operations.*

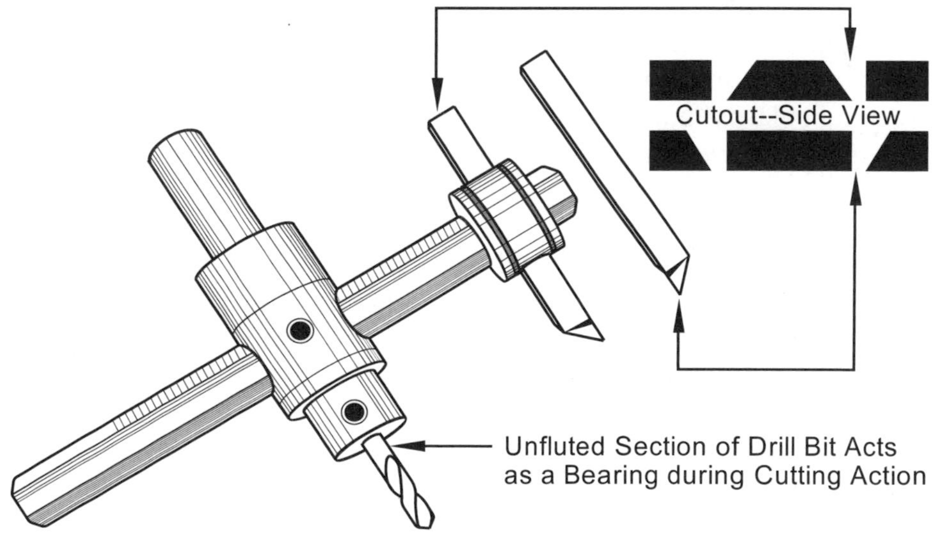

Figure 5–11. Flycutter with cutting tool set to produce a round hole with square edges. A mirror-image of the cutting tool in the flycutter will remove a circle with square edges and leave a hole with tapered edges.

- *Hole sawing* is trepanning with a cylindrically-shaped cutting tool. The tool, which usually has saw teeth around its cutting edge, is for cutting metal, plastic, drywall and wood. Saws for ceramics, marble, concrete, stone and glass have one or more carbide teeth or diamond grit to do the cutting. Starrett and Greenlee industrial-grade hole saws offer dramatically better performance and durability than bargain brands sold in home improvement centers. See Figure 5–12.

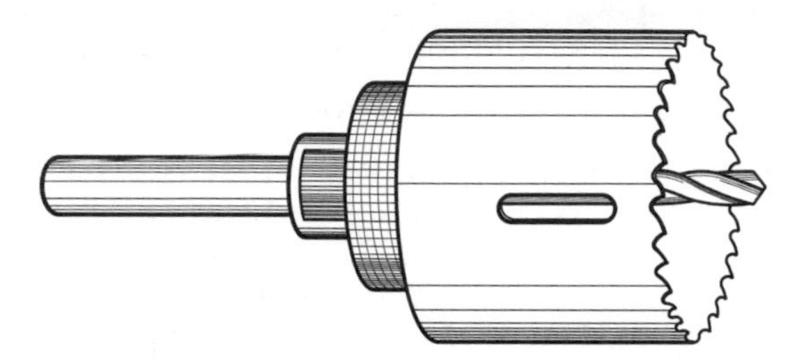

Figure 5–12. Hole saw.

  Most ¼ and ⅜-inch hand-held electric drills turn too fast at their lowest speeds for hole saws causing them to heat up and loose their temper. Hole saws work very well in a drill press using a low spindle speed.
- *Reaming* enlarges an existing hole to a precise diameter and improves wall surface finish. It can be performed equally well in a drill press, lathe or milling machine. See *Chapter 4 – Grinding, Polishing, Reaming & Finishing.*
- *Boring* enlarges the diameter of an existing starter hole with a single-point lathe tool. It can be performed in a drill press, but is more often done in a lathe or milling machine. Boring offers precise diameter control, perpendicularity and smoother walls than drilling. See *Chapter 7 – Turning Operations* and *Chapter 8 – Milling Operations*.

Center Drills

What do center drills look like and why are different designs used?

- The *NC spotting and centering drill,* Figure 5–13 (A), is popular because it is the least complicated center design, is easy to sharpen and may be sharpened many times. When using spotting drills for carbide twist drills, they should always have a flatter angle than the carbide drill point angle so the chisel edge of the twist drill makes contact with the work first, not the drill edges. For example, use a 120° spotting drill for a 118° carbide twist drill or a 140° spotting drill for a 135° carbide twist drill. Spotting drills are available with point angles of 60º, 82º, 90º, 118º and 120º.
- *Plain-type combination drill and countersinks,* Figure 5–13 (B), may be used for accurately starting holes and countersinking them, but they are

also essential for making properly shaped holes for lathe centers. The pocket created by the drill tips provides a space for lubricant that keeps the lathe centers from frictional overheating. They are available in many materials and point angles, but the most common HSS version with a 60° point is suitable for the majority of operations. These drills are also referred to as *plain-type combined center drills* to distinguish them from other center drill designs described next.

- *Radius-type center drills,* Figure 5–13 (C), provide longer tool life and higher productivity than the plain-type because of their greater chip clearance and coolant flow. Their long bearing surfaces are desirable during high-precision machining operation because they keep the work from shifting axially and their curved surfaces provide a better geometric fit to lathe centers than the plain-type center drill when cutting tapers by shifting the lathe tailstock off center. See *Chapter 7 — Lathe Operations.*
- *Bell-type center drills,* Figure 5–13 (D), offer two advantages. In a production setting, a quick visual inspection confirms that the center hole has been cut to minimum depth whenever the step can be seen. Also, when parts are jumbled around during production handling, the step or outer edge of the center hole is below the workpiece surface and protected from damage.

Most center drill designs are available in HSS, Cobalt HSS and solid carbide. See *Chapter 7 – Turning Operations* for more on center drills.

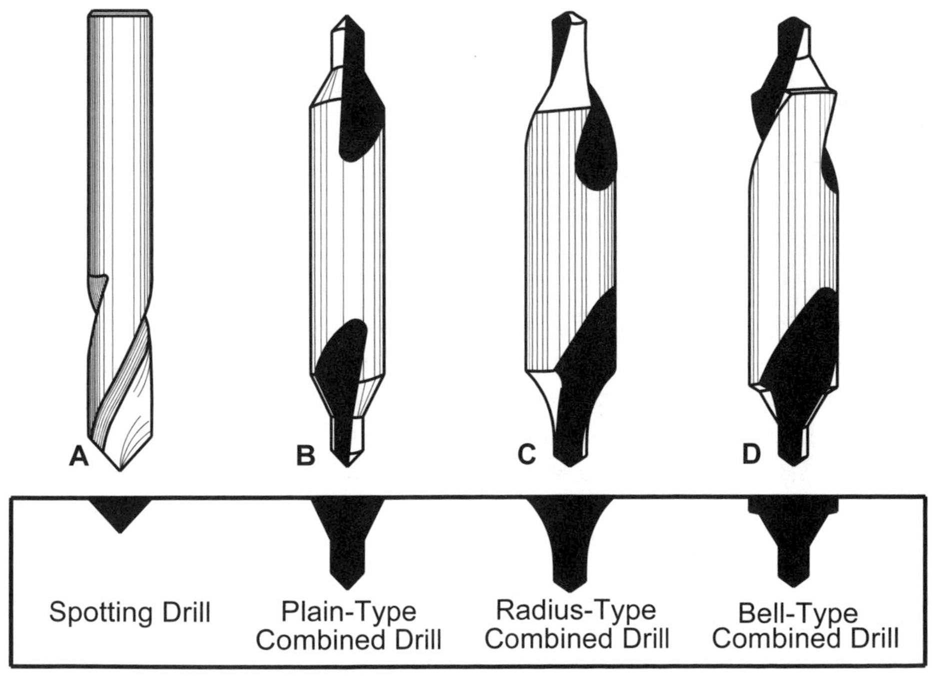

Figure 5–13. Center drill designs.

Snapped Center-Drill Tip Recovery

You have snapped off the tip of a center drill inside a valuable part. How can this part be saved?

The situation is shown in Figure 5–14 (A). The solution lies in the ruined end of the center drill. Remove the broken center drill from the drill press or lathe, and use a small abrasive wheel in a hand-held air or electric grinder to form a slot, Figure 5–14 (B). The slot should be just wide enough to fit around the broken tip. Finally, use this modified center drill to re-drill the center hole. When the slot in the tip of the center drill fits *around* the snapped tip, the tip will be released, Figure 5–14 (C & D), and a new, undamaged center drill can complete the drilling, Figure 5–14 (E).

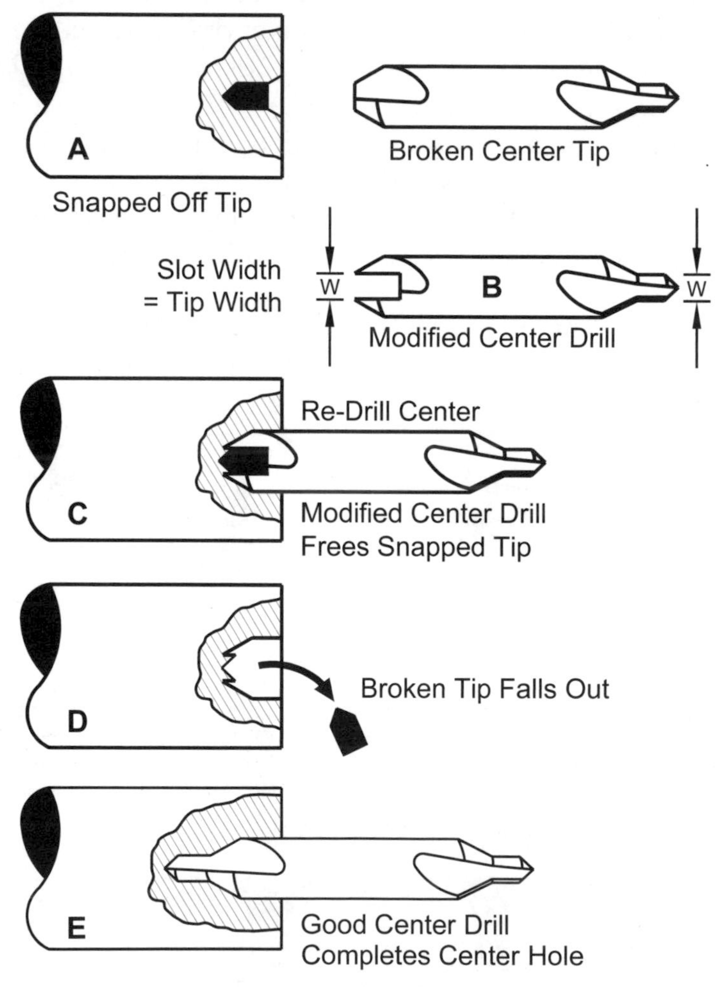

Figure 5–14. Recovering from a snapped center drill tip. If the work will be turned between centers, it will have to be refaced to provide access to a tapered center hole for the lathe center.

Section II – Twist Drills

Drill Nomenclature

What are the parts of a twist drill?

See Figure 5–15. The most common jobber-type twist drills have 118° point angles and the more expensive split-point drills have 135° point angles. Because Cobalt HSS and HSS bright finish drills look alike, some Cobalt HSS drills have a step cut into the end of their shank to indicate they are Cobalt .

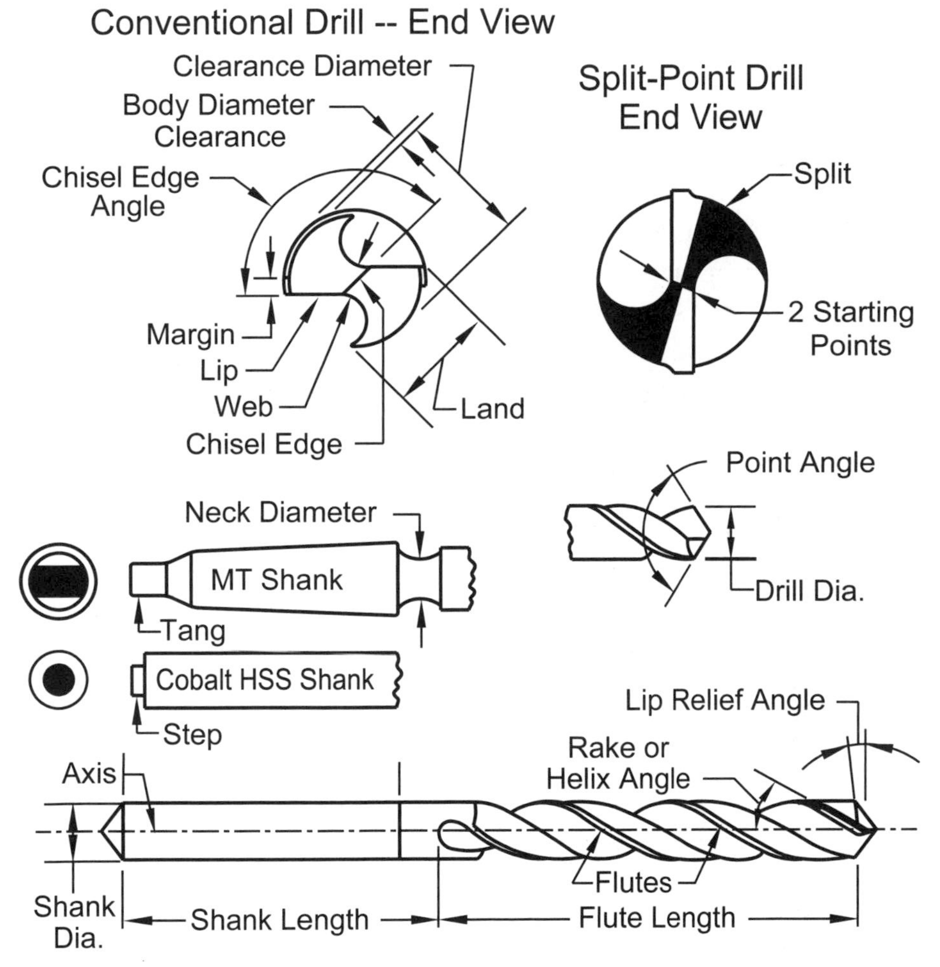

Figure 5–15. The parts of a twist drill.

Twist Drill Mechanics

How do twist drills work?

The process of drilling is complex because several different processes occur simultaneously and interact with each other.

- The chisel point both extrudes and cuts the metal under it. High axial forces on the drill are necessary to maintain this extrusion, not the lip cutting action.
- The workpiece metal may be softened by friction heating from chisel point rotation.
- The lips of the drill perform cutting action similar to other machining operations. However, it is complicated by the changing cutting angle along the lip, which goes from a negative to a positive rake from the center of the drill to its outer edges. The speed of the workpiece metal passing the cutting tool edge increases as the distance from the drill center increases. The cutting speed at the outer diameter of the lip can be four to five times faster than at its inner diameter.
- Metal removed from the hole slides along the flutes and out of the hole. Lubricant works its way down to the cutting face through the flutes.

Speeds & Feeds

What factors affect drilling speed and feed selection?

There are many factors including:

- Workpiece material.
- Drill material and coatings, if any.
- Cutting fluids, if any.
- Rigidity of the drill and drill press.
- Rigidity of the workpiece and its jig or fixture.
- Quality of the hole.
- Depth of the hole.

How are *speeds and feeds* determined?

Speeds and feeds have been established by experiment just like those for lathe and milling machine cutters. They are not exact because there are so many variables, but they do provide a good starting point. Remember that the speeds and feeds listed in handbooks are based on the optimum trade off between maximum cutting tool life and minimum machine time in a production environment. *These do not apply in R&D labs and prototype machine shops.* Usually, only a few holes are needed, and such optimization is unnecessary. Getting a hole in the right place without excessive drill wear is usually the objective. Drill speed is the rate that workpiece material and the cutting tool bit pass by each other during cutting. For drill bits, this speed is slowest toward the center of the bit and fastest at the outer diameter. Speed at the outer diameter is used in the cutting speed calculations. Given the recommended cutting speed for a given material and the drill diameter, we can determine the drill spindle speed just as for lathe and milling machine tools with the formula: $N = \frac{4V}{D}$

Where: N = Spindle speed of the drill press in rpm.

V = Cutting speed in feet/minute at the drill's outer diameter.

D = Diameter of the drill in inches.

Another way to determine drill spindle speed is to use the chart most drill presses have affixed to them that shows recommended speeds for common materials and a range of drill diameters. This makes calculations unnecessary.

Remember that these figures are for HSS drill bits. Carbide and coated bits operate at 2 to 3 times higher rpm. Adding lubricant can increase speeds by as much as an additional 50%.

What are typical feeds used for drilling?
See Table 5–1 to determine the drill feed in thousandths of an inch per revolution.

Drill Diameter (inches)	Drill Feed (thousandths of an inch/revolution)	
	Soft Materials	Hard Materials
< ⅛	3	1
⅛–¼	6	2
¼–½	10	4
½–1	15	7
> 1	25	10

Table 5–1. Typical drill feeds by drill diameter.

Drill Sizing Systems

What are the four different ways twist drills are sized?
They are:

- *Numbered drills*, also called *wire gage sizes*, have size increments between each drill based on the ANSI preferred number series: R80. This increment is about 3%. The size increments in the R80 series are based on: $10^{N/80}$ where N is any number from 1 to 80. This factor, $10^{1/80}$, is about 1.0292, or just under 103%. Numbered drills are usually furnished in a drill index with sizes from #1 to #60. Another smaller indexed set with

#61 to #80 drills is also available, but much less popular since the drills are relatively small and only needed for fine work. These eighty drills complete the usual numbered drill series. However, the set of eighty has been extended from #81 to #97. These drills are very small and are not commonly stocked. *Numbered drills remain important because the smaller numbered screw sizes (#1 to #12) require them for tap and clearance drills for the proper thread percentage.*

- *Letter drills* are another graduated series that continues on from where the largest number size, #1 at 0.2280-inches diameter stops. Each letter down the alphabet is about 2% larger than its predecessor. They are needed as both tap and clearance drills for medium-size inch-based threads and fasteners.
- *Fractional-inch drills*, known as *Imperial drills* in Europe, run from 1/64 to over 3-inches diameter. Even larger twist drills are available, but they require so much torque that medium-sized machine tools cannot drive them. Depending on the material, somewhere above 1-inch diameter machine tool power limits force a switch from drilling to boring and trepanning. Fractional drills provide both clearance and tap holes for large fractional-inch fasteners. Unlike number or letter drills, which have a 3 or 2 percent increase between sizes, fractional-inch drills come in *fixed size increments* of 1/64-inch for drills under 1-inch diameter and 1/16-inch increments for larger diameter drills. A complete set of 115 pieces includes 29 fractional, 60 numbered and 26 alphabetical drills.
- *Metric drills* are available from 0.3 to 76 mm and, like the fractional-inch drills, have fixed size increments. These drills are needed for metric fasteners.

See Table 5–2 and Figure 5–16 for comparison of drill sizing systems.

Drill Sizing Series	Minimum Diameter (inches)	Maximum Diameter (inches)	Increment or Step Size between Drill Sizes
Number	0.0059	0.2280	about 3%
Letter	0.2340	0.4130	about 2%
Fractional-Inch	0.0156 (1/64)	3.000+	Increments of: < 1-inch by 64^{ths} of an inch > 1-inch by 16^{ths} of an inch
Metric	0.118 (0.300 mm)	2.992 (76.000 mm)	Increments of: < 10 mm by 0.1 mm > 10 mm by 0.2 mm

Table 5–2. Drill sizing systems comparison.

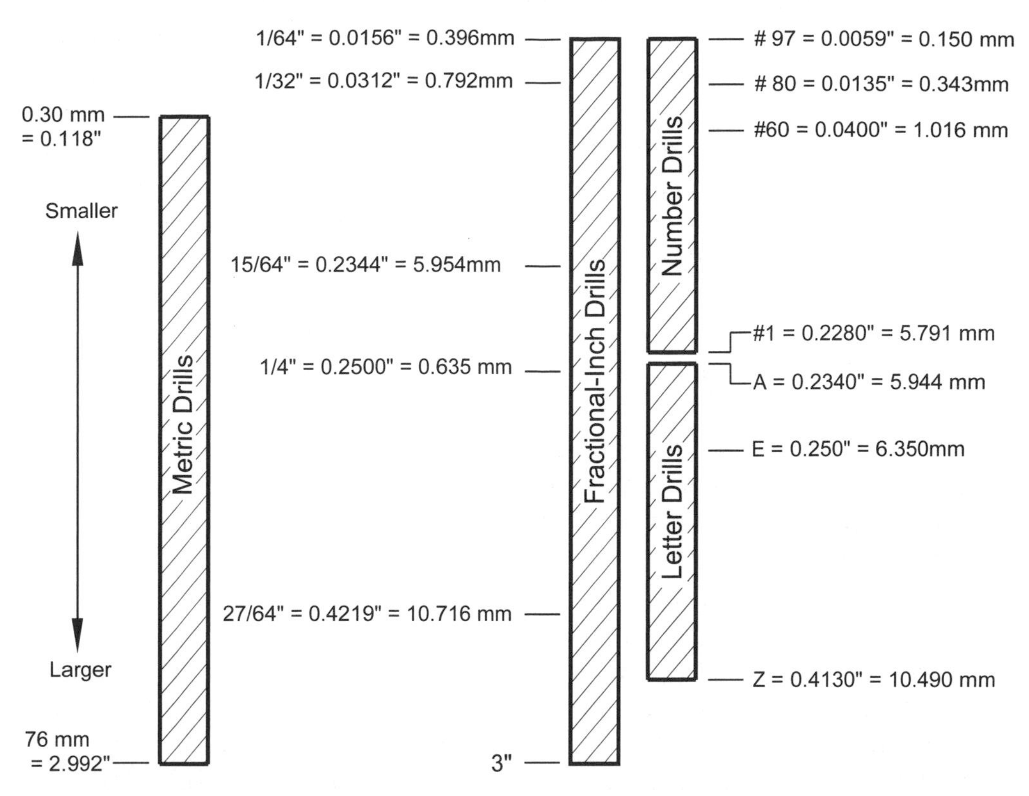

Figure 5–16. Size ranges for the four drill sizing series.

See *Appendix Table A–3,* pages 487–488, for decimal equivalents of fractional-inch, number, letter and metric drills.

Twist Drill Designs

What are the common *twist drill designs* and their uses?

- Twist drills come in many designs. The most common are *right-hand jobber length drills* with a medium helix or twist. These drills turn clockwise, viewed looking from the shank end. *Left-hand twist drills* have two uses. In production, multi-spindle drilling machines use them in some spindles, which turn counterclockwise to simplify the machine's design. They are also used in broken screw extractor kits because the force and vibration of drilling serves to loosen a right-handed screw or bolt, not tighten it. *High-helix drills*, sometimes called *parabolic deep-hole drills,* are designed to drill holes 10 to 15 times their diameter. Their flutes are designed to reduce chip friction up the drill and increase coolant flow down them. However, they are for soft materials like aluminum, white metal and other medium-tensile strength materials. See Figure 5–17.

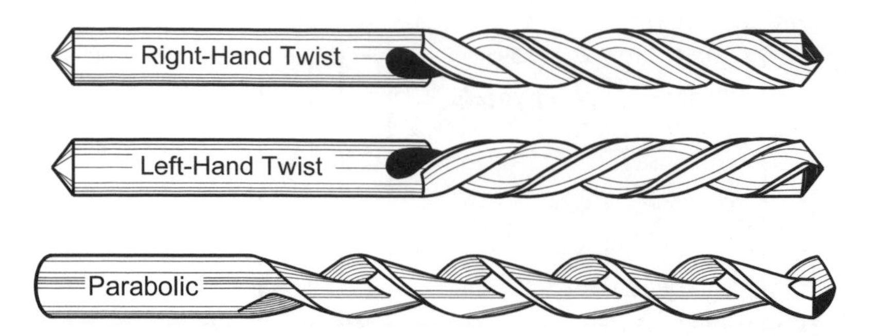

Figure 5–17. Common twist drill designs: right-hand standard (top), left-hand standard (middle), right-hand high-helix parabolic (bottom).

- *Reduced shank drills*, also called *Silver & Deming* drills (named for an early American woodworking machinery company that vanished at the beginning of the 20th Century), *blacksmith drills,* and in Canada, *Prentice drills*, permit the use of larger diameter drills than would usually fit in a given size chuck, Figure 5–18. Standard shank diameters are ¼, ⅜, ½ and ¾-inch diameter. They are available in inch and metric sizes and with a variety of point angles and anti-wear coatings. Remember to slow down the drill for larger diameters.

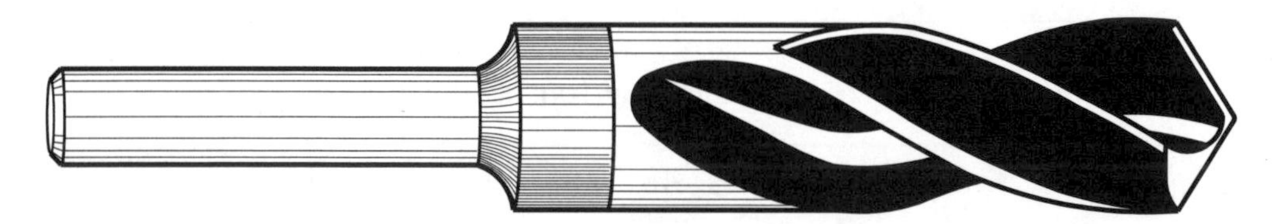

Figure 5–18. Reduced shank drill.

- *Step drills,* Figure 5–19, offer the advantage of drilling two different size holes at once. They are often used to drill and countersink holes for cap screws in a single operation, but also work better than plain twist drills for *chain drilling.* See *Chapter 12 – Other Shop Know-how.*

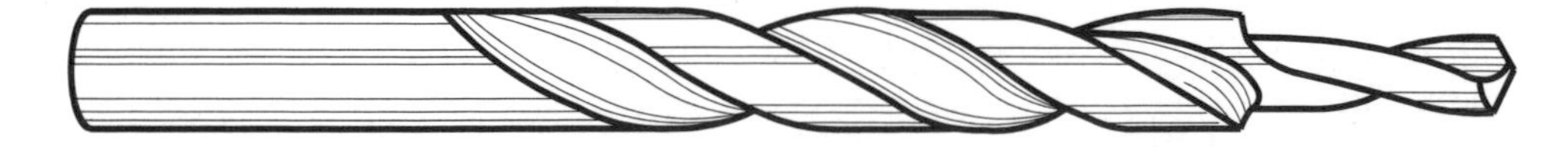

Figure 5–19. Step drill.

Drill Lengths

What are the standard lengths for twist drills?

They are:

- *Screw machine,* also called *stubby* – 2½ inches.
- *Jobber* – 4 inches.
- *Aircraft* (two sizes) – 6 and 12 inches.
- *Extended,* also called *long boy* – 18 inches.

See Figure 5–20.

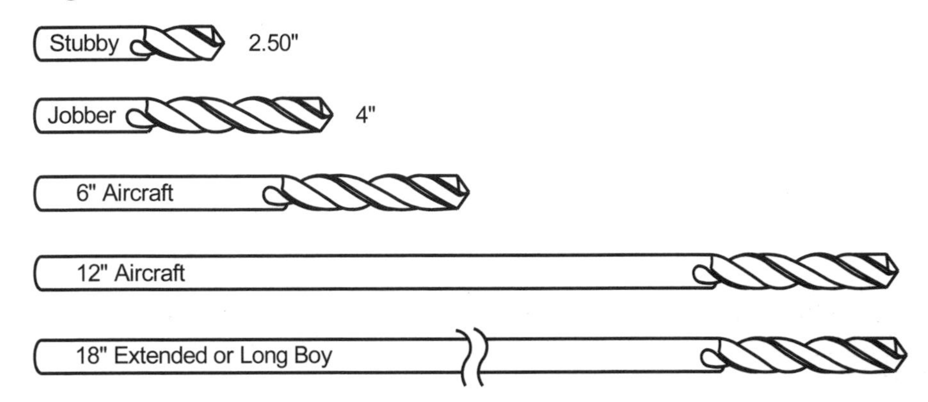

Figure 5–20. Standard twist drill lengths for ¼-inch diameter drills.

Drill Materials & Coatings

What metals are used for drills and what are their advantages?

The metals are:

- *High-speed steel*, abbreviated HSS and stamped on the drill shank, is the most common drill bit metal and is suitable for typical shop drilling.
- *M42 cobalt HSS alloy* costs about twice as much as HSS in an equal size and will last ten times as long as HSS in the same application. Cobalt bits are for drilling work-hardened, high-strength stainless and manganese steels, titanium and Inconel®. They are also useful for drilling hardened materials, like bolt heads. This alloy is heat resistant and maintains its cutting edge above 1000ºF which allows higher cutting speeds (SFM).
- *Solid tungsten carbide* bits last about three times as long as HSS when drilling zinc, brass, rubber, plastics and other abrasive, low tensile strength materials. They are easily damaged by shock and are not for use in steel. Be especially careful to avoid chatter with solid tungsten carbide countersink bits since they are brittle and fracture easily.
- *Titanium nitride (TiN)* and *titanium aluminum nitride (TiAIN)* are expensive, high-performance coatings. To gain the benefits of tool bits with these coatings, machine tools using them must run at 50 to 100% higher speeds than similar uncoated tools. Running *coated tools* at the same speeds as *uncoated tools* is likely to be counterproductive. Most of the time, they are neither needed, nor cost effective in the prototype shop. See Table 5–3.

Drill Material & Finish	Price (HSS Cost = 1 ×)	Performance
HSS (M2) with black finish	1 ×	—
HSS (M2) with bright finish	1 ×	Good chip ejection for low carbon and other soft materials
HSS (M2) with black finish left-hand twist*	2–3 ×	Same life as right-hand drills; used for broken screw or bolt removal
Cobalt HSS (M42)	2 ×	—
TiN coated	1.7 ×	—
TiAIN coated	2.6 ×	—
Solid Tungsten Carbide (mostly for production)	9 ×	For low-tensile abrasive materials like zinc, brass, plastics
Brazed Tungsten Carbide tip	4 ×	For abrasive, non-ferrous materials, not for steel

* Included to show price differential for left-handed drills.

Table 5–3. Approximate cost and performance comparison of straight-shank ¼-inch jobber length twist drills.

Does the twist drill's finish—black or bright—on a HSS twist drill make a difference in a non-production setting?

Manufacturers claim a difference in the ability of different finishes to slide chips along their flutes, but in most applications there will be no noticeable difference. Similarly, the added-cost of TiN and TiAIN coatings will not make much difference in a non-production setting, and the coating's most valuable area (on the end of the drill) is lost during the first sharpening.

Point Designs & Angles

Why are different *point angles* used on twist drills?

In general, softer materials, like plastics, drill better with a more pointed 90° drill, and harder materials drill better with a less pointed 135° one. The 118° pointed drills are a compromise between these two extremes and are good for general purpose shop use. Until recently most drill bits had tip angles of approximately 118°. More common now is 135° with a split point. This newer design requires less axial cutting force because it reduces the size of the chisel point and has less tendency to "walk" or "skate" when starting a hole without a center punch mark. Figures 5–15 and 5–21 shows these different point designs. Split-point drills are slightly more expensive because they have two additional surfaces to grind. Every shop should have drill sets of this design.

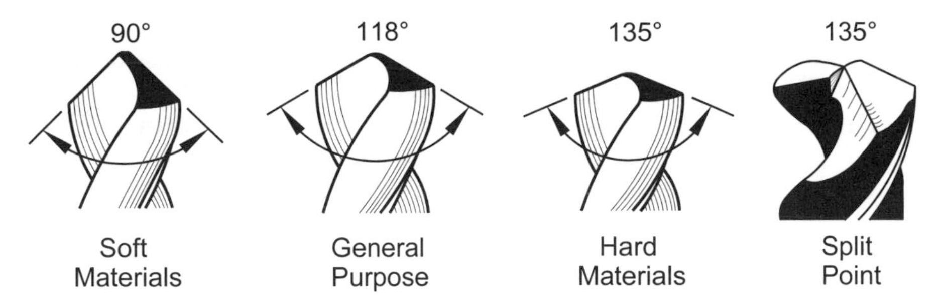

Figure 5–21. Point angles and split points.

Shanks

What is important to know about drill *shanks*?

- There are two designs: the straight shank and the tapered shank. Small- and medium-sized drills, up to ½-inch or 13-mm diameter, use straight shanks which fit in Jacobs and Albrecht drill chucks. Larger drills require more torque and use a tapered shank for a better grip under load. They fit in a Morse taper (MT) in a drill press, lathe or milling machine. The flattened end of the drill shank, called a *tang,* fits into a rectangular slot in the back end of the MT. This insures that the drill turns with the taper and does not spin in it. See Figure 5–15 for a comparison of shank shapes.
- Most drill shanks are not hardened. They can easily be shortened for tight quarters, or be turned down in a lathe to fit smaller chucks. However, a smaller diameter shank gives the chuck a poorer grip than a full-size shank, so remember that the drill may spin in the chuck if you treat it as a full-size shank. If this happens, you will want to remove thc burrs that result on the drill shank. Any time you "spin" a drill and raise burrs on its shank, the drill is compromised and cannot be used for precision work.

Deep-hole Drilling

How is *deep-hole drilling* defined?

It is defined as holes that are three or more hole diameters deep. When using twist drills on deep holes, reduce speed and feed. See Table 5–4. Chip removal is the biggest problem when deep-hole drilling because friction from the chips will heat up the tool bit and anneal it. Although use of parabolic drills provides good chip removal in soft materials like aluminum, hard materials like cast iron produce chips that tend to pack the drill flutes and prevent smooth chip flow. This requires "peck" drilling where every two or three diameters the drill is withdrawn momentarily and the chips removed from its flutes. A stiff brush may be needed for this. Depending on the material, lubricant may be added to cool the drill and reduce the tendency of metal to weld to the drill.

Hole Depth-to-Diameter Ratio	Speed Reduction from Recommended Drilling Speed (%)	Feed Rate Reduction (%)
3	10	10
4	20	15
5	30	20
6–8	35–40	25

Table 5–4. Speed and feed reductions for deep-hole drilling.

Nearly all holes smaller than 0.050-inches (1.3 mm) diameter are really deep holes and may have chip clearance problems requiring peck drilling. In production operations, there are drill designs that inject coolant under pressure down through holes *inside* the drill to clear the chips. Other drills use carbide inserts in spade bits that provide more chip clearance. Sometimes holes are drilled upside down to let gravity help remove chips.

Section III – Other Cutting Tools

Multi-Diameter Step Drills

What are *Multi-diameter step drills* and what are they used for?

They are a one piece HSS stepped-drill that cuts up to 12 different size holes in 1/16-inch diameter increments depending on how far the drill enters the work, Figure 5–22. They work equally well in drill presses or hand-held electric drills. Because they cut one diameter at a time, relatively little drill pressure is needed. Although they can cut stock up to ⅜-inches (9.5 mm) thick, they are primarily for thin materials like sheet metal, copper, brass, aluminum, wood, plastic and laminates. Unlike a twist drill in thin metal, these drills leave a hole nearly free of burrs. In smaller sizes, they are self-starting, but larger sizes require an initial starting hole. They can be resharpened without special tooling. These drills make holes from ⅛- to 1⅜-inches (3.2- to 34.9-mm) diameter.

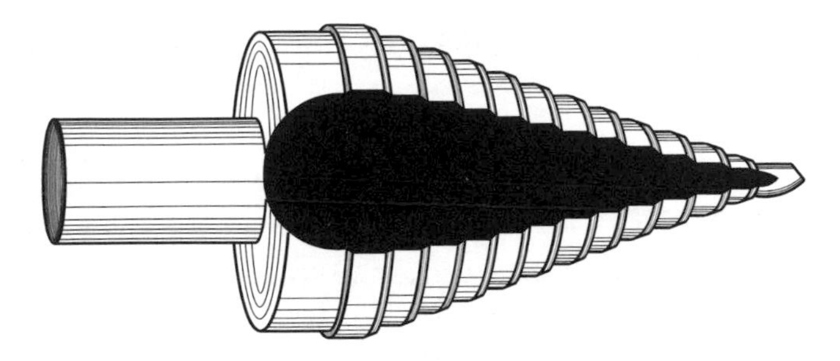

Figure 5–22. Multi-diameter step drill.

Straight Flute Die Drills

What are *straight flute die drills* for?

They are used to drill difficult materials that have been hardened up to R_c60 such as nickel, chrome, titanium, and high temperature alloys including stainless steel weldments. Their straight flutes provide edge strength, diameter control and good surface finish. Sometimes these drills are used on sheet brass and aluminum because the work does not tend to climb upward on the drill flutes. Straight flute die drills are limited to shallow holes, no more than 3 × drill diameter. See Figure 5–23.

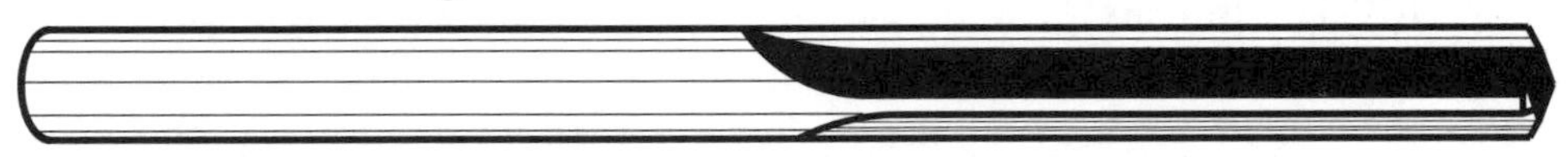

Figure 5–23. Straight flute die drill.

Masonry Drills

Where are *carbide-tipped masonry drills* used and how do they work?

Masonry drills put holes through brick, cement, concrete and the softer stones usually for anchor holes or to run wiring. The carbide insert brazed onto its steel body does the drilling and the flutes remove the resulting powder, Figure 5–24. Masonry drills run from ⅛- to 1-inch diameter. Although these drills work well in a drill press, most applications call for drilling work in place, so hand-held electric drills, typically ⅜-inch or larger, are usually used. Here are some precautions to their effective use:

- Use a slow drill speed to avoid burning up the drill bit. Stop to let the bit cool occasionally.
- Considerable pressure on the drill is usually required; they do not self-feed.
- When drilling concrete and the drill stops removing material, yet continues to rotate, the drill has hit aggregate (stone) inside the concrete. Use a star drill and hammer and break the individual piece of stone so drilling can continue. On shallow holes, an old center punch will work.

Usually only a few hammer strikes are needed. Using a hammer drill will completely avoid this problem.

- When drilling vertically down, material powdered by drilling will accumulate in the hole bottom and retard drilling. Stop drilling and use compressed air to remove it. *Be prepared for a dust cloud.*

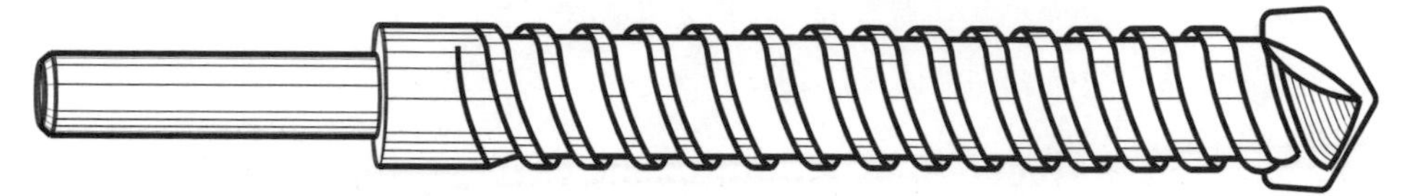

Figure 5–24. Carbide-tipped masonry drill.

Wood Bits

What are the most common hole-making tools for wood?

- *Spade bits* are the simplest and least expensive wood-drilling tools. They are available from ¼- to 1½-inches diameter. To work properly, they must be sharp and pressure must be applied to the bit, as they are not self-feeding. If the wood smokes, the drill is dull. Stop drilling and sharpen it with a file. Do not file the outer edges of the bit, as it will change the hole diameter. Be sure to slow down the spindle speed for larger drill sizes. Extensions are available for hard-to-reach spots and extra-deep holes. Several extensions may be put together in series for even more length. See Figure 5–25.

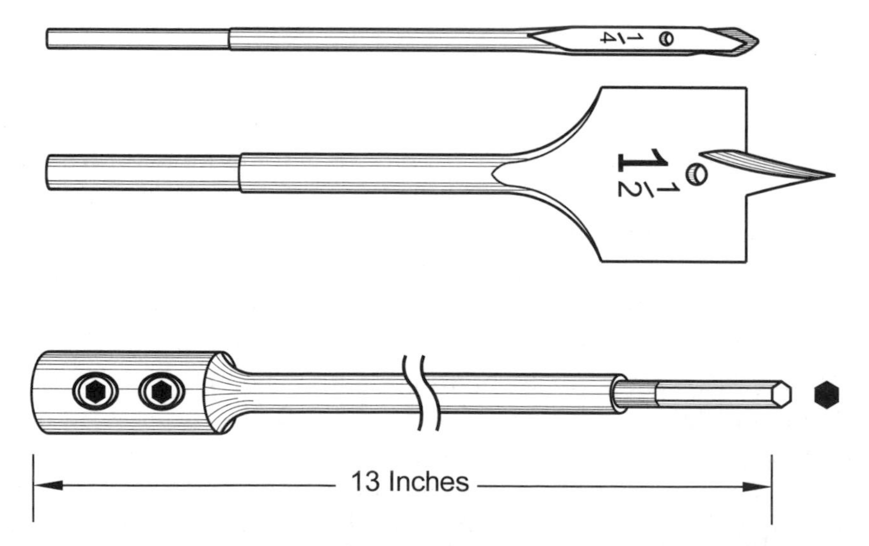

Figure 5–25. Spade bits (top and middle) and an extension for them (bottom).

- *Power helix bits*, also called *auger* and *ship auger bits,* Figure 5–26, are usually 8- or 18-inches long, although there are some longer versions available up to 29 inches. Unlike the traditional brace-driven auger bits, which have a double helix and two end spurs, they have a coarse single spiral, one spur, and a self-feeding central screw point. They may have round, hex or the old four-sided taper shanks. They originated in the boat building industry, but are used for electrical construction work in buildings, on power poles and for timber-framed buildings. They are designed for fast cutting, produce rough, splintery holes and are not for fine work. They require a lot of torque, so larger bits require a ½-inch electric drill. Sizes for these bits range from ¼- to 2-inches diameter.

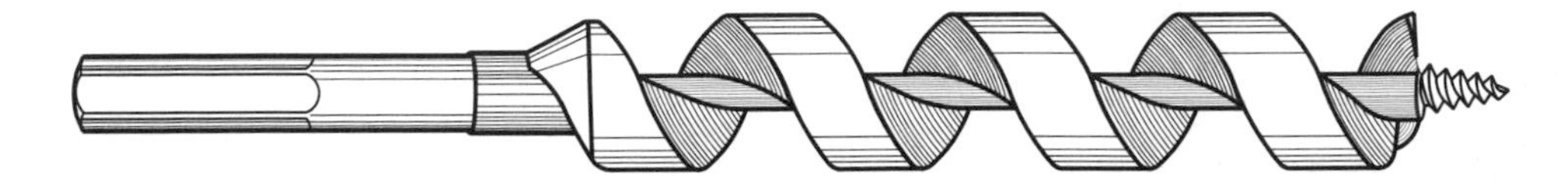

Figure 5–26. Power helix bit.

- *Brad-pointed, or spur- pointed, wood drill bits*, Figure 5–27, cut a cleaner hole in wood than power helix bits or twist drills and are ideal for drilling dowel holes. Their center point keeps the bit from wandering and the side spurs insure a smooth hole. They are available from ⅛- to ½-inch diameter in HSS.

Figure 5–27. Brad-pointed, or spur- pointed, wood drill bit.

- *Forstner bits,* Figure 5–28, make smooth-sided, flat-bottomed holes through hard wood, even knots, while holding tight diameter tolerance. They also drill well into end grain and thin veneers without splitting the wood. Using a drilling jig, they can make straight, angled or overlapping holes. They are often used for *pocket holes* to conceal hardware, especially hinges and fasteners. They are limited to wood and very soft plastics. Their long, rigid shank often is captured by a drill jig for guidance. They are available in diameters up to 3½ inches in HSS, carbide and titanium coatings. Forstner bits should be used in a drill press or in a drilling jig powered with a hand-held electric drill. Avoid using them freehand in an electric drill. Guides and drilling jigs are available for work outside the drill press. Some Forstner bits have no center spur and must be used in a drill press. They may be resharpened with a file and an Arkansas stone. Remember to use slow speeds to avoid burning the bit, 2400 rpm for ⅜-inch bits, just a few hundred rpm for 3-inch ones.

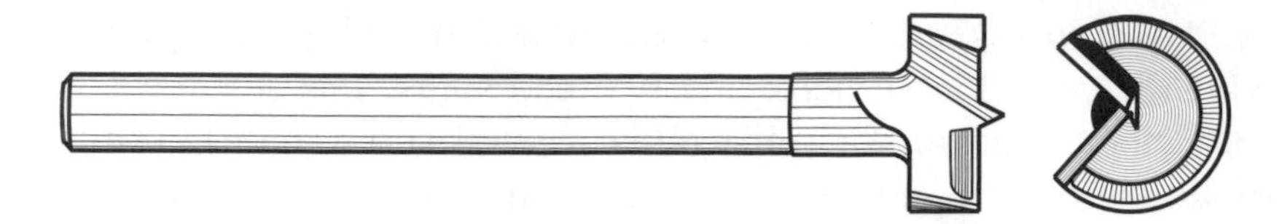

Figure 5–28. Forstner bit.

Section IV – Tool Holders

What are the two ways drill presses hold drills?

They hold smaller drills in *chucks*, and larger ones in Morse tapers (MTs) inside their spindles. While drill chucks depend only on friction between the chuck jaws and the drill shank to turn the drill, the Morse taper spindle traps the rectangular tang on the drill locking the drill to its spindle. It cannot spin as drills in chucks sometimes do.

Chucks

What are the two common chuck designs for drill presses?

- *Jacobs®-type chucks,* also called *key chucks,* are 3-jaw drill chucks tightened with a chuck key and are the most common chuck design, Figure 5–29 (left). Today, Jacobs Chuck Manufacturing Company makes both the traditional key chucks and the newer keyless chucks.
- *Albrecht chucks,* also called *keyless precision chucks,* are frequently used in production because their keyless design makes hand tightening possible. This is more convenient and efficient when changing drills. They provide better concentricity for small drills than do key chucks. They also have excellent gripping power, but are more expensive. Albrecht brand keyless chucks are available with diamond-impregnated jaws for additional holding power. Once you use a keyless chuck, you will not want to go back to the key design, Figure 5–29 (right).

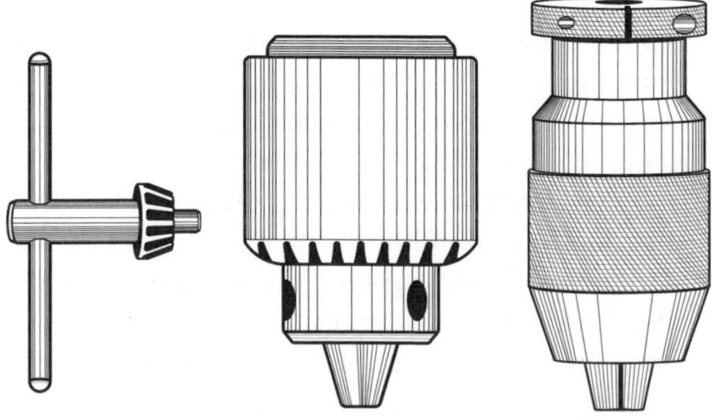

Figure 5–29. Jacobs-type chuck with key (left) and Albrecht chuck (right).

What three methods secure drill chucks to the drill press spindle?
The chuck:

- Has an integral taper and fits directly into the Morse taper in the spindle, Figure 5–30 (left).
- Fits onto an arbor and the arbor fits into the MT. The taper on the chuck end of the arbor is usually a *Jacobs taper* (JT) which is shorter and steeper than the MT, Figure 5–30 (center).
- Fits directly onto the JT on the end of the drill press spindle, Figure 5–30 (right).

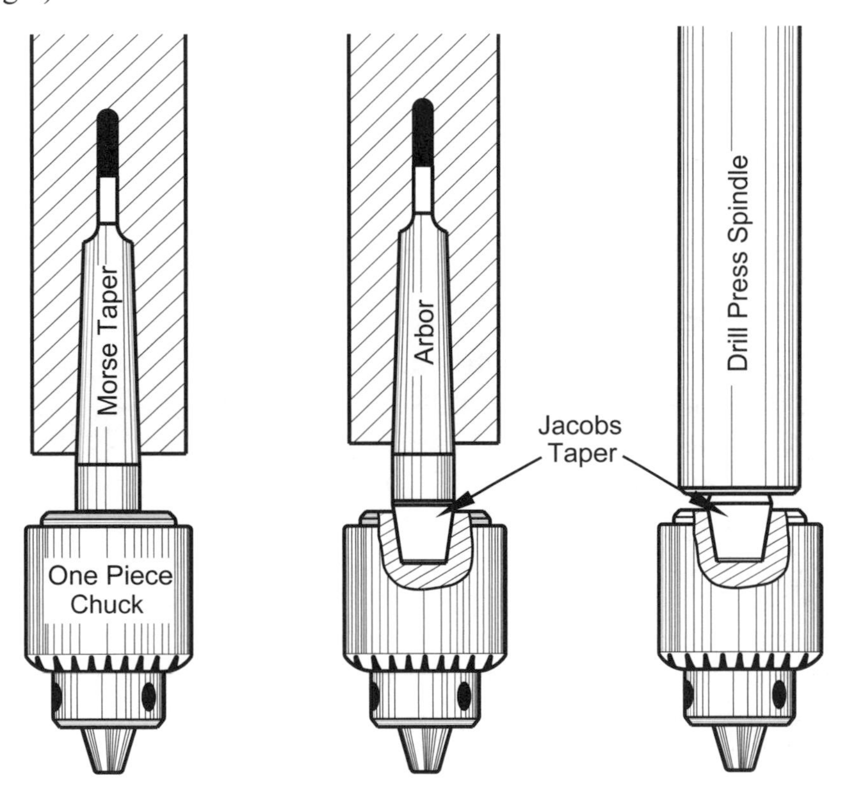

Figure 5–30. Securing chucks to the drill press spindle. Both the MT and JT are self-holding, meaning they lock up and do not release when axial force is reduced.

How are drill chucks on JTs removed from their arbors?
Special pairs of wedges are forced between the chuck and the arbor by lightly tapping the wedges. This dislodges the drill chuck. Do not allow the loosened chuck to strike the worktable. See Figure 5–31.

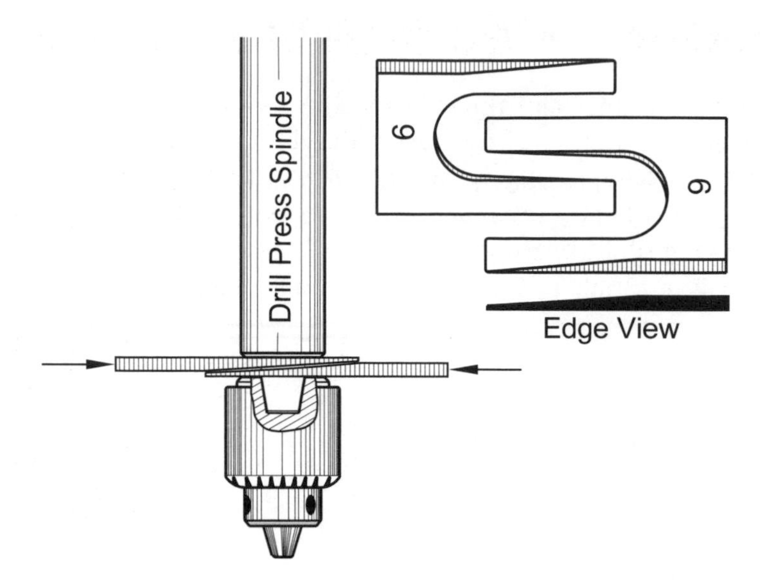

Figure 5–31. Using wedges to remove a chuck from its arbor.

Morse Taper Shank Drills

How do *MT-shank drills* fit into drill press spindles?

They fit into the spindle just as MT chucks do. See Figure 5–32.

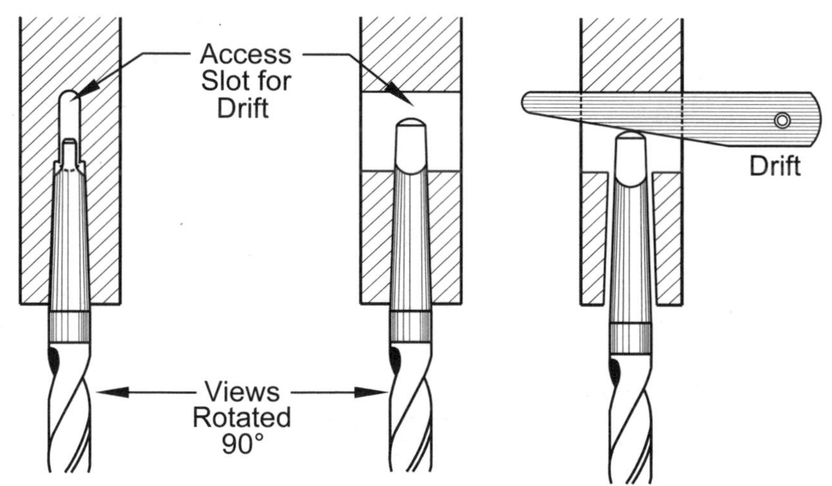

Figure 5–32. Fitting and removing MT-shank drills from spindles.

Figure 5–32 (right) shows how a *drift*, a special-purpose wedge, is used to release the shank from the spindle. Usually just a few light taps are needed. Place a piece of wood under the drill or chuck and raise the drill table to within a ½-inch of its end. This will prevent the tool from falling far and striking the work table when released. Be sure to match the drift size to the MT size. Like MTs, drifts have size numbers from 1 to 5.

Section V – Work-Holding Methods

What are the typical methods used to hold work in drilling machines?

- *Drilling vises,* Figure 5–33, are similar to milling machine work table vises, but are made to lower tolerances and so they may cost as little as 25% of the cost of a milling vise. Because most drilling work is less precise than milling operations, this is not usually a problem. However, a milling vise could also be used on a drill press table.

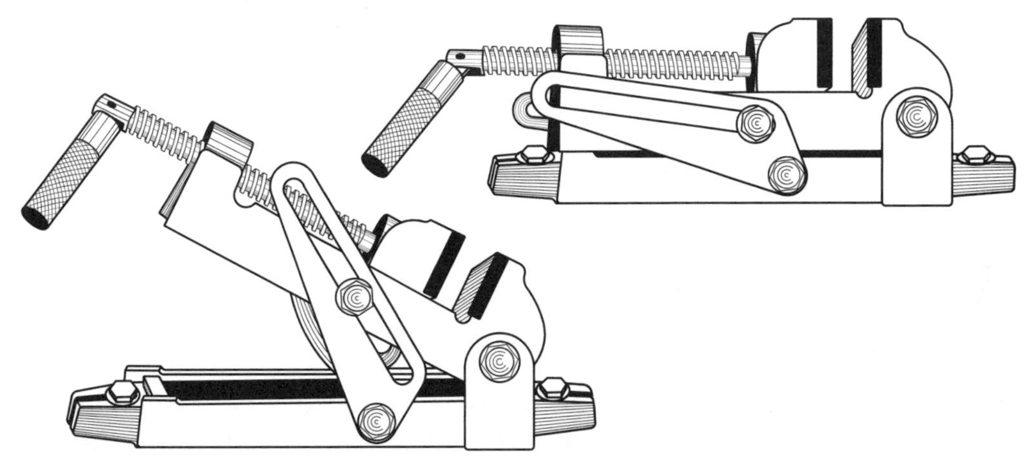

Figure 5–33. Drilling vise flat (top) and at an angle (bottom).

- *Locking hold-down clamps* bolt though the drill press table, Figure 5–34. They are quicker and easier to use than conventional C-clamps and just as secure. They are usually used in pairs.

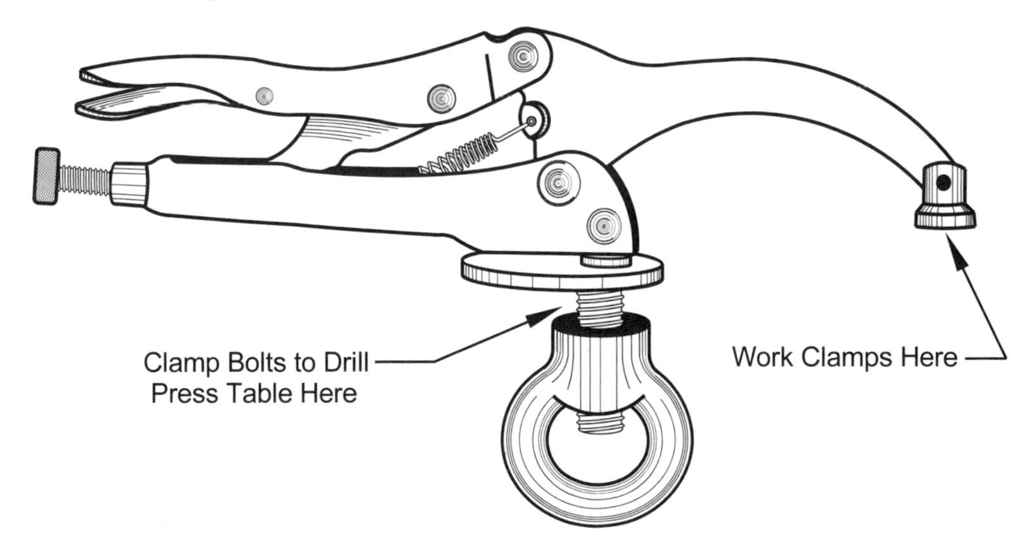

Figure 5–34. Locking hold-down clamp for the drill press.

- *Table clamps*, Figures 8–39 and 8–40, *Chapter 8 – Milling Operations*.
- *V-blocks,* Figure 8–47.
- *Angle plates,* Figures 8–49 through 8–51.

These last three methods to hold drilling work have a much higher degree of accuracy when used in milling machines and are covered in detail in *Chapter 8 – Milling Operations*.

Section VI – Practical Drilling Methods

General Drilling Procedure

What are the steps for drilling a hole?

1. Mark the hole location initially by scribing it, then mark it with a prick punch, finally enlarge the prick-punch mark with a center punch. This is essential on a curved or cylindrical surface.
2. Select the shortest drill possible for rigidity. This will increase drill life, concentricity and on-axis location. Check its sharpness and verify its diameter with a drill plate or micrometer. Remove any burrs on its shank with a file or abrasive stone.
3. Determine the spindle rpm and recommended feed rate from tables in *Machinery's Handbook*, and set this spindle speed on the drill press. It is more common to over speed twist drills than to over feed them. Over speeding generates excessive heat at the cutting edges, dulling the tool and possibly work-hardening the workpiece.
4. Mount the drill in its chuck and insure it is tight. Be sure to remove the chuck key. After making sure the drill will not touch either the drilling table or the work, "bump" the motor on momentarily to make sure the drill is concentric in its chuck, not trapped between two jaws of the chuck.
5. Secure the work to the drilling table, in a drilling vise or in a fixture. Use a wiggler to locate the center-punch mark exactly under the drill press axis. Wiggler use is explained in *Chapter 8 – Milling Machines*. Use a center drill to enlarge the center punch mark when high accuracy is needed. Make sure the drill will pass through the center hole in the drill press table, or that the work is spaced off the table so the table will not be drilled. Use a magnifier to check drill alignment for small diameter drills.
6. Decide what lubricant to use, if any, and have it handy.
7. Turn on the drill press motor and begin drilling.
8. When the drill hole depth is about 3 drill diameters deep, withdraw the drill and clear chips from the drill flutes and the hole. Use a stiff bristle brush, not your fingers. Repeat as needed.
9. Just before the drill breaks through the back side of the work, the drill will tend to stop cutting, dig in and lock to the work, spinning it if the work is

not well secured. To avoid this, ease up on drilling pressure to allow the drill to finish cutting its way through the work.

Thin materials often need support from below to prevent the drill from dishing the work around the hole. Plywood or Masonite® works well for this. Adding a top sheet of this material can prevent ripping thin, fragile materials. Clamp the resulting sandwiched layers around the hole.

Checking Drill Diameter

How can you check a drill's diameter?

There are three ways:

- By reading the size stamped on the shank.
- By measuring over the drill margins with a micrometer, however if the drill margins are worn, measure the shank diameter. Remember, if the drill margins are worn, the drill will not cut to size.
- With a drill check plate.

Good practice is to check every drill's diameter *before* use. See Figure 5–35.

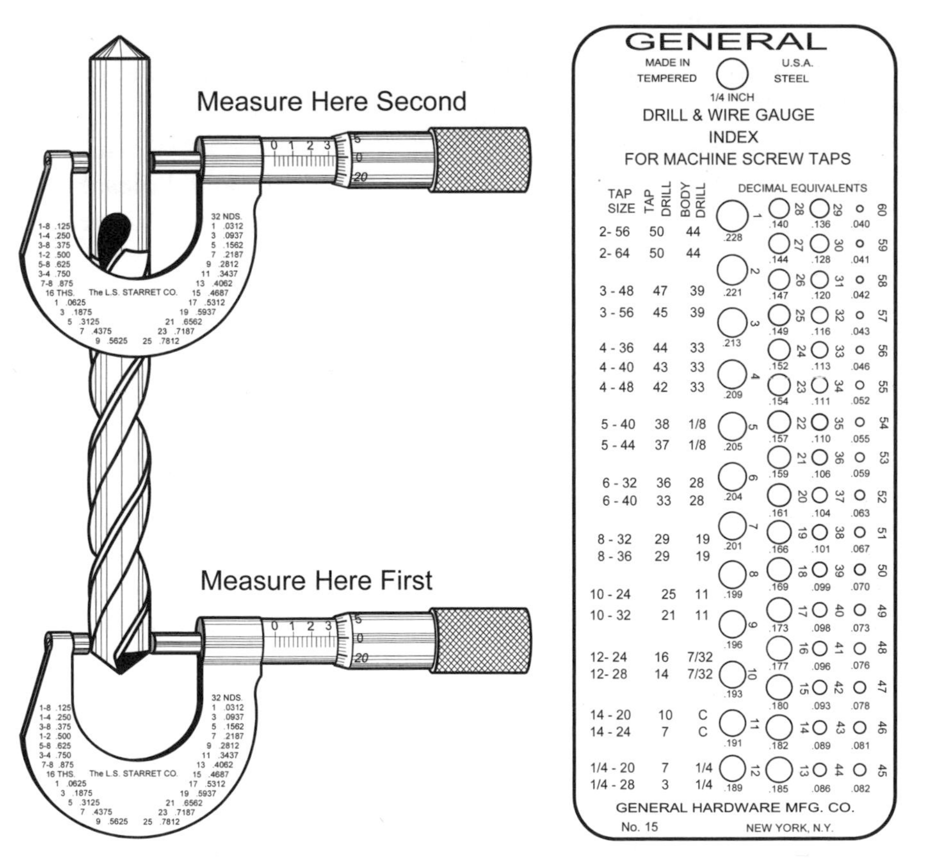

Figure 5–35. Checking drill diameter with a micrometer or drill-check plate.

Correcting a Drill's Position

When a drill drifts from the desired center, as in Figure 5–36 (A), how can it be brought back on center?

Cut a series of V-grooves with a cape or diamond-point chisel on the side of the hole to which the drill must move, as in Figure 5–36 (B), and begin drilling again to see if the drill has been drawn into place, Figure 5–36 (C). Repeat if needed. This method works only when the drill has *not* reached the depth to cut its full diameter. However, if a prick punch, center punch and center drill are used, this correction method is rarely needed.

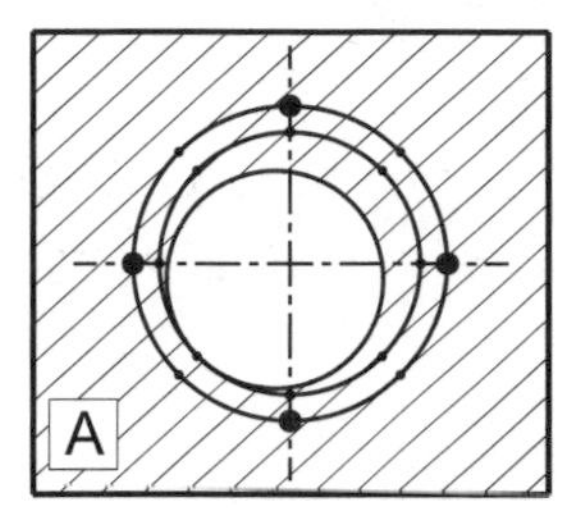

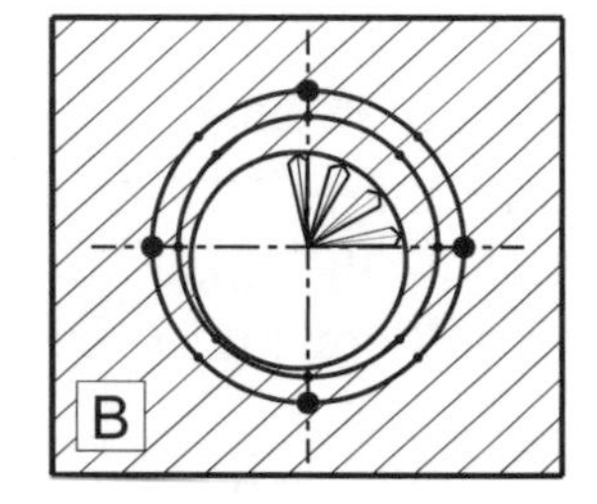

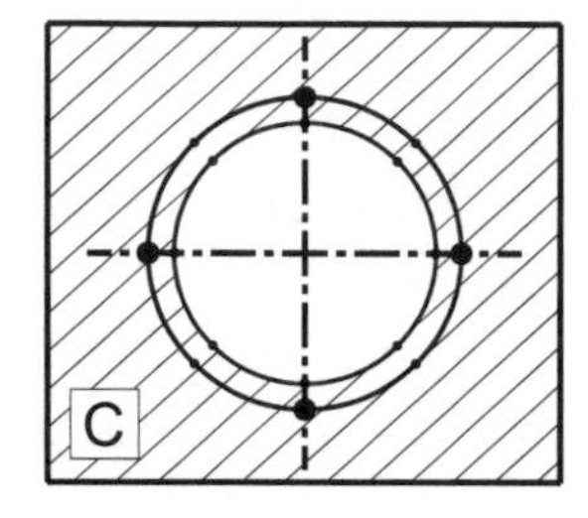

Figure 5–36. Correcting a drill's position.

How do you accurately transfer hole position from one part to an identical part?

There are three tools to do this:

- *Transfer punches* are used when both parts have through holes. See Figure 5–37. These are commercially available in fractional-inch, letter, number and metric sizes, but are easily shop-made from drill rod.

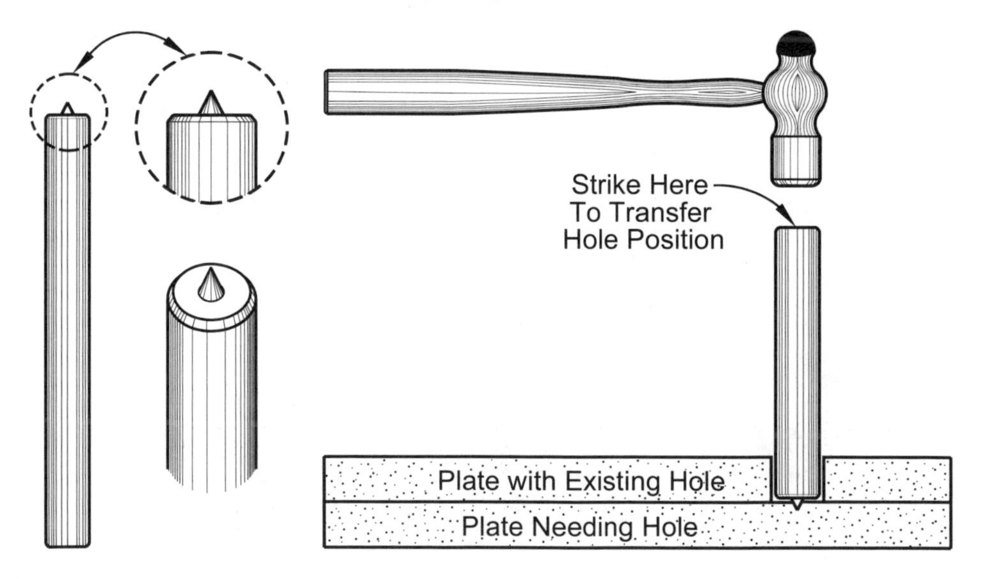

Figure 5–37. Transfer punch structure and use.

- *Transfer plugs* and *transfer screws,* Figure 5–38, perform the same function as transfer punches, but for blind holes. Plugs are used for plain holes and screws are used for threaded holes. Transfer screws have a hex cut in their tops so they can be adjusted for depth. Both are used like transfer punches and are available in inch and metric sets.

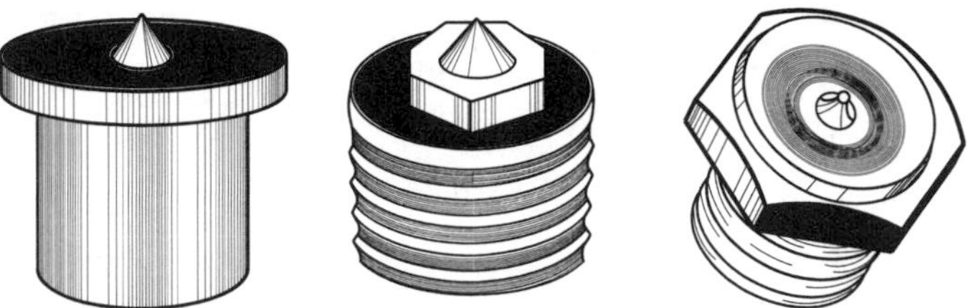

Figure 5–38. Transfer plug (left) and transfer screws (center and right). The transfer screw on the right also provides the outline of the hole in addition to the center punch mark.

What do *drill bushings* do?

They provide accurate initial drill positioning and on-going mechanical restraint during the entire drilling process, Figure 5–39. Available in many sizes and designs, they are hardened steel and ground to precise dimensions. They can enable the off-center drilling of cylinders and are also useful in precisely aligning reamers. They should be pressed into the drilling fixture or drawn in with a bolt, not hammered in. Hint: Position the bushing high enough over the workpiece so chips do not enter the bushing.

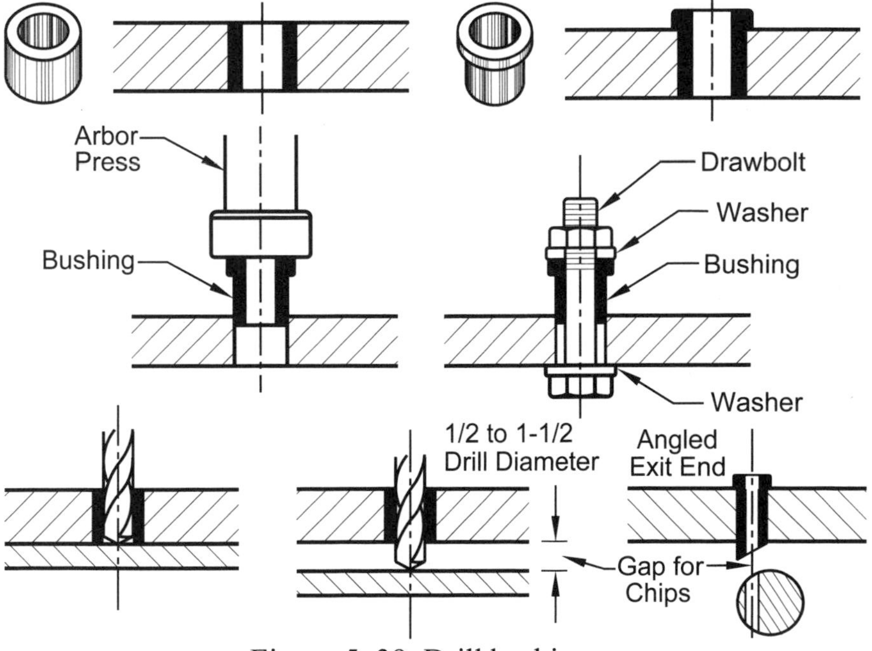

Figure 5–39. Drill bushings.

Drilling Fixtures

Why use a *drilling fixture*?

Drilling fixtures insure exact drill positioning, provide axial drill support to prevent the drill from bending, and also hold the work securely. As an example, Figure 5–40 shows a fixture for drilling bolt heads for safety wires. The drill is fully supported axially by the bushing and the threaded bushing jams the bolt head, preventing it from turning. Different size holes in the bushings allow the use of different size drills.

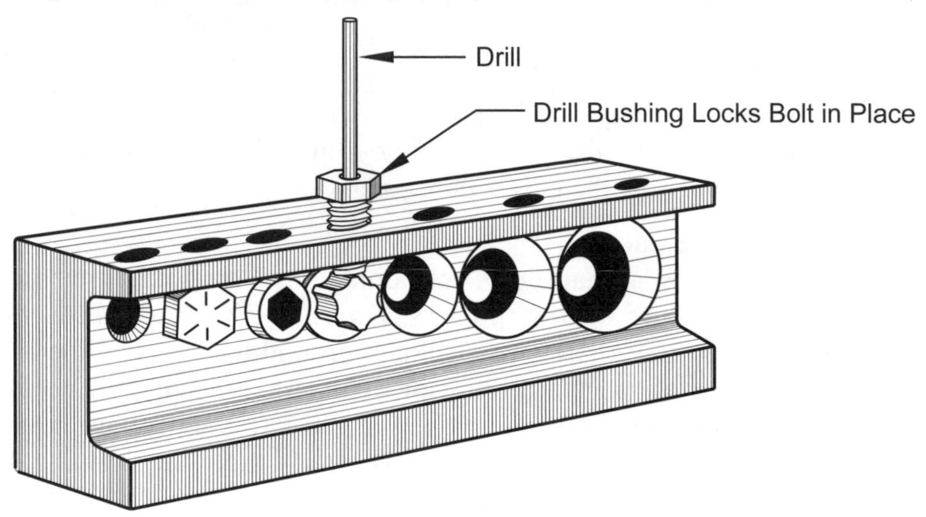

Figure 5–40. Commercial fixture for drilling holes in bolt heads for safety wires.

Coolants

What is the purpose of *coolants* when drilling?

They:

- Cool the drill and remove heat from the cutting site.
- Reduce friction at the cutting site.
- Minimize the tendency of chips to weld to the drill lips.
- Aid removal of chips.
- Improve hole wall finish.

What coolants should be used when drilling metals?

Typical drilling coolants are:

- Soluble oil
- Kerosene
- Cutting oil

While certain pairs of metals and coolants are especially effective, adding almost any type of oil improves a drilling operation. However, cast iron and other brittle materials produce small chips that tend to pack when lubricated, so do not use coolants with them. Instead, use compressed air, but be aware of the eye hazard of flying chips.

Pilot holes

Why does drilling *pilot holes* decrease stress on the drill and drilling machine?

The chisel point of the drill both extrudes and cuts the workpiece metal and requires most of the axial force on a drill to make it work. A pilot hole eliminates the need for these chisel point operations and so reduces the axial force. Use a pilot hole whose diameter equals the thickness of the drill's web, Figure 5–41. Using a much larger pilot hole concentrates cutting forces on too small a part of the drill lips and may cause them to chip. Pilot holes are not as effective for 135° split-point drills as their chisel points are very small to begin with and do not require much force.

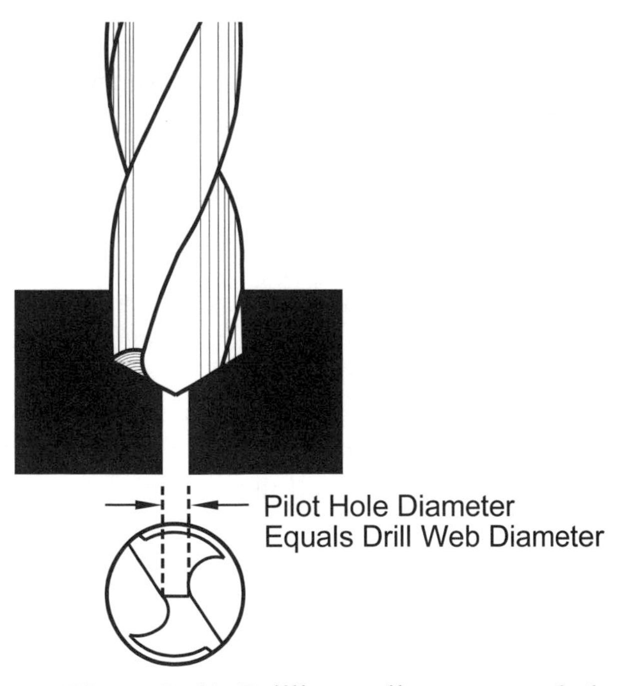

Figure 5–41. Drilling a pilot or starter hole.

Drilled Hole Size Accuracy

How does the diameter of drilled holes compare with drill diameter?

The drilled hole diameter in most materials will be *larger* than the drill and ranges from one or two thousandths of an inch for a 1/16-inch diameter drill to

8 or 9 thousandths of an inch for a 1-inch drill. Hole diameter is influenced by many factors including work material, drill length, drill sharpness, length and shape of the chisel edge, coolant, machine used and setup rigidity. But the biggest single factor depends on the accuracy of the drill point. Also, holes can be undersized when workpiece materials have a high coefficient of thermal expansion such as light metals and plastics.

Sharpening Twist Drills

Can twist drills be resharpened?

Yes, they can, but doing so requires professional-grade drill sharpeners costing between $1100 and $2500. Larger drills can be sharpened by hand, but this takes much practice. Small drills, say under ¼-inch diameter, cannot be reground freehand.

Owning professional drill-grinding equipment is not cost effective for small shops and regrinding drills without such equipment is always unsatisfactory. Sending drills out for sharpening is probably the best approach for small shops. A lot of high-quality drills can be bought for the cost of a good drill sharpener. Remember that in typical, undemanding applications (not drilling hardened materials) a drill will usually last for several hundred holes.

Making Extra-Long Drills

Sometimes only a jobber-length drill is available and a longer drill is needed. What can be done?

Here are three solutions:

- If the drill diameter is relatively large, chuck it in a lathe with the point-end toward the headstock. Cut off the drill shank square, center drill it and then drill it for a piece of drill rod. Drill shanks are usually not hardened and are easily machined. For maximum joint strength there should be between 2 and 3 thousandths of an inch space for solder around the drill rod. Too close a fit and the solder will not enter the clearance space, too great a space and the joint strength is reduced. See Figure 5–42 (A, B & C). Finally, silver solder the drill and drill rod together. Grind a small flat on the drill rod to let air escape and solder enter the joint.
- Trim the shank of the twist drill so it is square, center drill the twist drill shank, then drill the shank to accept the drill rod extension. Grind a flat on the drill rod for solder entry, Figure 5–42 (D).
- Another option when long reach is needed for a smaller diameter drill and hole depth is shallow, is to silver solder the twist drill into a hole in a length of drill rod. Grind a flat as above. See Figure 5–42 (E).

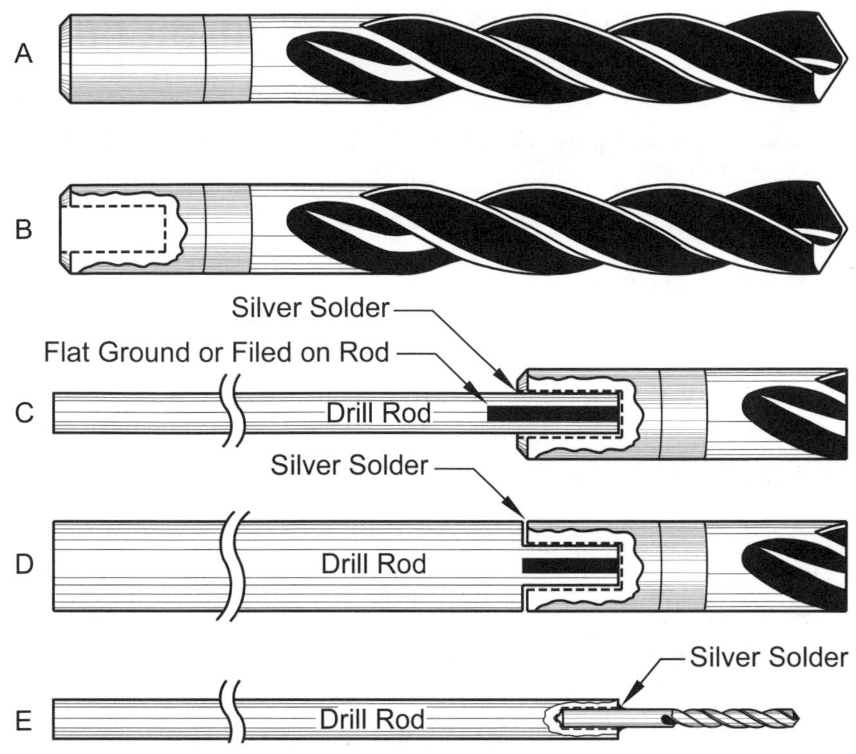

Figure 5–42. Extending drills with drill rod.

Another Drill Extension Method

How do you make a drill extension with a drill, steel rod and a file?

See Figure 5–43.

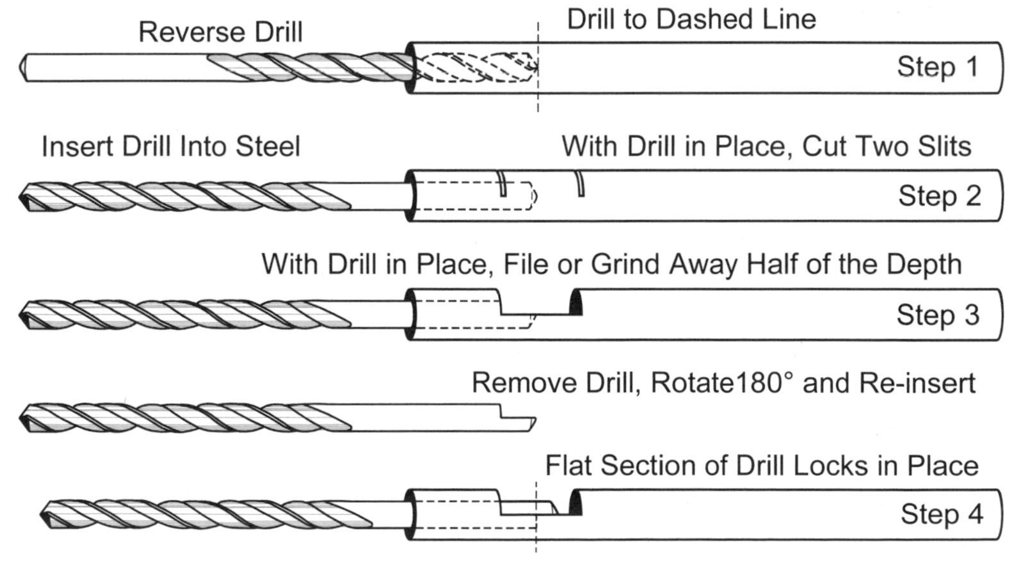

Figure 5–43. Making a drill extension without brazing or soldering.

Section VII – Drilling Non-metals

Rubber

How can rubber parts be drilled?

Cooling them in dry ice hardens the rubber and makes it possible to use a twist drill. Brass tubing with a sharpened cutting edge is also sometimes used.

Acrylic Plastics

What special twist drill designs work well for acrylic plastics, also called Plexiglas®?

Drills with a 60° point angle, a 0° rake and a blunted point, Figure 5–44, prevent the drill from biting into the plastic too fast, as is the case when using a regular twist drill. They make a clean hole and do not break out (create a ragged-edged hole) on the backside when drilling completely through the work. Modify a regular twist drill when drilling plastics.

Figure 5–44. Drill for plastics.

Glass

Cutting holes in glass is a common R&D lab requirement. What are the best methods of doing this?

The first choice is using diamond-coated cutting tools. Use a blunt cylinder-shaped diamond drill for the smaller holes, Figure 5–45 (top left). Trepan larger holes with a hollow bit, Figure 5–45 (top right). These bits must be flooded with a hose or immersed in water while cutting. A drill press or portable hole-drilling machine is necessary to rigidly hold the drill in its vertical position. Freehand drilling produces poor results.

Lacking diamond-coated bits, copper or brass bits and abrasive slurry will also work, but more slowly. Small holes use a solid "drill" and the larger holes are trepanned with a slotted-brass tool, Figure 5–45 (bottom). These tools are easily made in the shop. The glass in the hole area is solidly supported underneath by a rubber or vinyl pad, then a putty dam is built around the hole to contain the abrasive and its liquid carrier. Silicon carbide or aluminum oxide abrasives in water or oil are used. Like diamond-coated tools, a drill press or drilling jig is needed. These tools cut glass because they trap abrasive on the cutting surface and drag it across the glass wearing it down. Do not use excessive pressure; let the abrasive do the cutting.

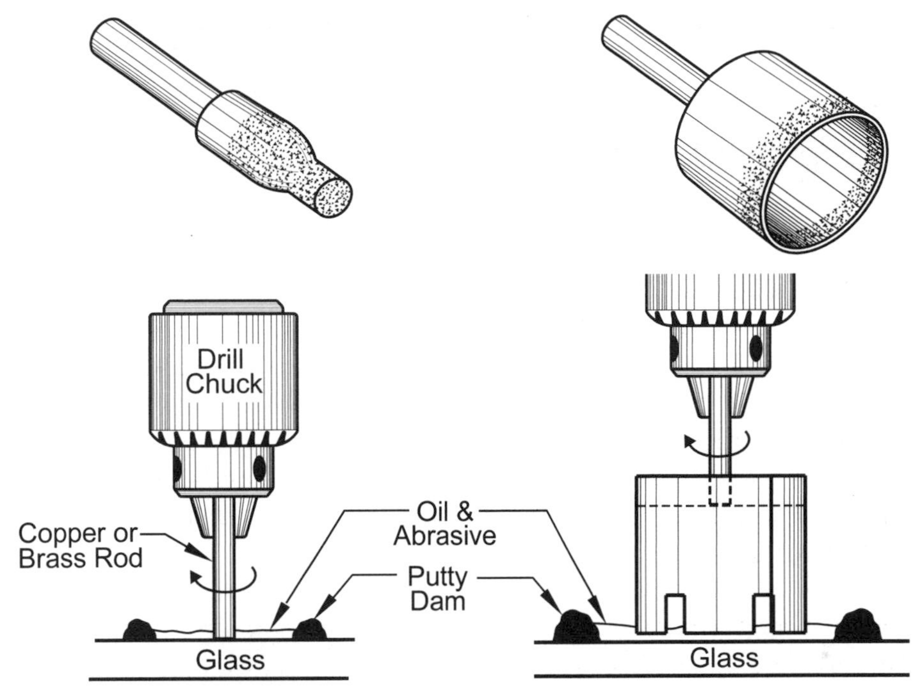

Figure 5–45. Drills for glass: diamond-coated bits (top) and copper or brass bits running in an abrasive-liquid mixture (bottom).

Section VIII – Other Hole-Making Tools

Whitney Punches

What are *Whitney punches* and why are they useful?

These are hand-held punches, Figure 5–46, for thin steel, aluminum, copper, brass, leather and soft plastics. Stainless steel is too hard and will damage the punch. Whitney punches have cast iron frames and make holes from 3/32 to 9/32-inches diameter. Metric size punch and die sets are also available. The advantages of the Whitney punch are:

- No electricity is needed.
- Throat depth permits punching as far as 1½ inches from the edge of the work.
- Adjustable depth gage aids putting holes parallel to the workpiece edge.
- Punch and dies are easily changed without tools.
- Punch has sharp center point to locate hole position precisely by feeling for scribe lines or center punch marks.
- Punched holes are burr-free.

There are larger punch models available with larger hole sizes and deeper throats. Imported clones are also available.

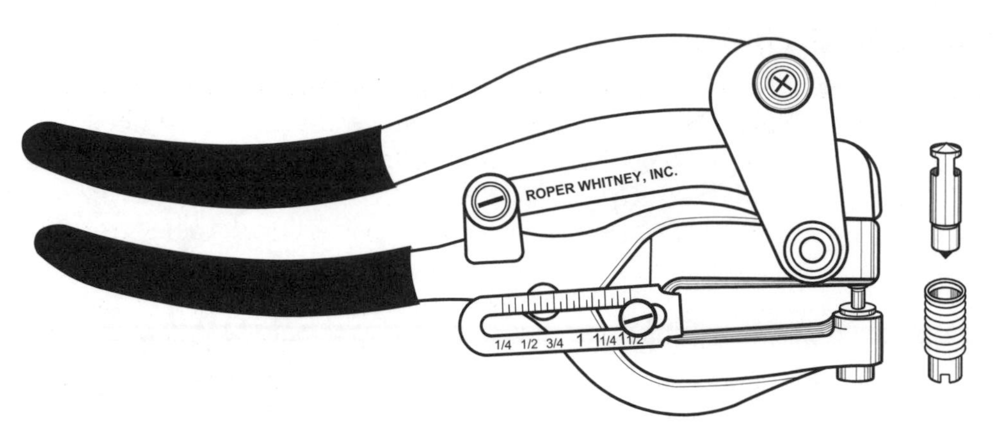

Figure 5–46. Whitney punch (left) and its punch and die (right).

Arched & Osborne Punches

How do *arched* and *Osborne punches* work?

These punches work like cookie cutters except they have hardened steel cutting edges and must be struck with a hammer, Figure 5–47. They are for cutting washers and sealing rings in many materials including foam, cork, linoleum, cardboard, leather, rubber, fiber, plastic, gasket material and lead. They are not for use on metals other than thin aluminum foil. Always punch through the work into a soft base material. End-grain wood works best. They may be resharpened with a file or a grinding wheel.

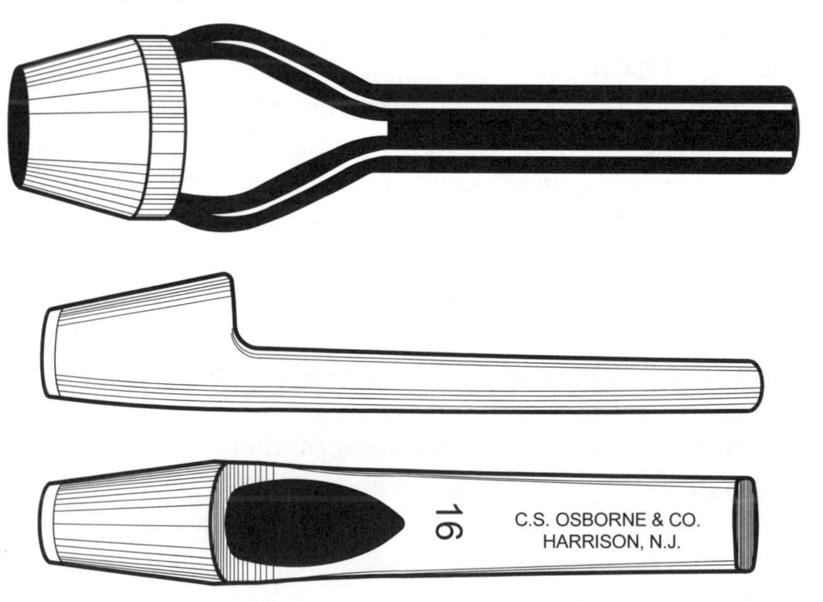

Figure 5–47. Arched punch (top) and two views of an Osborne punch (middle and bottom).

Chassis Punches

What are *chassis punches* and what are their advantages and drawbacks?
Chassis punches, also called *radio punches*, are used for punching holes from ½- to 5-inches (13 to 127-mm) diameter in mild steel up to 10 gage (0.135-in or 3.5-mm) thickness, Figure 5–48. Naturally, they also work on softer materials. Each "punch" consists of an upper die, a punch and a draw bolt. Most punches cut circular holes, but some cut square holes. Punched holes are relatively burr-free. After drilling a starter hole, typically ¼-inch diameter, the punch set works by pulling the punch into its die with a bolt turned by a wrench. There are also battery-, pneumatic- and hydraulic-powered accessories for production work.

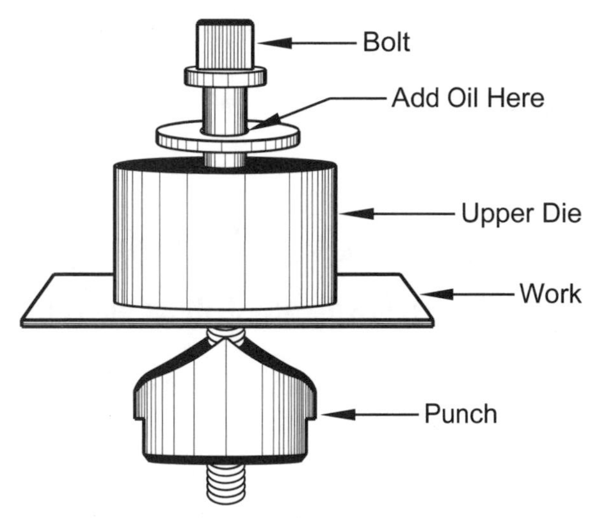

Figure 5–48. Chassis punch.

Gasket Cutter

How does the *gasket cutter* work?
It is similar to a flycutter, but a gasket cutter has two blades so it cuts the inner and outer dimensions of a gasket or washer at the same time, Figure 5–49. Also, the blades are more like knives than lathe tools. These blades may be resharpened on a stone. Use a very slow spindle speed and feed the tool slowly. Gasket cutters are for soft materials only.

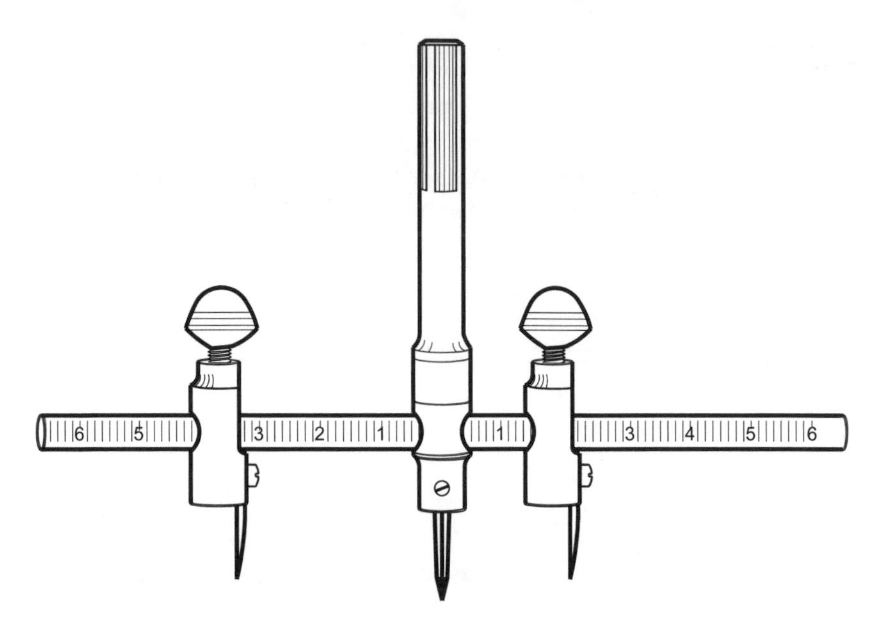

Figure 5–49. Gasket cutter.

Section IV – Drilling Safety

What hazards are associated with drilling operations, and how can they be avoided?

- The biggest single drilling danger comes from workpieces that are not adequately secured to the worktable and spin with the tool bit. While all shapes of work can be dangerous, pieces of sheet metal that seize to the drill become spinning knives. Always secure work in a secured vise, clamp it to the worktable or put it in a fixture.
- Allow the spindle to stop on its own. Do not attempt to stop it with your hand.
- A particular hazard when drilling an aluminum bar not secured to the drill press table arises from drilling chips. Initially the chips come out of the hole on the drill flutes, but as the hole gets deeper, they jam up at the bottom of the hole, heat from friction, and expand. When you attempt to withdraw the drill to clear it, the drill lifts the work *up* from the table and the end you are holding remains *down* on the table. Either the drill snaps or you let go the work and the work spins on the drill. At a minimum, the drill gets bent. Clamping the bar avoids the problem.
- Be sure to remove the chuck key *before* turning on the drill motor.
- Make sure the work is spaced off the drilling table or positioned so the drill will go through the table's center hole, not into the table itself. Place the work on a sacrificial piece of wood or metal or up on parallels.

Chapter 6

Threads & Threading

You can never plan the future by the past.
—Edmund Burke

Introduction

Screw threads were used in the time of Plato, about 500 BC, in grape and olive presses. About one hundred years later, Archimedes was credited with inventing a water pump based on the screw to irrigate crops and remove water from ship bilges. Later the Romans used this pump to dewater mines. Until the English instrument maker, Jesse Ramsden, developed the first satisfactory screw-cutting lathe in 1770, all screw threads were cut by hand. This limited most threads to large wooden ones for presses and clamps until 1800 when Henry Maudslay, a talented machinist, produced a large screw-cutting lathe.

The development of steam engines, trains and machine tools created a demand for threads in the form of nuts, bolts and leadscrews. But lack of standardization was a great obstacle to the widespread use of threaded fasteners since each workshop had its own fastener designs and they were not interchangeable.

To overcome these problems, Joseph Whitworth collected sample screws from a large number of British workshops, and in 1841 proposed that the thread angle be standardized at 55° and that the number of threads per-inch should be standardized for various diameters. His proposals became standard practice in Britain in the 1860s.

In 1864 William Sellers of Pennsylvania, an engineer and machine tool builder, independently proposed another standard based on a 60° thread, and set thread pitches for different diameters. This was adopted as the U.S. Standard, and subsequently developed into the American Standard Coarse Series (NC) and the Fine Series (NF). In Continental Europe, several different thread standards emerged, but German and French standards based on the metric system and a 60° thread prevailed and metric threads were established.

Section I – Thread Basics

Thread Uses

What are four important uses for threads?

Threads are used to:

- *Actuate* other mechanical components, like leadscrews on lathes and wing flaps on aircraft.
- *Measure distance* as with micrometer threads.
- *Adjust and hold length* as with tie rods and turnbuckles.
- *Fasten* as with nuts and bolts.

Parts of a Thread

What are the basic parts of a thread?

- *Major diameter*, the outer or largest diameter.
- *Minor diameter*, the smallest diameter.
- *Pitch*, the distance between adjoining threads.
- *Form,* the profile or shape of a thread.

See Figure 6–1.

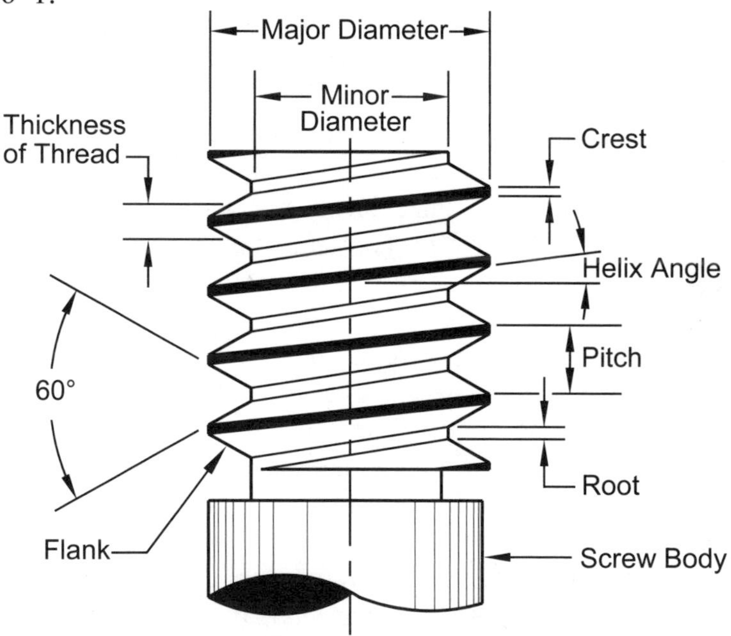

Figure 6–1. Parts of a thread.

Basis & Characteristics of Threads

What is the basis of threads?

Screw threads are really wedges wrapped around a cylinder, Figure 6–2.

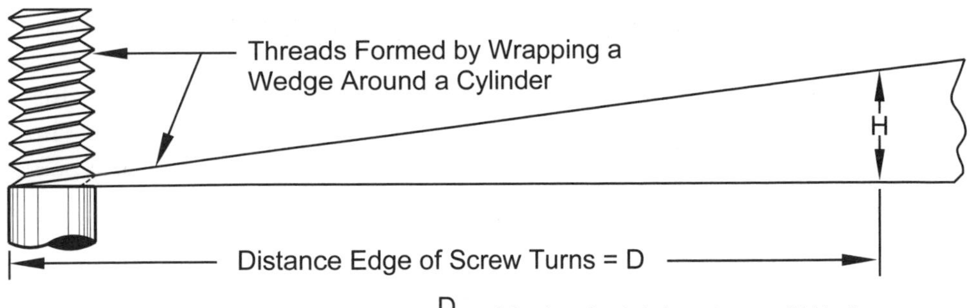

Figure 6–2. A screw thread and the wedge that formed it.

What are the special characteristics of screw threads?
Screw threads have the same mechanical advantage as the wedge that formed them. Figure 6–2 shows a wedge and screw thread with a mechanical advantage of about 7, the ratio of the distance the screw turns around the cylinder to the distance the screw moves vertically. This mechanical advantage enables screw jacks to lift heavy loads.

Measuring & Identifying Threads

How are threads measured?
Here are several ways to measure threads:

- *Use a screw-thread micrometer,* as in Figure 6–3, which differs from a regular micrometer in having a cone-shaped spindle and a mating anvil. These just fit over threads and measure the outside diameter (or major diameter) of the thread less the depth of one thread. This method works well for threading operations because the work does not have to be removed from the lathe to take the measurement.

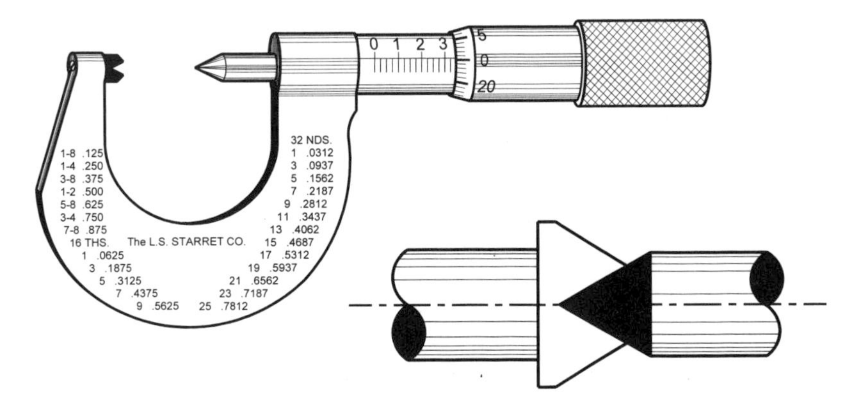

Figure 6–3. Screw-thread micrometer (top) and positions of its anvil and spindle corresponding to the zero position (bottom).

- The *three-wire method*, Figure 6–4, uses a regular micrometer placed over three same-size wires of a specific diameter for the thread being measured. Entering the thread dimensions into the formula that follows yields the diameter over the wires for a full-depth 60° American National Thread. This method works well for lathe threading operations since it can be done while the work remains in the lathe. Here is the formula:

$$M = D + 3G - \frac{1.5155}{N}$$

Where: M = Measurement over the wires
D = Major diameter of the thread
G = Diameter of the wires
P = Pitch = 1/N
N = Number of threads/inch (tpi)

$$\text{The largest wire size} = G_{Largest} = \frac{1.010}{N} = 1.010\ P$$

$$\text{The best wire size} = G_{Best} = \frac{0.57735}{N} = 0.57735\ P$$

$$\text{The smallest wire size} = G_{Smallest} = \frac{0.505}{N} = 0.505\ P$$

Note: G must be no larger or no smaller than the sizes shown above.
Any wire size between the largest and smallest may be used.
All wires must be the same size.

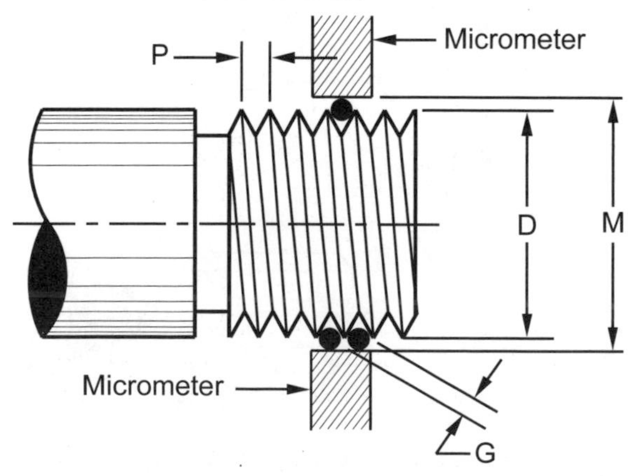

Figure 6–4. Three-wire method.

- The *thread triangle method*, Figure 6–5, measures the distance over a pair of steel triangles between the thread roots. A reference chart is then used to determine thread depth. This measuring method also works well for lathe threading while the work remains in the lathe. Triangles are awkward to use, hard to position and are to be avoided, if at all possible.

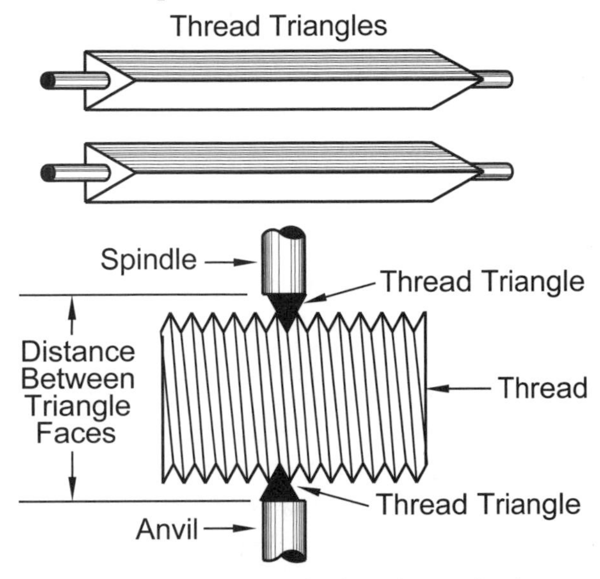

Figure 6–5. Thread triangle method.

- *Thread ring gages* and *plug gages*, as in Figure 6–6, are commercial master thread gages for external and internal threads. These are highly accurate and often used in pairs of *Go* and *No-Go* dimensions. Work may have to be removed from the lathe to measure external threads between centers.

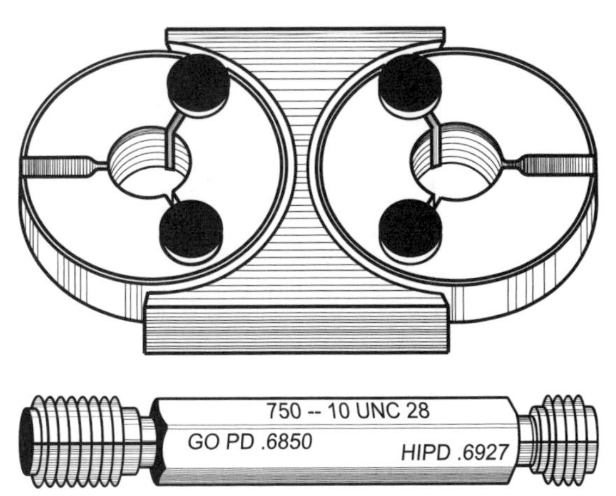

Figure 6–6. *Go/No-Go* thread ring gages in a holder (top) and *Go/No-Go* thread plug gages (bottom).

- An *optical comparator*, which projects a magnified shadowgraph of the part on a ground glass screen, is useful to determine the shape and dimensions of a thread, but not useful for checking work while it is in the lathe. These are also excellent for checking the shape of a cutting tool.
- Using *commercial screws or nuts of the matching size* are not as accurate as a plug or ring gage, but will often be adequate. Also, in the case of unusual threads, this method is used because gages may not be available.
- Using *the mating part itself* is often the best choice for checking threads, particularly in a non-production or spare parts situation.

Not all measurement methods work for all types of threads. For example, internal and tapered threads cannot be checked with a thread micrometer or with the three-wire method, but plug gages will work fine. Using the mating set of threads, if available, or a commercial threaded fastener of the matching thread, are low cost, effective alternatives to the fancier tool room methods.

What tool quickly determines a fastener's thread pitch?

A *screw pitch gage* quickly, easily and accurately identifies thread pitch. Starrett offers three different types for:

- *American National Unified 60°-inch threads.* These are available to measure pitches from 2¼ to 80 threads per inch, see Figure 6–7.

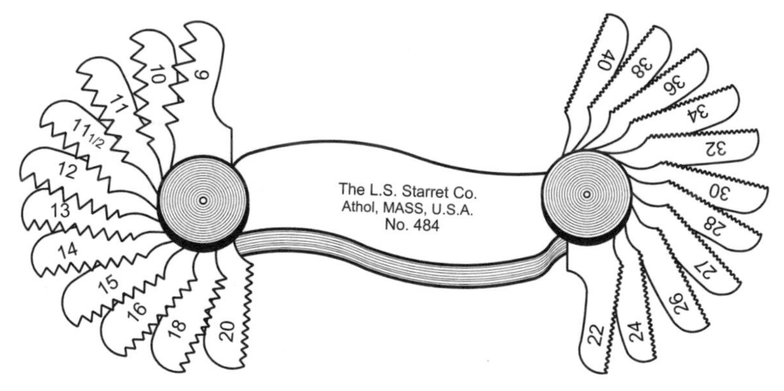

Figure 6–7. A screw pitch gage.

- *Metric threads* from 0.25 to 11.5 mm pitch.
- *Whitworth Standard 55° threads* (obsolescent and relatively uncommon).

Identifying Right- & Left-hand Threads

How can it be visually determined whether a screw, bolt or stud has *right-* or *left-hand threads*?

See the sloping lines in Figure 6–8. It does not matter which way the fastener is pointing, head to the left or head to the right, a right-hand thread always slopes upward to the left. Right-hand threads tighten when turned clock-wise as viewed from the head-end of the fastener.

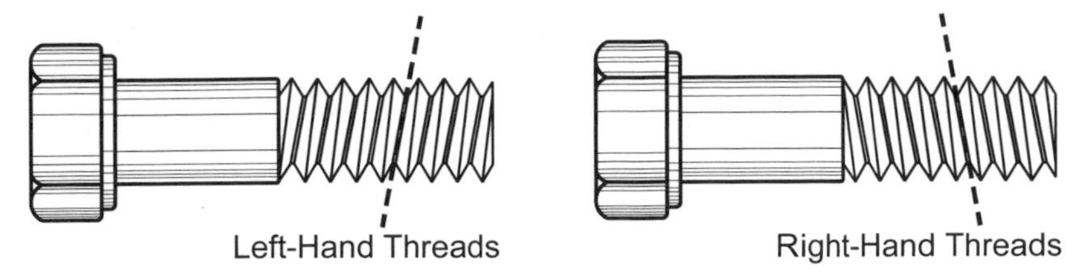

Figure 6–8. Visual identification of left- and right-hand threads.

Where are *left-hand threads* used?
Here are some applications:

- Pairs of right- and left-hand threads are used to adjust the length of mechanical linkages, Figure 6–9. If the end threads were the same hand, the ends would not be drawn together or apart; the center nut would merely move along both threads *without* changing their relative position. Turnbuckles work similarly.

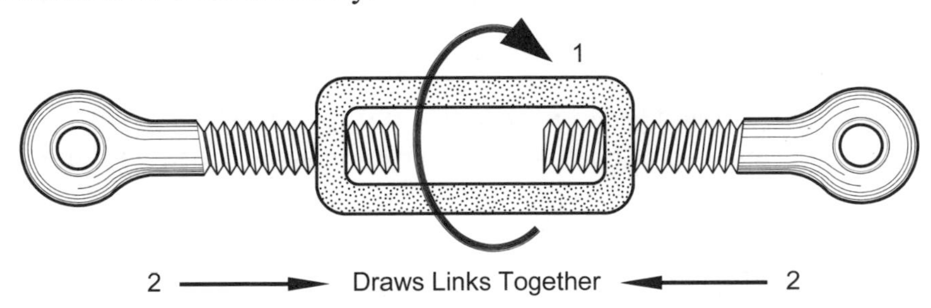

Figure 6–9. Turnbuckle uses right- and left-hand threads.

- Some lathe and milling machine leadscrews use left-hand threads to eliminate a direction-reversing (idler) gear and simplify machine design.
- Left-hand fittings on propane and acetylene gas handling equipment prevent mix-ups. *Fuel fittings are left-hand and oxygen fittings are right-hand. Nuts on fuel fittings have notches in the edges of their flats.*
- Nuts on the left side of a bench grinder spindle are left-hand so they will tighten with use.

Section II – Thread Forms

Overview

What are the main thread families?

- *Unified threads*, inch-based, popular in the U.S.; there are both a coarse and fine series.

- *Metric threads*, millimeter-based, more popular outside the U.S.
- *Machinery threads*, inch and metric, transmit large axial forces.
- *Pipe threads,* both inch and metric; there are several thread designs, some for liquid tight applications and some for mechanical applications where no liquid seal is needed.

There are many other threads for uses as diverse as microscopes, scientific instruments, watches, fire hose fittings, garden hoses and light bulb bases. See *Machinery's Handbook* for complete specifications.

Unified Threads

How did the *Unified Thread System* originate?

To overcome thread incompatibility problems they experienced with fasteners during WWII, in 1948 the U.S., U.K. and Canada established a new thread standard, the *Unified Thread System.* In the compromise the British accepted the 60° thread angle (in place of 55°) and the Americans accepted rounded roots and optionally rounded crests. The new Unified Thread System fasteners continued to fit fasteners manufactured under the older systems.

What does the *Unified Thread profile* (shape) look like?

Unified Threads and ISO Metric Threads have the same profile, Figure 6–10.

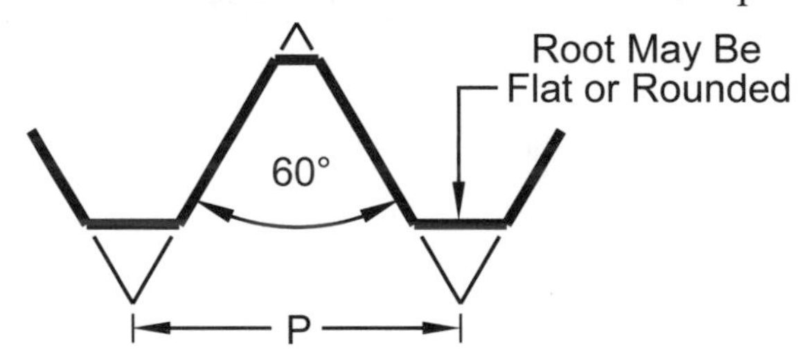

Figure 6–10. Thread profile for both Unified Threads and ISO Metric Threads.

What are the most common *Unified Thread Series*?

The two most common series for fasteners are:

- *Unified Coarse (UNC)*
- *Unified Fine (UNF)*

Table 6–1 shows the range of sizes, from about 1/16 to 1⅜ inches, that makes these two fastener series the most popular inch-based threads. The left-hand column shows the proper way to identify these fasteners on workprint callouts. Three common ¼-inch diameter fasteners are ¼-20 UNC, ¼-28 UNF and ¼-32 UNEF.

Thread Size Threads/Inch Series Designation	Major Diameter (inches)	Comment
0-80 UNF	0.0600	Smallest in series, #0, no matching UNC
1-64 UNC	0.0724	Next larger screw in series, #1 64 tpi for UNC and 72 tpi for UNF.
1-72 UNF	0.0724	
2-56 UNC	0.0854	——
2-64 UNF	0.0854	
3-48 UNC	0.0983	——
3-56 UNF	0.0983	
4, 5, 6, 8, 10 Number Sizes Continue – Diameter Increases ↓		
12-24 UNC	0.2150	Largest numbered screw in series, #12.
12-28 UNF	0.2150	
¼-20	0.2489	Fractional-inch sizes and bolts begin at ¼".
¼-24	0.2489	
Fractional-inch Sizes Continue – Diameter Increases ↓		
1⅜-6 UNC	1.3726	Largest Fractional-inch size and in series.
1⅜-12 UNF	1.3726	

Note: Numbered screw sizes are based on #0 size = 0.060 inches diameter; each larger size number increases by 0.013-inches diameter.

Table 6–1. UNC and UNF fastener series and their major diameters.

Are there other common members of the *Unified Thread Series*?
Yes, there are several. The most common are:

- *Unified Constant Pitch Series* consists of the 8-thread (8-UN) for diameters larger than 1 inch, 12-Thread (12-UN) for diameters larger than 1½ inches and 16-Thread series (16-UN) for diameters larger than $1^{11}/_{16}$ inches. These fasteners are for boilers, structural steel and machinery.
- *Unified Extra Fine Thread Series (UNEF)* have finer threads than the UNF. For comparison, here are three #12 screws from three different Unified Thread Series with their thread counts: UNC has 12 tpi, UNF has 28 tpi and UNEF has 32 tpi.

There are many other Unified Thread Series for specialized applications. See *Machinery's Handbook,* Thread Section for details. There are 319 thread sizes listed in the U.S. Standards for Unified Threads covering a range in diameters from 0.06 inches (1.5 mm) to 6 inches (152 mm), and pitches from 80 to 4 threads per inch.

Metric Threads

Where are *metric threads* used?
Metric threads are the most common threads everywhere but the in U.S. All countries are now in various degrees of transition to metric threads. Most EU countries are completely metric. Other countries, such as Canada, India and Australia are about half metric. The U.S. is about one-fifth metric. Many U.S. products and industries have been metric for a long time, while automotive companies have been in transition for more than twenty years. Most cars made in the U.S. contain a mixture of Unified and metric threads.

How are metric threads defined?
The *ISO Metric Thread Standard* defines them. Although metric threads have the same profile as Unified threads, Figure 6–10, very few fasteners with metric threads fit Unified thread fasteners of the opposite sex. Fasteners from these two systems are usually incompatible because few fasteners from either system have both the *same* diameter and the *same* pitch.

Metric series threads include 332 thread sizes covering a range from 1 to 300 mm in diameter and 0.2 to 6 mm in pitch. Like Unified threads, there are both coarse and fine threads in the same diameters. In general, common metric fasteners cover a similar range of sizes as Unified threads.

While many metric fastener diameter and pitch combinations are defined, experience has shown that relatively few fastener combinations meet the needs of most applications. To reduce inventory and increase parts compatibility, a series of *preferred fasteners* have been designated and are shown in Table 6–2.

Ranking Choice	Metric Fastener Size
First	M2, 2.5, 3, 4, 5, 6, 8, 10, 12, 16, 20, 24, 30, 36, 42
Second	M3.5, 14, 18, 22, 27, 33, 39, 45
Third	M15, 17, 25, 40
Avoid	M7, 9, 11, 26, 28, 32, 35, 38

Table 6–2. Preferred metric fastener sizes.

Metric threads are designated by the letter M followed by the nominal major diameter of the thread and the pitch in millimeters. For example, M10 × 1.0 indicates that the major diameter of the thread is 10 mm and the pitch is 1.0 mm. The absence of a pitch value indicates that a coarse thread is specified. For example, stating that a thread is M10 indicates a coarse thread series is specified with a diameter of 10 mm (giving the thread a pitch of 1.5 mm).

Miniature Threads

What threads are used in watches and instruments?

There are two series, both based on the 60° thread form:

- *Common Standard Miniature Instrument Thread Series*, an older inch-based series, which supplements the UNF in fastener diameters below 0.060 inches, Table 6–3. They are obsolescent and should not be used.

Designation	Major Diameter (inches)	Threads per inch
0000–160	0.021	160
000–120	0.034	120
00–112	0.047	112
00–96	0.047	96
00–90	0.047	90
UNF Thread Series Begins Here		

Table 6–3. Common Standard Miniature Instrument Thread Series.

- *Unified Miniature Screw Thread Series*, a metric-based series, covering fourteen diameter fasteners from 0.30 to 1.5 mm. See Table 6–4.

Designation	Major Diameter	Pitch mm	Approximate Threads per inch
0.30 UNM	0.30mm (.0118)	0.080	318
0.35 UNM	0.35mm (.0138)	0.090	282
0.40 UNM	0.40mm (.0157)	0.100	254
0.45 UNM	0.45mm (.0177)	0.100	254
0.50 UNM	0.50mm (.0197)	0.125	203
0.55 UNM	0.55mm (.0217)	0.125	203
0.60 UNM	0.60mm (.0236)	0.150	169
0.70 UNM	0.70mm (.0276)	0.175	145
0.80 UNM	0.80mm (.0315)	0.200	127
0.90 UNM	0.90mm (.0354)	0.225	113
1.00 UNM	1.00mm (.0394)	0.250	102
1.10 UNM	1.10mm (.0433)	0.250	102
1.20 UNM	1.20mm (.0472)	0.250	102
1.40 UNM	1.40mm (.0551)	0.300	85

Table 6–4. Unified Miniature Screw Thread Series. (UMS)

Machinery Threads

What are *machinery threads* and what are their common forms?

Machinery threads are used in machine tools, jacks, clamps and presses. They are much stronger than other thread forms in resisting axial force, Figure 6–11. Their common forms are:

- *Square threads* are mechanically efficient, but difficult to cut.
- *Acme threads* are almost as mechanically efficient as square threads, but easier to cut and often seen on lathe and milling machine leadscrews.
- *Buttress threads* are relatively easy to cut and are much stronger in one axial direction than the other.

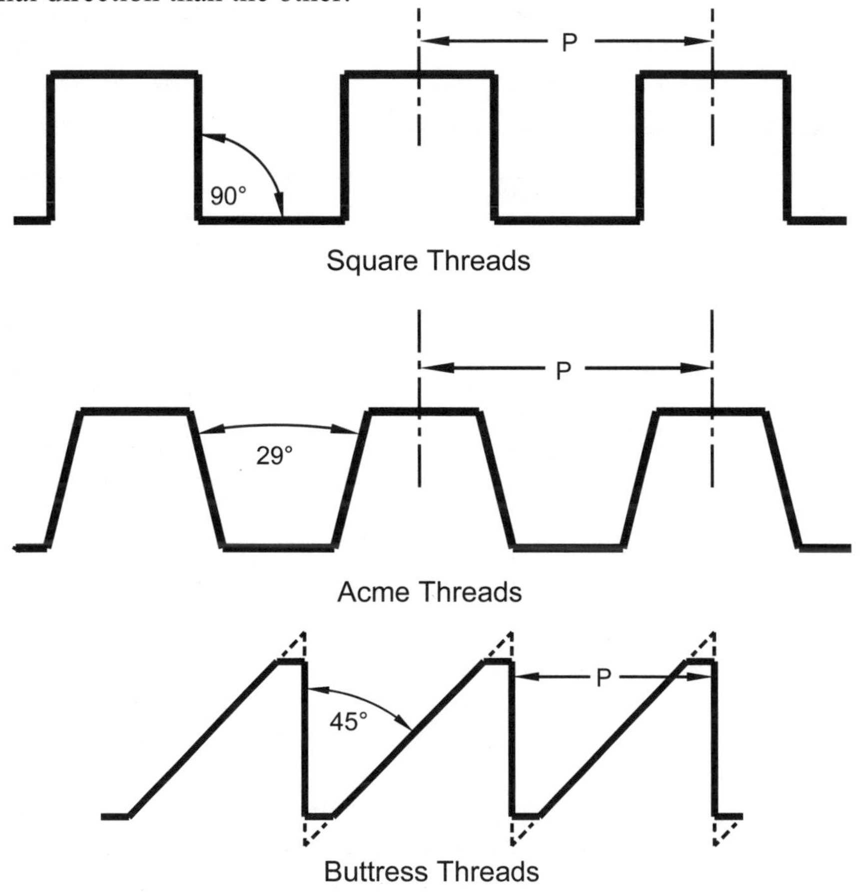

Figure 6–11. Machinery threads.

Pipe Threads

What are the common *pipe thread forms* and how do they differ from other threads?

The three most common forms are:

- *American National Standard Taper Pipe Threads (NPT)* are used for electrical conduit, railings and low-pressure piping. Teflon® tape or pipe joint compound is required to make a liquid-tight seal.
- *American National Standard Dryseal Pipe Threads for Pressure-Tight Joints (NPTF)* have the same thread form as NPT, but its major and minor diameters are adjusted so that a crushing of the threads takes place when they are tightened to make a mechanical seal. NPTF threads are used in more critical applications where the possibility of leakage from joint compound failure must be excluded.

Both thread forms are visually identical and have a taper of ¾-inch per foot. See Figure 6–12.

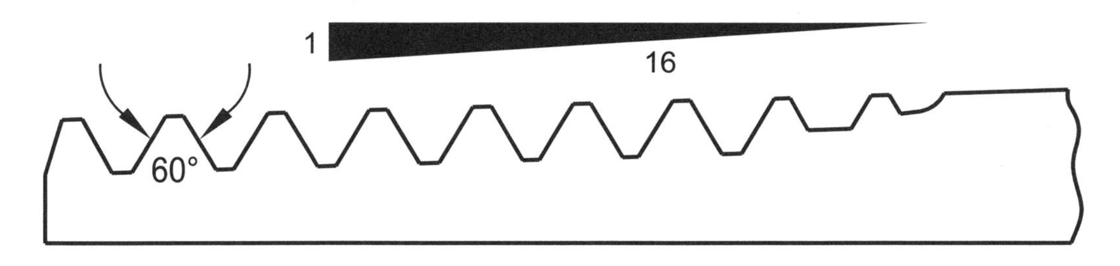

Figure 6–12. NPT and NPTF pipe thread profile.

- *American National Standard Straight Pipe Threads for Mechanical Joints (NPS)* have the same thread form as NPT except the male and female threads are parallel, not tapered. These threads are for mechanical applications such as electrical conduit, railings and lamp parts where no liquid seal is required.

Coarse & Fine Threads

The same diameter fasteners are available in both *coarse* and *fine threads*, for example, #8-32 and #8-36 screws. Why is this done?

Fasteners are available in coarse and fine threads because for a given diameter fastener, as thread pitch increases, thread depth decreases, and the cross sectional area of the fastener able to bear tensile loads increases. For inch fasteners, UNF threads offer 6 to 13% *more* cross sectional area and so *more* tensile strength than UNC fasteners of the same size. In soft materials like brass and aluminum, deeper coarse threads give the internal threads in nuts and tapped holes more female thread depth to hang onto. See Figure 6–13. These conclusions are summarized in Table 6–5. Metric fasteners are also offered in fine and coarse threads.

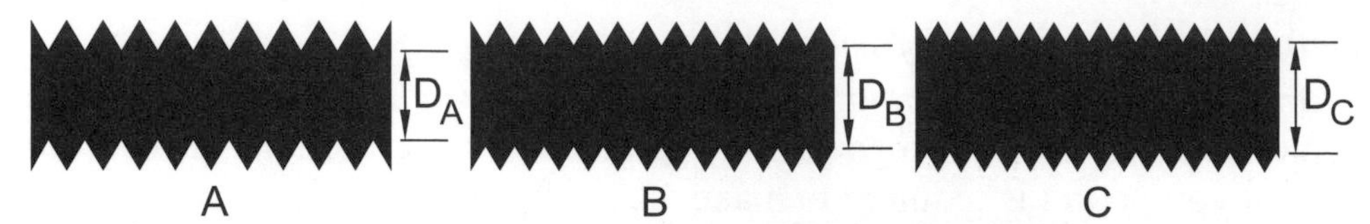

Figure 6–13. Increasing thread pitch increases core diameter of the fastener.

Characteristic	**Coarse Threads**	**Fine Threads**
Lower cost.	✓	
Most commonly used.	✓	
Speed of assembly.	✓	
Resistance to cross threading and handling damage.	✓	
Used in lower tensile strength materials like cast iron, mild steel, and softer materials like brass, aluminum, magnesium and plastics.	✓	
Used where length of engagement is short. That is, just a few threads of the screw are used as in thin-wall tubing.		✓
Used where wall thickness requires a fine pitch to avoid damage.		✓
Used where smaller lead angle is desirable, as when the fastener controls an adjustment, where more mechanical advantage is needed, or for more precise tightening.		✓
Used where resistance to stripping of both internal and external threads is equal to or greater than the load carrying capacity of the screw or bolt.		✓
Strength of male threads in same size fastener.		✓
Strength of female threads in same size fastener.	✓	
Less likely to seize in temperature applications and in joints where corrosion forms.	✓	
More easily tapped in brittle materials or materials that crumble easily.	✓	
Fatigue Resistance.	No difference.	

Table 6–5. Comparison between coarse thread and fine thread fasteners of the same diameter.

Section III – Applying Female Threads

Female vs. Male Threads

What makes applying *female threads* different from applying *male threads*?

Female threads can only be applied by tapping or lathe threading, and small threads only by tapping. Male threads can be applied by single-point lathe threading, die threading, rolling or grinding processes. NC-controlled milling machines can apply both male and female threads.

Taps

What do we mean by the word *tapping*?

Tapping cuts *internal* threads in a predrilled hole using a tool called a *tap*.

What are the most important terms used to describe a tap?

In Figure 6–14, they are:

- *Chamfer* – Tapering the threads at the front end of each land by cutting away and relieving the crest of the first few teeth. This distributes the cutting action over several teeth and makes starting the tap easier.
- *Core diameter* – The diameter of the circle tangent with the bottom of the flutes. Deep flutes reduce the core diameter and weaken the tap.
- *Flutes* – The longitudinal channels formed in the tap to create cutting edges on the thread profile. They provide chip space and cutting fluid passage.
- *External* and *Internal Center* – Small taps usually have pointed ends and external centers, on which they are held during manufacturing, because they are too small for center holes, Style 1 in Figure 6–14. Larger taps have internal centers because there is room to drill cone-shaped center holes, Styles 2 and 3.
- *Flank* – The sides of the thread. The sections between the root and the crest.
- *Land* – The threaded section between the flutes.
- *Rake* – The angle between the cutting face of the land and the tangent to the radius of the tap. It is the same as the rake on a lathe tool.
- *Square* – The four flats machined on the end of the shank by which the tap is turned by a tap wrench.
- *Style* – International specifications define three styles or shapes for *Standard Taps,* the most common tap designs. The shape of the shank end, and whether the shank is smaller than the threads, determine its style.

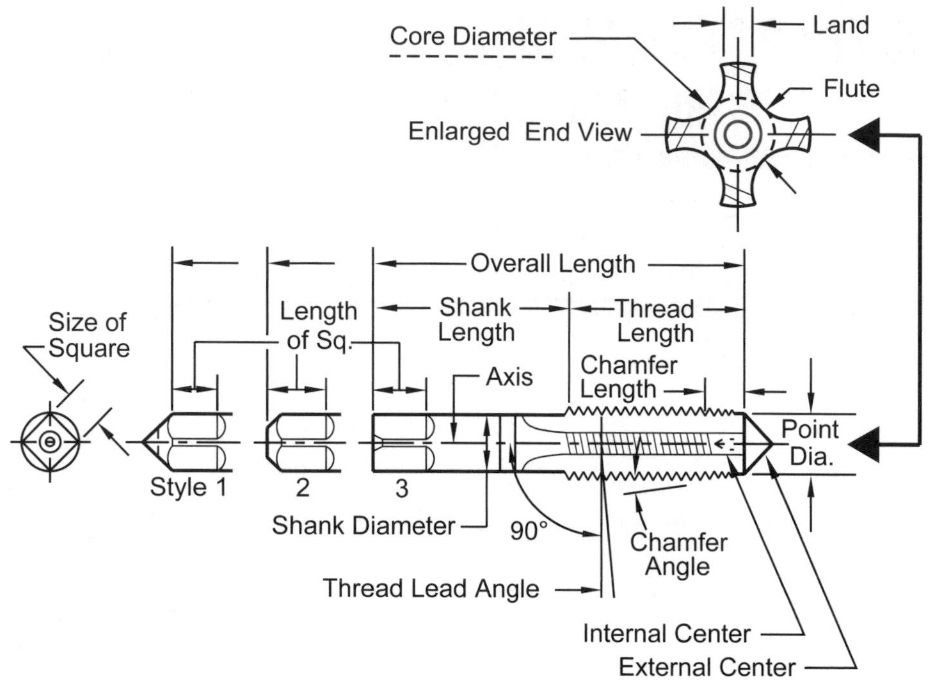

Figure 6–14. Tap terms.

Taps are furnished in different *chamfers.* Why?

Because the *taper tap* has the greatest chamfer, it starts threading straighter and most easily and so is often used to begin tapping. When initial threads are established, the *plug tap* can complete the tapping on through-holes. If it is necessary for the threads to extend to the bottom of a blind-hole, the *bottoming tap* puts in the last few threads. Other than chamfer length, the three types of taps are the same. See Figure 6–15.

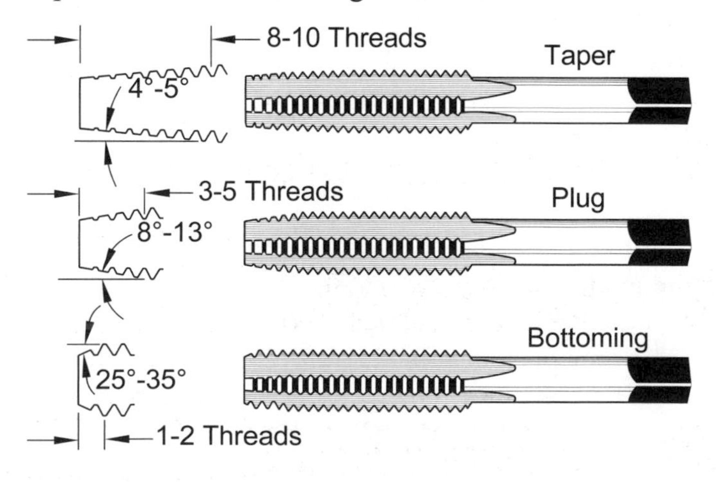

Figure 6–15. A set of standard taps.

What materials are used for taps?
Usually taps are HSS or Cobalt HSS. Some taps are solid tungsten carbide. Tap stock is hardened and then threads are applied by grinding.

Special Tap Designs

There are other tap designs. What are they and when are they used?

- *Spiral point taps,* also called *gun taps,* are similar to standard taps except for a spiral slash ground in their point. This ejects thread cuttings *ahead* of the tap and avoids clogging in the flutes. They work well in through-holes, their two-flute design makes them stronger because of their increased cross section, and they require less turning torque than standard taps.
- *Spiral fluted taps* have right-hand helical flutes of 25 to 35° for medium carbon steel, alloy steel and stainless steel, and 45 to 60° for aluminum, copper and cast steel. They draw cuttings out of the tap hole and bridge gaps caused by other intersecting holes or keyways. They also draw stringy-chip materials out of blind holes.
- *Forming taps,* also called *roll taps,* have no flutes except for the lubricating grooves. Because they do not perform cutting but extrude material from the sides of the tap hole into the threads, they require a larger tap drill. The selection of tap drill size is more critical than for standard taps. Consult the manufacturer for these sizes. The work metal grain structure is displaced *along* the thread profile making the threads stronger and more fatigue-resistant than cut threads. They are also smoother and have a burnished appearance. Forming taps are useful in blind holes where removing cuttings is a problem. Roll threading, a process to apply male threads, has similar advantages and is discussed later in this chapter under the heading Other Threading Methods.

See Figure 6–16.

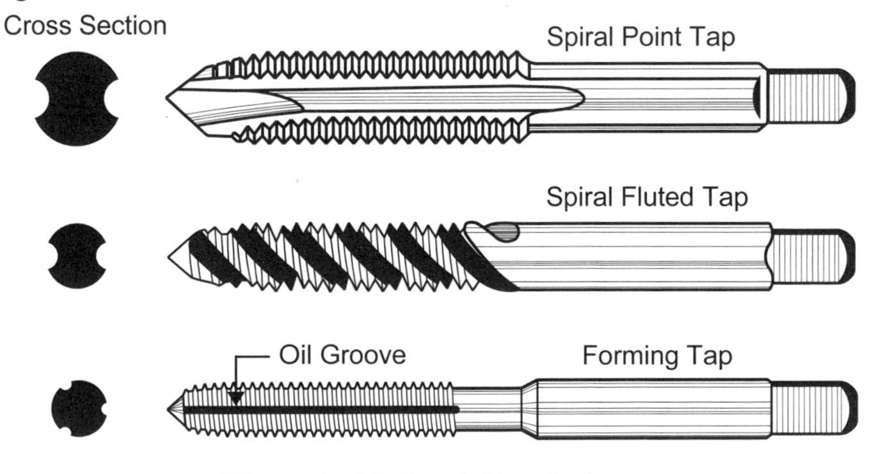

Figure 6–16. Special tap designs.

- *Oversized taps* compensate for heat-treat shrinkage when plating will reduce thread diameter, or in plastics, which tend to spring back when the tap is removed. They are most often 0.005 inches oversized. Standard and spiral point designs are also available.
- *Pipe taps* cut tapered threads for liquid-tight seals, Figure 6–17. Tapered reamers must be used to remove excess metal and to establish the taper of the pipe threads prior to tapping. There is too much material for the pipe tap to remove from the taper by itself.

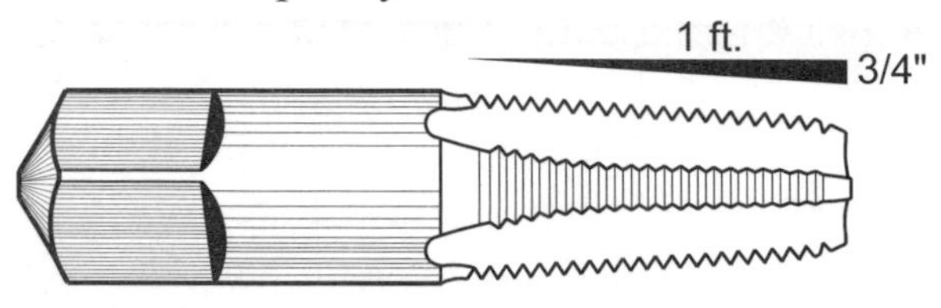

Figure 6–17. Pipe tap.

Tap Wrenches

What tools are used to hold and turn hand taps?

There are both *T-handle* and *straight-handle hand-tap wrenches*, Figure 6–18 through Figure 6–20. T-handle wrenches have a center hole that can be used for tap alignment. Larger taps have a center hole in their shanks left over from manufacturing. These holes can be used for tap alignment. Extension tap handle wrenches are available for pulley threads and other hard to reach threads. Cheaper tap wrenches, Figure 6–18, have an integral, 4-jaw collet cut into the body of the tap wrench itself. This design fits only one tap shank size well; all other sizes wobble.

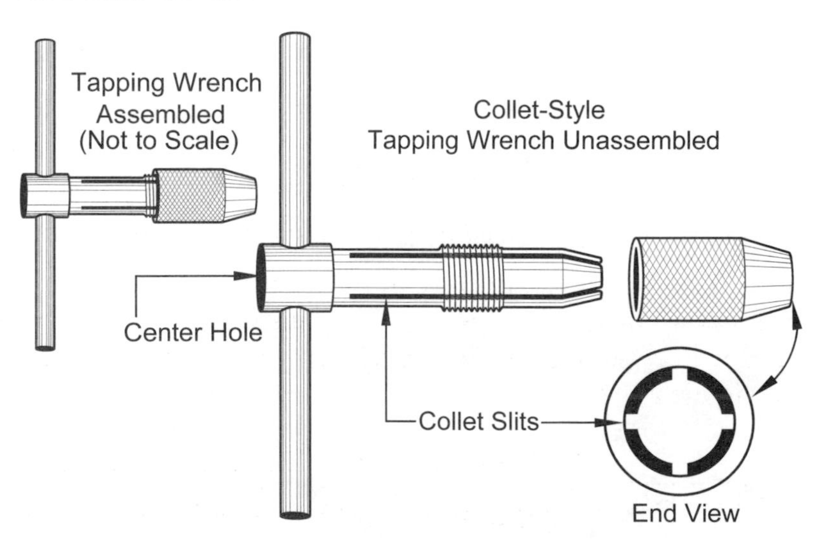

Figure 6–18. Collet T-handle hand-tap wrenches.

A better tap wrench design, Figure 6–19, has two jaws kept parallel by a spring and holds a wider range of tap shanks straight and secure.

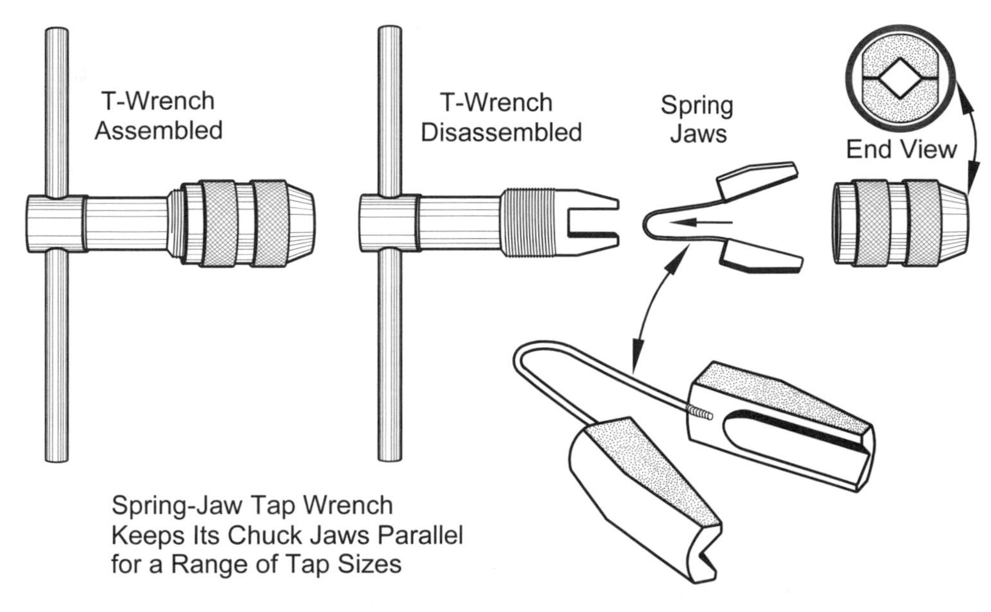

Figure 6–19. Spring-jaw T-handle hand-tap wrenches.

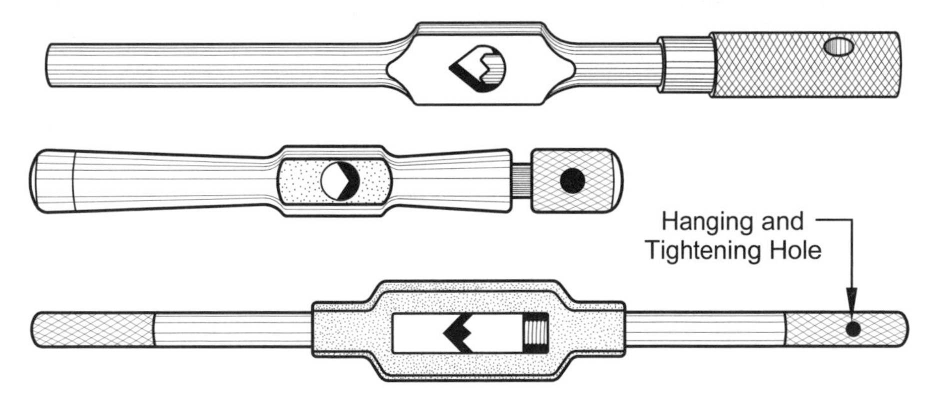

Figure 6–20. Three straight-handle hand-tap wrenches.

Small taps may also be held in *pin vises*, Figure 6–21, which put less stress on the tap than tap wrenches. Because they are turned with fingers, small taps in a pin vise provide a more sensitive feel of the tapping process and reduce the chances of tap breakage.

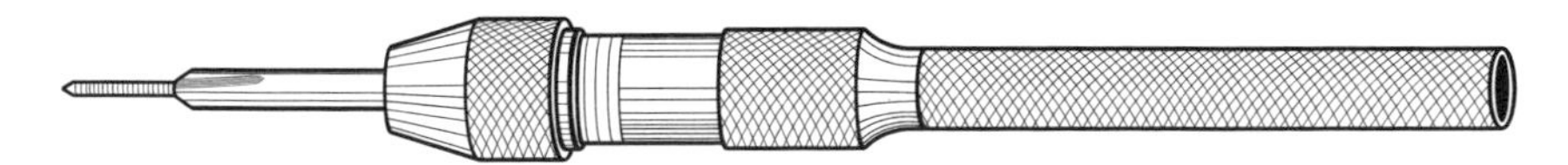

Figure 6–21. Small tap in pin vise.

Thread Engagement

How is *thread engagement* defined and why is it important?

Thread engagement measures how much flank-to-flank contact exists between threads. Thread engagement in Figure 6–22 (left) is nearly 100%. Because most threads have rounded or truncated crests and manufacturing allowances, thread engagement in commercial fasteners is usually between 70 and 75% as in Figure 6–22 (right). Even at 60% thread engagement, thread resistance to shearing off is 2.5 × stronger than the tensile strength of the fastener itself when used with a nut of the same quality. *There is little gained by high thread engagements which greatly increases the risk of tap breakage.*

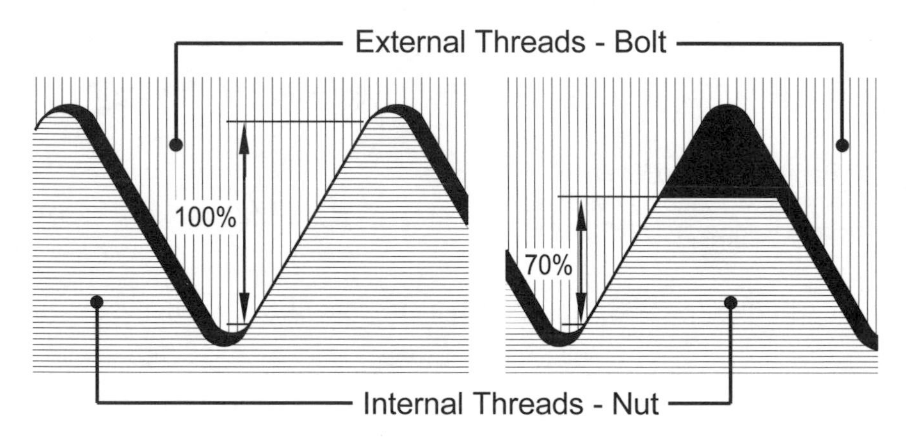

Figure 6–22. Thread engagement.

Tap & Clearance Drill Sizing

What are *tap drills* and how can their size be determined?

Tap drills make holes for thread tapping. Tap drill sizes appear in drill indexes, drill check plates. See Figure 6–23 and *Machinery's Handbook*. However, these tap drill sizes were established in the 1950s and 1960s when holes made by twist drills had a tendency to run oversized. Since today's drills produce holes much closer to their actual size, tap drill sizes listed in most tables tend to run small and produce tapped holes with high thread engagement. This leads to high tap breakage. Tap drill sizes in tables should be treated as a starting point rather than the optimum drill size. Small changes in hole diameter produce large changes in engagement and tapping torque, particularly in small fasteners. In fact, if tapping torque to produce 60% thread engagement is 100 torque units, then 70% engagement requires 200 torque units, and 80% engagement requires 300 units. There is little point to drilling undersized tapping holes and risking tap breakage when increasing thread engagement above 60% achieves little in additional thread strength.

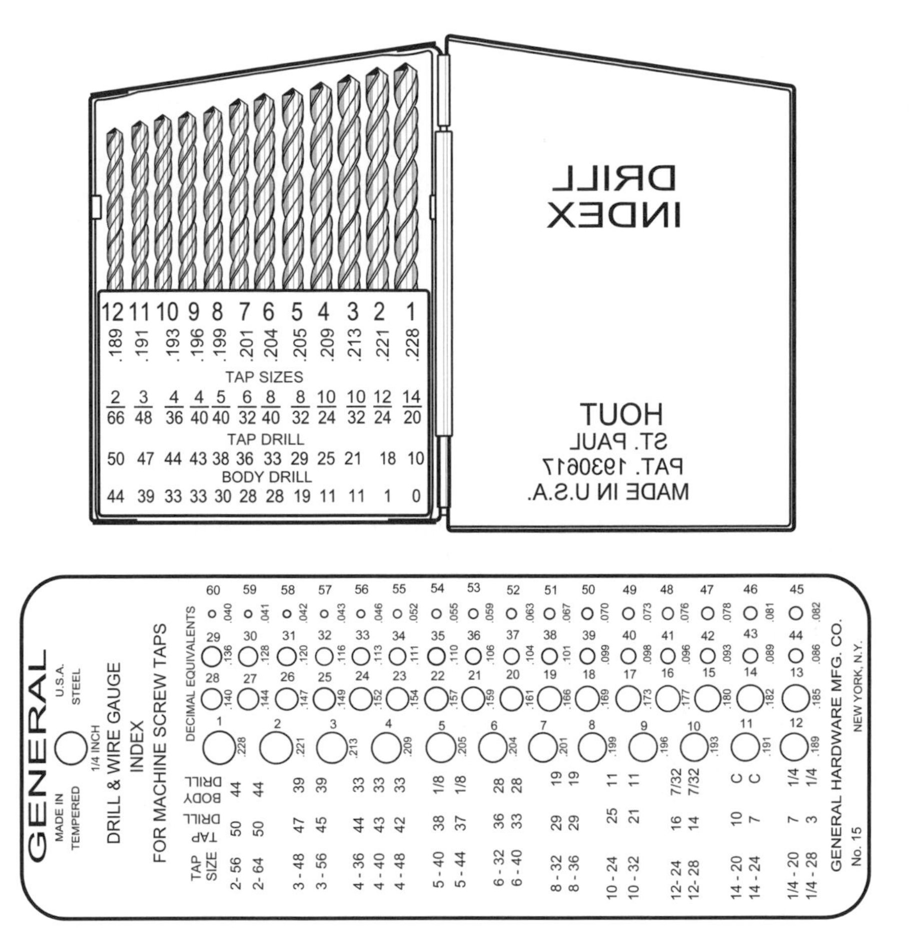

Figure 6–23. Tap and body drill sizes are usually embossed on the first leaf of a drill index (top) and on drill-check plates (bottom).

What are *clearance drills* used for?

Clearance drills provide space around a threaded fastener, Figure 6–24. The tapping holes they drill are usually 10% larger than the diameter of the fastener that passes through them. Clearance drill sizes are listed with tap drills sizes.

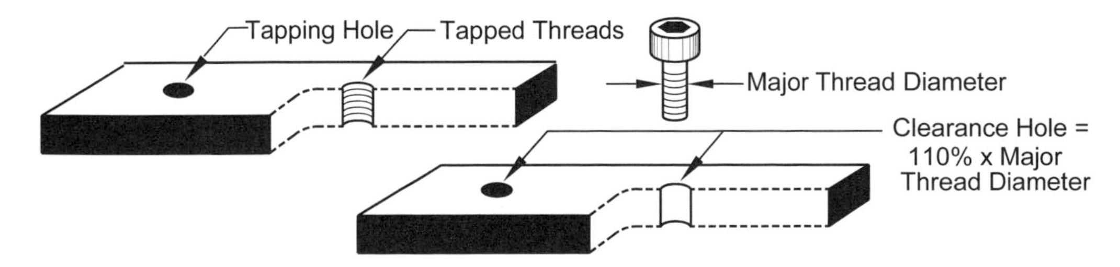

Figure 6–24. Clearance and tapping holes.

How can the accuracy of a tapped hole be checked?
Use a *Go/No-Go gage.* Never use a screw or bolt because they will not allow you to verify the size of the thread produced, but only indicate that the specific fastener used will fit into that particular hole. Figure 6–6 (bottom) shows a *Go/No-Go* thread plug gage.

Tapping Procedure

How is hand tapping performed?
Here is the hand tapping sequence:

1. Choose the correct taps and tap wrench.
2. Determine the correct *tap drill size* for the threads being tapped.
3. Drill the tapping hole perpendicular to the work surface.
4. While not absolutely essential, chamfering the tap hole (using a countersink to clean up any burrs and put a slight taper on the edge of the hole) starts the tap more easily and prevents raising work metal around the edges of the tap hole. In critical, fine work, the tap hole may also be reamed to exact size.
5. Apply tapping lubricant to the hole and tap threads. Use:
    - No lubricant on cast iron or brass.
    - New Rapid Tap™ or Tap Magic® on steel.
    - A9® or Tap Magic® Aluminum on aluminum.
6. Insert the tap into the hole perpendicular to the work surface.
7. Apply moderate downward pressure on the tap and turn it about two turns clockwise (for right-hand threads) to get the tap started.
8. Remove the tap wrench and, using a square, check that the tap is perpendicular to the work surface. Check this at two positions around the tap 90° apart.
9. If the tap is not square with the work surface, back it out, reposition it perpendicular to the surface, and start the threads again. Recheck for squareness.
10. Insert the tap into the hole. Turn the tap clockwise a quarter-turn and then backward a half-turn to break the chip. Do not apply downward pressure after the tap has been started. Use a smooth even rotary motion, as taps are brittle and easily broken.
11. After several turns, remove the tap, clear metal cuttings from the tap and hole, add fresh lubricant, and continue tapping.

What are the best methods for getting the tap hole drilled square to the work?
Using a drill press, lathe or milling machine for the tap hole will produce the best results. The least accurate method is using a hand-held drill.

When tapping, what methods can be used to hold the tap square to the work surface to reduce the chances of tap breakage?

There are several methods:

- *Shop-made steel tapping blocks*, Figure 6–25, are inexpensive and fit into tight spots. Machine a rectangular steel block and drill holes to hold the tap vertical to the work.

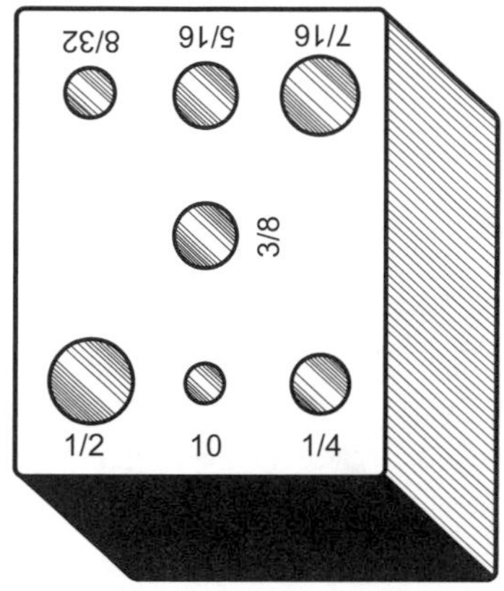

Figure 6–25. Shop-made steel tapping block.

- *Hand tappers*, Figure 6–26, are commercial fixtures that hold taps and keep them square to the work. They are effective and inexpensive, but cannot be used on large parts or in tight quarters.

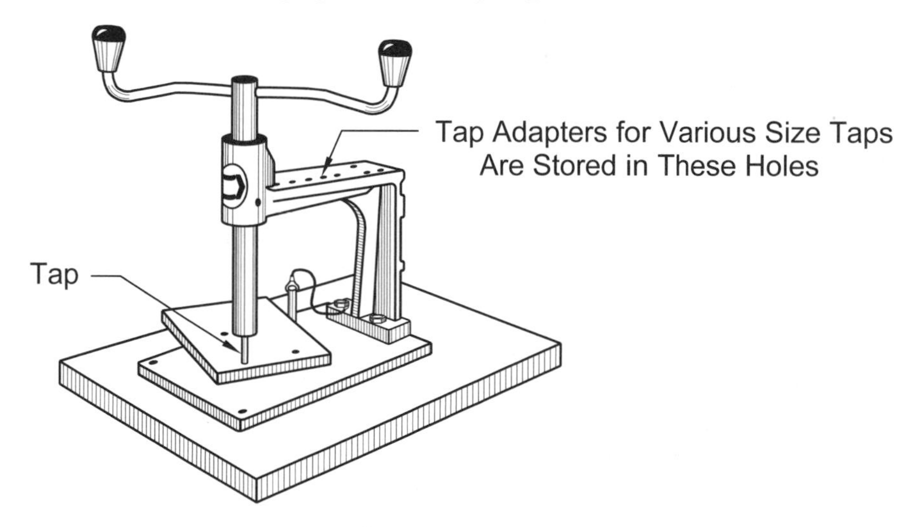

Figure 6–26. Hand tapper.

- *Spring-loaded tap guides*, Figure 6–27, are inexpensive and work well. They have reversible plungers: pointed on one end for large taps with center holes, and with a cone-shaped depression on the other end for small taps with pointed shanks.

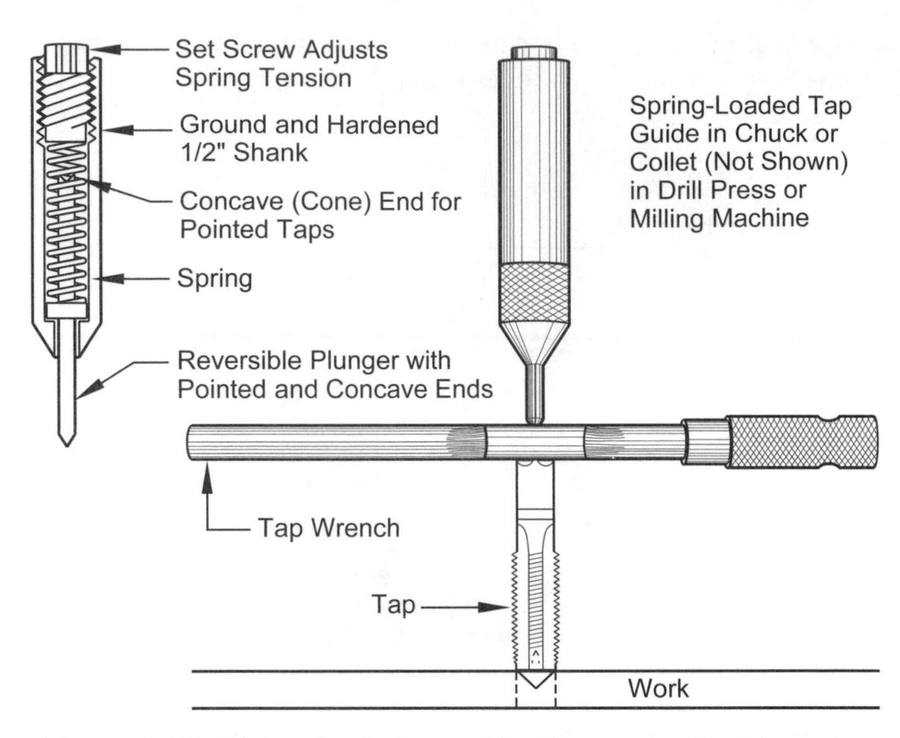

Figure 6–27. Spring-loaded tap guide. These can also be used to position a part accurately for drilling in the drill press.

- *Drill press or milling machine chucks with shop-made turning disk*, Figure 6–28. Taps may also be held vertical by loosely holding them in the chuck of a drill press or milling machine. A small, knurled, shop-made aluminum disk turns the tap and provides good feel to the machinist.

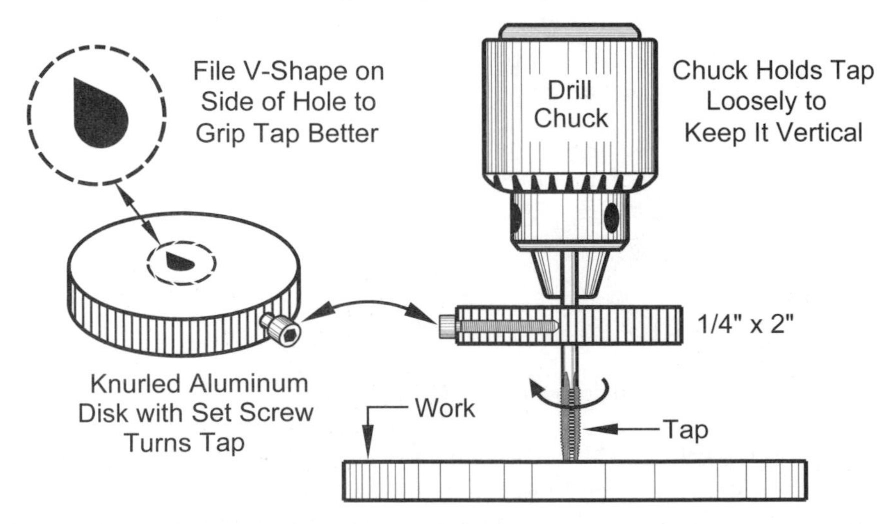

Figure 6–28. A chuck with a shop-made turning disc to hold a tap vertical.

Tapping Blind Holes

How can the chances of tap breakage in blind holes be reduced?

Put solid band saw blade lubricant in the blind hole first. As the tap moves into the hole, the solid lubricant will be displaced with the cuttings. See Figure 6–29.

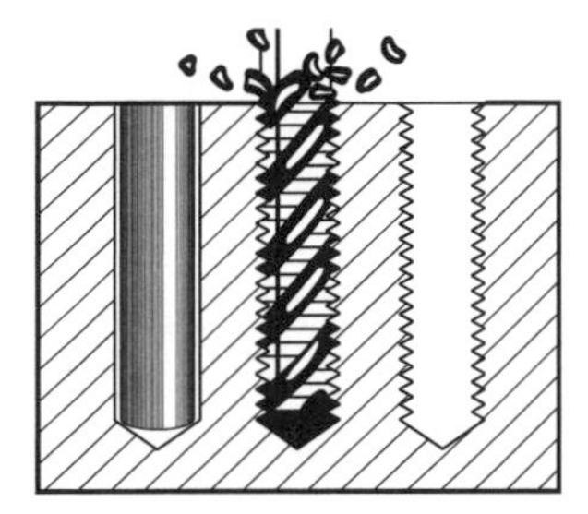

Figure 6–29. How solid lubricant removes cuttings from blind holes.

Section IV – Applying Male Threads

Dies

What are *threading dies* and how are they used to apply threads?

Threading dies are the female version of taps. Like taps, they cut threads because their own threads are interrupted and the resulting land faces do the cutting. One side of a threading die has an internal chamfer to ease starting on the work. This side is usually marked with the die thread size. You must start threading on this face of the die. Dies have a slit so the die handle can pinch them to adjust their size, and some dies themselves have screws for size adjustment, Figure 6–30. Some dies are fixed in size; these are for *re-threading,* the repair of existing damaged threads. Dies for hand threading are held in *die handles*, also called *die stocks,* Figure 6–31.

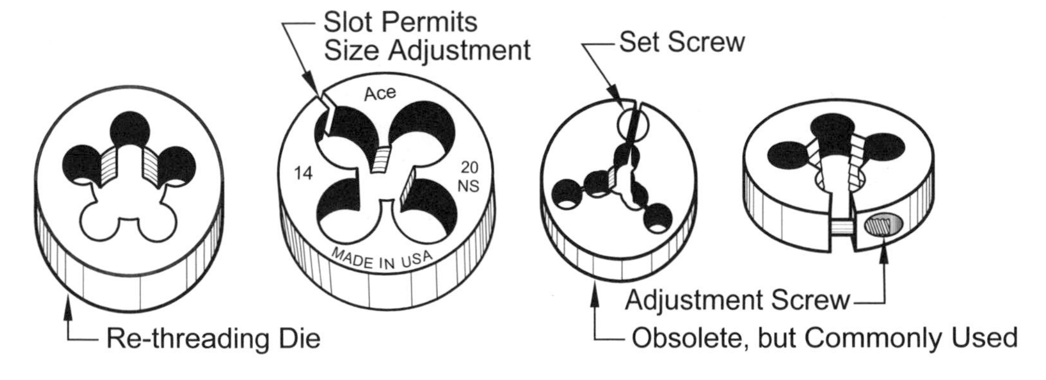

Figure 6–30. Threading dies have a means of adjusting the diameter of threads cut; re-threading or thread repair dies are solid and do not.

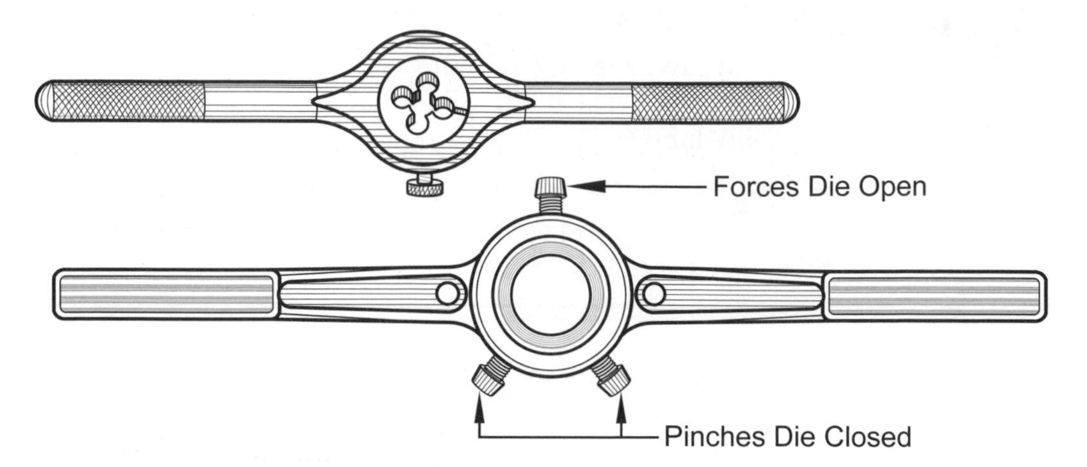

Figure 6–31. Die handles, also called die stocks.

What is the procedure for die cutting threads?

1. Select round stock the *same* diameter as the desired thread: A ¼-20 UNC thread requires ¼-inch diameter stock.
2. Choose the proper die and secure it in its die stock.
3. Grind, turn or file a chamfer on the end of the work to make starting the die easier.
4. Secure the work solidly in a vise.
5. Fit the tapered, or starting end, of the die over the work. Usually the face of the die with the thread size and maker's name is the starting end.
6. Apply lubricant to the top of the die. It will lubricate the die threads and the work.
7. Press down on the die or die handles and turn the die clockwise (for right-hand threads).
8. Reverse die rotation every one or two turns to break up the chips. The larger the diameter of the work, the more frequently the chips must be cleared.
9. Remove any burrs from the finished work with a fine-cut file.

Because of its starting chamfer, the die will not cut threads up to a shoulder. What can be done?

Turn the die over so the *chamfered* end threads face out of the work. Then the die can cut threads down to the shoulder.

The threads cut with a die are rough and torn. How can better threads be cut with a die?

Thread quality can be improved by using a die with a screw or pinch adjustment and cutting the threads in two or more passes. Open up the die for the initial pass and gradually close it on subsequent passes. The initial pass should just make a fine trace on the work. Check the minor diameter with a thread micrometer or dial caliper to reach the desired finished size. By observing the number of turns needed on the die adjustment screw to reach finished thread depth, threading cuts on subsequent workpieces can be made in roughly equal depths. Opening up the die at the end of cuts will also improve thread quality because it prevents tearing when the unchamfered end of the die reverses into the threads. These steps are needed only for fine work. Most die threading is done in a single pass.

Other Threading Methods

What other methods are used to apply male threads and what are their advantages?

- *Roll threading* presses two or more hardened steel dies with the desired thread form against a rotating fastener blank, transferring the thread profile to the blank. This process produces stronger, more fatigue-resistant fasteners and smoother threads than fasteners with machined threads, Figure 6–32. For screws and bolts up to 1-inch diameter, heads and threads are cold formed. Larger bolts and high-strength screws have hot-forged heads and cold-rolled threads. Threads are cut or tapped only on fasteners which are too large for thread rolling which is faster and less expensive than cutting them.

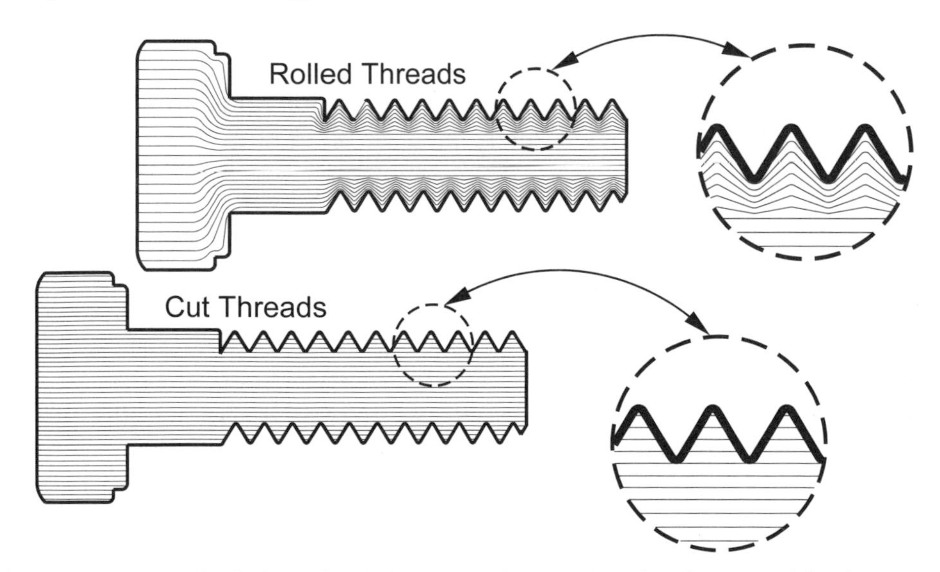

Figure 6–32. Rolled threads deforms and extrudes the fastener blank's metal to match the thread (top), while machined or cut threads (bottom) removes metal from the thread.

- *Grinding threads* offers high accuracy, the ability to machine hard materials and a smooth thread finish. Precision measurement tools, gages and master threads are frequently ground. Most threading taps and micrometers have ground threads.
- *Lathe threading* with single-point tools is covered in *Chapter 7 – Turning Operations*.

Thread Fit Classifications

Why are there different fit classifications for threads and what are they?

Different applications require different degrees of fit, or tightness, between male and female thread pairs. Since there is a great difference in production costs, it makes sense to match cost and fit with application.

- *Classes 1A and 1B:* For work of rough commercial quality where loose fit for spin-on-assembly is desirable.
- *Classes 2A and 2B:* The recognized standard for normal production of the great bulk of commercial bolts, nuts and screws.
- *Classes 3A and 3B:* Used for high-quality work where a close fit between mating parts is required.
- *Class 4:* Obsolete, not used.
- *Class 5:* For a wrench fit. Used principally for studs and their mating tapped holes. A force fit requiring the application of high torque for semi-permanent assembly.

Bolt Pre-Tensioning

Why are bolts pre-tensioned or "torqued up"?

A tightened bolt stretches, acting like a spring, and clamps the work together. How much the bolt is tightened determines this clamping force, Figure 6–33. Bolts are pretensioned:

- To increase friction between the clamped layers to prevent them from moving with respect to each other.
- So that the clamped layers cannot move with respect to each other, so the bolt is never put into shear.
- To prevent the nut and bolt pair from loosening.
- Because without proper pre-tensioning, bolts experience fatigue forces. If the fluctuating forces on bolts remain below the pre-tensioning force, the bolt does not *see* them and is not subject to fatigue forces that could eventually destroy the bolt.
- If the bolt clamps a gasket, adequate tension is required to make a seal.

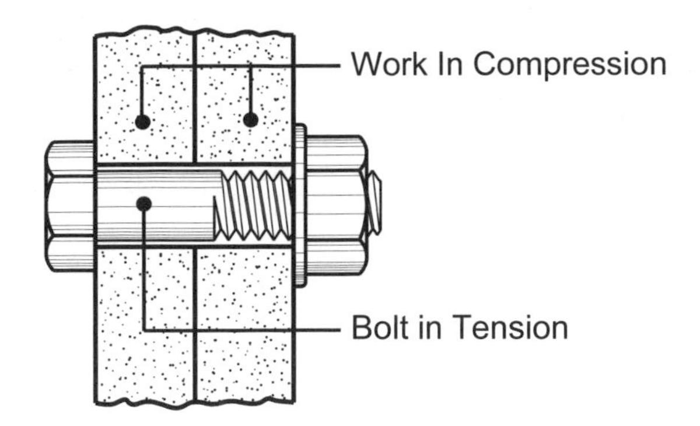

Figure 6–33. Bolt tightening.

How is pre-tensioning done?
Using a torque wrench is the most common pre-tensioning method. Although this tool accurately measures *total* applied torque, the important torque needed to stretch the bolt is just 8 to 15% of the total torque. The remaining torque overcomes thread and nut-face friction, which varies depending on lubrication and thread conditions. The large variations in these two torques prevent accurate measurement of the stretching torque itself. No matter how accurate the torque wrench, bolt stretching torque is at best an estimate. Drilling a hole down through the center of the bolt nearly to its base and measuring bolt stretch though this hole is very accurate and is done on critical engine parts. There are also good ultrasonic methods as well as torque-indicating nuts.

Washers

Why are washers used under nuts, bolts and screws?
Washers distribute forces from fasteners over a greater surface area to prevent *embedding*. This is when the fastener sinks into the work, allowing the fastener to lose its pre-load and possibly loosening. Second, washers provide a smooth surface for the bolt or nut to turn against when tightened so sufficient torque can be applied without rupturing the fastener. If high-strength bolts are used, matching hardened washers must also be used. *Non-heat treated washers will yield.*

Section V – Removing Frozen or Broken Fasteners & Taps

Frozen or Broken Fastener Removal

What are the common methods for removing frozen or broken threaded fasteners?

Here are some approaches and cautions. Take the least destructive removal steps first. Often several steps are needed, like heat cycling, tapping and overnight penetrant. Be persistent and try alternatives. The choice of removal methods should take into consideration where the fastener was snapped—above or below the part's surface, and what caused the failure—rust in the threads, excessive loads or over-torquing.

- Begin all fastener removal efforts by cleaning the screw or bolt head, the stem and the area around the stem where it enters the work. If the fastener has a slot or Phillips cross, clean them. Cleaning is best done with a stainless steel brush and acetone. Cleaning lets you see the problems better, gives screwdrivers a better grip on the slots, and opens a path to the threads for penetrants.
- Often a *tap disintegrator (TD)*, also called a *bolt disintegrator*, *metal disintegrator* or *spark eroder*, is the best way to remove a frozen or broken fastener. This device works on the same principle as Electrical Discharge Machining (EDM) and removes the fastener quickly and without damage to the work. It works for all fastener metals. Some TDs are stationary while others may be brought to the work. When using TDs, the workpiece must be immersed in water or oil. Many machine shops have these machines and offer this service. For high-value workpieces, this is an excellent approach and presents minimum risk to the part.
- On stripped out Phillips head screws, use a die grinder (Dremel® tool) with an abrasive disc to cut a new slot. Be aware that the grinding will heat the fastener. Then use a flat-blade screwdriver in the new slot.
- On socket head cap screws whose hexes have stripped out, use a SOCK-IT-OUT tool, Figure 6–34, or make one from a larger Allen wrench using a bench grinder. Hammer the SOCK-IT-OUT tool into the recess. If the Allen wrench itself strips out, grind down the damaged end to expose fresh hex area.

Figure 6–34. SOCK-IT-OUT tool for removing Allen-head screws and bolts with damaged recesses.

- If the work can withstand heat, temperature cycle the fastener and surrounding areas by heating with a torch or electric heat gun. Heat small screws with a soldering iron. Do not give up until you have temperature cycled 5 to 10 times. Cool with ice or compressed air after each heating cycle. Even raising fastener/workpiece temperature 200 to 300°F may be helpful, but raising the fastener to a dull red, if the work will not be damaged, is even better. Other alternatives are heating with a butane pencil torch to limit damage to the surrounding area. If flame damage to the surrounding workpiece area is a problem, heat to red-hot the end of a piece of metal held in a Vise Grip® pliers, then apply the hot end to the top of the fastener.

- Use penetrating oil and allow it to soak in overnight. Be aware that many proprietary products, which formerly contained hydrofluoric acid and other aggressive chemicals, worked well, but have been replaced by ineffective, but government-approved hydrocarbon oils with the same names.
- Sometimes medium to light tapping on the fastener, in combination with other steps, works. Apply shock to help loosen the fastener by striking a screwdriver inserted into it or directly to the fastener itself. Do not apply so much force as to damage or snap off the fastener head. Consider tapping on the side, head or stem of the fastener with a center punch. Apply the center punch force off center, tangentially and counter clockwise to encourage the fastener to unscrew.
- Socket head cap screws are hardened and almost impossible to drill out. Consider removal by grinding them out with a diamond bit in a die grinder. After removing the bulk of the fastener, the remaining metal in the threads may be picked out with a scriber. Tap disintegrators are also excellent for removing socket head cap screws.
- Severely rusted fasteners cannot be removed by force. The three forms of iron oxide (rust) occupy a greater physical volume than the steel it came from, locking the fastener solidly in place.
- Frozen and rusted nuts can be removed by cutting them in half with a die grinder and an abrasive wheel. Nut crackers, which use a screw to drive a hardened wedge into the side of a nut while holding the other, are also effective.
- When the fastener or tap is snapped off flush or below the surface, GTAW (TIG) welding works well:
    - Case I: If the fastener is *flush* with the work surface, a nut is placed over it and weld metal is used to join the fastener and nut. Turning the nut backs out the fastener.
    - Case II: If the fastener is *below* the workpiece surface, weld metal is fused to the fastener, and the hole is filled with weld metal. Then a nut is added as in Case I. Skilled weldors can remove broken fasteners which are 5 to 8 diameters below the surface. Weld metal added inside the fastener hole will not adhere to the sides of the hole and ruin the

threads because welding only takes place in line with the electrode, not to its sides. Tilting the GTAW torch inside the nut welds filler metal to the threads of the nut. See Figure 6–35.

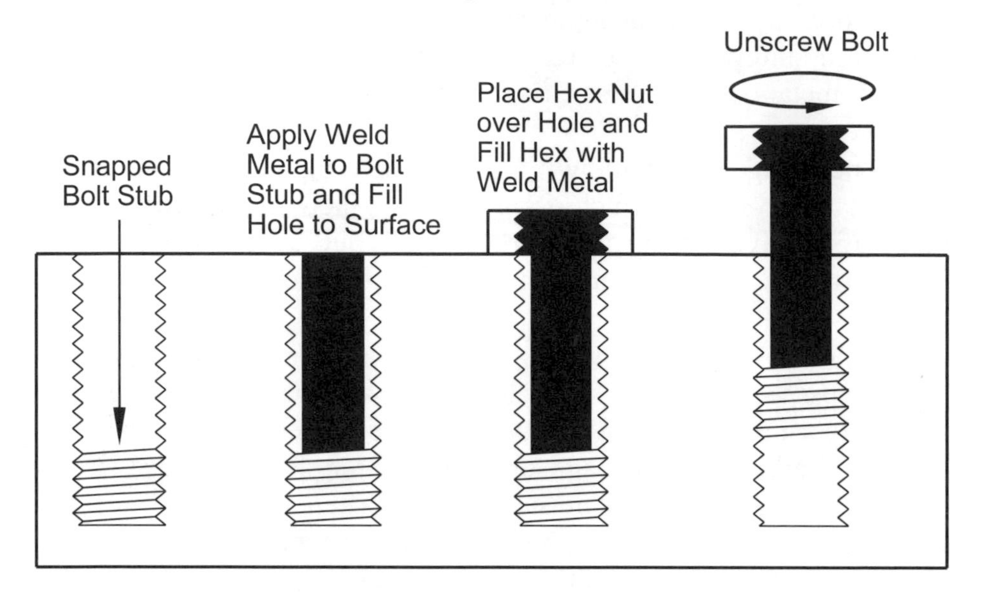

Figure 6–35. Welding a nut onto a broken bolt.

- If the head is snapped off but the stem projects above the work surface, use Vise Grip® pliers to grasp the stem and unscrew it. Flats can be filed on the sides of the stem to improve your grasp.
- If the fastener has broken off flush or below the work surface and the fastener is *not* hardened, try a screw and bolt extractor, which come in several designs, Figure 6–36. Center punch the fastener at its center. Even better than center punching to center the drill is to use a diamond tool in a die grinder. This makes a cone-shaped depression in the top of the fastener so the drill will self-center. Except for very small fasteners, begin with a ⅛-inch diameter twist drill and drill down about ½ inch. Use drilling lubricant. *A left-hand drill is better than a right-hand one because drilling forces tend to unscrew the fastener.* Drill progressively larger holes until the drill hole is about half the diameter of the fastener. Reverse the drill for left-hand drills. Then, insert the largest spiral-point extractor, tap it in place so it bites into the fastener, and unscrew the fastener. Use a tap wrench to turn the extractor. Add penetrant and try to worry it out. Too much force will snap the extractor off in the drill hole and no more drilling can be done because extractor metal is hardened. Better to remove the extractor and drill the hole larger. Do the drilling in a drill press or milling machine, if possible.

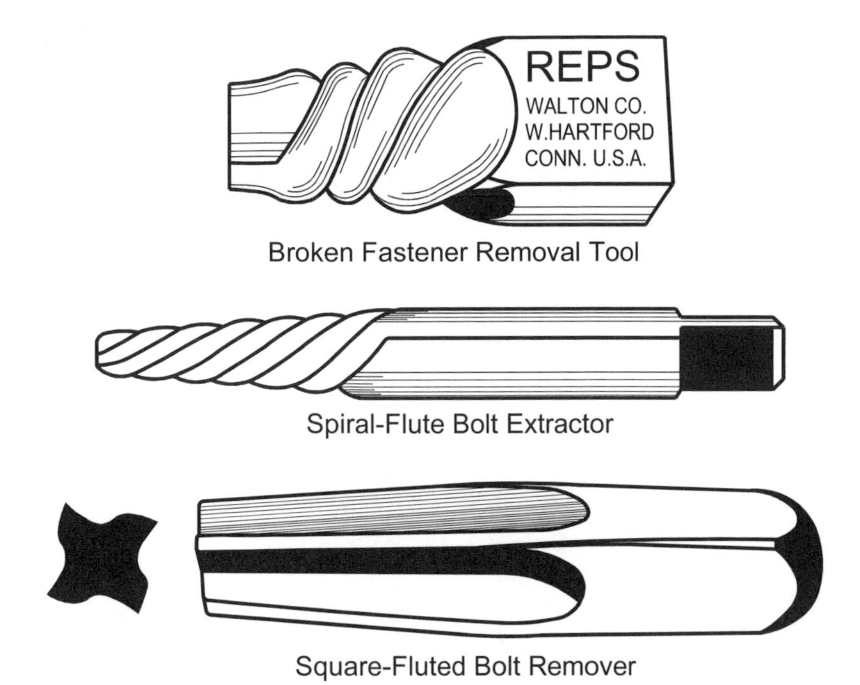

Figure 6–36. Broken bolt and screw extractors.

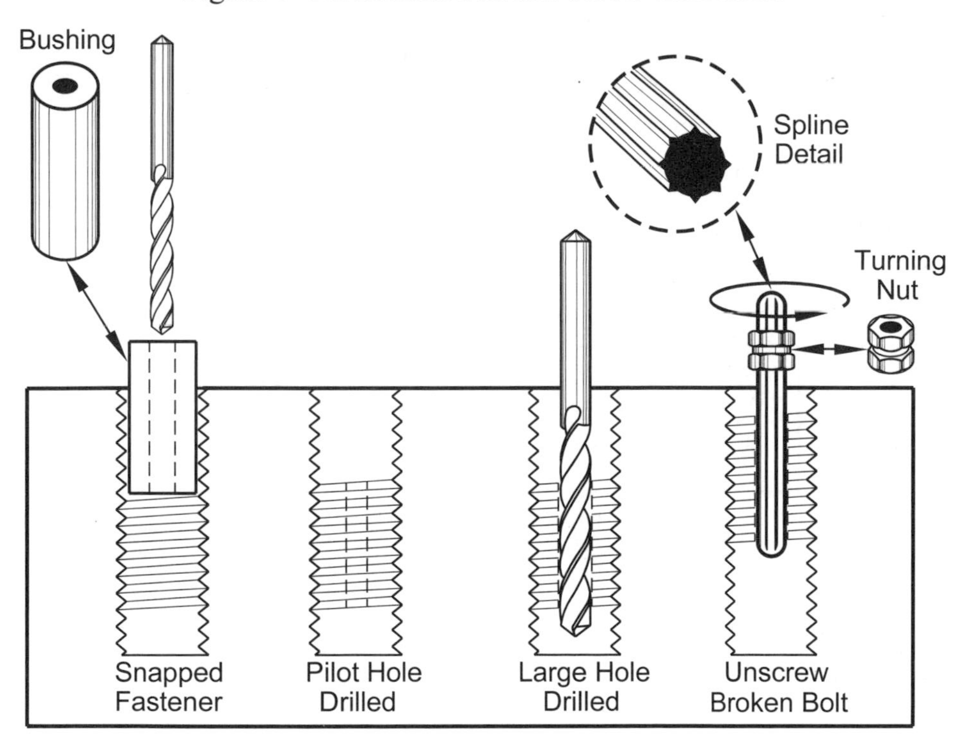

Figure 6–37. Broken bolt removal tool using a centering bushing and spline wrench.

- Another alternative for unhardened fasteners is a commercial product which uses a drill-centering bushing, two twist drills, spline wrench and a spline turning nut, Figure 6–37. Torque on the spline wrench is not concentrated at the point it enters the fastener hole like a screw extractor, so the wrench is less likely to snap off. The spline wrench must be tapped into the fastener hole.

Broken Tap Removal

What methods can be used to remove a broken tap from its tap hole?

- *Tap removal tool*, Figure 6–38, is a good choice for a first removal try. Too much force will break off fingers on the tool, so go lightly. If the tap does not come out easily, try another method.

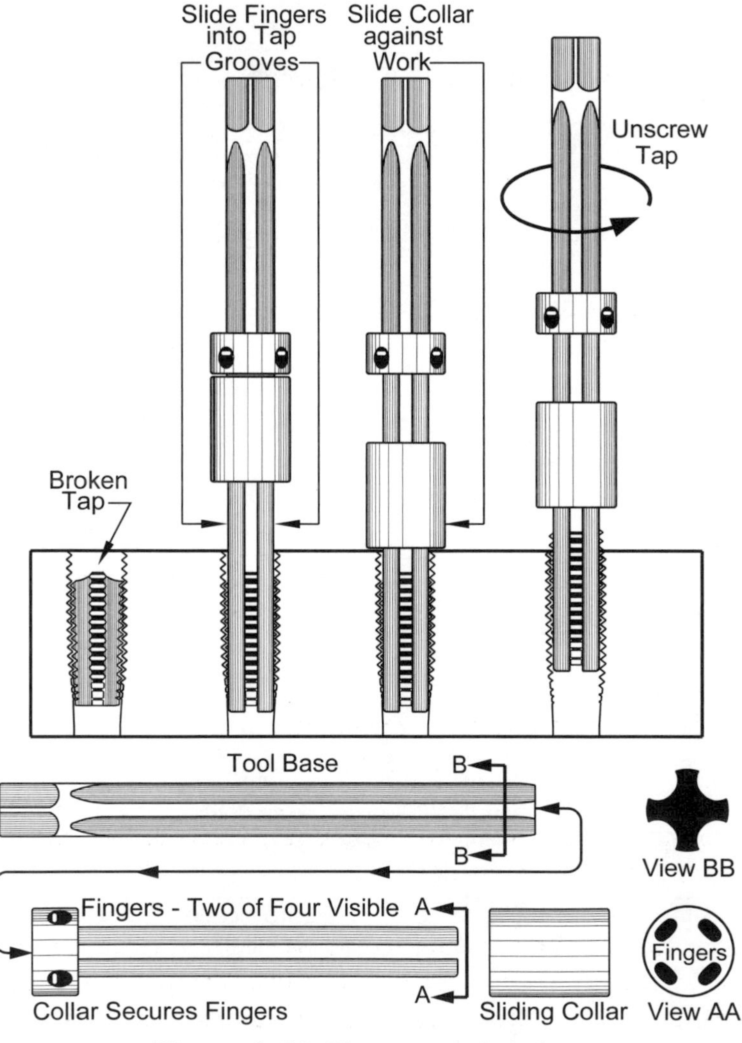

Figure 6–38. Tap removal tool.

The removal steps are:

1. Remove any loose tap material in the tap hole with tweezers or compressed air.
2. Tap removal tools usually have four fingers; two can be removed for taps with just two grooves. (Three-fingered tap removers are rare.)
3. Slide the fingers of the tap removal tool down into the tap grooves.
4. Slide the cylindrical collar down against the workpiece to hold the fingers against the tool base.
5. Back out the tap. You may need to twist the tap back and forth to worry it out rather than unscrew it in one operation.

- *Nitric acid* dripped into the hole dissolves the edges of the tap, reducing its diameter and easing removal. If the hole becomes too large and the threads are damaged, use a Heli-Coil. Be sure to wash the acid out completely.
- *GTAW (TIG) welding* works well when taps snap off flush or below the surface.
- *Tap disintegrators,* discussed at the start of Section V, work well on taps.

Section VI – Repairing Damaged Threads

Thread Repair Options

What repair options are available if threads tapped in a casting or machined part have stripped out?

- The stripped threads can be drilled out and new threads tapped for the next larger size fastener.
- Install a Heli-Coil® screw thread insert, which makes a stronger, more durable repair.
- Install an E-Z Lok insert, which works particularly well in plastics and pressed wood, as well as ferrous and non-ferrous metals.

Heli-Coils®

What is a Heli-Coil and how is it installed?

A Heli-Coil consists of an 18-8 stainless steel wire helix, Figure 6–39. The wire forming the helix has a diamond-shaped cross-section with 60° ends. One edge of the diamond fits into the tapped threads in the work and the other forms a new thread for the fastener.

Heli-Coils are sold in kits containing the proper size twist drill to remove the old fastener's threads, the tap for new threads, the Heli-Coils themselves and an insertion tool. There are several different insertion tool designs, but all have a shaft that extends down through the Heli-Coil to turn the tang at the base of the Heli-Coil. Turning this tang causes the Heli-Coil to behave like a clock spring. When it is wound, its diameter decreases. This allows the Heli-Coil to be easily threaded into its new threads.

To install the Heli-Coil:
1. Drill out the damaged threads.
2. Tap the hole to accept the Heli-Coil insert.
3. Wind the Heli-Coil insert into the tapped threads with an installation tool.
4. Snap off the Heli-Coil tang so it does not interfere with fastener insertion.

The result is a stainless steel thread exactly the same size as the damaged thread. The insert has more depth and more area of the base metal to grip and will not wear out when fasteners are removed. Many aluminum and magnesium castings have Heli-Coils installed initially. Heli-Coil inserts are available in inch, metric, pipe and spark plug threads.

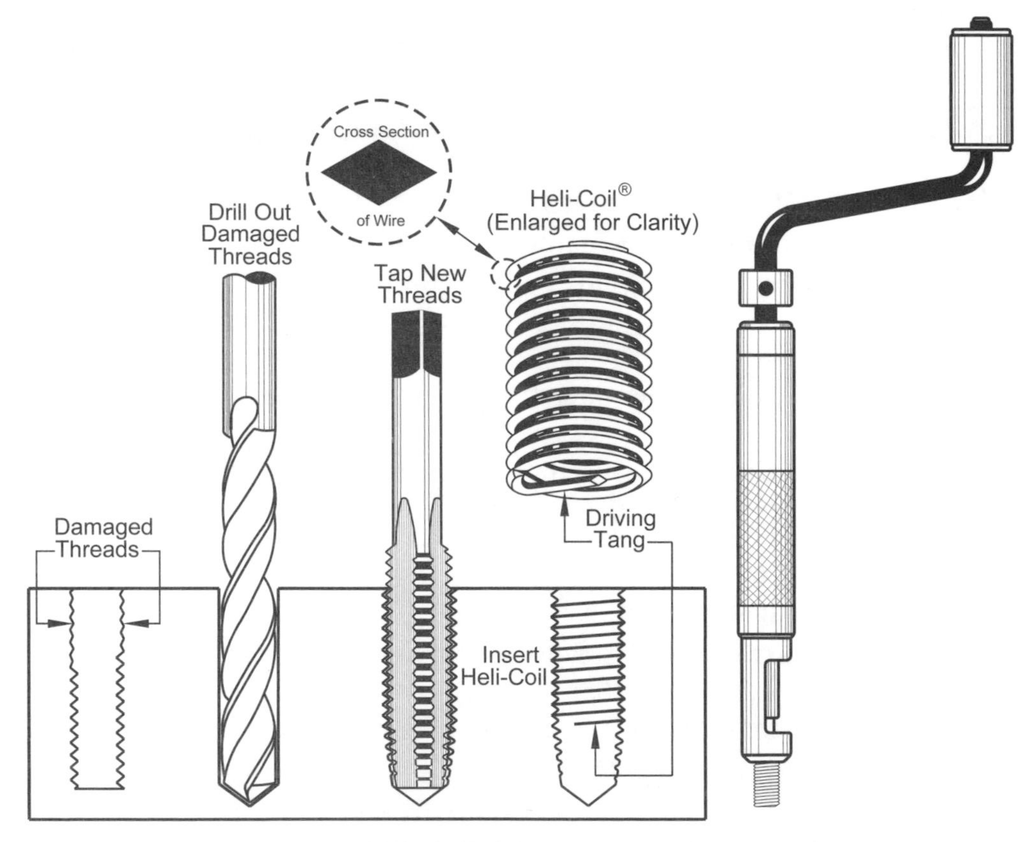

Figure 6–39. Heli-Coil thread replacement steps (left) and installation tool (right).

E-Z Lok Thread Inserts

How do E-Z Lok thread inserts differ from Heli-Coil inserts and how are they installed?

E-Z Lok thread inserts consist of a one-piece steel fastener with *two* sets of threads, Figure 6–40. The external threads secure the insert in the work, and the internal threads replace the damaged threads. The external threads on the insert contain a thread-locking adhesive or a nylon pellet that locks the insert in place. They are simple to install:

1. Drill out the old damaged threads to make a new hole for the insert.
2. Thread the new hole with a tap for the insert.
3. Screw in the insert using either a special purpose tool or a nut and bolt pair as in Figure 6–40 (left).

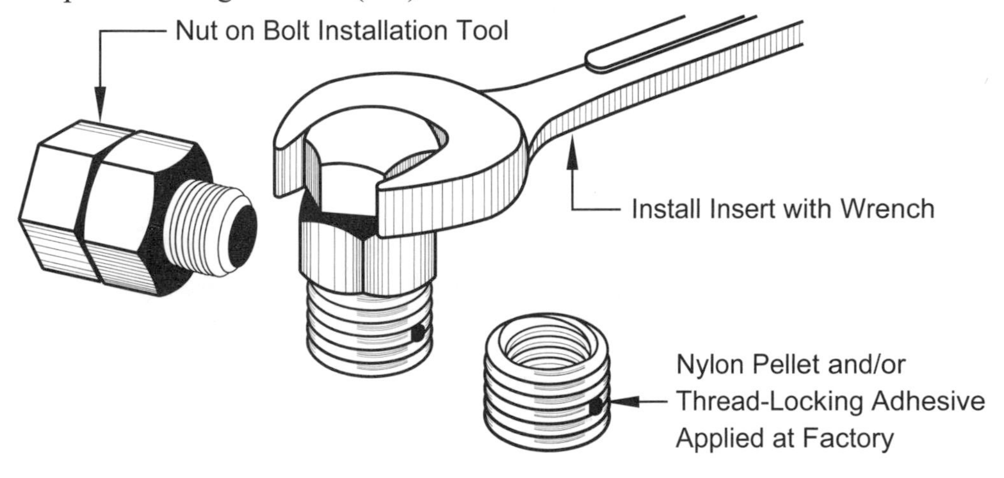

Figure 6–40. E-Z Lok thread insert.

Other Thread & Fastener Repair Options

Besides cutting an Acme thread on a lathe to repair a broken machine rod, what other options are available?

- If the rod is for a non-critical application, like a jack or clamping mechanism where a 0.020- to 0.035-inch error in 3 or 4 feet of thread length is permissible, most industrial tool supply companies sell suitable rods.
- If the rod must be accurate over its length, there are four choices:
  - Apply Acme threads to the rod stock using a lathe.
  - Purchase threaded rod specifically made for leadscrew or other measurement applications.
  - Purchase the exact replacement part.
  - Obtain an identical used part from a scrapped machine.

You are repairing a firearm and need a replacement screw. Other than ordering one from the factory, what other options do you have?

The screws in firearms may be Unified Extra Fine, Unified Fine, metric or Whitworth. The particular size, stem length and shape of many screws are specific to their function in that particular firearm: There are two options:

- Make a screw from the beginning.
- Begin with *unthreaded* screw stock (a slotted screw head on an unthreaded shank). Reduce the shaft diameter to the proper finished screw diameter, trim it to length, thread and blue it. Although the screw can be case hardened, most firearms frames are heat-treated, and most screws are not. Starting with unthreaded screw stock is a much better option than making a screw from scratch. Gunsmith supply companies, such as Brownells, offer selections of this unthreaded screw stock.

Chapter 7

Turning Operations

When a true genius appears in the world, you may know him by this sign, that all the dunces are against him.

—Jonathan Swift

Introduction

The lathe is the oldest and most basic machine tool. Egyptians used primitive lathes at least 3000 years ago. Lathes work by holding and rotating a workpiece while a tool, whose position is controlled by the lathe, is held against the work. Lathes can cut cylinders and cones, both solid and hollow. Metal lathes can make truly round parts to tolerances of less than one-thousandth of an inch that permits the production of matched components such as axles and bearings, pistons and cylinders, and gun barrels and projectiles. Lathes are also important because of their ability to thread shafts, nuts and bolts. In addition, the lathe can turn a steel forging or cylinder into a one-piece crankshaft.

Not only does the lathe make parts required to build all other types of machine tools, it also makes components for engines, pumps, valves, vehicles and electric motors, and performs over a dozen different operations, more than any other machine tool. Because the versatile lathe makes essential parts for many of the devices that make modern life possible, you can see why 20% of all machine tools in the U.S. are lathes.

During their early development, lathes were used by furniture makers, wheelwrights, joiners, bell founders and pulley makers. The lathe has no single inventor, but evolved as dozens of incremental improvements were added over time. Before 1700, long periods passed when the lathe saw few improvements and was limited to turning work between centers. Over the next 200 years, the lathe evolved into its present form. The basic design of lathes built after 1900 differs little from those made today. The majority of improvements were made first in England and then in the United States. Six major developments brought us to the modern lathe:

- Introduction of the fixed tool holder to replace the use of hand-held tools (as in a wood lathe) was essential to making accurate cuts.
- Transition from human power to water wheels, steam and electricity increased the size of the lathe, the work possible and the cutting speeds.
- Replacement of wooden lathe frames with cast iron ones increased rigidity, accuracy and cut depth.
- Development of the leadscrew for thread cutting and power feeds.
- Addition of chucks and collets extended lathe use from merely turning between centers to facing, drilling, reaming, filing and a host of other capabilities.
- Replacement of carbon steel tool bits with high-speed steel alloys, and later replacing the high-speed steel with tungsten carbide cutting tools, dramatically increased cutting speeds and tool life.

While the manual lathe remains important today in making prototypes, one-of-a-kind units like huge turbine and generator shafts and spare parts, the advent of CNC lathes and machining centers has rendered the manual lathe uneconomical for most production. However, when just a few similar parts need to be made, a skilled machinist can produce the required parts before the program for a CNC machine can be written. But on large production runs, even in low-wage, third-world countries, a modern CNC machine is often more cost-effective than a manual lathe. For this reason, few factories exist today with row upon row of manual production lathes as seen in WWII training films.

Some older lathe models and their components appear in this chapter because many machinists are likely to own or have to work on one. Many excellent older machines remain in shops and labs today. Most do not differ significantly from new machines. There are four reasons so many of these older machine exist:

- Unless deliberately damaged or left to rust, lathes can have working lives of over 60 years. When properly lubricated, they wear little even in heavy service. The exception is when they are cutting cast iron and not carefully and frequently cleaned of chips which are very abrasive.
- Even in worn lathes, spindle bearings can be replaced and ways can be reground at a much lower cost than buying a new lathe; also, spare parts are available for most US-made lathes built during the last 50 years.
- The economics of CNC tools displaced many lathes from production use making them surplus.
- Many jobs can be accomplished on lathes with considerable wear.

While metal lathes are the most common, other specialized lathe designs also contribute to our modern world. Some grind and polish lenses, others carve out gunstocks or shoe lasts, or cut decorative patterns in wood, metal and plastic. Still others spin flat sheets of silver and brass into plates and pitchers or recondition automotive brake drums. Very specialized lathe designs, called automatic screw machines, begin with rod stock and make nuts, screws and other cylindrical metal parts at high speed and low cost.

This chapter:

- Shows how lathe work capacity is measured.
- Pictures and describes five different modern lathes that illustrate the classes of lathes you are most likely to use.
- Examines how lathes are constructed, looks at each part and describes its function.
- Describes and compares lathe cutting tool materials, holders and shapes.
- Investigates cutting tool feed rates and spindle speed, called "feeds and speeds," to determine their optimum settings.
- Looks at lathe accessories, the various devices that attach to the lathe to hold the work or guide the tool and add versatility.
- Presents step-by-step procedures for setting up a lathe and then performing specific operations.
- Reviews lathe-related safety issues.

Section I – Lathe Sizes

Specifying Lathe Size

How is *lathe capacity* measured?

Several dimensions are needed to get a complete picture of a machine's capacity:

- *Swing over the bed* is the largest diameter work that will clear the ways. Note: European manufacturers designate lathe size by the distance from the lathe center to the ways, the radius, or half the dimension we call swing over the lathe bed.
- *Swing over the cross slide* indicates the largest diameter work that can be turned between centers. This dimension is always less than the swing over the bed.
- *Maximum distance between centers* indicates the maximum length of work that can be turned. This figure takes into account the distance of the lathe bed taken up by the tailstock.

- *Bed length* is often mentioned, but is much less useful in comparing work length capacity of two lathes than the maximum distance between centers. This is because some beds extend under and support the headstock. This portion of the bed length is not usable compared with the same bed length in front of the spindle nose. This dimension can be misleading.

See Figure 7–1.

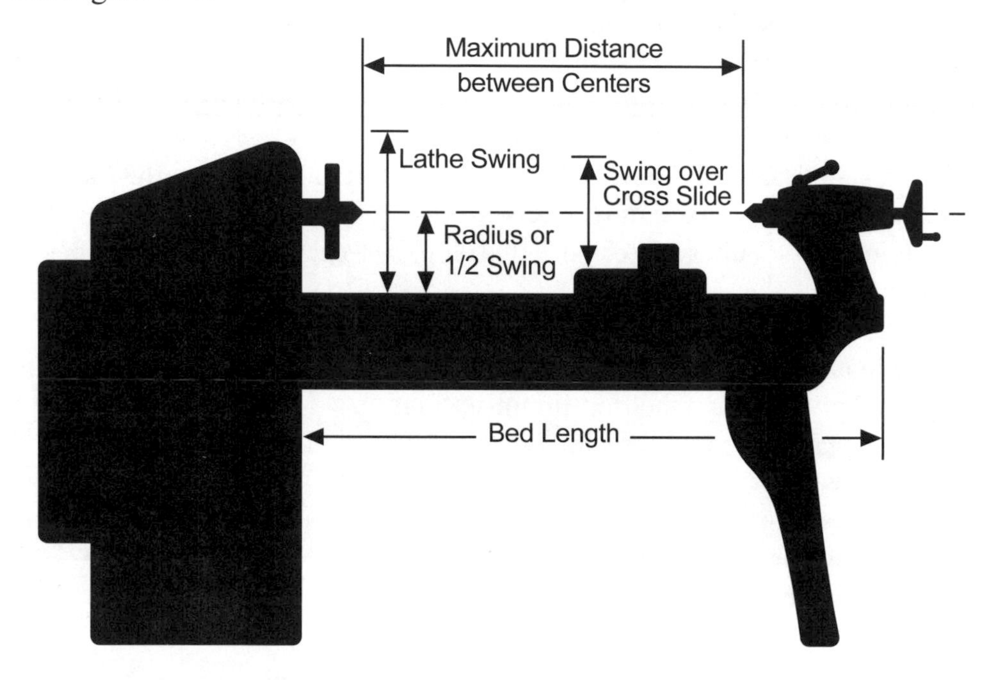

Figure 7–1. Lathe size measurements.

Is horsepower important when evaluating a lathe for an application?

Yes, horsepower limits the maximum depth of cut and thus maximum metal removal rate. The maximum usable lathe horsepower and metal removal rate is also limited by the:

- Strength of its power train—belts, pulleys and gears.
- Rigidity or stiffness of the machine. Cutting operations place large forces on the carriage and lathe bed, twisting them out of alignment, degrading accuracy and causing vibration or chattering between the work and cutting tool.
- Lathe cutting tool material, cutter size, shape and tool mounting fixture rigidity also determine how much of the available horsepower can be used. A high-speed steel lathe bit would burn up quickly when subjected to the heat generated by the production surface speeds and feed rates a modern carbide tool easily withstands. Metal volume removal rates are 2 to 6 times higher for carbide tools than for HSS ones.

Studies show that during the last century improvements in lathe design, work holding methods and, most importantly, tool bit materials have led to 100:1 reductions in the time needed to do the same turning task on a low-carbon steel workpiece.

What other factors are important in selecting a lathe?

- Although two lathes may have the same dimensions, the one with the more robust headstock, tailstock and carriage castings will be able to take bigger cuts, remove more pounds of metal per hour and perform more work in a given time while maintaining accuracy. A rough way to compare lathes is by their capacity to remove metal in pounds/hour. In general, the more a lathe weighs, the stiffer its castings. This would not be as important in a work situation where the machine is only used occasionally such as in a research lab, but would be critical in a production setting.
- While the same lathe design may be calibrated in either inches or millimeters and holds the same accuracy in either measurement unit, the lathe using the measurement units most often used in the shop is the better choice. The carriage, cross-feed and tailstock collars are usually calibrated in one or the other units, not both. Working on a machine calibrated in one set of units while using drawings dimensioned in another is possible, but it is difficult and leads to errors. Regardless of their measurement units, most lathes can cut both inch or metric threads simply by changing gears.

Modern Lathes

What are some examples of typical modern lathes?

Here are five quality lathes from very small to large:

- *Levin Precision Instrument Lathe*, Figure 7–2, makes parts for electrical and scientific instruments, medical devices and watches, so Levin lathe operators frequently work under stereo-microscopes. Work made on these lathes is usually small enough to fit under a dime and is machined in collets, but 3- and 4-jaw chucks are also used. Their standard ⅓-horsepower motors have 0–5000 rpm SCR drives with IR-drop compensation to improve speed regulation. Levin offers a wide range of collets, chucks, tool holders, turret tailstocks, milling attachments, grinding attachments and coolant systems, so their lathes can be used as a tool makers' lathe, turret lathe or micro-drilling machine. These are very fine lathes for miniature critical and demanding applications and probably the best in its size class. A Levin lathe with a bench, cross slide, DRO (Digital Read Out) and a good selection of accessories could easily exceed $20,000. The Levin website is www.levinlathe.com.

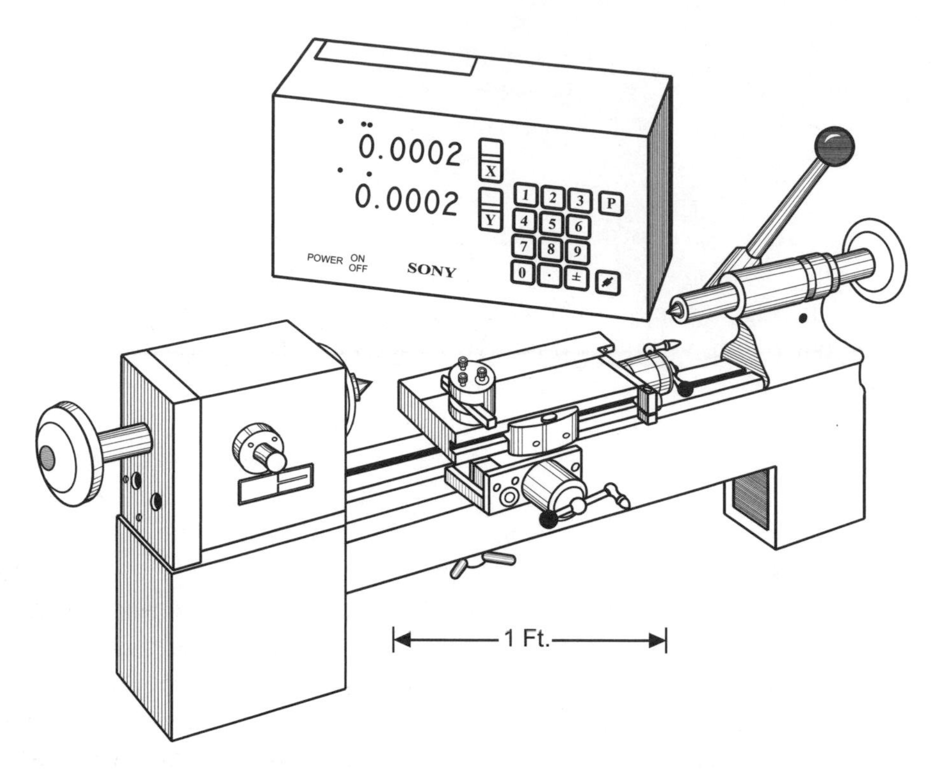

Figure 7–2. Levin Precision Instrument Lathe with Sony digital readout.

- *Sherline Miniature Lathe*, Figure 7–3, is an ideal machine for work too large for an instrument lathe and too small for a bench lathe. It easily makes 0.060-inch cuts in ¾-inch aluminum round stock and is capable of cutting mild steels or even stainless steel and titanium with carbide tools. Don't let its size fool you, it is definitely not a toy. A clever threading attachment cuts 36 unified and 28 metric threads. It holds tolerances of 0.0005 inches and has a TIR (Total Indicated Runout) under 0.0002 inches. An SCR speed control provides 70–2800 rpm spindle speeds. Lab technicians, prototypers, model makers, watchmakers and clockmakers use this machine. However, half of all Sherline lathes go into manufacturing applications where they are often used for small production jobs to free up more expensive machines. There are both inch and metric versions. Sherline offers dozens of accessories including quick-change tool holders, carbide insert cutting tools, digital readouts, chucks, collets, steady rests and adjustable-zero handwheels. Nearly every attachment seen on a large lathe is available for a Sherline. A basic machine costs about $575 and a good selection of accessories adds $1000 to $2000 more. See www.sherline.com for more detailed information, including many ingenious techniques and accessories developed by lathe users. Sherline Products Inc. manufactures these machines in Vista, CA 92081.

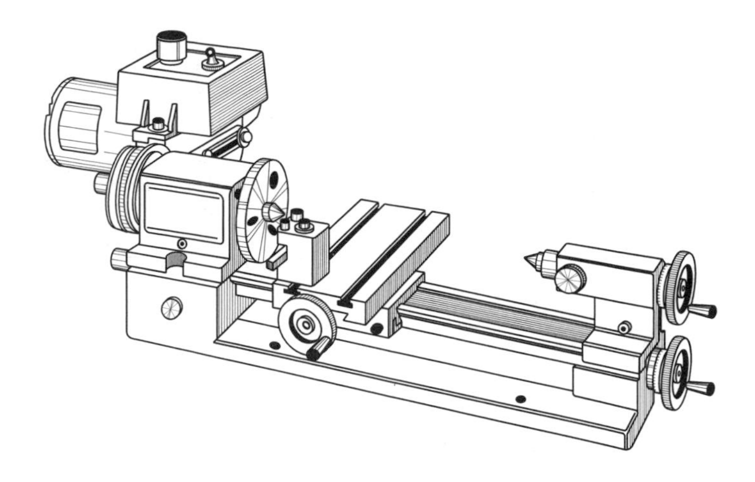

Figure 7–3. Sherline Miniature Lathe.

- *Myford Super 7 Bench-top Lathe*, Figure 7–4, is a high-quality, smaller, precision lathe for research labs, industrial prototypers, model makers and advanced hobbyists. Models are available in either inch or metric units. It can be fitted with a quick-change gearbox, which is both highly desirable and recommended. A basic Myford Super 7 on a metal stand costs about $9500. Spindle speeds are changed by moving belts. Necessary tooling could add another $4000 to $6000. While these items—chucks, collets, Morse taper adapters, tool holders, live centers, face plates and a dozen other items—do not come standard with the lathe, they are essential to performing tasks beyond turning between centers. See www.myford.com for more information.

Figure 7–4. Myford Super 7 Bench-top Lathe.

- *Clausing 10-inch Lathe*, Figure 7–5, is a typical, medium-sized industrial lathe. This is a rugged, mature, tested design, which was in production for decades in the U.S. with only minor changes until replaced by foreign-built models from the United Kingdom and Spain. There are thousands of these lathes in shops and labs around the world. Because this lathe is so typical of similar machines in its class and likely to be encountered by the reader, a detailed labeled view is shown in Section II – Parts of the Lathe, Figure 7–7.

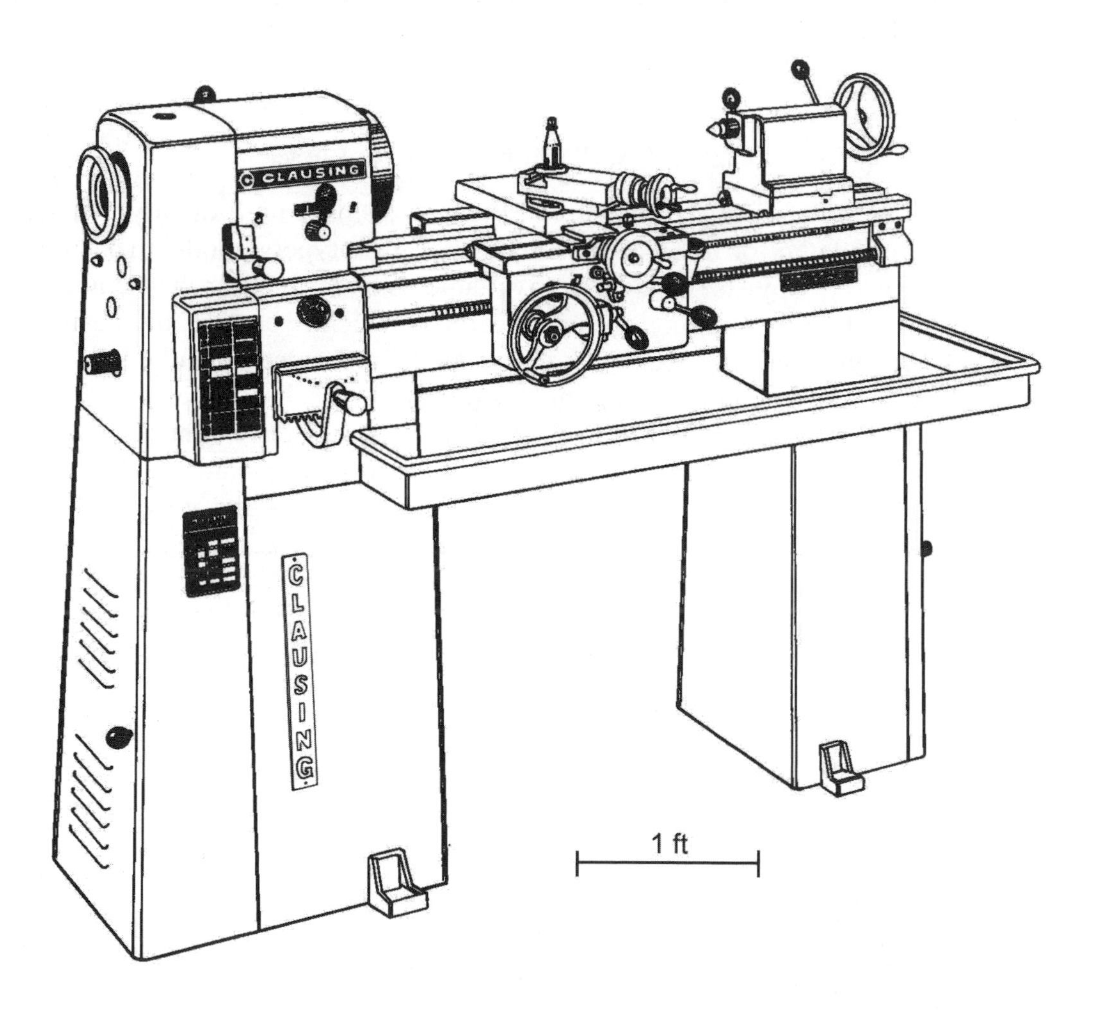

Figure 7–5. Clausing 10-inch Lathe.

- *Clausing/Colchester Geared Head 15-inch Lathe*, Figure 7–6, is a high-quality, medium-large industrial-grade lathe. This model is the smallest member of a family of similar designs offering swings of 15, 18 and 21 inches and distance between centers from 50 to 120 inches. Guaranteed maximum spindle runout is 0.0001-inches TIR. All headstock gears are ground and hardened. Lubrication of spindle bearings, gears and shafts is by a pumped oil bath. The apron gears are hardened and also run in an oil bath. Their three-phase motors range from 7.5 to 15 horsepower depending on the model. Some models offer variable-speed drives, also. High-end models offer constant surface speed: the motor automatically adjusts spindle speed with the turning diameter so the SFM remains constant and optimized. These lathes weight between 3300 and 8000 lbs. depending on the model. The Clausing website is www.clausing-industrial.com. While this lathe is made in the United Kingdom, Clausing Industrial, Inc., Kalamazoo, Michigan 49007, distributes and supports these machines in the United States.

Table 7–1 lists the specifications of these five lathes for easy comparison.

Figure 7–6. Clausing/Colchester Geared Head 15-inch Lathe.

	Levin Instrument Lathe Model 1213-02	Sherline Miniature Model 4000 Lathe	Myford Super 7 Plus Lathe	Clausing 10-inch Lathe	Clausing/Colchester Geared Head 15-inch Lathe
Swing	4	3.5	7	10	15.7
Swing over Saddle	3.5	1.88	4⅛	5½	9.7
Distance b/w Centers	3.5	8	18½	24	50
Bed Length	12	15	27	30	—
Spindle Hole	—	0.405	1.026	1.024	2.125
Spindle Taper & Nose Mount	3C, D & WW Collets	#1MT & ¾-16 thread	#4MT & 1⅛-12 thread	#4MT & 1¾-8 thread	#4MT & D1-6
Tailstock Mount	D Collet	#0MT	#2MT	#2MT	#5MT
Speeds	0–5000	70–2800	25–2500	52–1700	25–2000
Threads	—	36 inch & 28 mm[1]	Yes[2]	40 inch	56 inch & 51 mm
Longitudinal Feed	—	—	Yes[2]	0.0006–0.04	0.0007–0.048
Horsepower	⅛	⅓	¾	¾ or 1	7½
Manufacturing Origin	Culver City, CA, USA	Vista, CA, USA	Nottingham, UK	Kalamazoo, MI, USA	West Yorkshire, UK Kalamazoo, MI, USA
Dimensions (L x W x H)	12 x 4 x 9	23 x 10 x 8	42 x 27 x2 5	68 x 44 x 28½	80 x 50 x 45
Weight	25	24	270	700	3300
Cost	8500	600	9000	N/A[3]	19,000

Note: Dimensions in inches, speeds in rpm, weights in pounds, and costs in dollars.
Cost is the approximate base price of least expensive machine and does not include tooling, accessories, freight, installation, or taxes.
[1]Threads are applied with hand-powered attachment.
[2]Change gears used, a quick-change gear box is optional.
[3]Not in production.

Table 7–1. Lathe comparison chart.

Section II – Parts of the Lathe

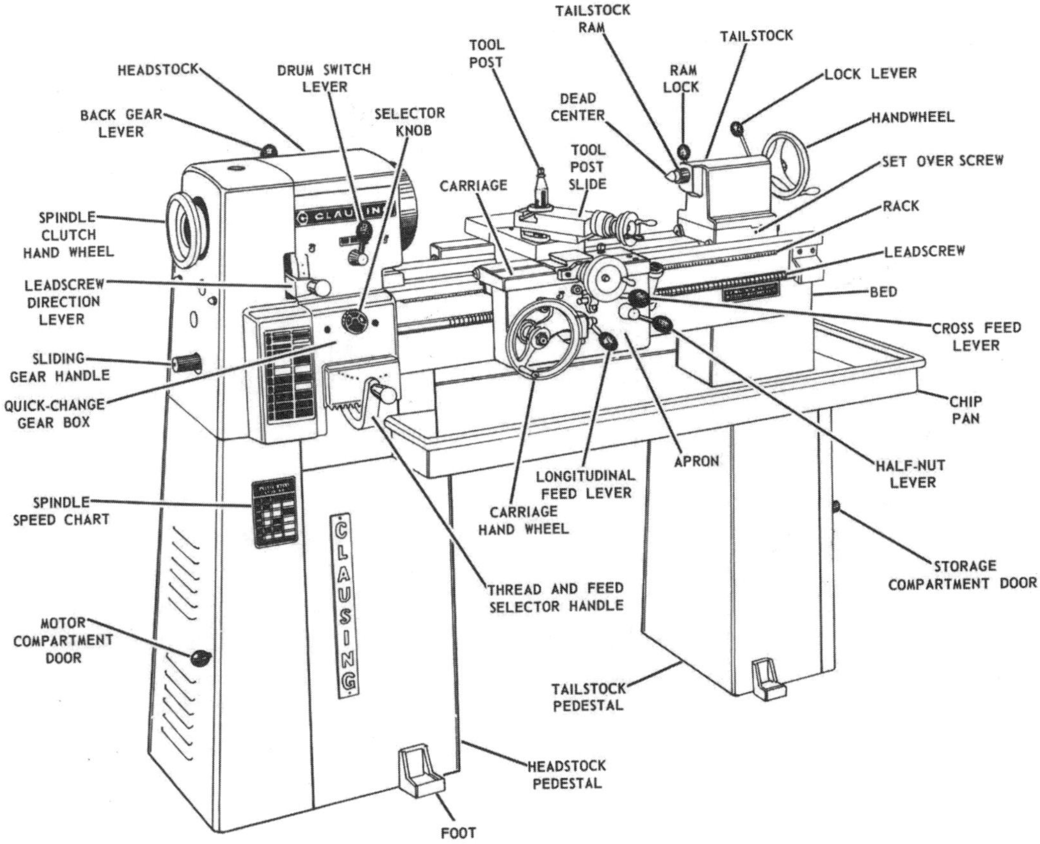

Figure 7–7. Clausing 10-inch Lathe.

Bed

What is the function of the lathe bed and what are common designs?

The lathe bed provides a rigid foundation for the entire machine and holds the headstock, tailstock and carriage in alignment. Lathe beds are usually fine-grain cast iron. The machined and ground surfaces of the bed on which the carriage and tailstock slide are called *ways*. Quality lathes often have induction or flame-hardened ways to minimize wear. Cast into the beds are reinforcing ribs to increase rigidity against cutting tool forces. These forces tend to throw the headstock, carriage and tailstock out of alignment. To provide better carriage and tailstock alignment, one or more inverted Vs are usually part of the ways design. Figure 7–8 shows two bed castings. The left one bolts to the headstock and the right one has an integral base for the headstock. Both are excellent designs.

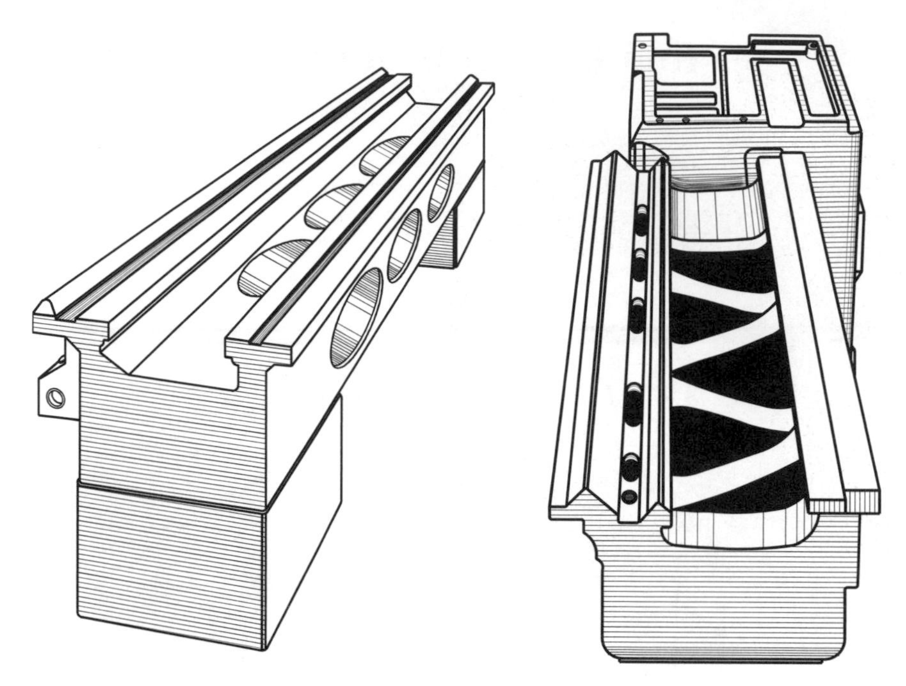

Figure 7–8. Typical lathe bed castings.

Ideally, the center axes of the headstock and tailstock lie on the same line. The tool path as the carriage moves along the ways, is also parallel to the headstock-tailstock centerline. Failure to achieve this alignment prevents turning a perfect cylinder and all work becomes cone-shaped.

What is the frosted pattern sometimes seen on the ways of older lathes?
This pattern results from hand scraping to achieve extreme flatness and to provide minute surface channels to retain an oil film. This film reduces wear and increases long-term machine accuracy. Hand scraping is the mark of a quality older machine. Lathes made today usually have ground and flame-hardened ways which lack this frosting.

Are there other lathe bed designs?
Yes, there are three other designs used when additional swing space is needed to turn extra-large work: *Gap-bed lathes, removable gap-bed lathes* and *sliding bed lathes*. On gap-bed lathes the ways stop short of the headstock, Figure 7–9 (top) and on others a section of the ways near the headstock is removable, Figure 7–9 (middle). On some very large lathes, called sliding-bed lathes, the entire lathe bed slides back to open up a gap for the work, Figure 7–9 (bottom). Many lathe owners do not like to remove the gap bed section because they believe lathe accuracy will bc compromised when the section is replaced which is probably often true.

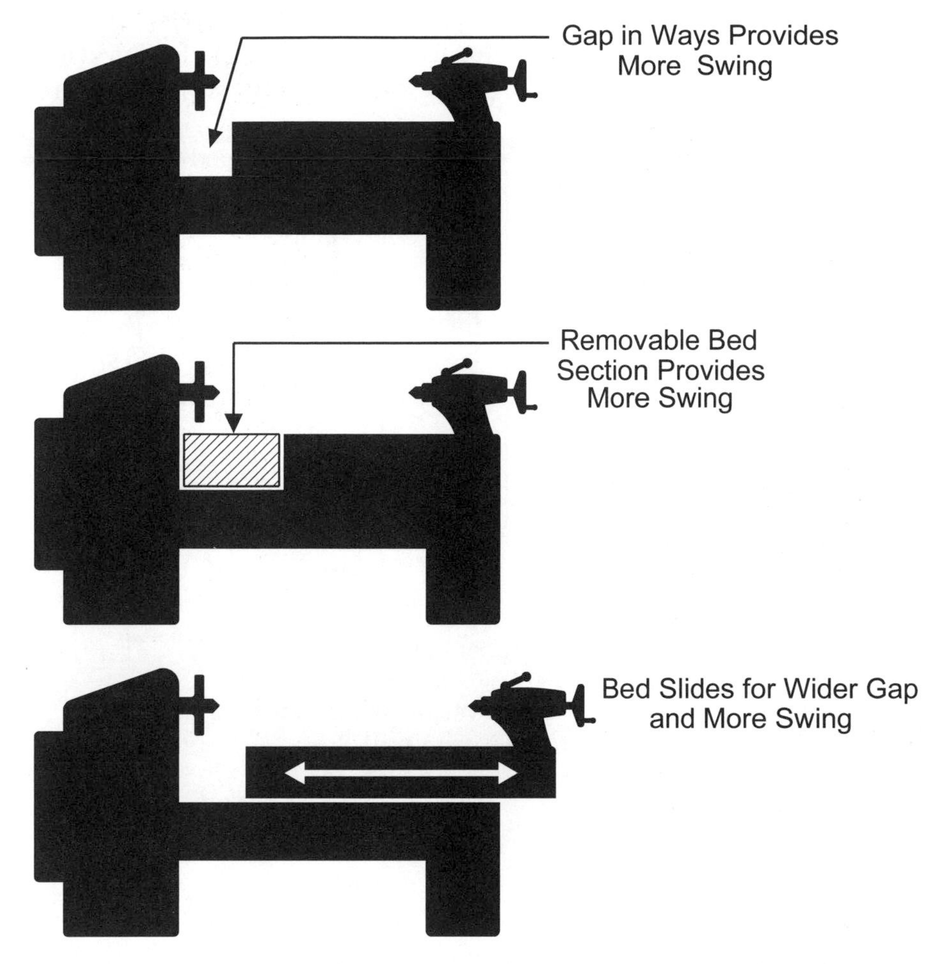

Figure 7–9. Lathe bed designs provide additional swing by increasing the gap between the headstock and the ways.

Headstock

What does the *headstock* do?

The headstock has several functions:

- Supports and aligns the spindle and its bearings so the axis of the headstock remains coaxial with the tailstock and parallel to the ways. Like the bed, the structure of the headstock resists cutting tool forces acting to force it out of alignment. In most lathe designs, it is permanently and rigidly connected to the lathe bed or part of the same casting.
- Contains and supports belts, pulleys and gearing that couple the lathe motor to its spindle and provides a range of spindle speeds.
- Sometimes the headstock also provides support for the lathe motor.

What is the function of the *spindle* and how is it constructed?

The spindle is turned by the lathe motor using a belt and pulley or gears on a gear-head lathe. It holds a face plate, center, chuck or collet, Figure 7–10.

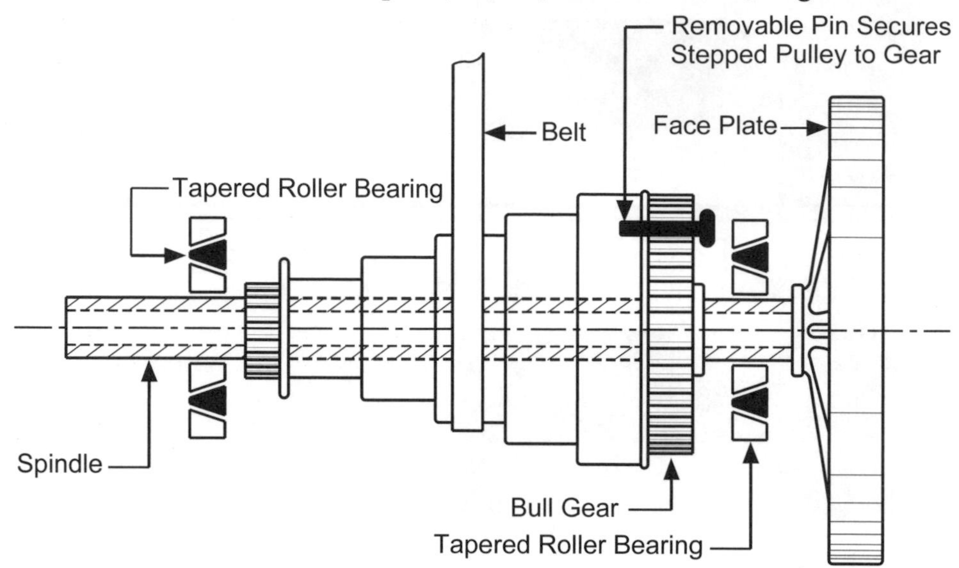

Figure 7–10. Spindle cross section showing bearings, drive belt, stepped pulley and bull gear.

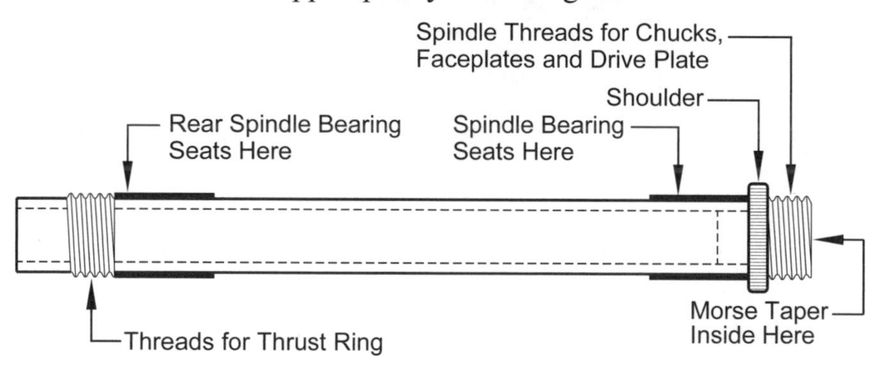

Figure 7–11. Threaded lathe spindle construction.

These attachments hold the work itself. Made of hardened, heat-treated steel, the spindle is usually in the shape of a hollow cylinder. A hole through its entire length permits the headstock to hold a long piece of rod or bar stock internally while its exposed end is machined. Ball or tapered bearings in the headstock usually support the spindle. Spindle surfaces are ground for smoothness and concentricity. Although Figure 7–11 shows a threaded spindle nose, there are other nose designs shown in Figure 7–12. In older lathes, the spindle surface itself forms a bearing against bronze or babbit bearings.

How does the spindle support the actual work?

The spindle holds centers, face plates, collets or chucks and they hold the work. The spindle does not support the work directly.

- Internally–a taper *inside* the spindle nose holds the Morse taper lathe centers and collets. Many small lathes use a MT in their spindles, but many medium-sized lathes use a 5C taper design.
- Externally–the spindle end has one of four designs to hold face plates and chucks:
    - The *threaded-spindle nose* is the least expensive design, Figure 7–12 (A). The face plate or chuck screws onto the spindle nose threads. Because these threads wear with use, the TIR will increase over time. This design is self-tightening when the lathe is run forward, but the chuck will unscrew if the spindle is reversed. This is acceptable and common on smaller lathes, but not suitable for larger production lathes where high-speed reverse capability is needed. The two most common spindle nose threads are 1½-8 and 2¼-8.
    - *Type D, cam-lock spindle nose*, is less prone to wear than a threaded spindle, but more expensive, Figure 7–12 (B). A short taper on the nose aligns the work-holding attachment and a series of studs on the back of the attachment fit into holes in the spindle nose. These studs are locked into place by six cams located radially around the spindle nose. This is the most common spindle connection used on modern, quality lathes and allows high-speed spindle reversal.
    - *Type L, long-taper key spindle nose*, Figure 7–12 (C), is a rugged design, accurate and less prone to wear than the threaded spindle. Like the cam-lock spindle nose, it permits running the spindle in reverse. This design is found on quality older South Bend and Clausing lathcs.
    - *Type A, American Standard lathe spindle nose,* is another excellent design, Figure 7–12 (D). This design is found on large older lathes.

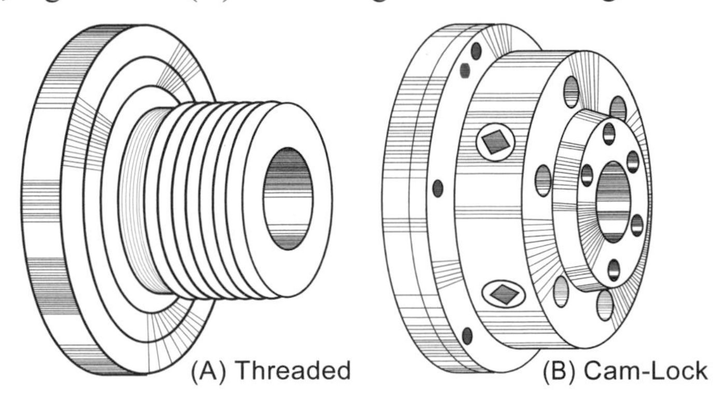

Figure 7–12. Common lathe spindle nose designs.

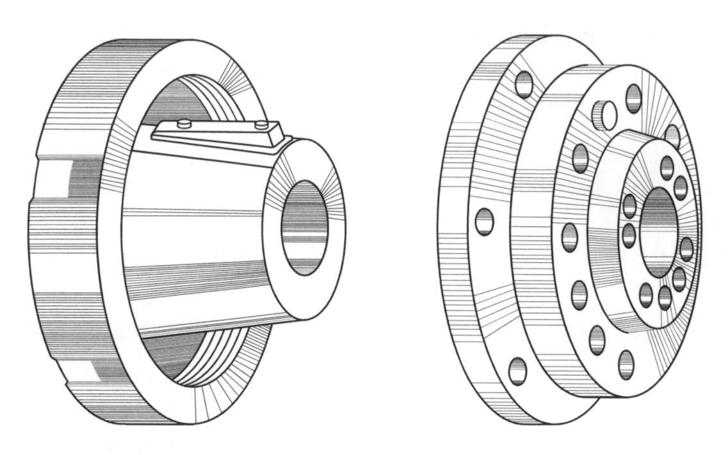

(C) Long-Taper Key (D) American Standard Short Taper

Figure 7–12. Common lathe spindle nose designs (continued).

How are chucks attached to the spindle?
On threaded-spindle lathes, the chuck has a mating female thread that screws onto the spindle. Some chucks for threaded spindle-lathes have threads cut directly into the back of the chuck. This is common on cheaper chucks and necessary on smaller chucks like those for Sherline lathes. Other chucks use a threaded adapter plate that is fastened to the chuck with several screws. Use of an adapter plate allows 3-jaw chucks to be set (or reset) to run true. The downside of threaded spindle designs is that they allow chucks to unscrew if the lathe runs backward and they lose accuracy over time due to wear.

On all other spindle designs, there is an adapter fitted to the back of the chuck that mates with the spindle nose.

Tailstock

How is the *tailstock* made and what are its functions?
The tailstock is usually a casting which slides along the ways. A locking mechanism called the *clamp bolt* or *binding lever* secures it to the ways and prevents it from moving, Figure 7–13.

The tailstock:

- Holds a lathe center in its ram for turning or facing. The *ram* (also called the *tailstock spindle* or *barrel*) is driven in or out of the tailstock casting by a screw thread and handwheel, and locked into position, Figure 7–13. This forces the center against the work and keeps it there.
- Holds a drill or reamer in a chuck or directly in its internal Morse taper. The screw mechanism of the ram forces the tool into the rotating work.

- Although the tailstock barrel is usually concentric with the headstock, most tailstocks can be moved out of this alignment to cut tapers.

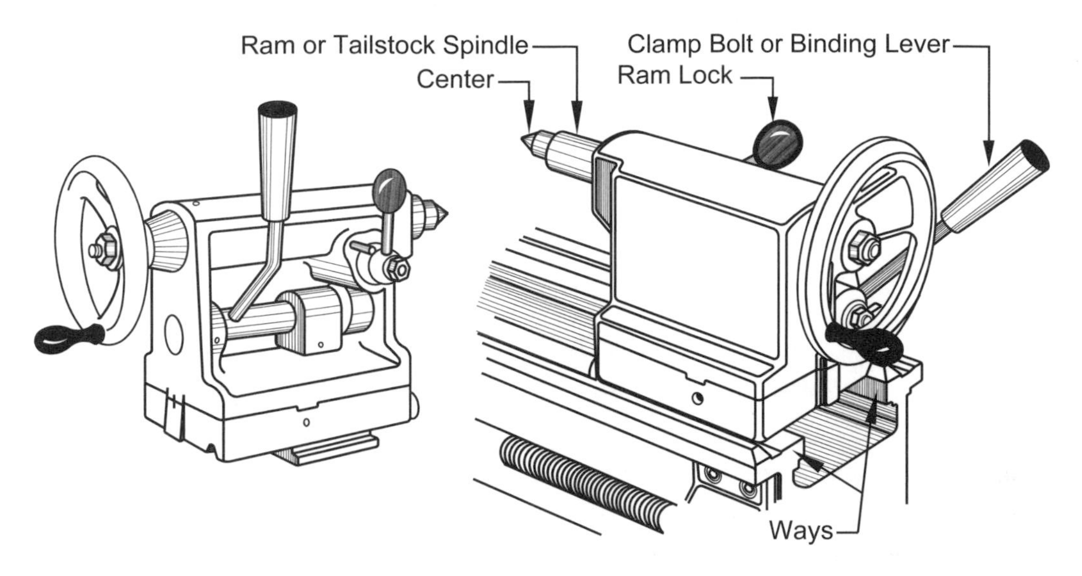

Figure 7–13. Tailstock holding lathe center: view of interior from back side of lathe (left) and front view sitting on the ways (right).

Carriage

What are the *carriage components* and what do they do?

The carriage has four components:

- *Saddle* is the H-shaped casting that rests on the ways. It forms the base of the carriage and supports both the cross slide and the apron. The underside of the saddle faces slide along the ways.
- *Apron,* which is the flat, vertical, rectangular plate on the operator's side of the saddle. Inside it holds drive mechanisms to move the carriage along the ways using manual or power feed. On lathes with cross feed—the ability to drive the cross slide at right angles to the ways—an additional mechanism inside the apron uses the leadscrew to drive the cross slide. The half-nuts mechanism inside the apron locks the carriage to the leadscrew for thread cutting. Controls for power feeds and threading are located on the apron face. The carriage locking screw binds the saddle to the ways. This insures the calibrations on the compound rest are accurate by preventing saddle movement.
- *Cross slide* or *top slide* is a casting on top of the saddle which holds the compound rest. The cross slide moves at right angles to the ways either manually by turning the cross feed handwheel or by using the power cross feed.

- *Compound slide,* which holds the tool holder. On most lathes the compound slide may be swiveled through 360° and locked in any position. This permits the tool to move across the work at any angle by turning the tool post slide handwheel. The compound slide has degree calibrations to simplify setting its angle. The compound slide does not have a power feed. See Figures 7–7 and 7–14.

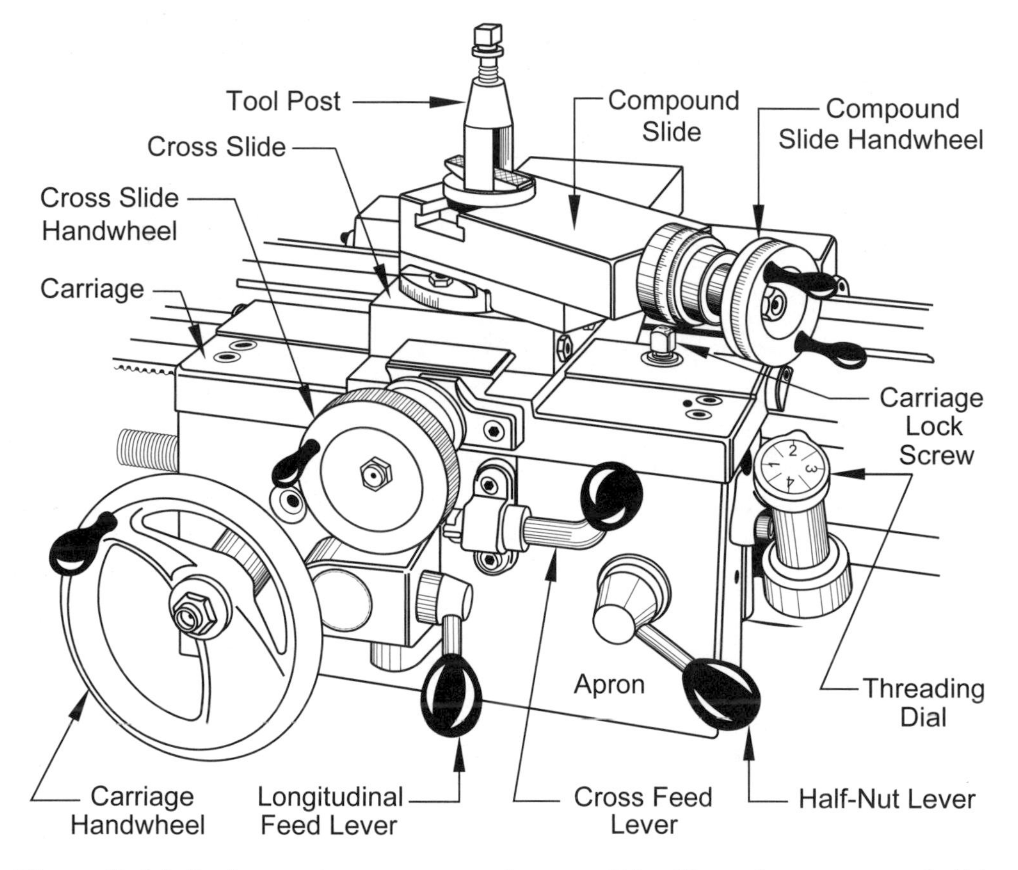

Figure 7–14. Lathe carriage, apron and cross slide. Keep the compound slide angle at 29–30° for normal turning so the compound and cross slide handwheels do not interfere with each other.

Graduated Collars

What are *graduated collars*, where are they found on lathes, and what is their function?

Graduated collars are cylinders that rotate with the handwheels for the cross slide and compound slide, Figure 7–15. Some lathes also have them on the tailstock ram handwheel. Divisions or graduations around the circumference of the cylinder show how much movement turning the handwheel causes. Graduations are in 0.001- or 0.005-inch increments for inch unit machines and

0.02-mm increments for metric machines. They permit the lathe operator to set the cutting tool accurately to remove a precise amount of material from the workpiece.

Figure 7–15. Graduated collars for a typical medium-sized lathe (left) and for a Sherline lathe (right).

There is a *locking screw* on each of the graduated collars in Figure 7–15. What is its purpose?

This screw permits the collar to turn freely on the handwheel shaft so it can be set to zero and then re-clamped to the shaft. Resetting the collar to zero eliminates the need of doing arithmetic when the collar is *not* set to zero to determine the final collar setting. This reduces the chances of error. Some collars have a friction mechanism instead of a locking screw. For these, hold the handwheel so it does not turn and then set the collar to zero.

The graduated collar indicates that there is 0.003 to 0.005 inches of backlash, or play, in the feed screw. Does this affect the accuracy of the calibration?

All screw feed mechanisms have *some* backlash, but there is an easy way to prevent it from affecting accuracy. If the operator backs the tool *away* from the work about one-quarter turn of the handwheel, then gently moves the tool *toward* the work until resistance is felt, backlash is eliminated. Then the proper increments on the collar are set. This action becomes a habit and is not a problem.

What can be done to provide finer tool settings with graduated collars?

Setting the cross slide at 30° to the work moves the tool one-half unit along the work for each one unit of infeed, Figure 7–16 (top). An even finer setting is obtained by setting the cross slide at 84.26° to the work which moves the tool one-tenth of a unit laterally for one unit of infeed, resulting in a 10× increase in sensitivity, Figure 7–16 (bottom).

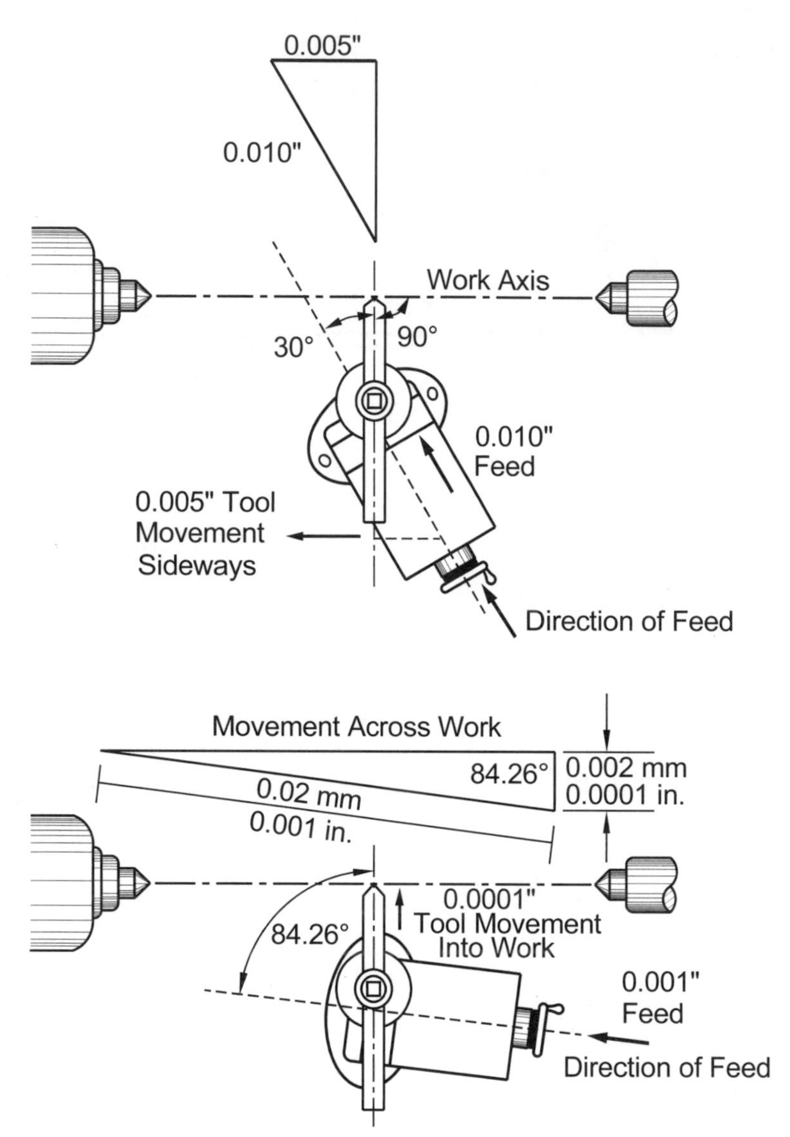

Figure 7–16. Using the compound and graduated collar for finer tool settings.

What is a lathe *gib*?

A gib is a wedge-shaped strip of metal or other material which can be adjusted to maintain a proper fit between movable surfaces of a machine tool. Properly adjusted gibs allow the two machined and dovetailed surfaces to slide by each other, but prevents them from wiggling with respect to one another.

Gibs allow slides to fit snugly, as if they were perfectly machined. They provide adjustments for wear to maintain this fit. Figure 7–17 shows how the prism-shaped gib removes play from the dovetail. A screw mechanism moves the gib until all slack is removed from the slide, then locks the gib in position.

Gibs are located on the side of the slide *away* from the force usually imposed by cutting action, so the metal-to-metal side of the dovetails see this load. There are other gib designs, but they all have the same function, to adjust dovetail tightness.

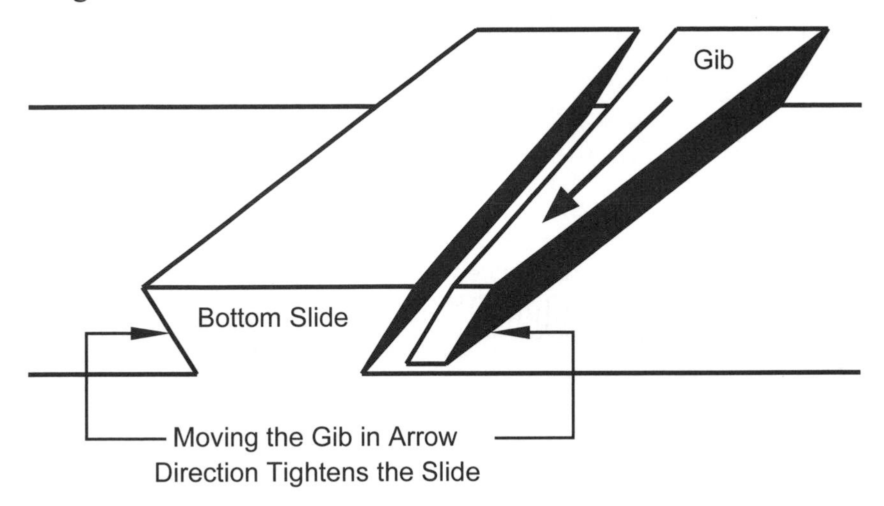

Figure 7–17. How a gib reduces play in a dovetail joint; the spacing between the upper and lower dovetail is exaggerated for clarity.

Leadscrew & Its Drive

What is the purpose of the *leadscrew*?

The leadscrew serves several purposes. It moves the:

- Carriage along the ways under power instead of by hand to make smooth, long cuts.
- Cross slide under lathe power to make cuts at right angles to the ways. The capability of the leadscrew to drive the carriage parallel to or across the bed is called a *power feed* and the different rates of movement are called *feeds*.
- Carriage in a precise ratio with spindle revolution to cut threads, Figure 7–18. Note that if the gears connecting the leadscrew and spindle have the *same* number teeth, the lathe will cut the *same* number of threads per inch on the work regardless of the work's diameter. The threads/inch cut on the work varies directly as the ratio of teeth on the leadscrew gear to those on the spindle. More teeth on the leadscrew gear put more threads/inch on the work, fewer teeth put fewer threads/inch on the work. The half-nuts on the carriage apron lock the carriage to the leadscrew. Opening up the half-nuts on the leadscrew disconnects the carriage from the leadscrew.

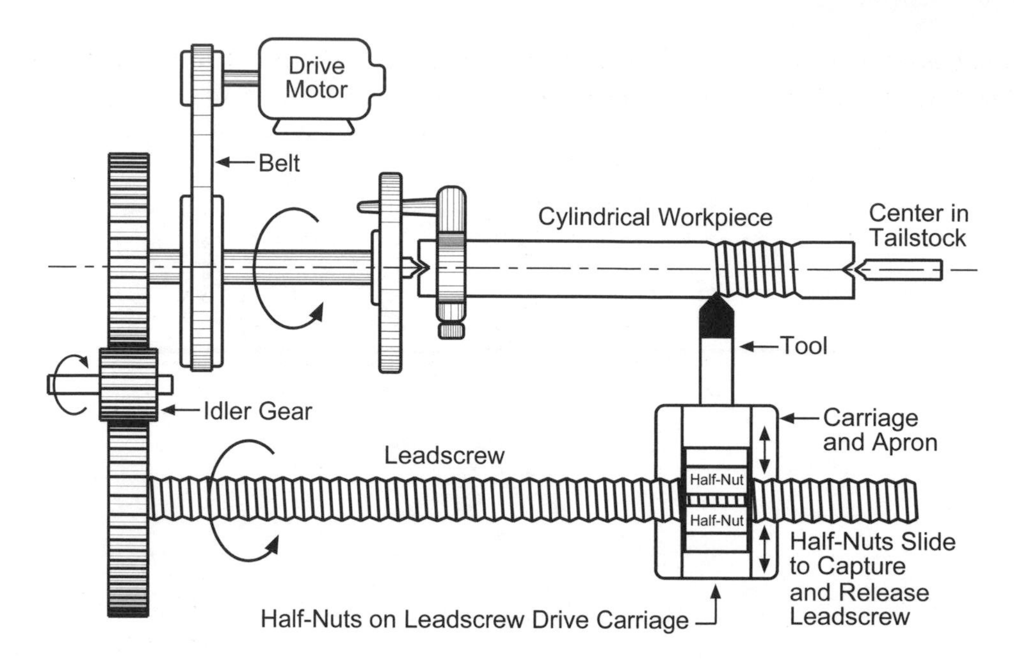

Figure 7–18. How a leadscrew drives the carriage and cuts threads.

What is the difference between using the leadscrew for power feeds and for thread cutting?

A power feed moves the tool along the work so slowly it makes overlapping cuts in the shape of a tight helix. This smooth, slow tool movement produces a fine surface finish. Just the opposite is true when using the leadscrew to cut threads: the leadscrew moves the tool so rapidly along the work that an open helix or threads result.

Where is the *quick-change gearbox* located and what does it do?

The quick-change gearbox sits on the left front side of the lathe below the headstock. Changing the position of one or more levers on the gearbox engages different gears inside it and provides a convenient way to select various speeds for auto feeds or single-point thread cutting. Older lathes do not have quick-change gearboxes so individual gears, sometimes called *change wheels*, must be selected and installed each time a different thread pitch is needed. See Figure 7–7 for the location of the quick-change gearbox.

When the quick-change gearbox is properly set up and the leadscrew fails to turn, what is wrong?

This lathe has a device, a kind of mechanical fuse, to protect the leadscrew and its gear train from overloads. If the leadscrew fails to turn when it has been engaged, this overload protection device may have been actuated. Many lathes use a shear pin, Figure 7–19, which must be replaced when it fails because of overload. Other lathes use a spring-and-ball clutch, which pops out during overload conditions. Check the lathe instruction manual.

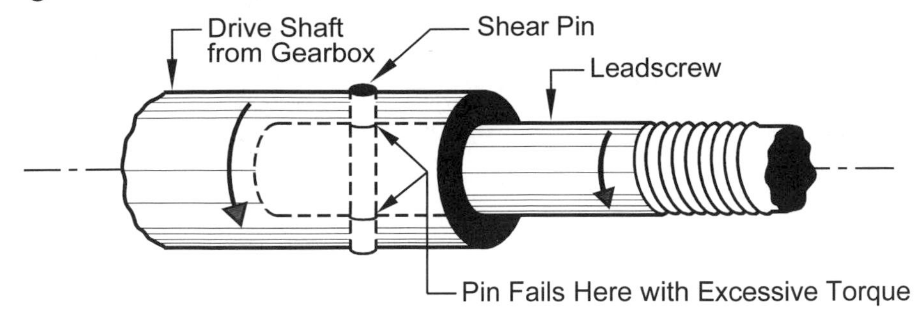

Figure 7–19. Shear pin for overload protection.

Motor and Motor Drive

What type motors are used to drive lathes?

Small lathes, such as jewelers' lathes and Sherline lathes, use DC motors with a variable-speed SCR drive. Medium-sized lathes, such as most Atlas, South Bend and other bench-top lathes, use single-phase induction motors. Large lathes, such as the Clausing Colchester machines, use three-phase induction motors. As costs have declined, variable-speed motor drives, also called variable-frequency drives (VFDs) and inverter drives, have been introduced for many lathes providing infinitely-variable speed control. These drives usually control three-phase motors, but some can control smaller single-phase motors. VFDs are small; a 5-horsepower unit is the size of a shoebox.

Countershaft

Where is the lathe *countershaft* located and what is its purpose?

The countershaft is located either behind the headstock on the back side of the lathe, or, if the lathe has a motor mounted inside the pedestal stand, underneath the headstock. The countershaft reduces the motor's rpm to one the lathe can use. Motors running on 60 Hz power turn at about 1750 rpm, much too fast for many lathe operations. The countershaft holds two intermediate pulleys and belts. Older machines use flat leather belts, while newer ones use either V-belts or toothed belts. These belts and pulleys reduce motor speed so the spindle turns at about 50 rpm, Figure 7–20. Belts are used between the motor and countershaft instead of gears because gears running at motor shaft speed would be very noisy. Not all lathe designs use countershafts. Many small lathes, such as Sherline and Levin with SCR speed controls, are belt-driven directly from their motors.

Backgear

What is the *backgear* for and how does it work?

The backgear is a clever arrangement of gears, which in addition to the reduction provided by the countershaft, provides low spindle speeds—25 to 50 rpm—and high torque. This enables the lathe to turn large diameter work without stalling. Low spindle speeds are also useful when cutting threads which run up to a shoulder because a slow spindle speed gives the operator time to stop the carriage before running the tool into the shoulder.

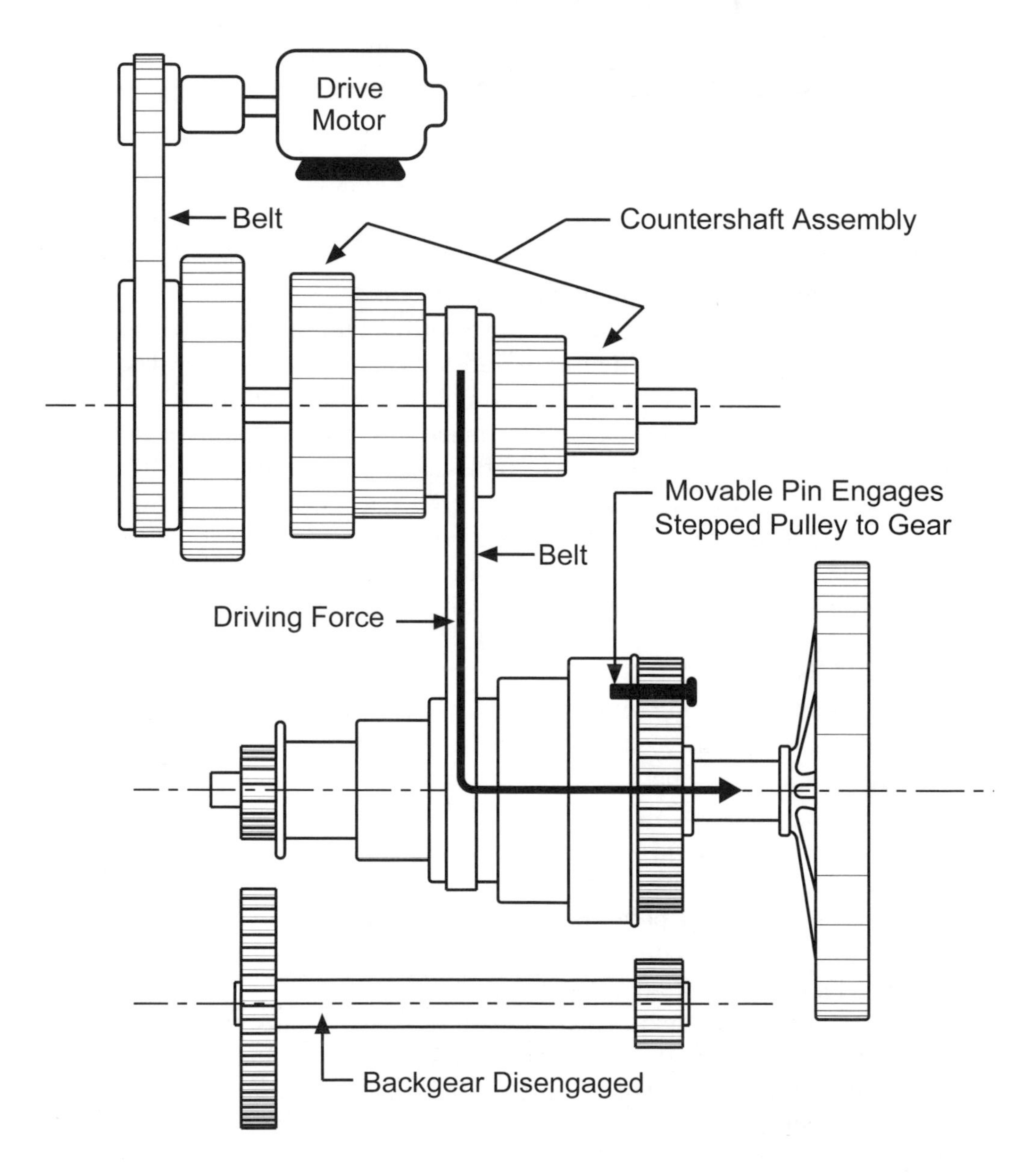

Figure 7–20. Countershaft and backgear: backgear disengaged.

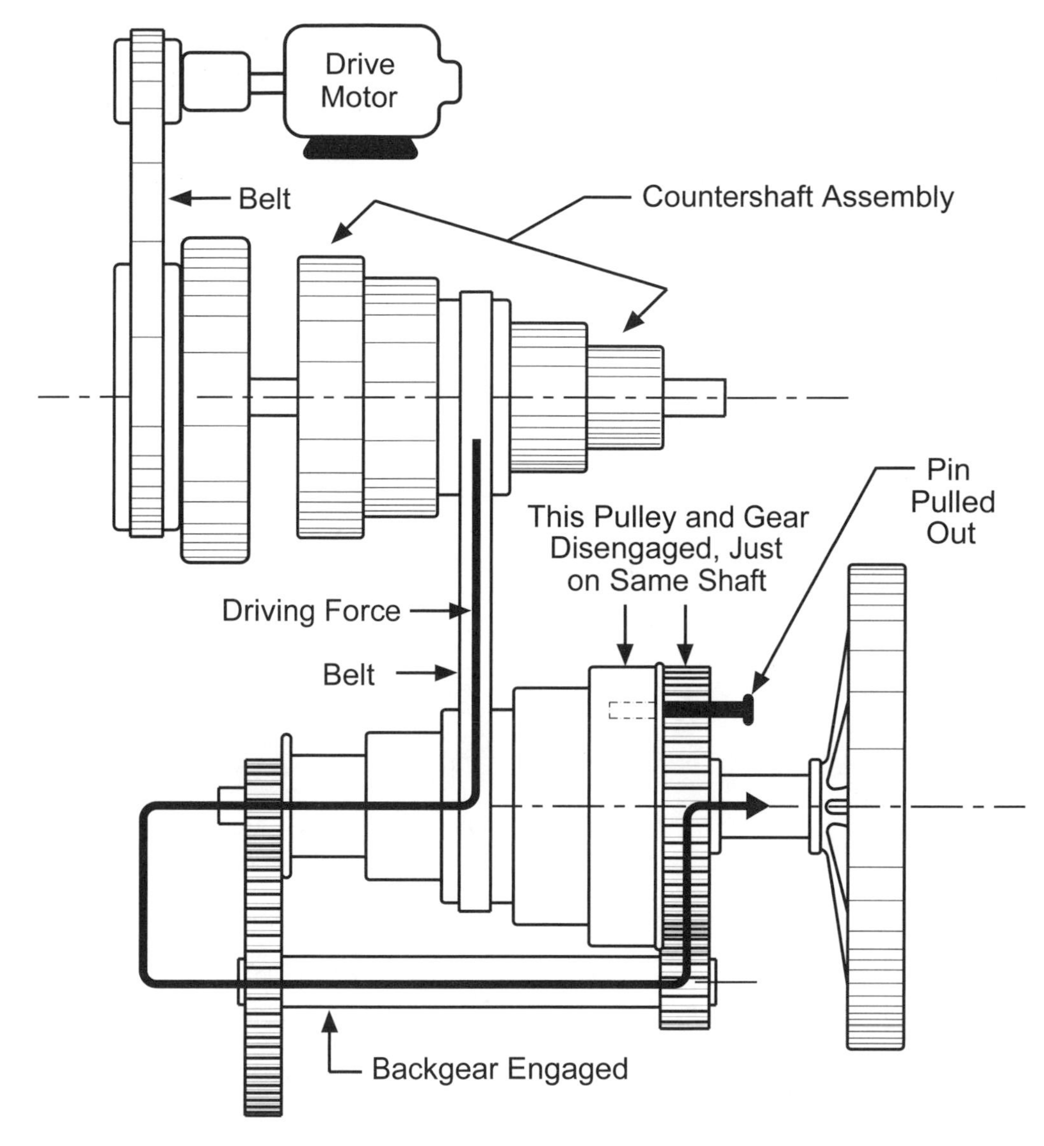

Figure 7–20. Countershaft and backgear: backgear engaged (continued).

Digital Readouts

What are *digital readouts (DROs)* and why are they useful?

DROs are electronic digital displays, usually mounted above and behind the lathe where they are easy to see, Figure 7–2. Regardless of the size lathe, DROs look much alike. They tell the lathe operator exactly where his tool is relative to the work. Once the DRO is zeroed with respect to the work, the operator can dispense with using a caliper or a micrometer to locate his position on the work. This is a great time saver. Some DROs can also display spindle speed.

Section III – Cutting Tools

Tool Posts

What are the common *tool post designs* used to hold lathe tools on the compound slide and what are their merits?

There are four common tool designs:

- *Traditional Rocker Tool Post,* Figure 7–21, holds a single cutting tool. Tool height is adjusted by rocking its curved base wedge in a concave hemispherical base and locking the tool in place with its top screw. The disadvantage of this post design is the need to readjust tool height each time the tool is changed. When doing work requiring multiple tool changes, a lot of time is lost making height checks and adjustments. This type of post was usually furnished with the lathe by its manufacturer. For extra rigidity, some machinists used this tool post without the bottom rocker base.

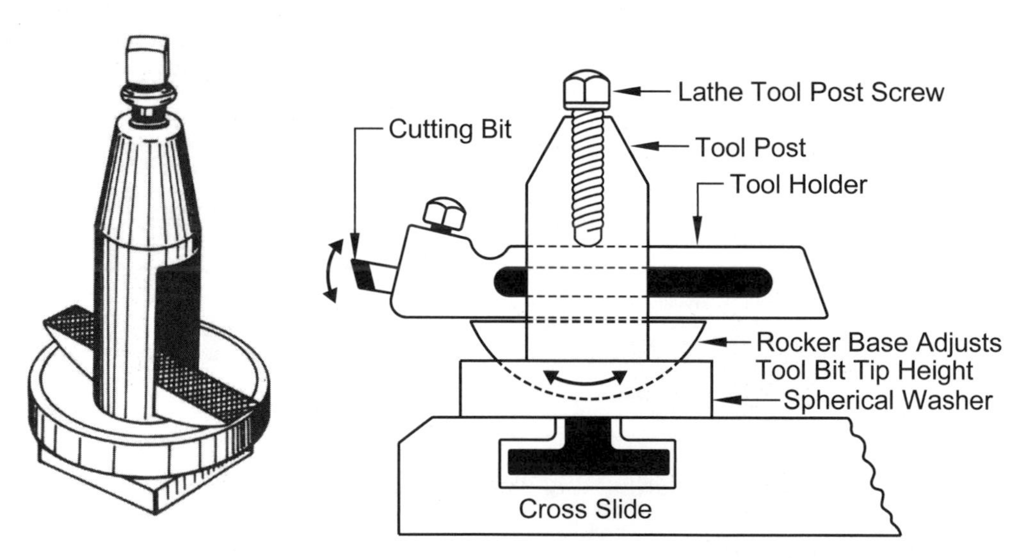

Figure 7–21. Traditional rocker tool post: without cutting tool (left) and side view of tool holder with cutting bit (right).

- *Open-sided Heavy-duty Tool Post,* Figure 7–22, also holds a single tool. Tool height is changed by adjusting the rocker or inserting shims underneath the tool. This design is rigid and excellent for the heaviest cuts, particularly with carbide tool bits. This type of tool post requires checking and shimming the tool height with each tool change.

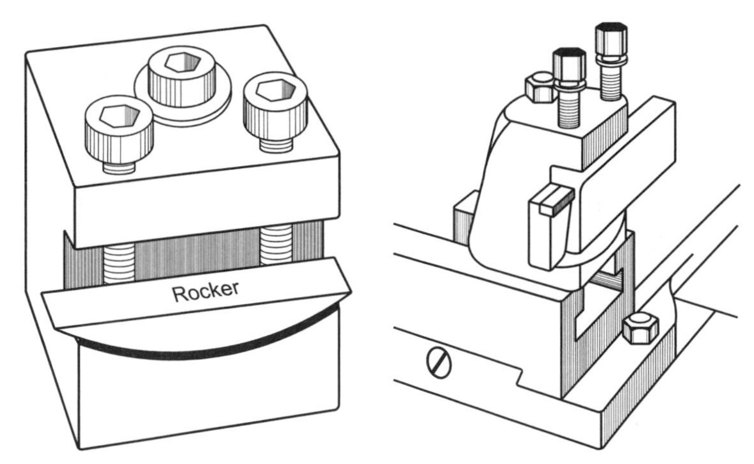

Figure 7–22. Open-sided heavy-duty tool posts: with rocker (left) and without rocker (right).

- *Four-sided Tool Post,* delivered with most lathes currently, Figure 7–23, is also called a *turret post*. It holds one tool in each of four sides. Tools are locked in place with set screws on the top of the tool holder. Any one of its four tools may be used without resetting tool height. Like the open-sided heavy-duty post, shims may be needed to set proper tool height when a tool is changed. This design is convenient for production work since it reduces tool change time. Some four-sided tool posts have a series of detents that permit repositioning of the tool within 0.001 inch.

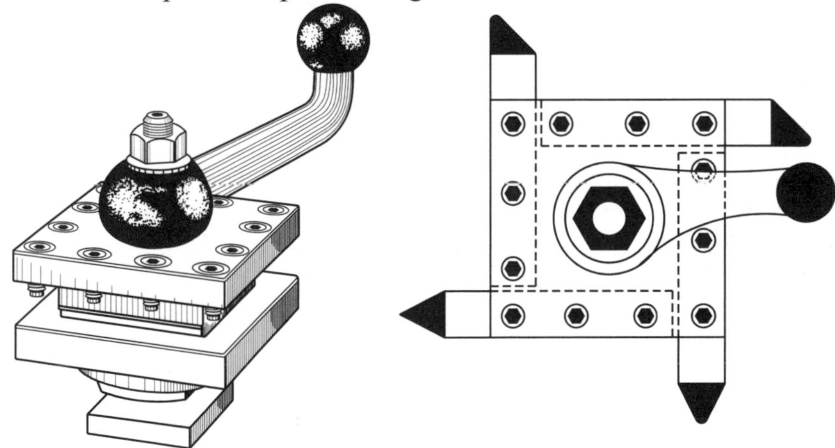

Figure 7–23. Four-sided tool post: without cutting tools (left) and top view holding four tools (right).

- *Quick-change Tool Post System*, Figure 7–24, offers the advantage of needing to set tool height just once for each tool in its holder because the tool height setting is preserved by its height-adjusting screw. These great time savers are easy to use and are common in industry.

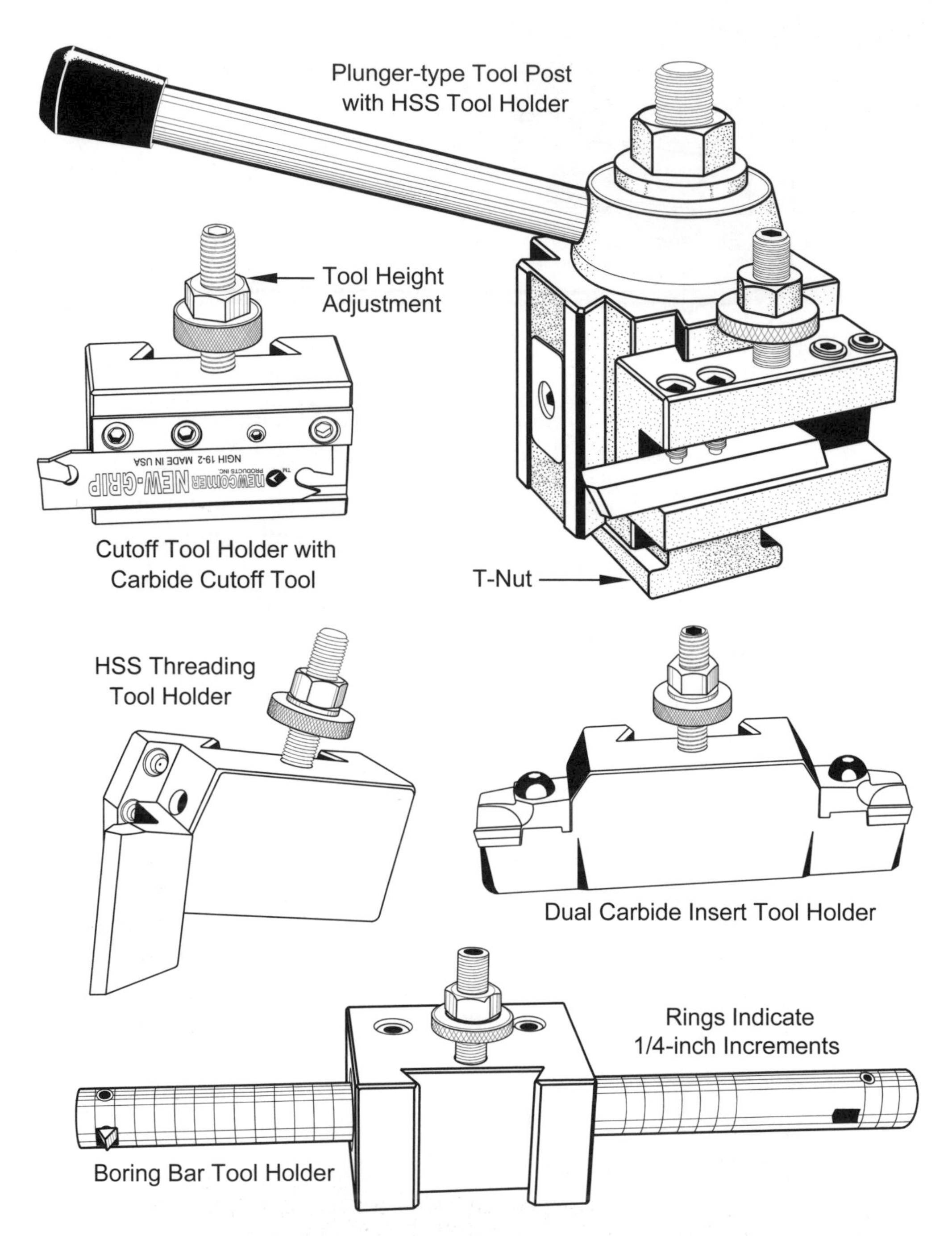

Figure 7–24. Quick-change tool post system. Aloris, Dorian, Phase II and budget imports for the same size lathes have the same dimensions and are interchangeable. There are more than a dozen different tool holders for this system.

Tool Materials

What materials are used for lathe tool bits and how are they designed?

- *Individual Steel Tool Bits*, Figure 7–25, are usually 2- to 3-inches long, square in cross section and made of HSS (high-speed steel) or Cobalt HSS. They are available in a wide range of sizes, from 1/16-inch to 1-inch square, both sharpened and unsharpened. The most common bit for medium-sized lathes is ⅜-inches square by 3-inches long. These bits are supplied hardened and no further heat treatment is needed. The advantage of these over carbide bits is that they are readily sharpened on aluminum oxide grinding wheels, easily formed into custom shapes, and provide a better surface finish on many materials. They remain in use by tool and die makers and in prototype work, but not in production operations.

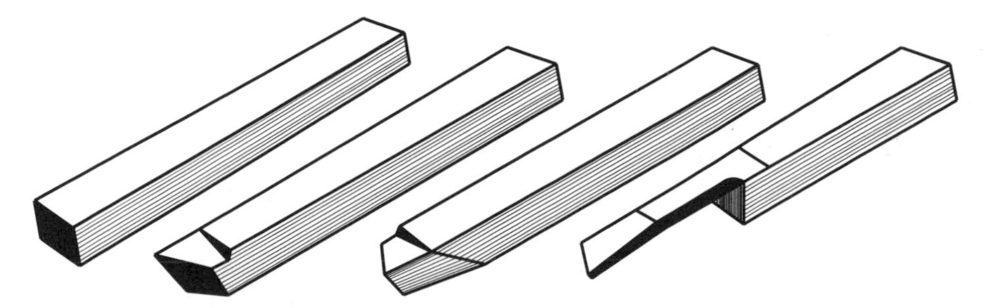

Figure 7–25. Individual HSS tool bits: unsharpened (far left) and sharpened (all others).

Smaller tool bits are held in tool bit holders that fit into the tool post, Figure 7–26. These now obsolete tool holders were introduced because HSS steel was expensive and they allowed the use of smaller tool bits. Today these tool bit materials are not expensive, and larger bits that fit directly into the tool post are most often used. However, because the cutters for these tool holders are smaller than cutters that fit directly into tool posts, they continue to be used because being smaller, they save a lot of grinding time, and sometimes they provide access to areas of a part that larger tool holders cannot reach.

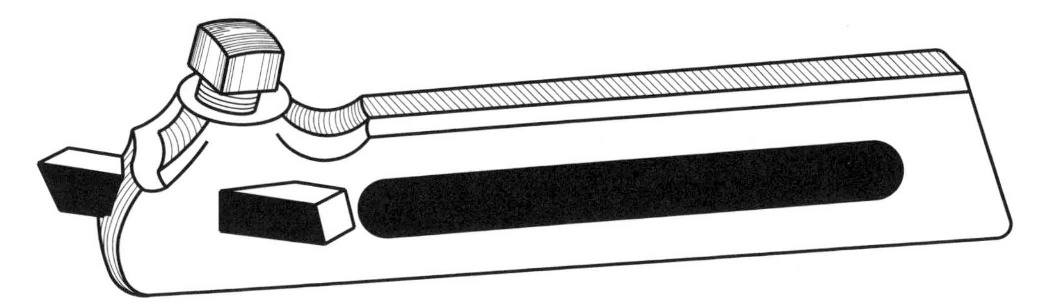

Figure 7–26. Common, but obsolete small tool bit holder.

- *Brazed Carbide Bits* consist of a small piece of tungsten carbide brazed onto the end of a square piece of carbon steel as in Figure 7–27. Only the carbide tip performs cutting. In a comparable size, a brazed carbide bit costs about twice as much as a HSS bit, but removes 5 to 10 times more metal at higher speeds. They also fit directly into a tool post like those used for conventional HSS bits and are available in similar shank dimensions as square tool steel bits. These tool bits cannot be ground and shaped to fit the particular turning or boring situation like HSS bits because they are too hard for aluminum oxide grinding wheels. They require a special "green" wheel for shaping and a diamond wheel for finishing. However, there are over 20 industry "standard styles" with different shaped carbide tips to cover most applications. Brazed carbide bits allow the use of carbide cutters without the expense of purchasing a matching tool holder that is needed for carbide inserts.

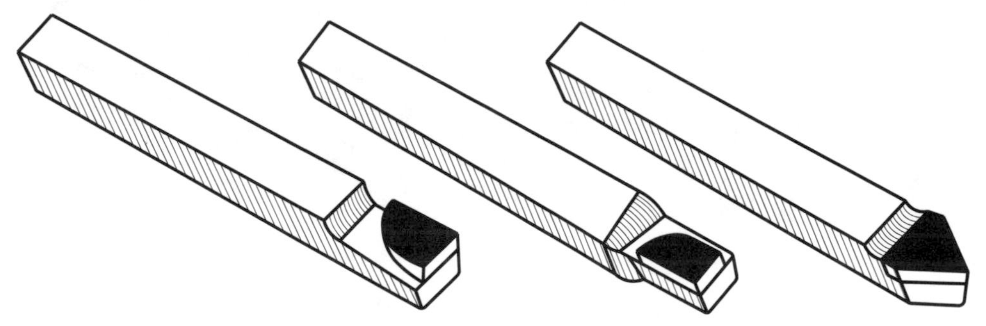

Figure 7–27. Brazed carbide bits.

- *Inserts* are individual pieces of tungsten carbide (or other materials) held in a matching insert tool holder by a small screw or clamp (or both), Figure 7–28. Carbide inserts are not usually reshaped and are used as supplied because the expensive holder would also have to be modified. Most inserts have multiple cutting edges accessed by rotating and/or flipping the insert in its holder. For example, a triangular negative-rake insert usually has six cutting edges, three on each side. Carbide insert cutting tools perform better than brazed carbide tools because the brazing process leaves residual stresses in the carbide chip causing them to crack sooner than the carbide inserts which are not brazed. The different thermal coefficient of expansions for steel and carbide cause this stress.

There are two reasons why the majority of production work is done with carbide insert tools:

- Inserts withstand much higher temperatures than HSS tool bits so cutting speeds can be 3 to 5 times faster than with HSS cutters.

- Because inserts are made to tight dimensional tolerances, their cutting edges appear in the same position relative to the tool holder and workpiece. This means the tooling does not have to be reset whenever the insert is changed, a very important issue in DRO and CNC machines. Resharpening would destroy this positional accuracy and is not usually done, although it is possible with a "green" or diamond wheel. Also, sharpening an insert reduces the clearance between the insert and its holder, often making it unusable.

Taken together, these factors lead to much lower costs in a production setting than with HSS bits.

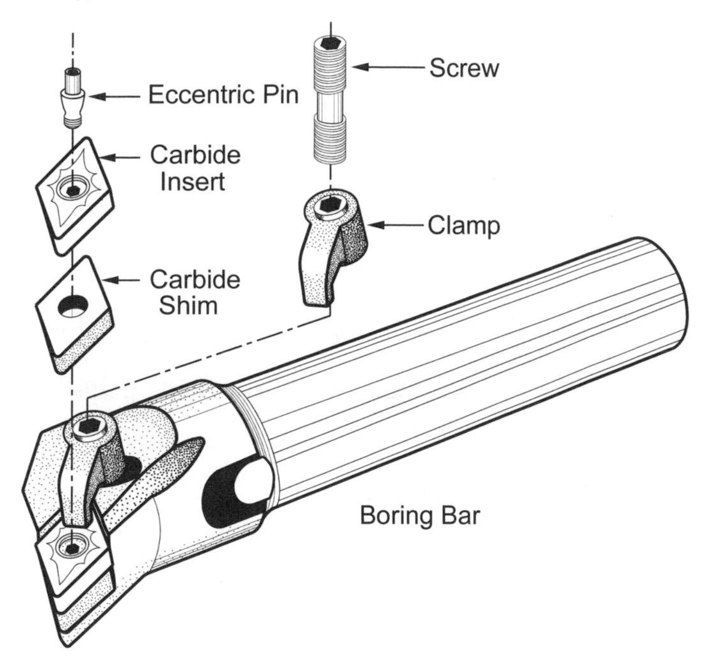

Figure 7–28. Carbide negative rake insert and its boring bar.

How are these three types of cutting bits sharpened?

Tool Bit Material	Sharpening Method
High-speed Steel (HSS)	Sharpened with an aluminum oxide grinding wheel.
Brazed Carbide	May be sharpened like tool steel except requires special "green" grinding and diamond wheels. Sharpening is not usually cost effective commercially compared with carbide inserts.
Carbide Insert	Disposable, not resharpened.

Table 7–1. Sharpening methods for three common tool bit materials.

Cutting Tool Nomenclature

What are the names given to the parts and angles of a lathe cutting tool?
See Figure 7–29.

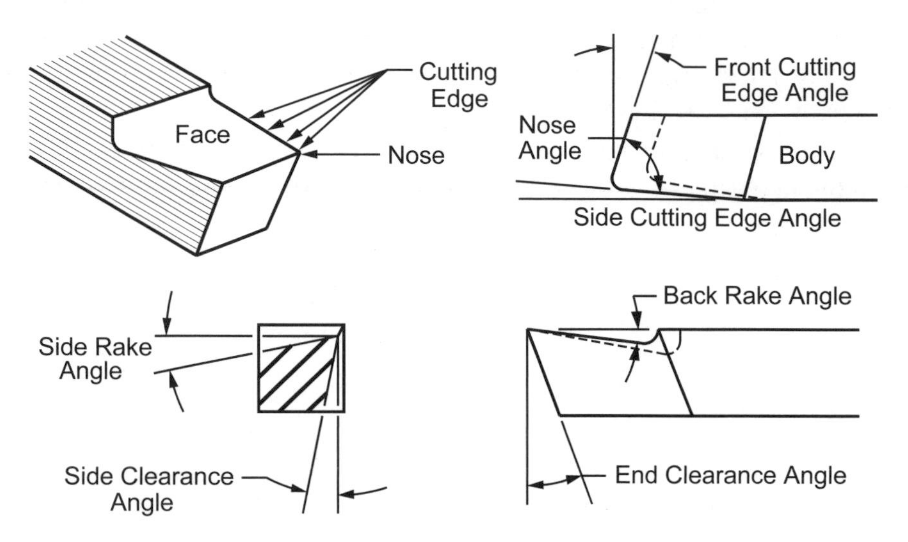

Figure 7–29. Lathe cutting tool bit nomenclature.

Where can these cutting tool angles be obtained for grinding HSS bits?
See Appendix, Table A–1.

Cutting Tool Mechanics

How do cutting tools remove metal from the workpiece?
As the tool enters the workpiece, material just ahead of the tool is deformed, sheared from the workpiece, and flows into the space above the tool in the form of a chip, Figure 7–30. The mechanical energy supplied to the lathe is converted to heat as shearing and deformation occurs and as the metal slides along the cutting tool surface. With coated carbide inserts, the tool bit also burnishes the part.

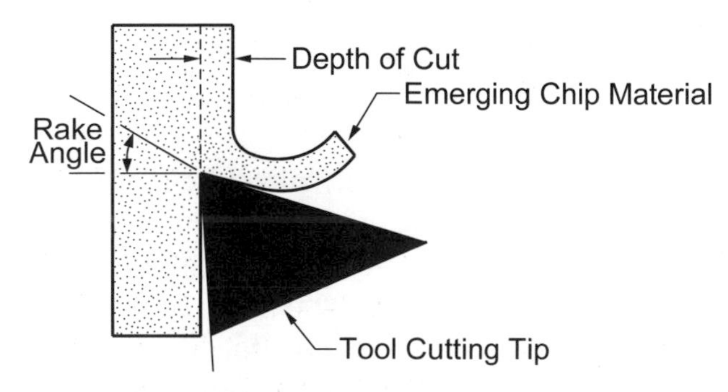

Figure 7–30. Single-point cutting tool action showing chip formation.

What factors determine how well a cutting tool works?

- The most important factor is tool shape, particularly the back rake and side rake angles, Figure 7–29. In general, side rake is more important because more cutting is done on the sides of tools than on their ends.
- The position of the cutting edge in relation to the workpiece is also very important and explained in this chapter under the section heading *Lead Angle*.
- Other factors which affect tool shape are relief or clearance angles—the taper on non-cutting tool surfaces—to prevent rubbing or dragging the tool against the work.

Positive & Negative Rake Angles

How are *positive and negative rake angles* defined?

- A positive rake angle tool has the chip moving *down* the top face of the tool bit.
- A negative rake angle tool has the chip moving *up* the top face of the tool bit.

See Figure 7–31.

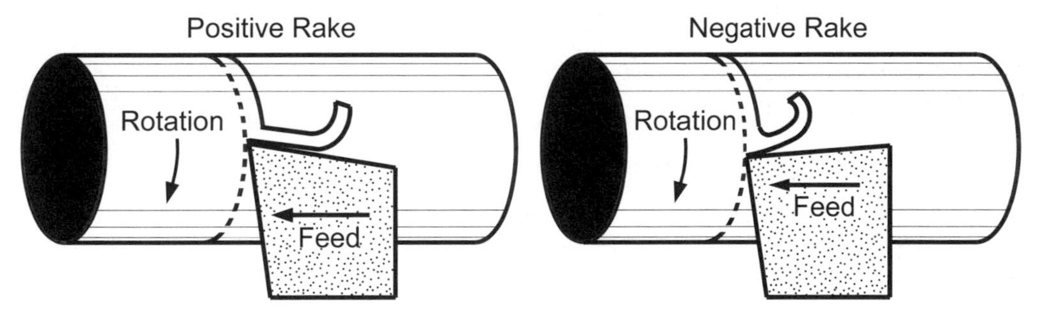

Figure 7–31. Positive and negative rake angles.

What are the advantages and drawbacks to positive and negative rake angles?

Positive rake angle tool bits, Figure 7–32 (A), are the most efficient shape for removing metal because they produce lower cutting forces and generate less heat than negative rake angle tool bits making the same cut. Use a positive rake angle tool bit when cutting forces must be minimized such as when machining thin sections, slender shafts or using a lathe with poor rigidity.

Negative rake tool bits, Figure 7–32 (B), also have several advantages:

- They have 90° side and end clearance angles, square ends, and still have adequate clearance. This results in a symmetrical insert which can be flipped over as well as end for end, yielding eight cutting edges on a square insert and two on a diamond-shaped or square one. Conversely, positive rake inserts must have a clearance taper on their sides and ends, Figure 7–32 (A), and so are not symmetrical and can only be turned end for end.

- Provide more tool surface to absorb and distribute heat, so higher cutting speeds are possible.
- Distribute the shock and force of the work hitting the tool onto the top surface of the tool, not its cutting edge as positive rake angle tool bits do. This prolongs tool life.
- Are particularly suited to making interrupted cuts and also work well on cast iron because the tough, abrasive outer scale hits the top surface of the tool, not its cutting edge.
- One drawback to using negative rake angle tool bits, besides the need for a rigid lathe structure, is that the chips leaving a negative rake bit come off as a continuous ribbon and are more likely to create a hazard for the operator than the individual chips made by a positive rake tool. However, many of these inserts have a built-in chip breaker.
- Only tool bits with carbide inserts are hard enough to work with a negative rake angle. HSS or alloy bits all must have a positive rake except for brass. See *Appendix Table A–1*.
- The positive/negative rake design, Figure 7–32 (C), offers the lower cutting force advantage of a positive rake design and the groove on the outside edge of the insert acts as a chip breaker. This chip breaker is needed because ductile materials produce a continuous, unbroken chip that will move off the tool rapidly creating a hazard to the machine operator. A negative rake tool holder is used with a positive/negative carbide insert.

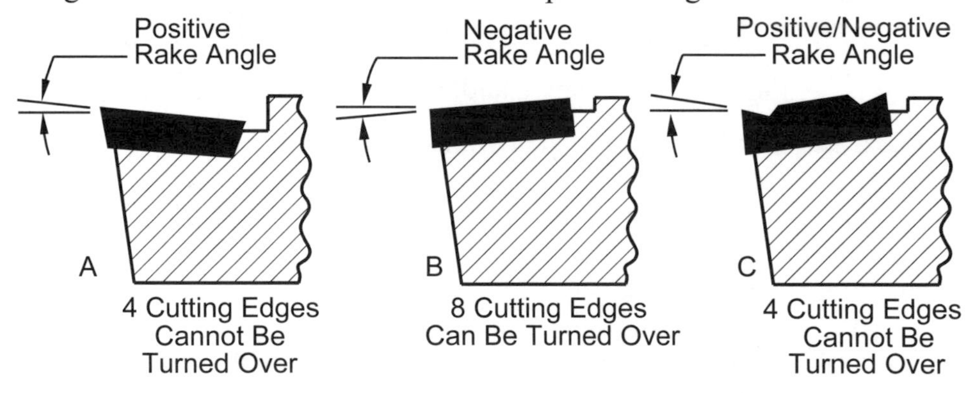

Figure 7–32. Square insert cutter rake angles.

In small lathes, a negative rake tool may not work well due to high cutting forces and lack of machine rigidity. However, in larger machines, they should be considered where increased cutting speed is important or when cutting cast iron, stainless steel or titanium.

Lead Angle

How is cutting tool *lead angle* defined?

Lead angle is the angle at which the side cutting edge of the cutting tool is positioned with respect to a plane perpendicular to the workpiece, Figure 7–33. Note that when the lead angle equals the side cutting angle, the tool axis is perpendicular to the work axis.

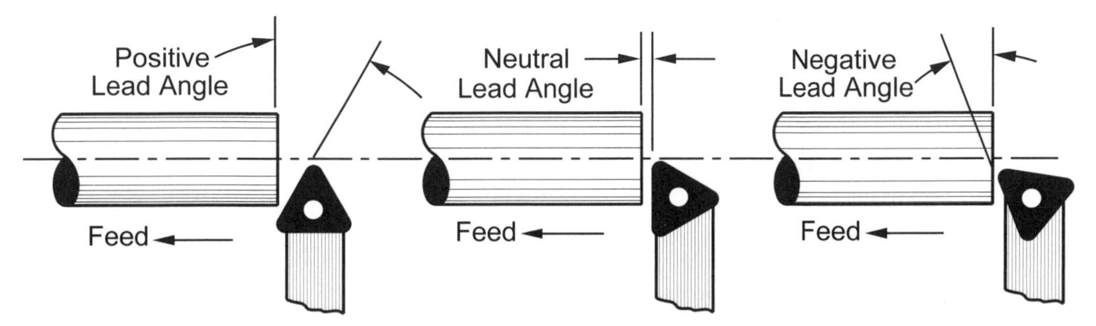

Figure 7–33. Positive, neutral and negative lead angles.

Why is it often advantageous to use tool bits with a *positive lead angle*?

A positive lead angle puts more of the cutting edge on the work. This has three effects:

- Although the volume of the material removed is about the same as with a neutral lead angle, the chip is thinner and more easily removed from the lathe, but more likely to chatter. See Figure 7–34.

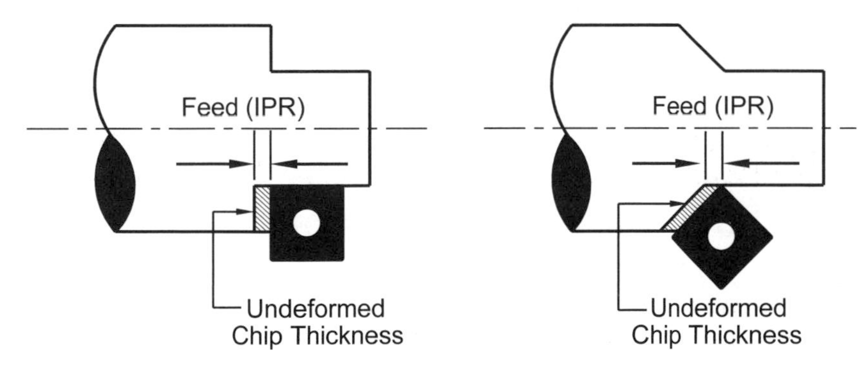

Figure 7–34. How a positive lead angle affects chip thickness.

- Cutting force is distributed over a longer length of the tool edge which means that tool life is increased and feed rate may also be increased. See Figure 7–35.

- It increases radial force and reduces longitudinal force on the workpiece which may become a problem with work that is thin or not well supported, Figure 7–35.

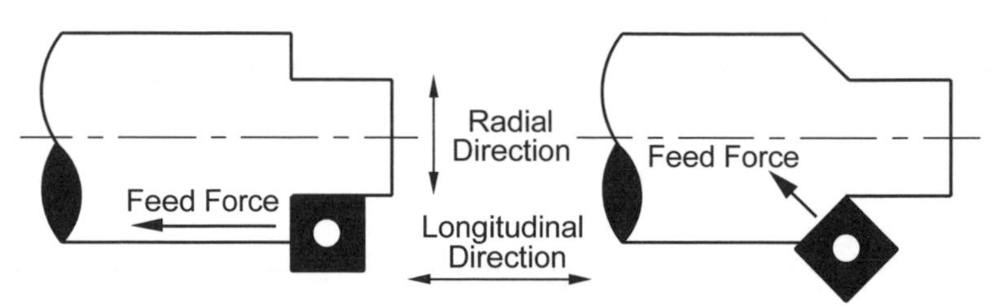

Figure 7–35. How lead angle affects longitudinal and radial cutting forces on the workpiece. The left-hand drawing shows the preferred tool setup for flexible, thin workpieces.

How does a positive lead angle protect the cutting tool bit?
A positive lead angle tool subjects the tool bit or insert to less shock when it first contacts the work because the work impacts the tool in the center of its face where it is stronger rather than on its end. See Figure 7–36.

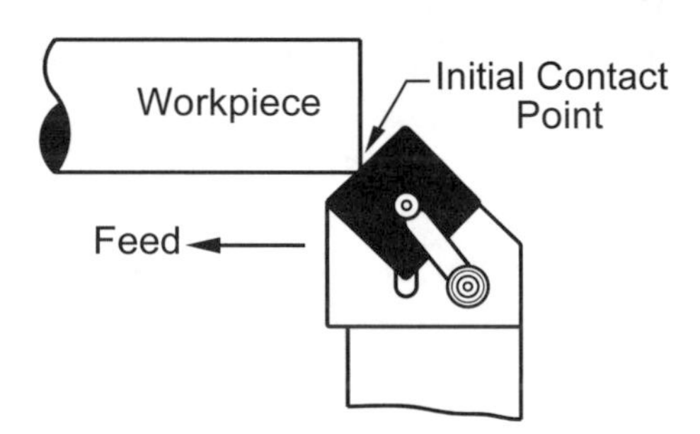

Figure 7–36. Positive lead angle protects the tool bit by contacting the work where it is strongest.

Cutting Tool Shapes & Functions

Why are different shaped cutting bits used?
The overall shape of the bit, usually as viewed from above, is determined by the kind of cutting it will do. Here are some examples of HSS bit shapes, Figure 7–37, and their functions:

- Tools labeled *left cut tools* cut only from left to right, that is, from the headstock end of the lathe toward the tailstock. *Right cut tools* cut only from right to left.
- A *round nose tool* with no back or side rake cuts in either direction. The round nose produces a smoother finish than a pointed bit.

- *Roughing tools* have small side relief angles to provide more material to support the cutting edge during deep cuts. This greater tool edge support reduces tool wear and damage during deep cuts.
- *Facing tools* have a shape to provide clearance for the center holding the work. Two tools must be used, one for each end of the workpiece.

Often the machinist will round the point on a tool after sharpening it to produce a smoother surface finish. In general, just a few tool shapes will meet most shop needs, but often a particular shape tool will be worth making in a production situation. Aluminum and brass cut better with tools shaped for them (Appendix I). Other specialized tools for cutting off, boring, forming and trepanning are covered later under their particular operations.

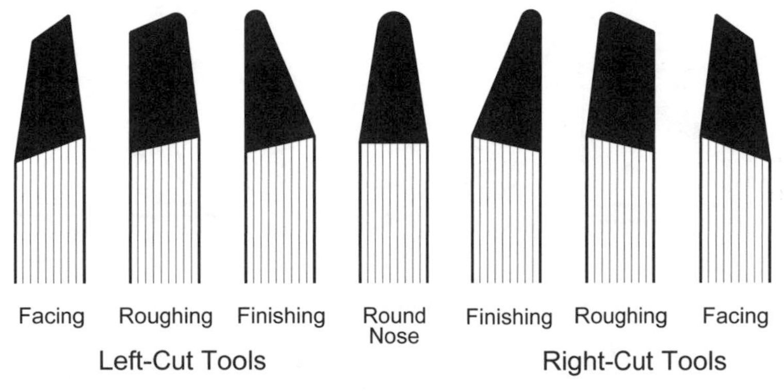

Figure 7–37. Common HSS tool bit shapes.

What selection of brazed carbide tool bits are commonly available?

Figure 7–38 shows a basic selection of commercial bits in standard industry styles. All bits shown are right handed, but are also available in their mirror images to cut from the left hand side.

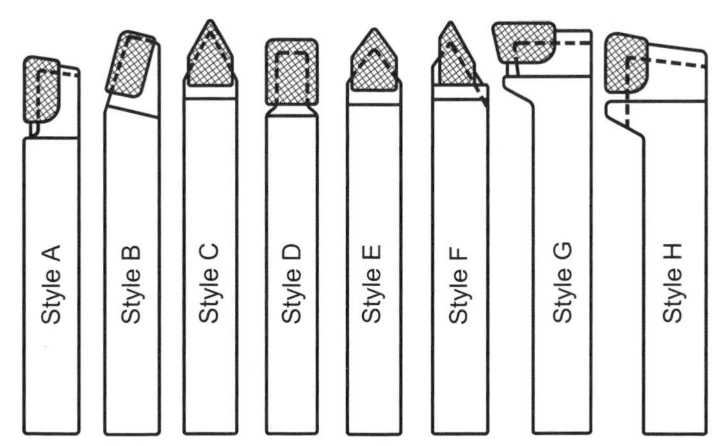

Figure 7–38. Some standard styles of brazed carbide tool bits.

The more complex issue is the shape of the cutting bit faces with respect to each other. On HSS and alloy bits the user may choose to grind or modify his own bits, but on carbide and insert bits this is not an option. Fortunately, there is a wide range of carbide tool bit shapes available. Also, there are literally hundreds of choices of inserts and insert holders. However, 95% of jobs can be done with the six shapes in Figure 7–39.

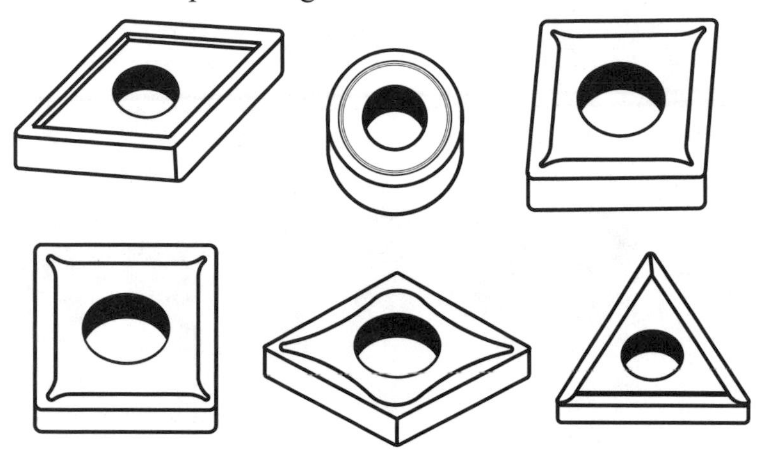

Figure 7–39. Most common insert shapes.

If just a few insert shapes are needed to do most jobs, why are there so many inserts offered in carbide cutting tool catalogs?
There are several reasons:

- In a production setting using CNC machines, which can quickly access many different tools and tool holders automatically, having just the right shaped tool can save a lot of cutting time and money.
- The number of inserts rapidly increases when the same insert size and shape is offered in different carbide grades and also with different wear-resistant coatings.
- Many inserts are for milling machine cutters, automatic screw machines, automatic profiling and coping lathes and other machine tools.

Installing Tool Bits

What details should be considered when installing tool bits on the lathe?
There are two:

- The length of the tool bit between the tool holder and the work should be minimized because longer lengths result in less tool rigidity and deflection of the tool under cutting force. This may lead to chatter and may also make taking small cuts impossible.
- The length of the compound overhang should also be minimized for the same reasons.

Best practice is to move the cross slide so both unsupported tool length and compound overhang are minimized, Figure 7–40.

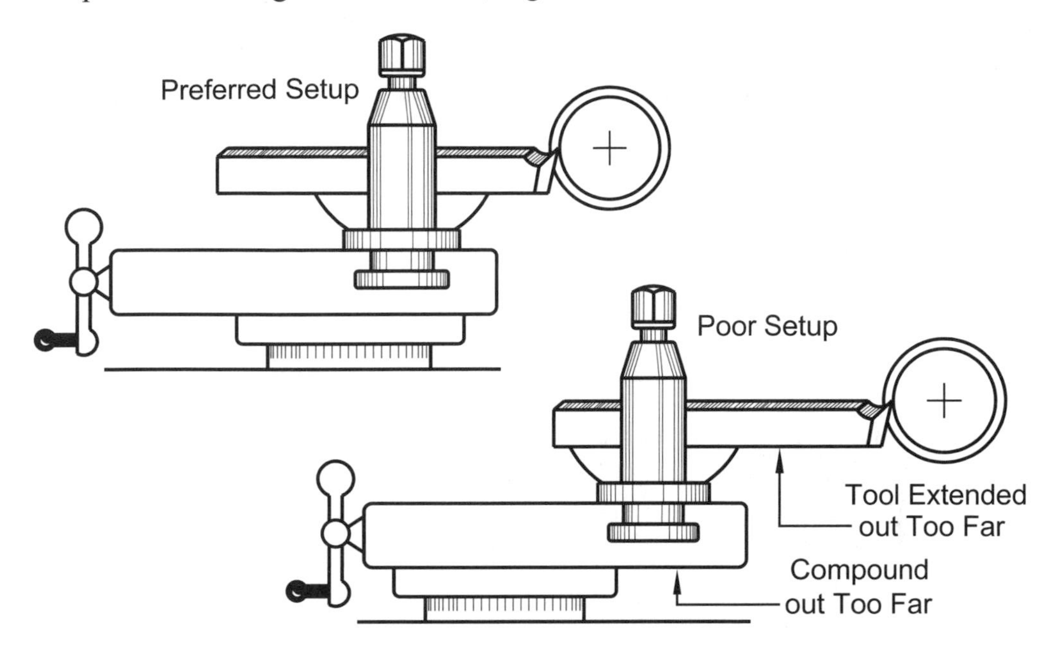

Figure 7–40. Location of tool bit overhang relative to the lathe work center affects cutting tool rigidity and the tendency to chatter.

Section IV – Lathe Accessories

Centers

What is the purpose of lathe *centers*?

Centers provide a pair of bearing points that allow work to rotate on the lathe headstock-tailstock axis, but prevent all other motion.

How are center holes made in the ends of the work?

A *center drill*, also called a *combination center drill and countersink* or *Slocombe bit*, makes these tapered holes with a bottom pocket, Figure 7–41 (A). The most common center drills and lathe centers have 60° points. This makes for an exact fit between the lathe centers and work center holes forming a nearly perfect bearing, Figure 7–41 (B). Center holes drilled too deep or too shallow will damage the center or the work, Figure 7–41 (C & D). The bottom end of the center hole contains a small pocket that must be filled with special high-temperature lubricant before inserting the lathe center to avoid overheating the center and the work. As the work turns, this lubricant makes its way into the tapered section where it is needed.

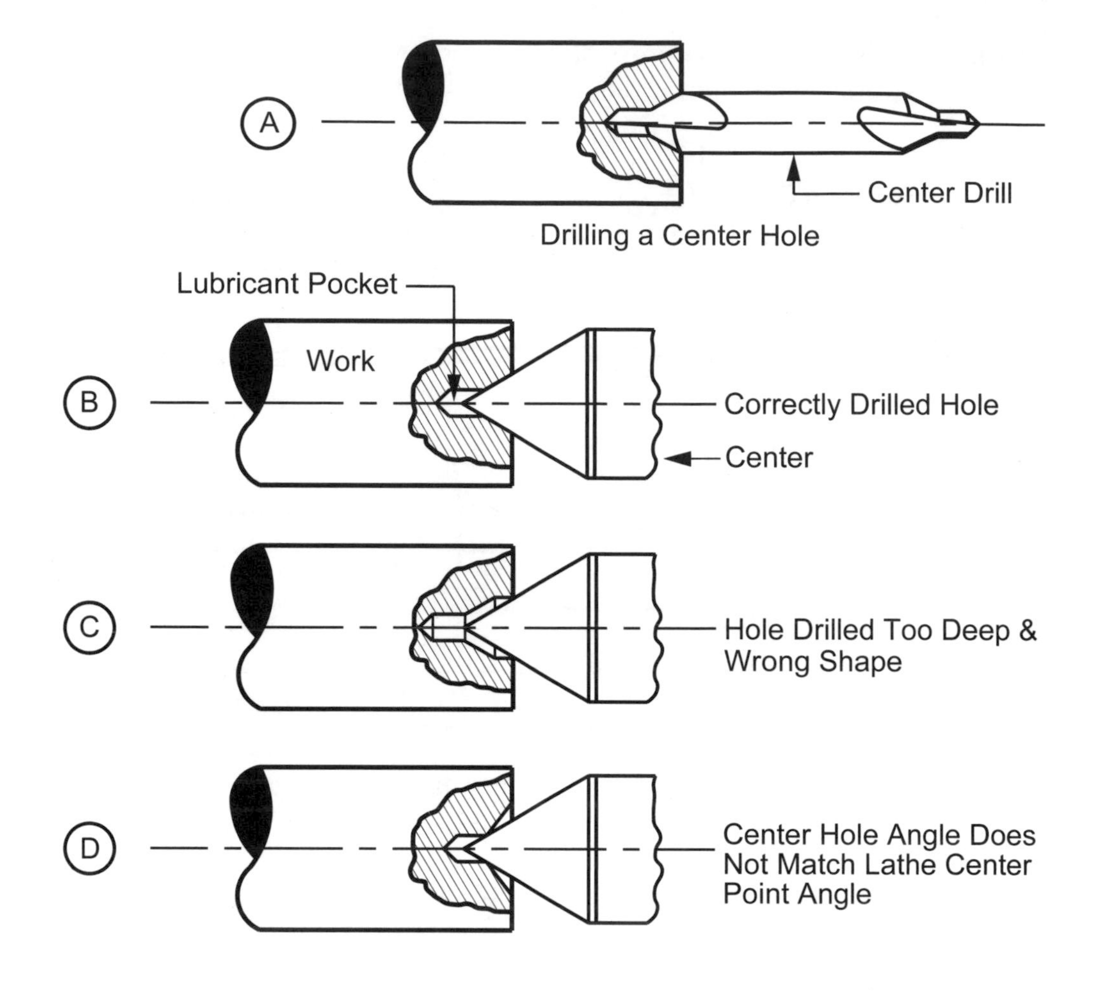

Figure 7–41. Drilling center holes.

How are center holes drilled?

Center holes are drilled by first locating the center point as described in *Chapter 1 – Measurement Tools, Layout & Job Planning.* Mark the point with a center punch, then drill the center hole with a combination drill. This is done either:

- With a drill press.
- In a lathe with the work held in a 3- or 4-jaw chuck.
- For long workpieces, in a lathe with the work held in a 3- or 4-jaw chuck and the tailstock end of the work in a steady rest.
- By holding the work in a lathe against a drill pad as in Figure 7–42.

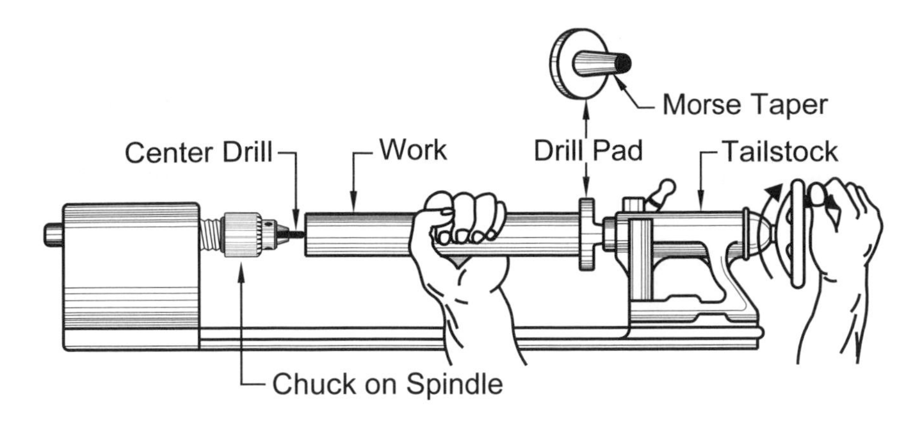

Figure 7–42. Drilling center holes in a lathe with a drill pad.

What size center drill should you select?

Use the largest center drill that will not make the work unsightly or interfere with its function. Table 7–2 lists combination center drill sizes, dimensions and suggested work diameters.

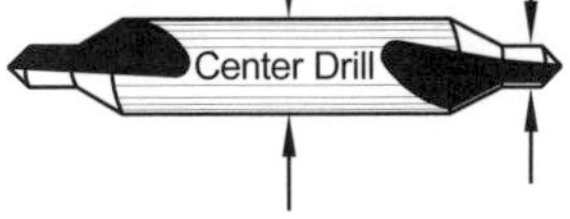

Combination Drill Size	Body Diameter (inches)	Point Diameter (inches)	Work Diameter (inches)
00000	0.125	0.010	—
0000	0.125	0.015	—
000	0.125	0.020	—
00	0.125	0.025	—
0	0.125	0.031	—
1–0	0.125	0.040	—
1	0.125	3/64	3/16–5/16
2	0.1875	5/64	3/8–1
3	0.2500	7/64	1¼–2
4	0.3125	1	2¼–4
5	0.4375	3/16	4–6

Table 7–2. Center drill sizing.

How are *centers* constructed?
Solid lathe centers are ground, hardened steel. They can be reground if their points are damaged. Some older centers, particularly those used in the headstock are not hardened, and can be repointed in the lathe. To determine if a damaged center is hardened, test it with a file. If the file bites into the metal, the center is not hardened. Most lathe centers today have 60° points, but other angles are also used. The angle on the points must match the angle on the center drill and the center hole in the work.

What are *live centers*?
Live centers resemble centers (or *dead centers*), but have one or more bearings between the center point and the end taper, Figure 7–43 (upper right). These bearings permit the center to turn *with* the work eliminating the frictional heating centers create against the work when they do not turn. Without this heating—and the lengthening of the work it creates—more tailstock axial force can be applied to the center securing the work so deeper cuts may be taken. A TIR of less than one thousandth of an inch is not unusual. A quality live center has two or more ball bearings just behind the center: one handles radial loads and the second, often a tapered roller bearing, handles axial thrust. Inside and near the end of the Morse taper, there may be an additional bearing. A typical MT3 has a work load rating of 750 pounds or more. *Using a live center instead of a dead one is always good practice and will produce more accurate work with fewer problems.*

Are there other lathe center designs?
Yes, there are several:

- *Bull-nose live centers* have a shallower end than a plain center to hold pipe, tubing or hollow shafts. Like other live center designs, they also have one or more bearings inside.
- *Spring-loaded centers* allow the work to expand lengthwise from the heat of cutting and yet maintain enough force on the center to limit sideways motion. When a workpiece is held too tightly between centers and has no expansion room, it will bow along its axis, ruining the work.
- *Half-centers* provide enough clearance to allow the cutting tool to face the end of the work without bumping into the center. They are only for facing since they do not provide as much support as a regular center, nor do they hold lubricant to keep them from burning up.

See Figure 7–43 for examples of common lathe centers. Morse taper adapters are often required to fit a center of one Morse taper size into the lathe spindle Morse taper of a larger size.

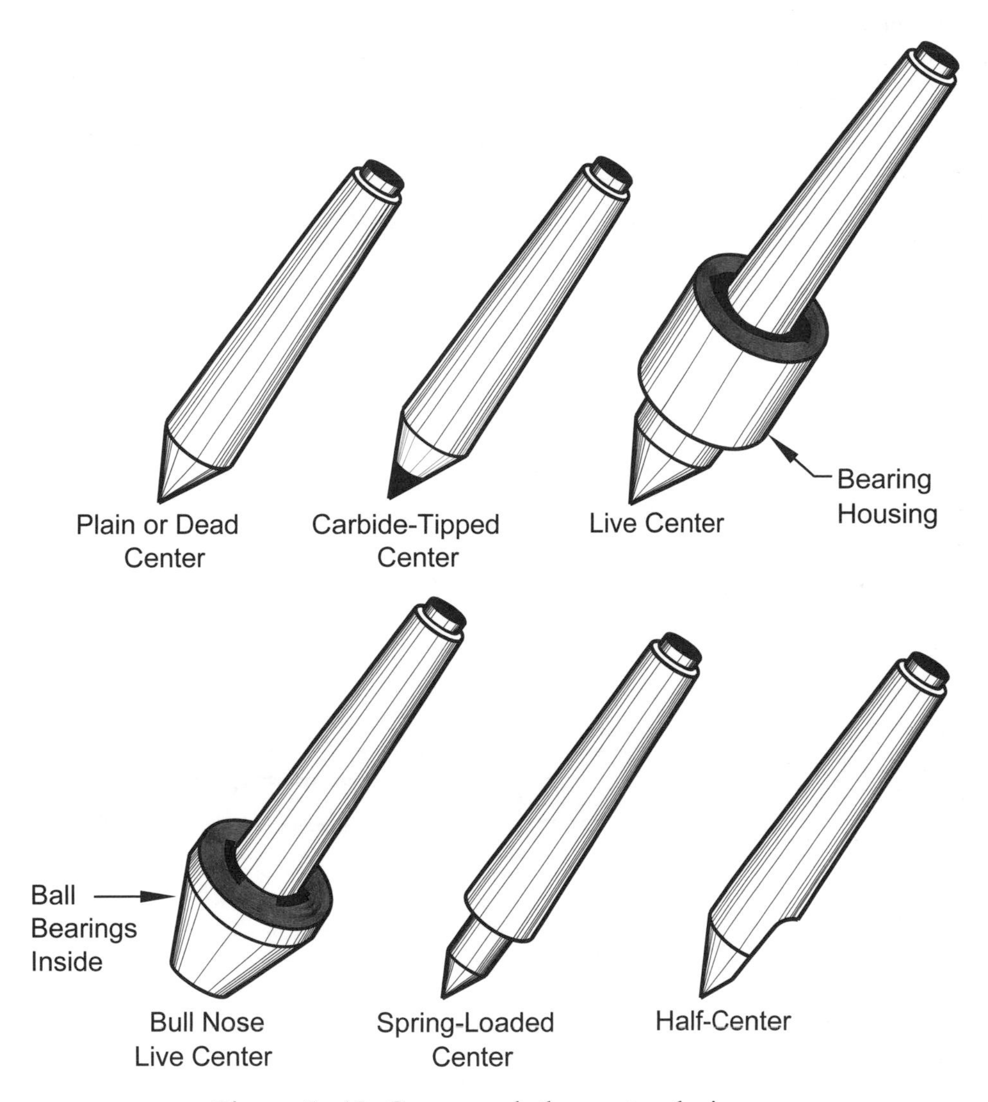

Figure 7–43. Common lathe center designs.

Installing Centers

How are lathe centers installed in the headstock and tailstock?

There are three easy steps to installing centers:

1. Clean debris from the center's Morse taper surfaces and the internal tapers of the headstock and tailstock.
2. If there are burrs on the Morse tapers, use a file or stone to remove them and wipe away any filings. Alternatively, use a strip of aluminum oxide abrasive cloth with a steel backing strip to polish the center.
3. Insert and seat the center in the headstock or tailstock the same way as it is installed in the spindle, Figure 7–44.

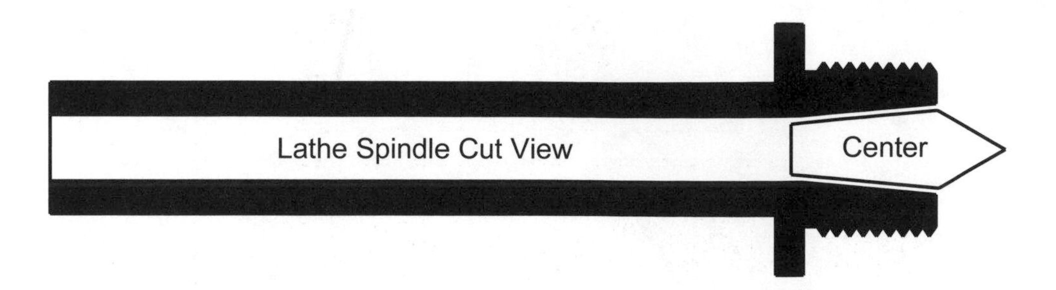

Figure 7–44. Lathe center installed in the spindle taper.

Removing Centers from the Headstock

How are centers removed from the headstock?

Start by holding a rag over the point of the center to protect your hand from the center's sharp point, then insert a *knock out bar* into the left end of the spindle hole and tap out the center, Figure 7–45.

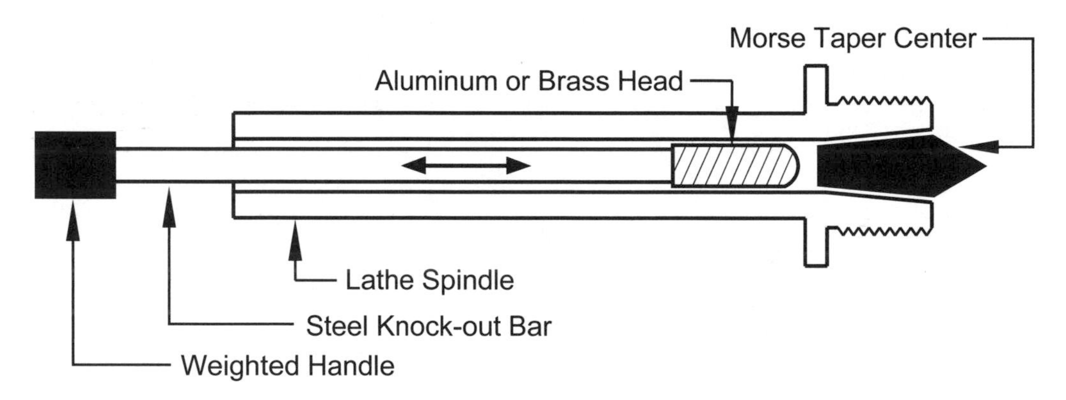

Figure 7–45. Removing lathe center from the headstock.

Removing Centers from the Tailstock

How are centers removed from the tailstock?

Figure 7–46 shows how the threaded shaft that moves the tailstock ram will push the center out of the ram's Morse taper when the spindle is cranked back into the tailstock. Be sure to catch the center as it is removed so it does not strike the lathe ways.

Morse taper drills without tangs will sometimes need an extension added on their tail ends so they will eject properly from the tailstock.

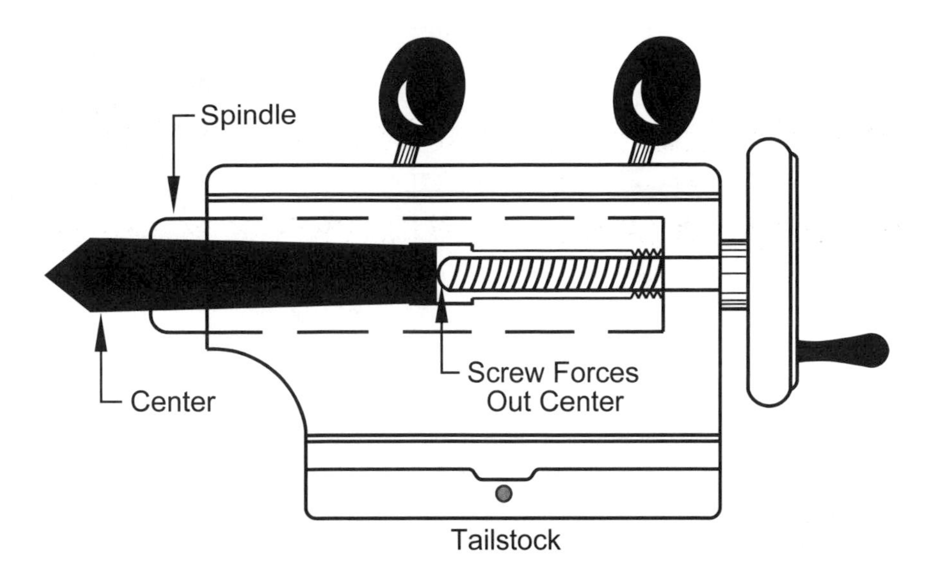

Figure 7–46. Removing center from tailstock.

Drive Plates & Lathe Dogs

How are *drive plates*, also called *catch plates*, used with *lathe dogs*?

Although lathe centers secure work for turning, they provide little traction to transmit spindle rotation to the work. Lathe dogs clamp onto the work and fit into the drive plate making the work rotate with the spindle. Drive plates, lathe dogs and centers are usually used together, Figure 7–47.

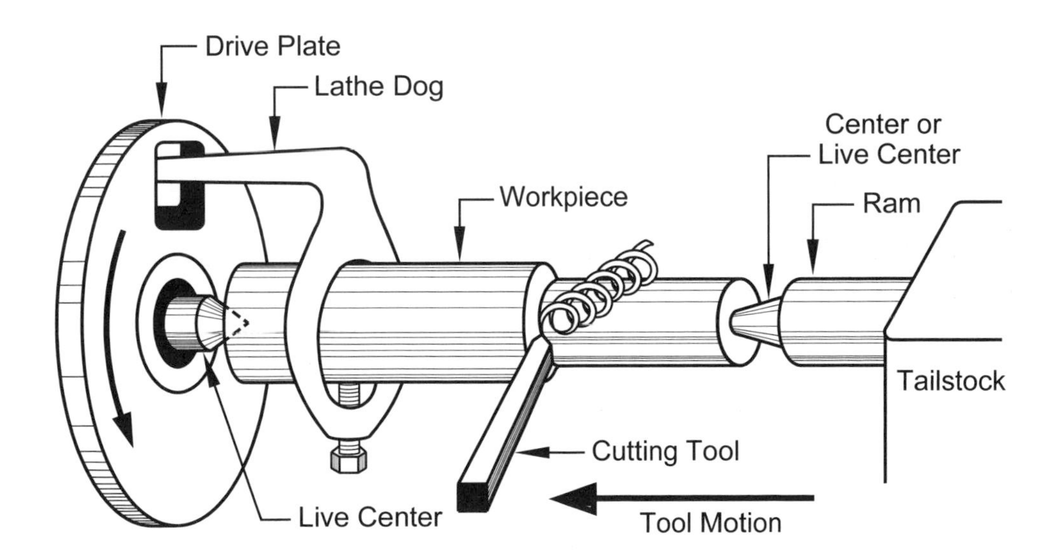

Figure 7–47. Turning work between centers with a drive plate and lathe dog.

What are the various designs of lathe dogs?
See Figure 7–48.

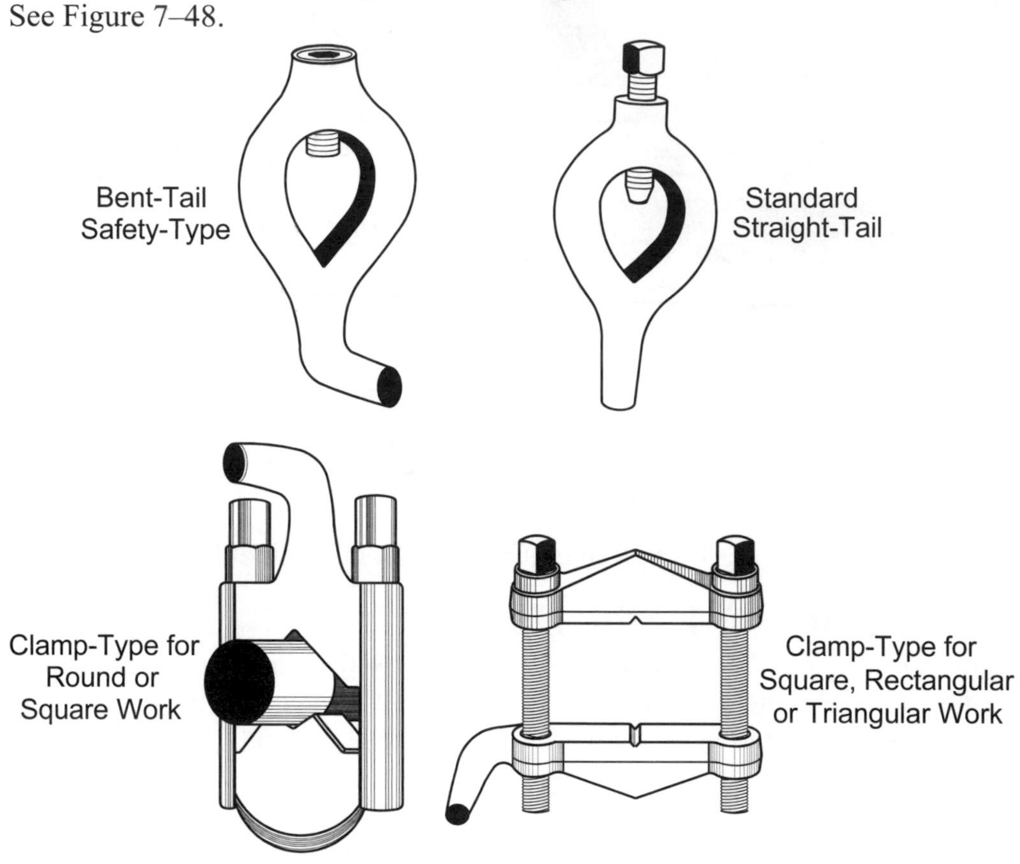

Figure 7–48. Lathe dog designs.

Face Plates

What is the difference between a *drive plate* and a *face plate*?
In many instances there is no difference in structure and the terms are used interchangeably. But there *is* a difference in function: Face plates hold and position work for turning, facing and other operations by having the work secured to it by bolts, brackets or adhesives. A drive plate does not hold work; it turns the work via a lathe dog. Figure 7–49 shows several face plate and drive plate designs and Figure 7–50 shows a face plate with a special lathe dog. Sometimes an angle plate is bolted to the face plate, and the work bolted to an angle plate to correctly position the work for facing or boring operations, Figure 7–51. Additional counter balance weights, such as washers or gears, are sometimes bolted onto the face plate opposite the workpiece and angle plate to balance the load and reduce vibration.

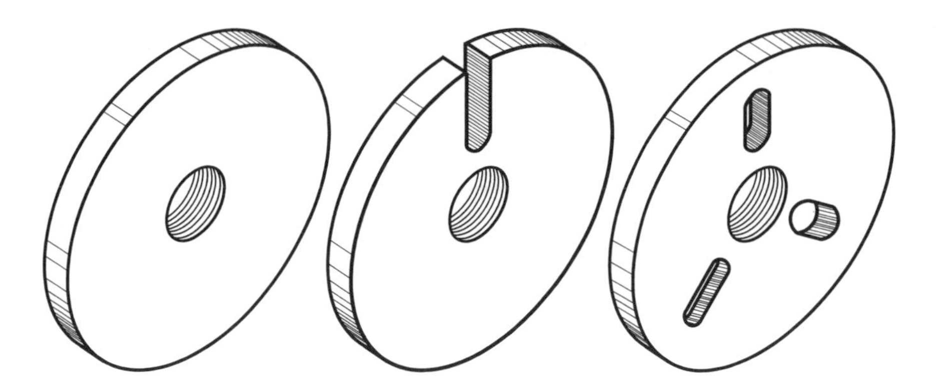

Figure 7–49. Plain face plate (left), drive plate (center), combination face plate/drive plate (right).

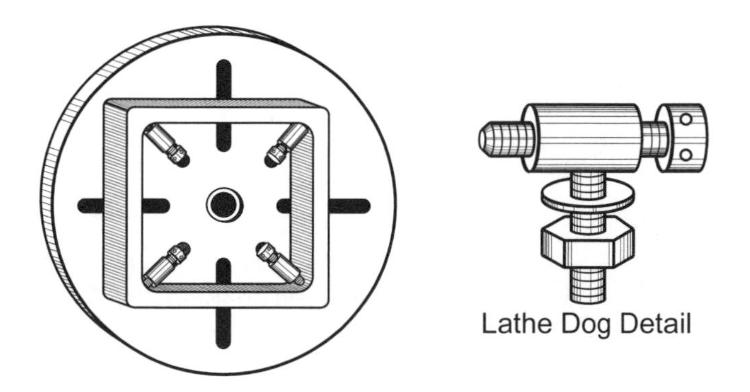

Figure 7–50. Face plate holding a square casting for facing (left) and detail of the lathe dogs used to secure the casting at its corners (right).

Figure 7–51. Angle plates bolted to face plates hold castings for facing and boring.

How can thin work be held on a face plate when the entire face of the work must be accessible?

Work can be held in place using either double-faced tape or an anaerobic adhesive such as Crazy Glue.® To achieve a solid bond, use carburetor or brake cleaner first to remove all residual oil from the work and faceplate. Keep the spindle speed down, take light cuts and stay out of the plane of the face plate. When the work is complete, heat the parts with a heat gun to release them. Figure 7–52 shows six washers adhered to a face plate so they can be thinned with a facing cut.

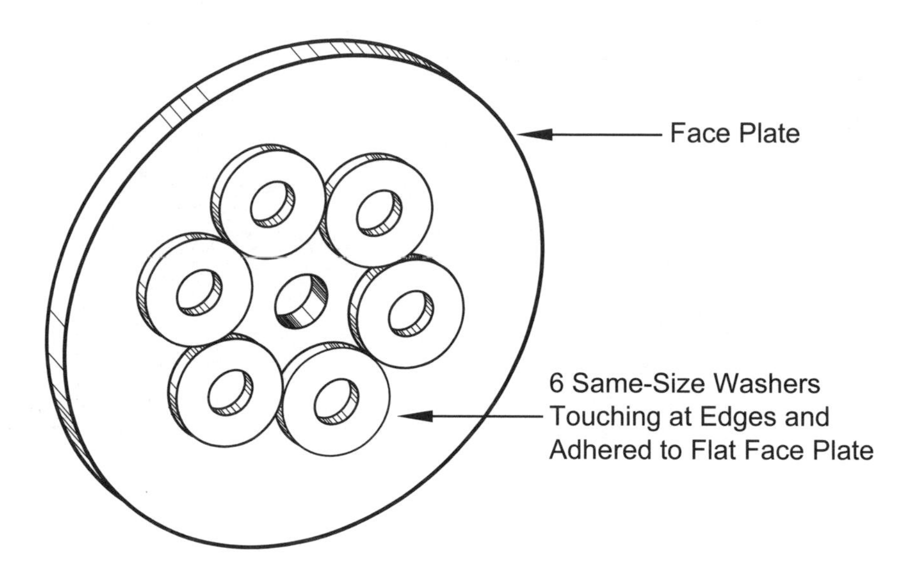

Figure 7–52. Washers to be thinned adhered to a face plate.

Drill Chucks

How are *drill chucks* used in the lathe?

Most often drill chucks are used to hold twist drills, center drills, taps or reamers in the tailstock as in Figure 7–53, but chucks are used in a couple of other ways:

- For small work, the spindle holds the chuck in the spindle taper and the chuck holds the work.
- For large work, the spindle holds the chuck and the drill and the work is clamped to the carriage.

Note that chucks are designed to handle *axial forces*, but have little strength for lateral or side-to-side forces. *Never use drill chucks to hold lathe centers or milling cutters since they cannot handle lateral forces.*

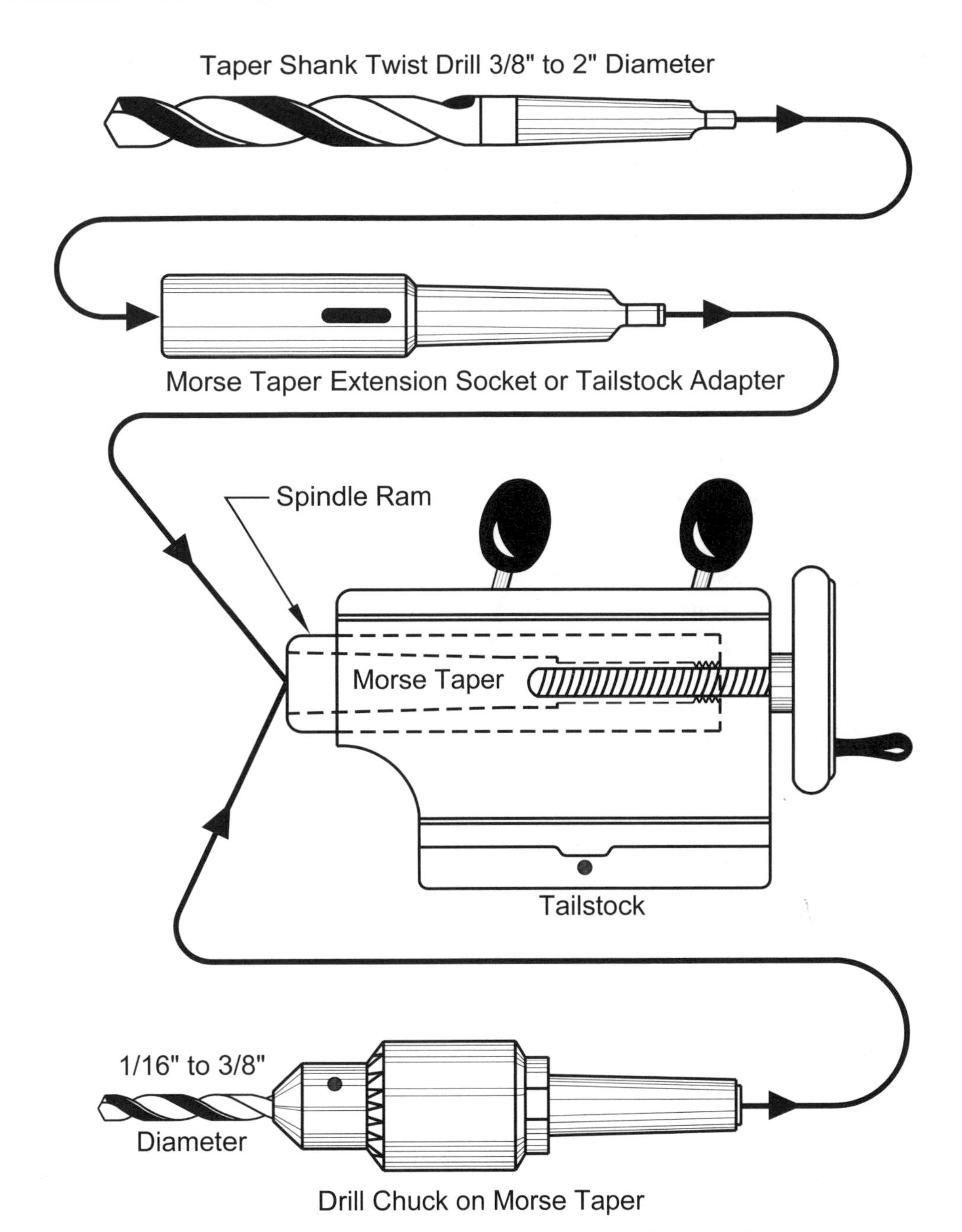

Figure 7–53. Methods to hold drills in the tailstock.

Are there other ways to hold large drill bits in the lathe?
Yes, Figure 7–54 shows an additional method. The drill is held in a commercial Morse taper tool holder. The holder has been drilled and tapped on its side for an anti-spin rod. This modification will not only prevent spinning, it prevents damage to the MT socket and the tailstock socket.

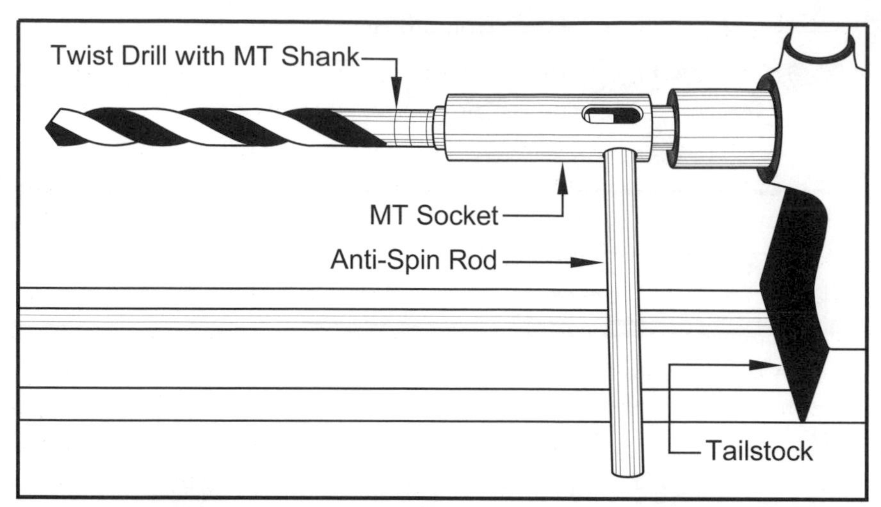

Figure 7–54. Method of holding large diameter drill bits in MT socket without spinning.

Are other types of drill chucks used on lathes?

Yes, hand feed drill accessories, as shown in Figure 7–55, are used to hold drills as small as 0.008 inches which are too small for regular sized drill chucks. Drills this size would drop out of a larger chuck. A hand feed drill provides the operator with a sensitive feel of the resistance on the drill. A regular chuck driven by the tailstock spindle ram would not provide this feel and drill breakage would be likely. The accessory shank fits into the tailstock and drilling operations are performed by grasping the free-spinning knurled knob and pushing it to drive the drill into the work. A spring between the chuck and the base shank draws the drill out of the work when the knurled knob is released. Alternatively, the hand drilling accessory can be held in a larger chuck, used in a drill press or in a milling machine.

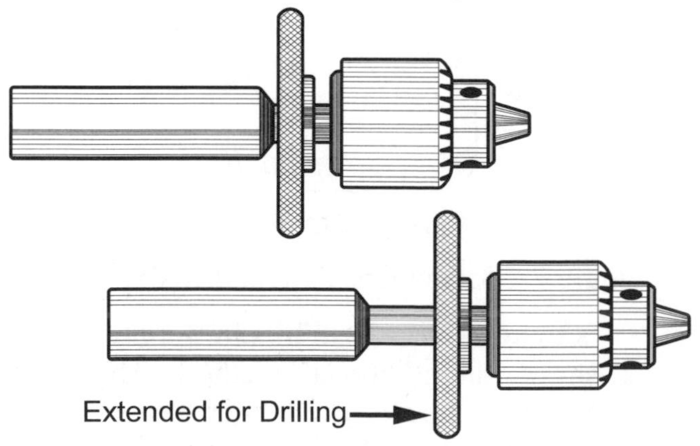

Figure 7–55. Hand feed drill accessory.

Lathe Chucks

How are *lathe chucks* used and what are the most common designs?
Lathe chucks fasten onto the spindle and hold work while rotating it. Although there are 2-, 3-, 4- and 6-jaw chucks, 3-jaw *self-centering, universal* or *self-acting* chucks and 4-jaw *independent* chucks, are the most common designs. Three-jaw chucks, as illustrated in Figure 7–56, are very convenient because they automatically center round work as the jaws are closed using its key. Centering is usually within a few thousandths of an inch. Four-jaw independent chucks, Figure 7–57, require each jaw be independently positioned on the work. They take more time to set up, but can center work much more accurately than a self-centering 3-jaw chuck and they also provide a better grip. A 4-jaw chuck is more useful for holding square, rectangular or eccentric work. Many chuck jaws are reversible for holding work either on its inside or outside edge as in Figure 7–58. Other chuck designs have two-part jaws and the upper jaws can be reversed to hold work either inside or outside. There are also magnetic chucks that have no jaws at all.

Chucks are available in a wide range of sizes from 2-inch diameter chucks for jewelers' lathes to 24-inch models for large lathes. They are usually of very similar design and differ only in size.

It is good practice to disassemble chucks periodically to remove dirt and metal chips, which increase wear and reduce accuracy. Add fresh lubrication when reassembling them.

Caution: On small chucks, like those for the Levin and Sherline lathes, it is easy to apply too much force while closing them and damage their workings. Large tightening forces are not needed on small chucks.

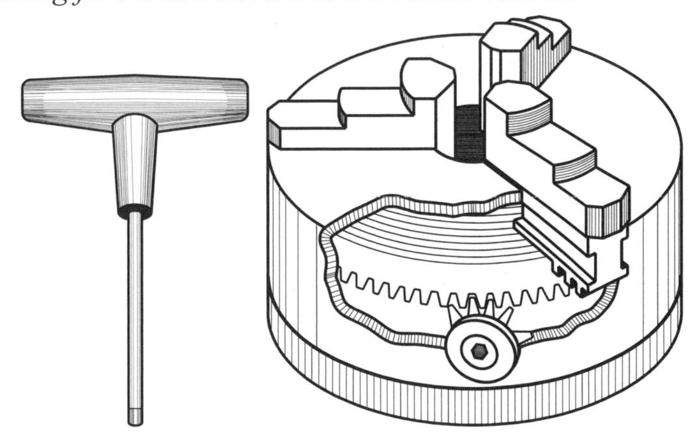

Figure 7–56. Three-jaw universal, self-centering chuck and its key.

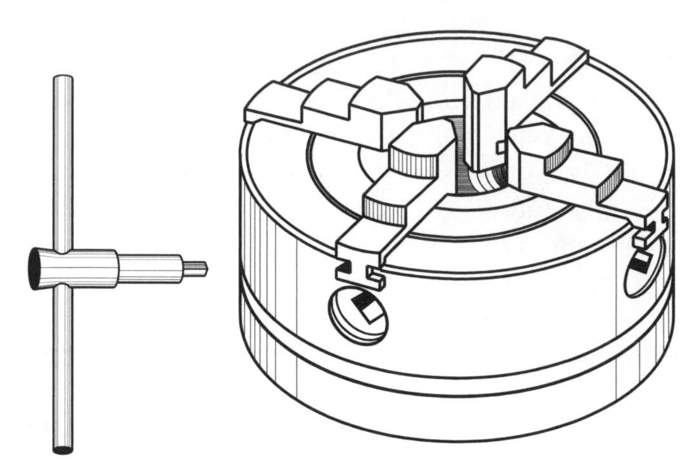

Figure 7–57. Four-jaw independent chuck and its key.

Important safety guideline: Do not permit the work to extend more than three times the work diameter beyond the chuck. Use a steady rest for longer work or use a center in the right end of the work.

How should work be placed in a chuck?

Figure 7–58 shows three ways to properly secure work in a chuck.

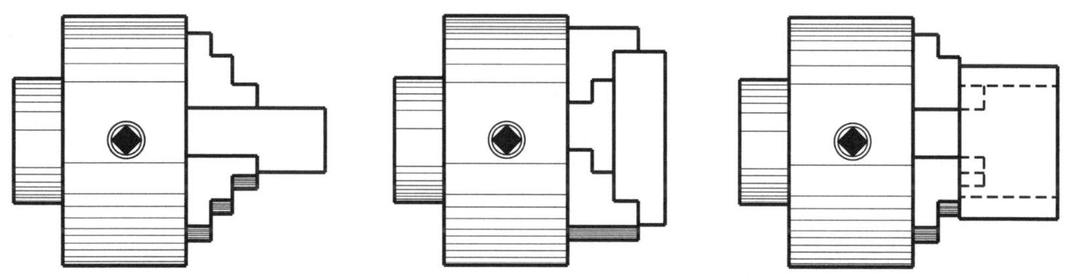

Figure 7–58. Smaller work held in normal jaw position (left), larger work held with jaws reversed (center), hollow work held with jaws inside the work provides access to the full length of the work (right).

Figure 7–59 (A) shows a proper match of chuck size and work size. Figure 7–59 (B) shows work extending more than three times the diameter of the work. Only very light cuts may be taken, but it would be much better to use a larger chuck or, if the work permits, put a center in the free end of the work. Figure 7–59 (C) is poor practice as the chuck has a poor grip and the work is likely to come loose. A better way to handle this job is to start with a longer piece of work so more length stays in the chuck and the excess is parted off when machining is done. The chuck in Figure 7–59 (D) has only a light grip on the work so only light cuts may be taken or the work will escape the chuck.

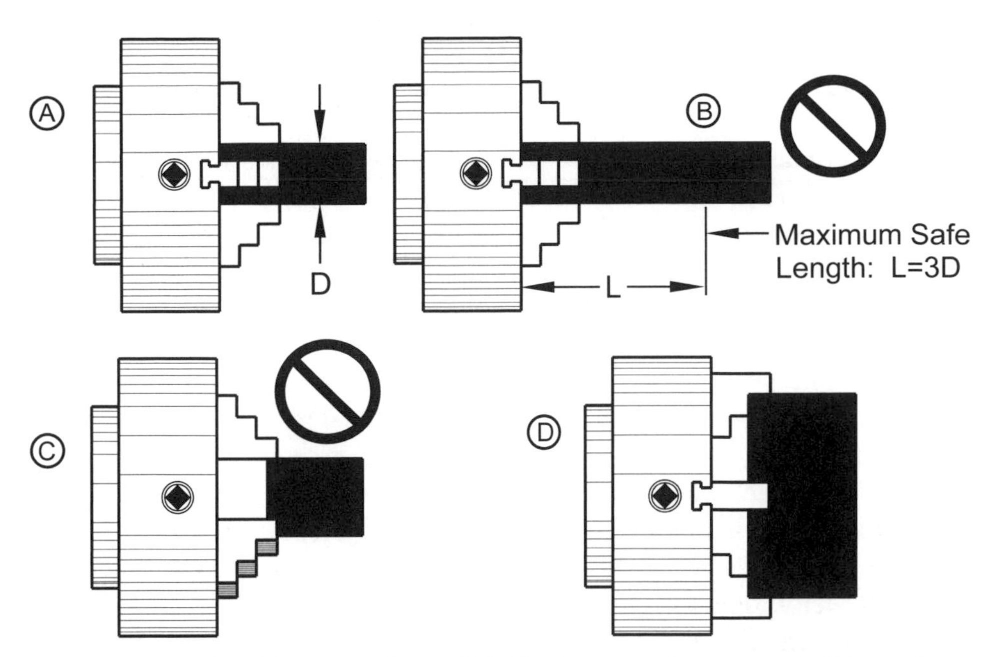

Figure 7–59. Chucking practices: (A) Work properly secured, (B) work too long for chuck, (C) work not deep enough in jaws for a secure grip and (D) marginal grip on work, take only light cuts.

Installing Heavy Chucks

How can heavy chucks be installed without risking damage to the ways?
Use a shop-made wood-block cradle, such as the one shown in Figure 7–60, which positions the chuck at the right height for mounting it onto the spindle. This block is also used to prevent the chuck from dropping off the spindle onto the ways when removing it. Notches across the bottom of the cradle align the chuck with the spindle. Very large chucks require a chain hoist.

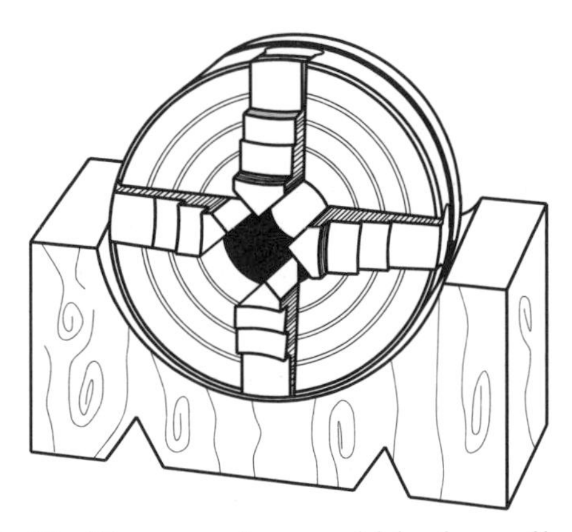

Figure 7–60. Shop-made wood block cradle supports large chuck for safer installation or removal.

Centering Work in 4-Jaw Chucks

How do you center work in a 4-jaw chuck using a dial test indicator (DTI)?

1. Use the concentric rings on the chuck face to center the work by eye.
2. Set up the DTI as shown in Figure 7–61 (A).
3. Align any one of the chuck's jaws with the axis of the DTI plunger and rotate the DTI scale to set it to zero, Figure 7–61 (A). The black dot on the chuck illustration is to make it easy to keep track of the starting jaw.
4. Rotate the chuck exactly one-half turn, read the DTI dial and adjust these two opposite jaws by one-half the DTI reading to center the work. Use the DTI to make this adjustment, Figure 7–61 (B). The work is now centered between these two jaws.
5. Rotate the chuck back and forth one-half turn to verify that the work is centered between these two jaws. If not, repeat steps 1 and 2.
6. Rotate the chuck one-quarter turn in either direction to put a new chuck jaw parallel with the DTI probe, zero the DTI scale, Figure 7–61 (C).
7. Rotate the chuck one-half turn in either direction, read the DTI and adjust these two jaws one-half the DTI reading so as to center the work between these two jaws, Figure 7–61 (D).
8. Verify the work is centered by slowly turning the work through a complete turn while observing the DTI. Repeat the above steps if needed.

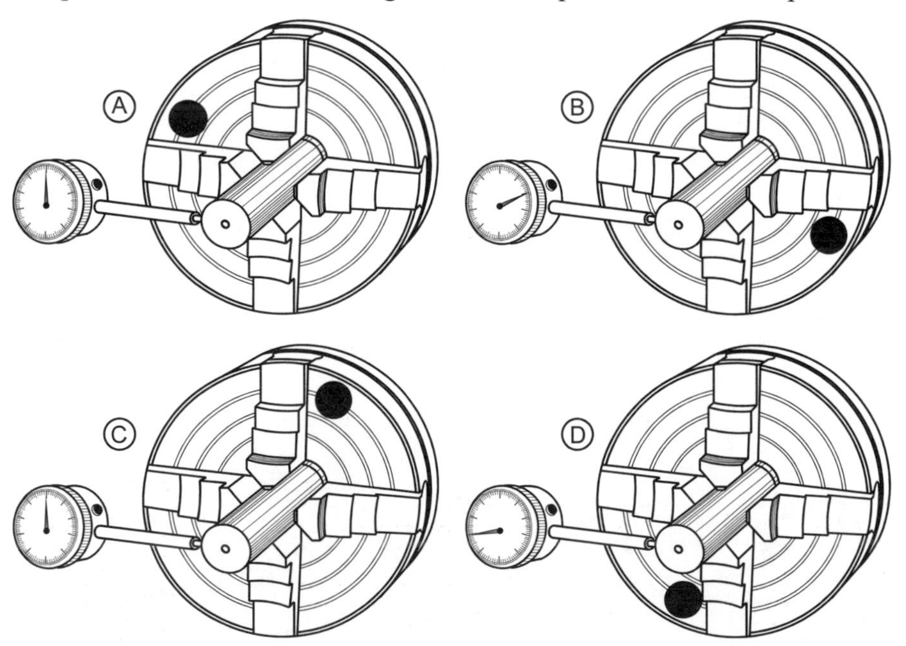

Figure 7–61. Steps to center round work in a 4-jaw chuck. The black dot is a reference point showing chuck rotation. On square bars, use the same centering process, but measure from the corner edges.

With practice you can center work to less than 0.001-inch eccentricity in under a minute. Many workpieces are not actually round and cannot be centered with precision. For example, many aluminum rounds are extruded from larger billets and are only roughly symmetrical.

Collets

What are *collets* and how do they work?

Collets are cylindrical devices, usually of steel, which grip either a workpiece or a tool and do so with a minimum of TIR (Total Indicated Runout). There are two styles of collets—the draw-in-to-close and the push-to-close. All collets work in the same way: by being pulled or pushed into a spindle with a tapered mouth that forces the head of the collet to collapse around the work or tool and grip it tightly.

Draw-in collets use a drawbar, which tightens on threads at the back of the collet pulling the front end of the collet, the end into which the work is fitted, into the adapter. This forces the collet to contract, gripping the work. See Figure 7–62. A series of three or more slits in the front half of the collet provide clearance for contraction. Collets are usually hardened and ground. Their advantage is that work can be repeatedly inserted and removed from collets while maintaining concentricity. The nose cap is used to force off the collet adapter for its removal.

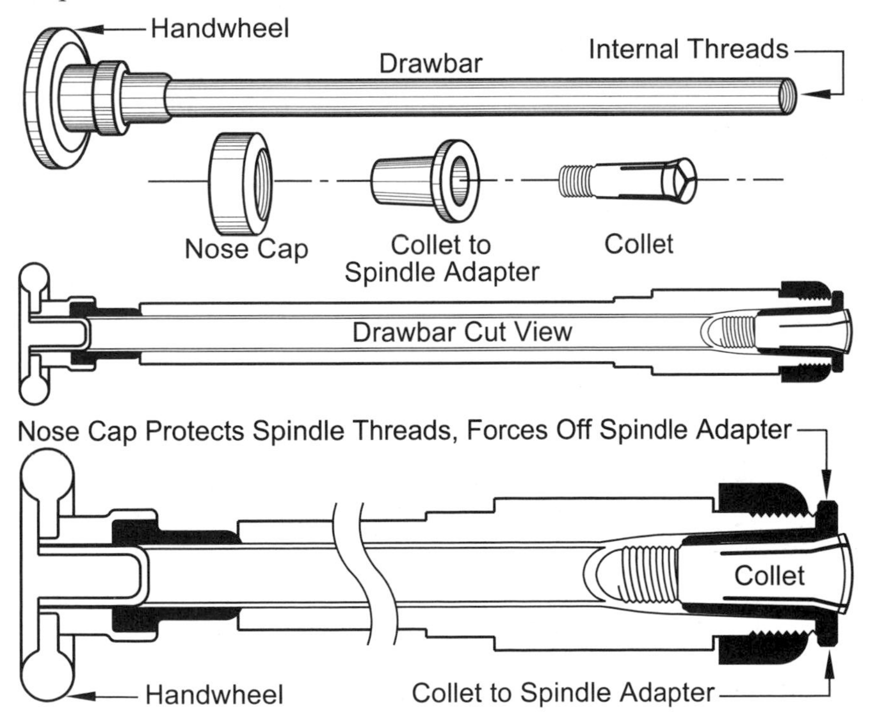

Figure 7–62. Drawbar, spindle adaptor, nose cap and collet.

What is the disadvantage of 3C and 5C collets?
The biggest disadvantage is that each collet covers only a small range of work diameters as in Figure 7–63. A well-equipped shop requires a set of 12 to 24 collets. Using a collet outside its size range will damage the collet and will not allow it to properly grip the work.

The collet design in Figure 7–63 permits long work to extend back through the length of the collet and into the spindle. Sometimes long stock extends out through the hole in the drawbar handwheel on the left side of the lathe and may rest in a V-shaped holder located on a stand several feet behind and to the left of the headstock. This is convenient when many short pieces of stock are to be machined. Each part is machined and cut off from the raw stock. The remaining raw stock is then advanced through the collet and locked in place for machining the next part. Using this method very little stock is wasted.

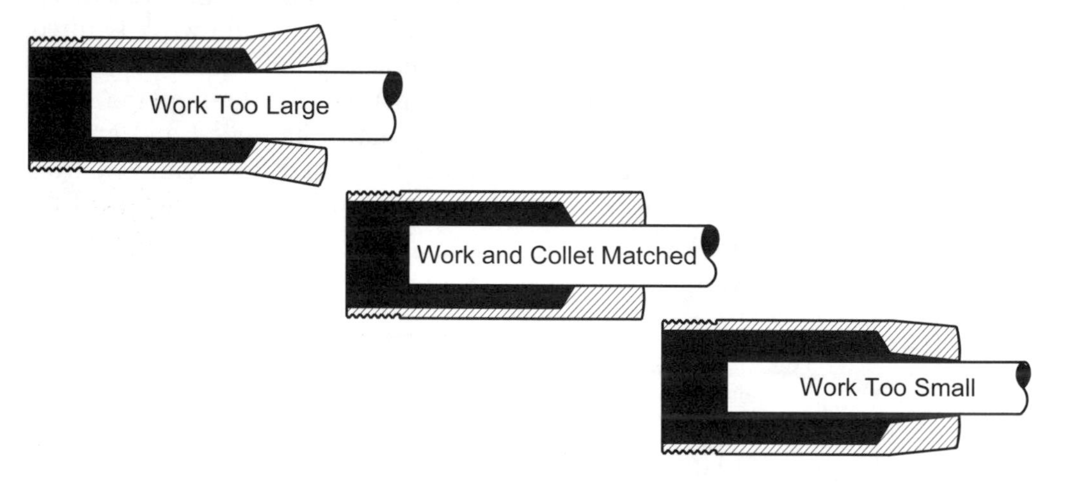

Figure 7–63. Collet sizing.

What are the most common collet designs?
There are several common designs:

- *Round Collets* are the most common. Made of hardened and ground steel inside and out, they fit into the spindle's Morse taper. As they are pulled into the spindle by the drawbar, they clamp down and grip the work. Their advantage is that they provide high clamping power and concentricity, TIRs of 0.0005 inches. Work can be removed and replaced while maintaining its center. Their drawback is that each collet handles only a narrow range of work diameters, typically 0.002 inches larger to 0.001 inches smaller than its marked size. The most common collet design for medium-sized lathes is Type 5C, seen in Figure 7–64.

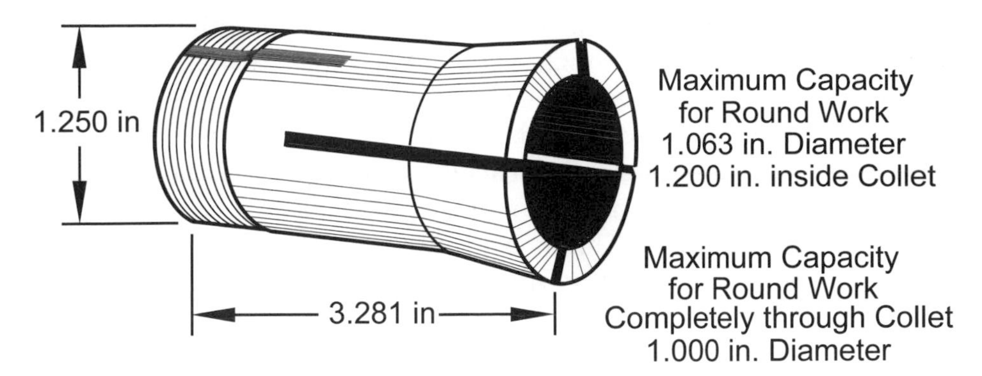

Figure 7–64. Type 5C round collet.

- *Hex* and *Square Collets* are just like round collets except for having a square- or hex-shaped hole instead of a round one. They are usually ground and hardened, Figure 7–65.

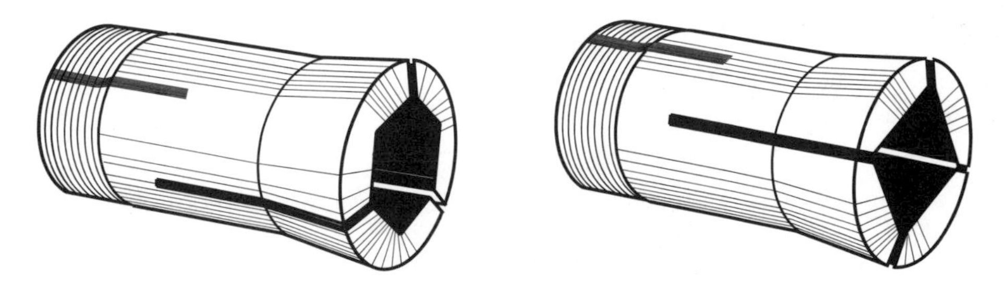

Figure 7–65. Type 5C hex and square collets.

- *Emergency Round Collets,* seen in Figure 7–66, are similar in design to round collets, and are made of *unhardened* steel, brass or nylon. They are made of these soft materials so you can drill or bore the size hole needed to fit the work at hand when you don't have a steel collet of the proper size. Emergency collets come with steel spacer pins that hold the slits open during boring and insure proper gripping action in use. These pins are removed and discarded when boring is complete.

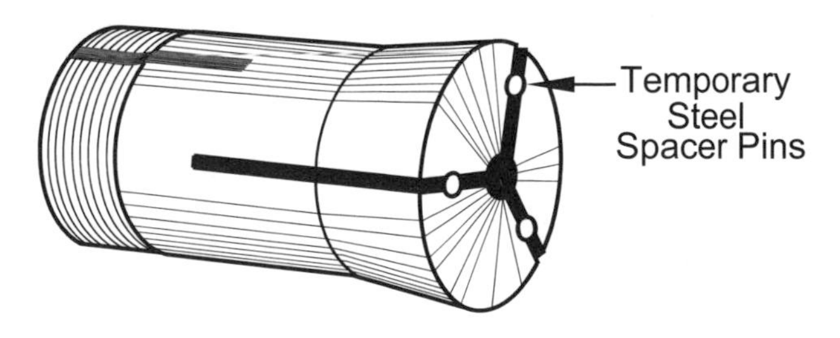

Figure 7–66. Type 5C emergency round collet.

- *Step Collets*, also called *step chucks*, have one or more internal steps for gripping the outside of round work, or a series of external steps for holding the inside diameter of cylindrical work. They are also supplied machined and ready to use or blank (unbored) for the user to customize. These may be heat-treated after machining. The advantage of a step collet is that it can hold work larger than the largest capacity collet which will fit inside the lathe spindle. This is because the work does not go inside the spindle but remains in the portion of the chuck in front of the spindle, as in Figure 7–67.

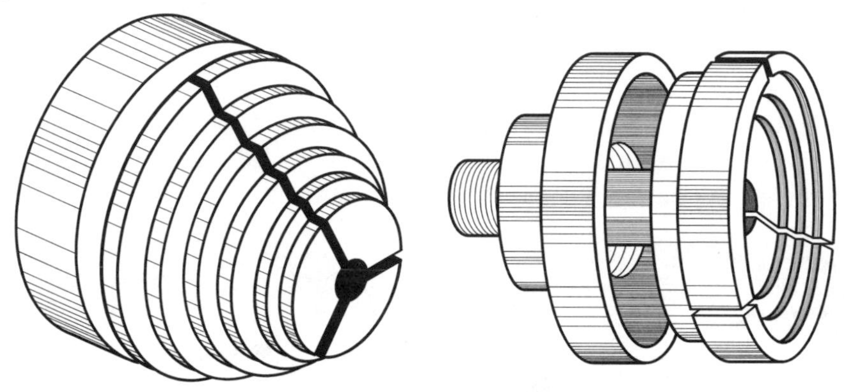

Figure 7–67. External (left) and internal (right) step collets.

- *Rubber Flex Collets* consist of rubber segments between steel ones as shown in Figure 7–68. This collet is a push-to-close type, using a threaded holder mounted to the spindle to compress and close the collet. These collets are useful because each size collet can hold a wider range of work diameters than conventional steel collet designs.

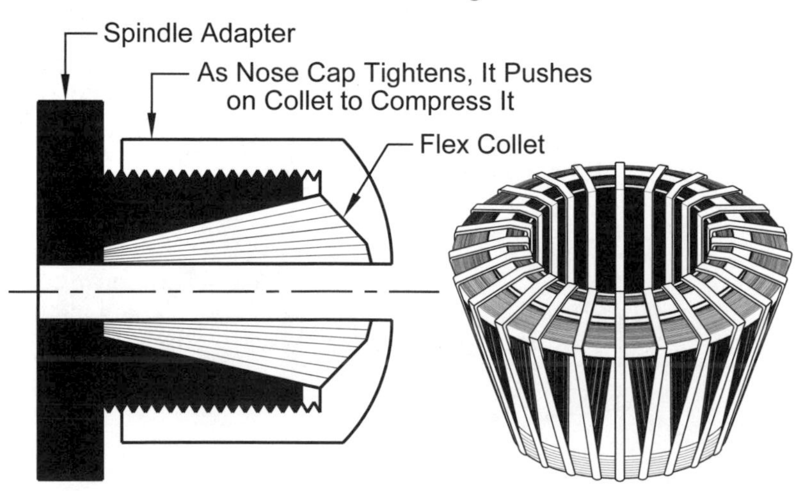

Figure 7–68. Flex collet: cross-sectional view (left) and overall view (right).

- *WW (Webster Whitcombe) Collets* are also called *wire collets* or *8 mm collets* after their outside diameter, Figures 7–69 and 7–70. Similar to the Type 5C round collet pictured in Figure 7–64, but much smaller, they hold work for jewelers and instrument makers in lathes like the Levin and Sherline models shown earlier in the chapter. These collets, like their big brothers, offer high concentricity. Unfortunately, each size collet handles work just 0.001 inches (0.02 mm) under or over its nominal size. Metric versions handle work diameters from 0.1 mm to 5.0 mm in 0.1 mm increments; English versions handle diameters from 1/64 to 3/8 inches in 1/64-inch increments. A hole extends through the collet and drawbar to hold longer work inside the headstock as with the 5C collets. Metric collets are supplied from Sizes 1 to 80. The diameter of metric collets in millimeters can be found by dividing the size number by 10. For example, a size 5 collet has a nominal diameter of 0.5 mm.

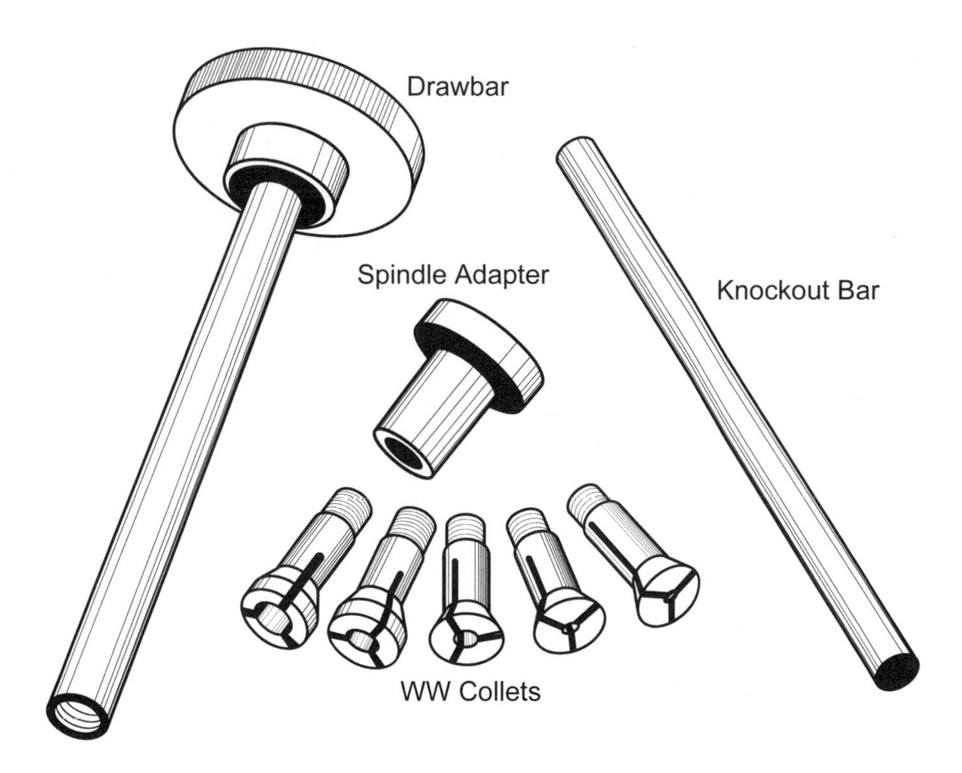

Figure 7–69. WW collets, drawbar, spindle adapter and knockout bar for a Sherline lathe.

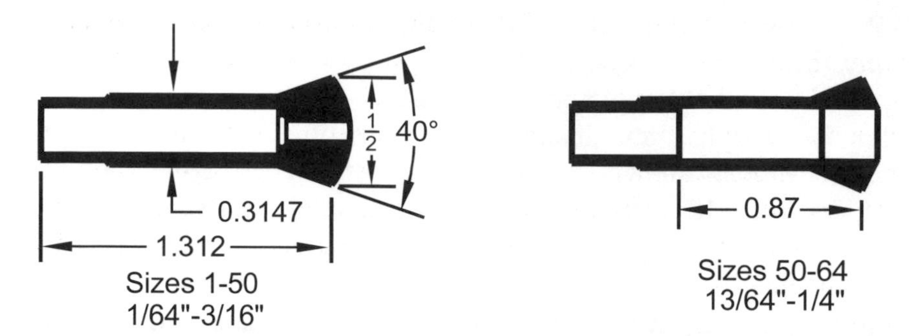

Figure 7–70. WW collets are available in both metric and inch sizes.

- *WW pot chucks*, Figure 7–71, are a close relative of the WW collets seen in Figure 7–70 and have the same *outside* dimensions, but handle work too large to fit completely *through* the chuck and into the headstock. They hold work in front of the collet. Unbored blanks for making emergency, or custom pot chucks, are available, Figure 7–72 (left). The user inserts a pin in the chuck blank, bores the chuck to size, then removes the temporary pin which holds the chuck to size for boring. Blank WW workholders for the user to customize are also available, Figure 7–72 (right).

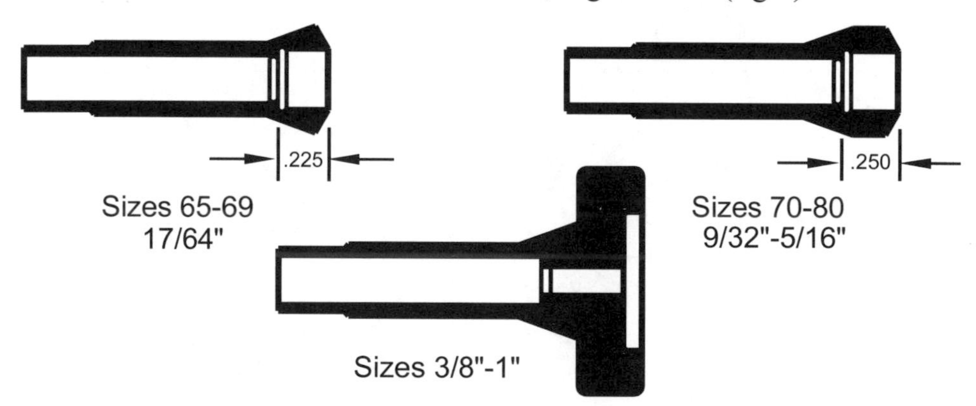

Figure 7–71. WW pot chucks in both metric and inch sizes.

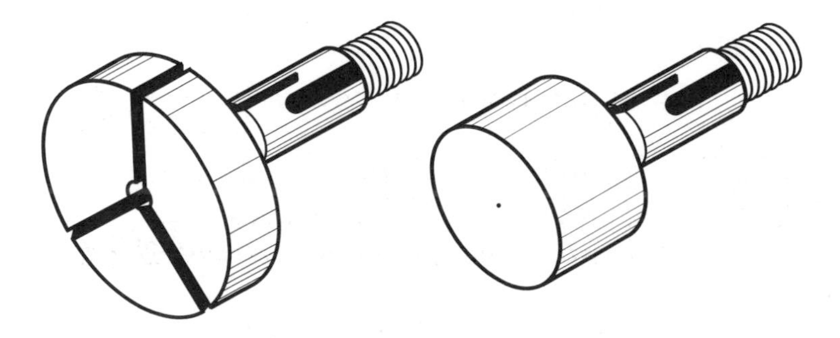

Figure 7–72. WW emergency pot chuck blank (left) and WW blank workholder (right).

What is the correct procedure for boring out an emergency collet?

1. Insert the collet in the spindle and make sure that the temporary pins supplied with it are in the face of the collet.
2. Adjust the drawbar so that the collet is firmly closed over the pins. Make sure the collet angle is fully seated in the spindle. If you can pull the pins out with your fingers, increase the drawbar tension until you cannot.
3. Rough and finish bore the collet. Remove the collet from the lathe, remove the pins and deburr it.
4. Thoroughly clean the spindle and collet, then replace the collet in the lathe. The collet is now ready to use.

What is a *collet stop*?

This simple device usually screws into the back internal threads of a collet and controls the depth a workpiece is held by the collet. This makes it faster and easier to make multiple identical parts. Stops may be purchased or shop-made. Figure 7–73 shows a collet stop for a 5C collet. There are extension tubes that accommodate longer workpieces; several of these extension tubes may be threaded together as well. For collets without internal threads, other collet stop designs expand inside the collet to hold themselves in place.

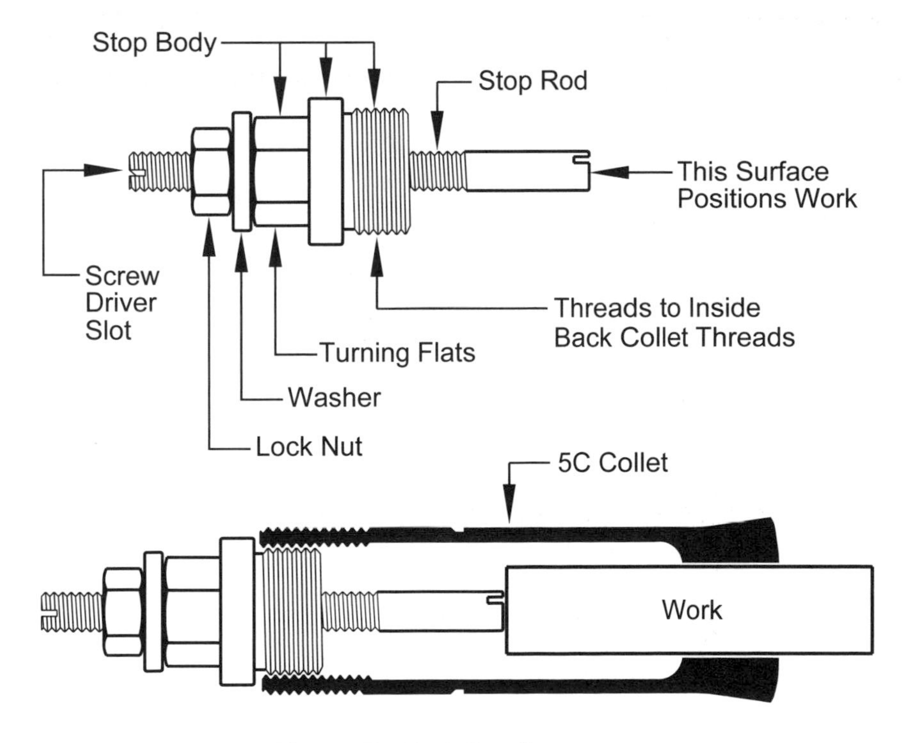

Figure 7–73. 5C collet stop.

5C-Mounted Chucks

What are the advantages of the 5C-mounted chuck in Figure 7–74?

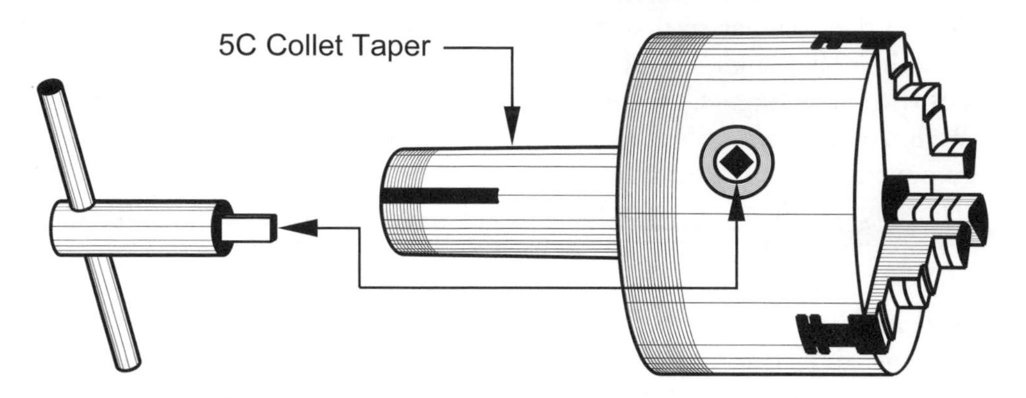

Figure 7–74. 5C-mounted 3-jaw chuck and its key.

- The chuck holds workpieces too large to fit in the largest 5C collet (these collets have a capacity of up to $1\frac{1}{16}$ inches in diameter), but does not require installing a large chuck.
- Concentricity can be preserved if the part must be removed from the lathe for some other operation. A 5C collet is often used in rotary-tables, dividing heads, collet indexers and collet chuck blocks.
- Switching between a collet and a chuck can be done without removing the collet adapter and collet closer from the spindle.

Both 3- and 4-jaw collet chucks are standard industrial tool catalog items and both domestic and imported models are available. They are typically 3- or 4-inch chucks furnished with two sets of jaws, internal and external. Two sets are needed because these chucks are too small to hold jaws machined for both internal and external holding.

Steady Rests

What is the function of a *steady rest*?

A steady rest, often called a *steady*, is essential when work which hangs out too far from the chuck needs extra support and a center cannot be used, Figure 7–75.

The steady rest clamps to the lathe ways with a T-nut. Steady rests are usually supplied by the lathe manufacturer to fit his machine's ways shape. Three adjustable brass or bronze guides, called *fingers,* support the work. Lubrication of the guides is critical. Some steady rest designs have ball bearing rollers instead of fixed fingers.

How is the workpiece set up in a steady rest?

Here is how:

1. Measure the diameter of the work where the steady rest will support it.
2. Turn a piece of scrap to the same diameter as the work. This piece may be a solid cylinder or a pipe, providing the walls are thick enough not to distort when clamped in the chuck. The scrap must be long enough to fit both in the chuck and pass through the steady rest at the same time.
3. Mount the scrap in the 3- or 4-jaw chuck. If using a 4-jaw chuck, center the scrap with a dial indicator.
4. Position and clamp the steady rest close to the chuck so the scrap passes through the steady rest.
5. Adjust the fingers to support the scrap securely, lock the fingers in place, and using a dial indicator, check to see that the scrap remains centered.
6. Release the top finger, remove the scrap, mount the work in the chuck, and move the steady rest along the bed to where it will support the work. Secure the steady rest to the lathe bed.
7. The bottom fingers are already in the correct location to support the work, now only the top finger must be positioned.
8. Use plenty of high-pressure lithium-based grease on the fingers. Some machinists also add the appropriate Never-Seeze® compound to match the metal of the workpiece to the lubricant as well.

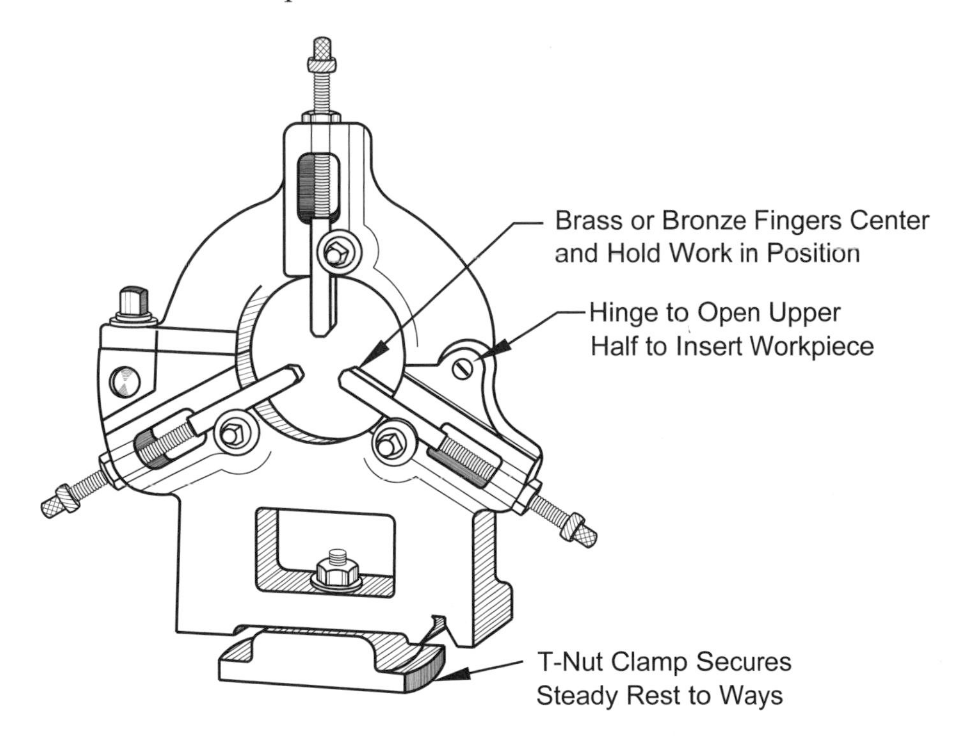

Figure 7–75. Steady rest.

Follower Rests

What is a *follower rest* and how does it work?

A follower rest is similar to a steady rest except it mounts on the lathe saddle just ahead of the cutting tool and prevents tool forces from bending the work, as in Figure 7–76. Most follower rests have just two brass work guides or rollers and are supplied by the lathe manufacturer. Sometimes a follower rest is used with a steady rest, Figure 7–77. As with a steady rest, lubrication on the fingers is important.

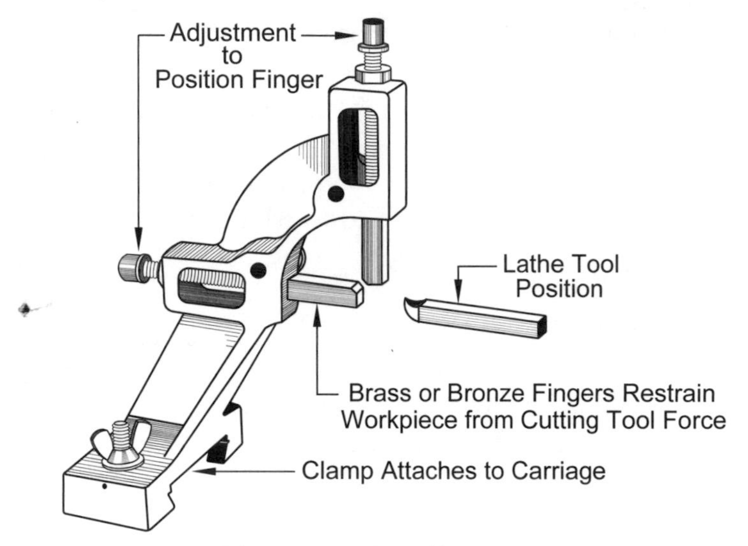

Figure 7–76. Follower rest.

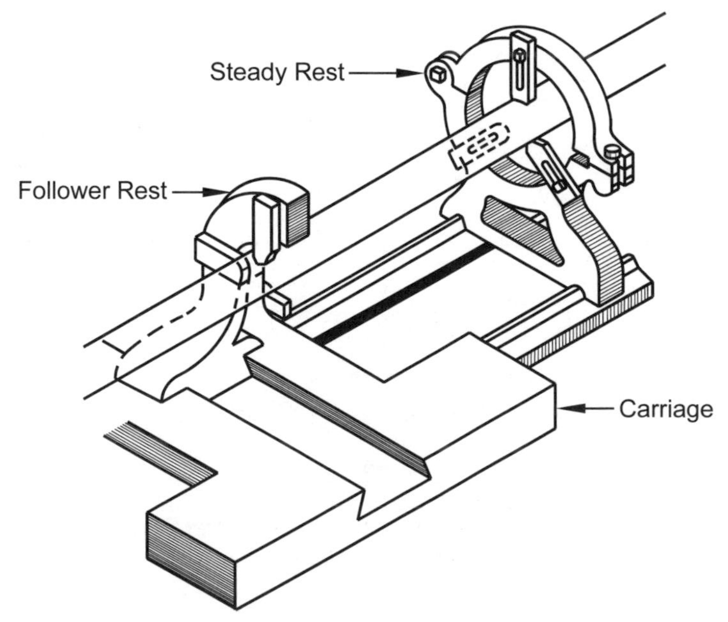

Figure 7–77. Steady rest and follower rest in use together.

Mandrels

What are *mandrels* and how do they work?

Mandrels are cylindrical hardened and ground steel work holders that fit through a hole in the work to grasp it. Some mandrels are tapered and forced through a hole in the work to grip it as in Figures 7–78 and 7–79. These mandrels are forced into the work using an arbor press and removed the same way. Others work by expanding *after* the work is slipped onto them. These expansion-types either are held only on the spindle end, as in Figure 7–80, or turn between centers, Figure 7–81. One advantage to using mandrels is that they leave no grip marks on the work while maintaining concentricity. Another is that they provide tool access to the full length of the work. Mandrel surfaces and work mounting holes should be cleaned of debris and lubricated before assembly.

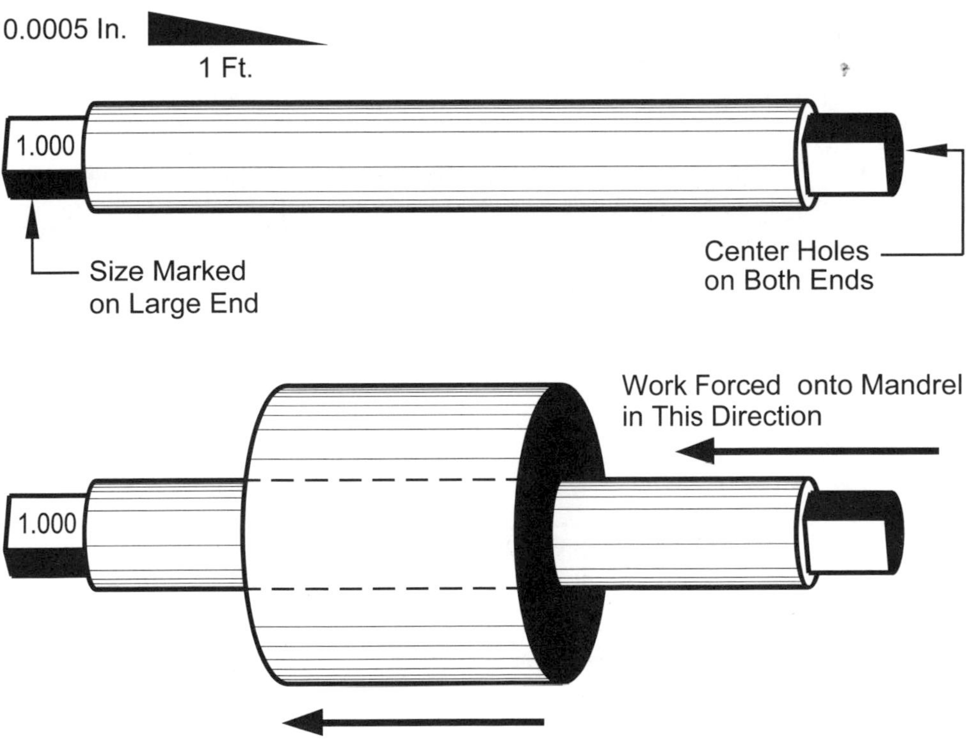

Figure 7–78. Tapered solid mandrel (top) and with work installed (bottom).

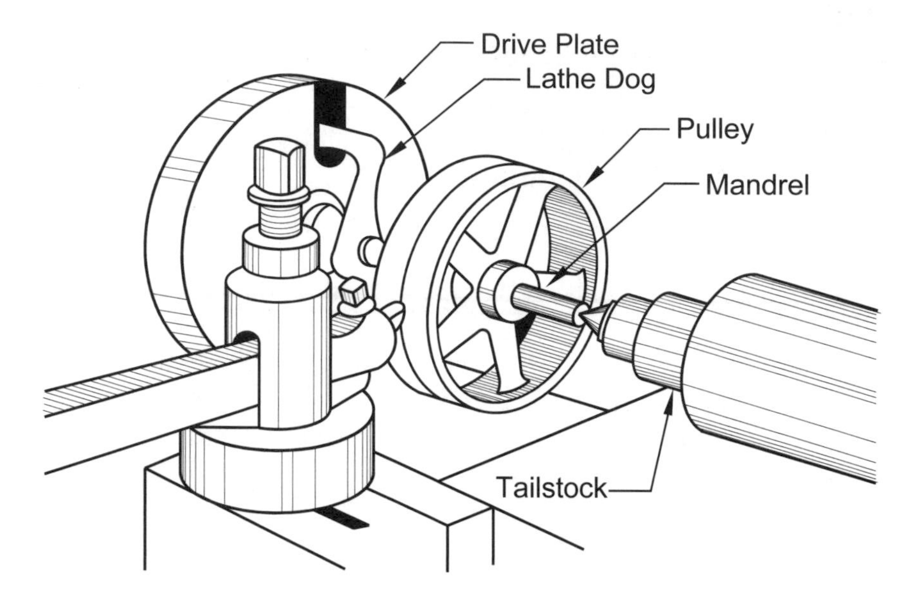

Figure 7–79. Machining a pulley mounted on a mandrel.

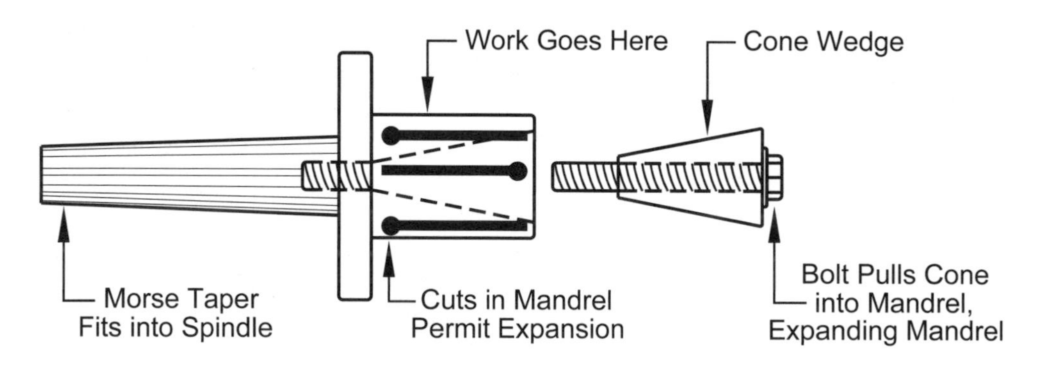

Figure 7–80. Expansion mandrel on a Morse taper.

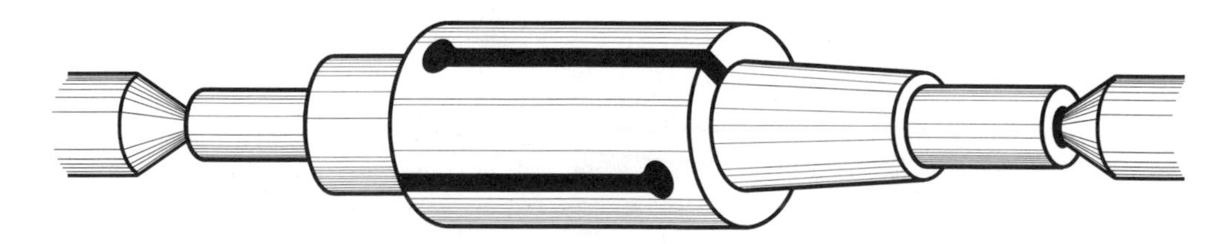

Figure 7–81. Expansion mandrel held between lathe centers provides for non-standard work diameters. In general, the more expansion slots, the better the mandrel will perform.

Arbors

What are *arbors* and how do they differ from mandrels?

Arbors are work holding fixtures, Figure 7–82. They differ from mandrels in that they grip by clamping the work from its ends rather than by gripping it internally as a mandrel does. Many times arbors are shop-made.

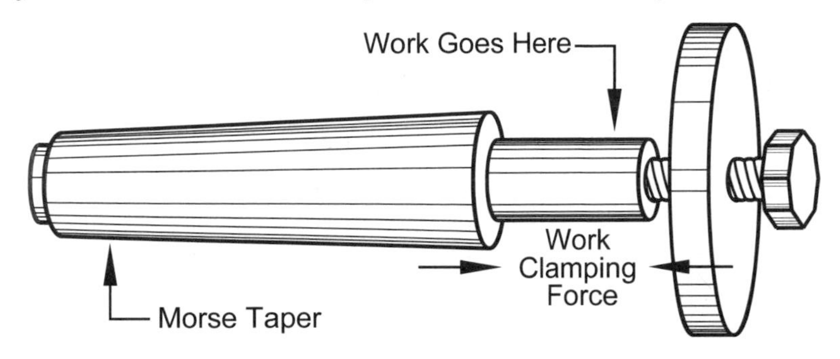

Figure 7–82. Arbor.

Taper Attachments

How do *taper attachments* work?

These accessories clamp onto the ways and link the cross slide to the taper attachment guide bar. Because of this connection, the cross slide follows the angle of the guide bar with respect to the axis of the lathe, and so, cuts a taper, Figure 7–83. It is easy to set accurate tapers using the calibrated scales.

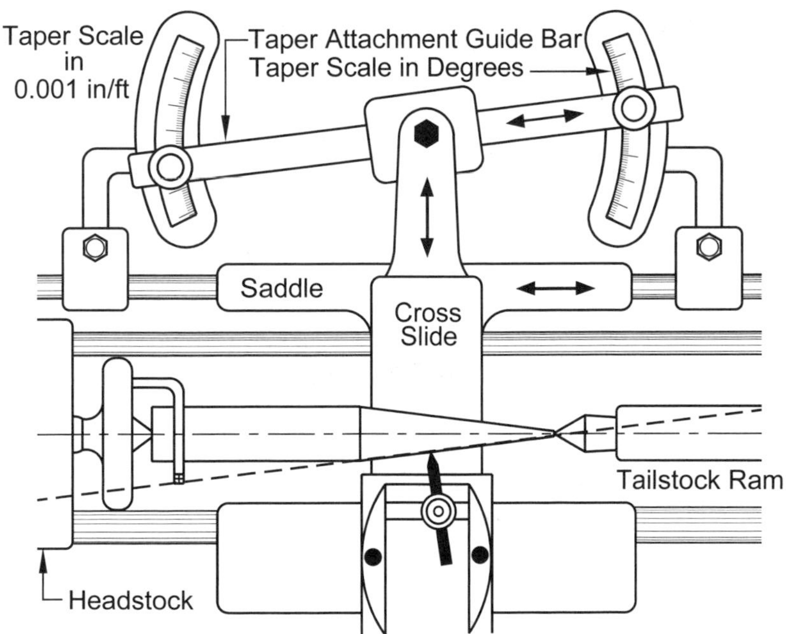

Figure 7–83. Taper attachment guides the cross slide to cut a taper.

Radii Cutters

What is a *radii cutter* and how does it work?

A radii cutter is a special tool bit holder that guides the tool around a fixed point. It cuts concave surfaces, convex surfaces, hemispheres and spheres. Some radii cutters have a horizontal axis of tool motion like the Sherline tool in Figure 7–84 (top). Other designs have a vertical axis, as in the Holdridge tool, Figure 7–84 (bottom). These tools hold tolerances in the thousandths of an inch range and make a wide range of products from bowling balls to prosthetic eyeballs, and from lenses to optical reflectors. Radii cutters bolt onto the cross slide or compound slide depending on their design. They are often used in plastic mold making. Figure 7–85 shows four shapes of radii that can be cut. In addition to making spheres and their sections, radii cutters also apply graceful curves to machine tool handles.

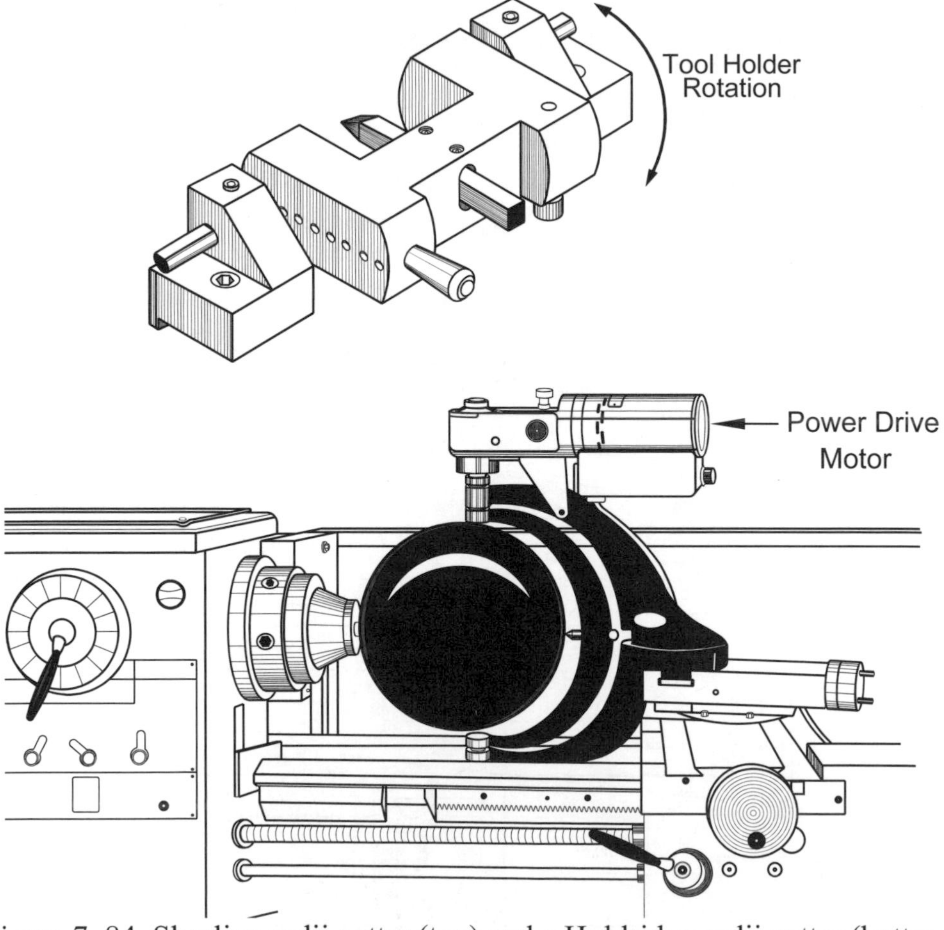

Figure 7–84. Sherline radii cutter (top) and a Holdridge radii cutter (bottom) with a power drive cutting a 12-inch sphere on a large lathe.

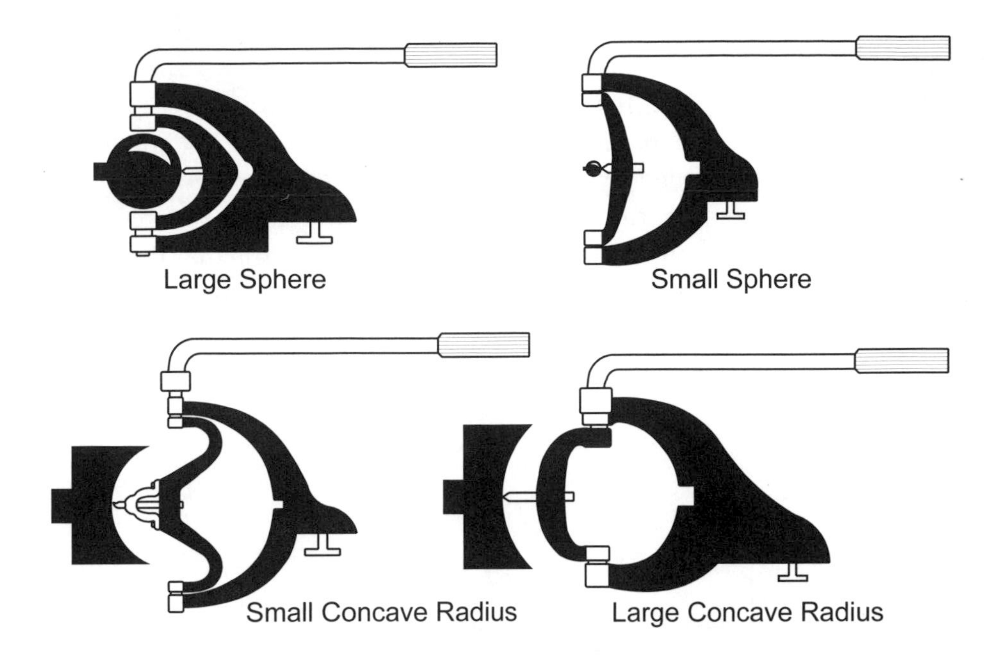

Figure 7–85. Four Holdridge radii cutter configurations for convex radius or sphere cutting (top) and concave radius cutting (bottom). These tools can be preset for specific diameters.

Spur Drivers

What is the purpose of a *spur driver*?

A spur driver is used to hold and drive a length of wood for turning. It has a Morse taper on one end and a small center and driving fins on the other, Figure 7–86. The spur driver fits into the Morse taper of the spindle. The work is held by the center pin of the spur driver and a center in the tailstock. The four fins on the spur driver's face dig into the work and turn the work with the spindle. No face plate or lathe dog is needed for this operation. The full length of the work is available for turning. Some plastics may also be turned using a spur driver.

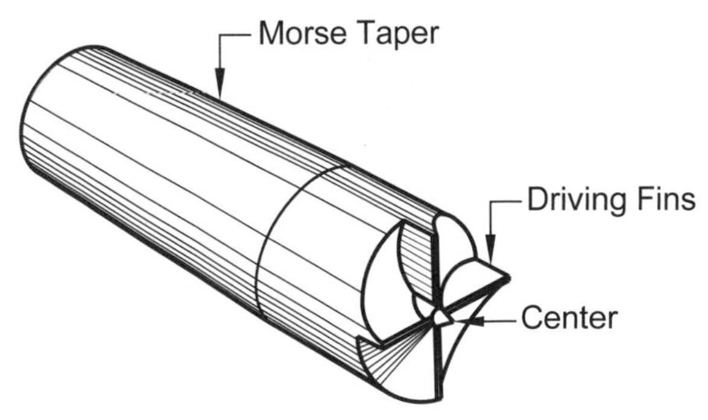

Figure 7–86. Spur driver.

Rear Tool Posts

What is the purpose and advantage of using a *rear tool post*?

The rear tool post, which mounts on the back end of the cross slide, permits parting off to be performed by cranking the cross slide *toward* the operator without changing out the front—or regular—cutting tool, Figure 7–87. Its advantage is that it is a major time saver. Its disadvantage is that it is harder to get cutting fluid onto the parting blade than on a top-mounted tool. The cutting tool must be mounted upside down to bring the work into the cutting edge.

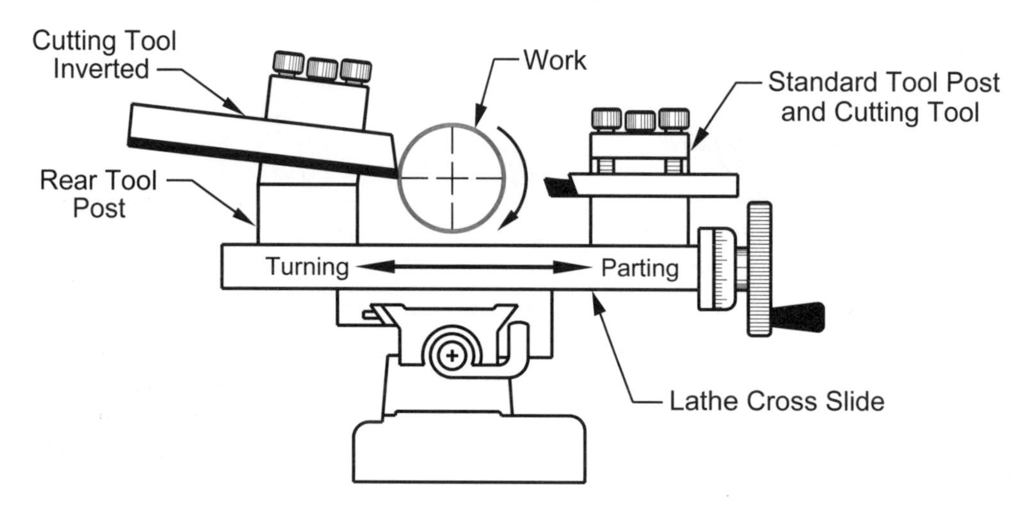

Figure 7–87. Rear tool post holds cutoff tool.

Section V – Speeds and Feeds

Chatter

The term *chatter* often appears in connection with machining operations. What action does this describe and why does it occur?

Chatter is an unwanted resonance or vibration between the cutting tool and the work. Chatter occurs with lathe tools, milling cutters, drills, reamers and countersinks. It is an unacceptable condition because it rapidly breaks down the tool's cutting edge and leaves a rough surface on the work.

Here is how chatter occurs in a lathe: As the work turns more rapidly, cutting forces on the work, cutter and their supporting structures increase enough to cause them to deflect, interrupting the cutting process, and storing energy in these deflected components like a spring. With cutting stopped, the deflected components spring back into place, releasing their stored energy and cutting begins again. This cycle repeats itself many times a second.

Lathe chatter may be caused by:

- Excessive spindle speed. Reduce the speed and increase the feed.
- Large lead angle, which causes cutting tool forces to deflect the work rather than apply them in line with the lathe axis. This is especially important when using a boring bar. Reducing the lead angle often eliminates chatter.
- Excessive cutting tool overhang.
- Dull cutting tool.
- Insufficiently supported work, such as a long, slender shaft, which deflects under cutting forces. Use a steady rest or follower rest.
- Loose headstock spindle bearings.
- Poorly fitting cross slide and compound.

Limiting Factors

If a lathe or other machine tool is not limited by its horsepower or tendency to chatter, what determines how fast metal can be removed?

The ability of cutting tool bits to withstand high temperatures generated by friction limits cutting speed. HSS loses its strength at about 1000°F, Cobalt HSS at 1300°F and carbide tools at about 1400°F. A tool run well below its softening temperature is not being used efficiently and is wasting machine time. A tool run above its softening temperature will have a shortened life as its cutting edge breaks down. When this happens, substantially more electrical energy will be needed as the tool dulls. Some large machine tools have meters to show motor horsepower demand, which indicates when tools are failing.

What factors determine heat generation in cutting tool bits?

There are three:

- *Cutting speed*, the speed at which the work surface passes the cutting tool. Cutting speed is measured in feet/minute (ft/min), sometimes called surface feet/minute (sfm), or meters/second (m/sec).
- *Feed rate,* the speed the carriage moves along the work parallel to its centerline measured in inches/revolution or millimeters/revolution.
- *Depth of cut* is measured in inches or millimeters.

How do heat-generating factors affect cutting tool life?

When the optimum cutting speed, feed rate and depth of cut are found and cutter bit life has been established, experiments have shown that doubling:

- Cutting speed above the optimum reduces tool life by 90%.
- Feed rate above the optimum reduces tool life by 60%.
- Depth of cut above the optimum reduces tool life by 15%

Operating cutting speed is the most important factor in tool life.

Metal Cutting Characteristics

What determines how easily a metal can be cut?

The most important factor is the metal's own microstructure. Another factor is any cold work that may have been done to the metal because cold work increases hardness. Usually metals with a similar microstructure machine in a similar manner, and those with a similar hardness show similar machining properties. Because harder metals are more difficult to penetrate, cutting them requires more horsepower, which generates higher temperatures on the cutting tool at any cutting speed. This makes it necessary to reduce the cutting speed for hard materials so that the cutting bit temperature does not rise so high it destroys the tool bit. This means that cutting bit materials that can withstand higher temperatures can run at higher cutting speeds and are more economical. For example, tungsten carbide tool bits have cutting speeds two to three times higher than high-speed steel (HSS) because of carbide's ability to remain hard at high temperatures.

What does *recommended cutting speed* for a material mean as shown in production engineering manuals such as *Machinery's Handbook*?

It is the most economical cutting speed to use in a commercial production setting. This speed is determined experimentally and takes into account fixed charges, tool-changing time, tool costs, sharpening costs and overhead.

Not all users will be able to use the recommended cutting speed because:

- Their lathe may not have adequate horsepower to maintain the recommended speed.
- The lathe and tool holder may not have enough rigidity to prevent chatter.
- The work holder may not be strong enough to sustain the forces seen at this speed.
- The physical properties of the their workpiece may differ significantly from the material used to establish the recommended cutting speed.
- Many users will not be in a commercial production setting where the economic costs used to develop the recommended cutting speed are valid.

Conclusion: These recommended commercial cutting speeds are most useful in a production environment, but can provide an approximate starting point for selecting the cutting speed in any shop.

What are some typical recommended *cutting speeds* for common materials?

Table 7–3 shows these speeds for HSS tool bits. Cutting speeds may be increased by a factor of 2 to 3 for carbide tool bits. Speed may be increased by up to 50% if cutting fluid is used. Cutting speed may vary within material lots and within sections of material.

Material	Roughing Speed		Finishing Speed	
	ft/min	m/min	ft/min	m/min
Low-carbon Steel	70–90	21–27	77–100	23–30
Medium- & High-carbon Steel	50–70	15–21	55–77	17–23
HSS	40–60	12–18	52–78	16–23
Stainless Steel	60–80	18–24	78–104	23–31
Aluminum	200–300	60–90	400–800	96–144
Brass & Copper	150–190	45–57	165–210	50–63
Cast Iron	60–80	18–24	102–128	29–38

Table 7–3. Cutting speeds for roughing and finishing with HSS tool bits.

How can this *ft/min*, sometimes called *surface feet/min (sfm) speed*, be used to set lathe rpm?

To set the lathe rpm, the diameter of the work must be taken into account because the speed at the tool tip varies with the diameter of the work. The smaller the diameter, the higher the spindle speed must be set to achieve a given ft/min surface speed. Here are the formulas:

For inch units:

$$\text{Spindle Speed (rpm)} \approx \frac{\text{Cutting Speed (feet/min)} \times 4}{\text{Diameter of the workpiece (inches)}}$$

For metric units:

$$\text{Spindle Speed (rpm)} \approx \frac{\text{Cutting Speed (meters/min)} \times 320}{\text{Diameter of the workpiece (mm)}}$$

Remember that these speeds are only guidelines, not exact speeds for your application. Choose the closest speed available on your lathe. If the work chatters, reduce the spindle speed and increase the feed. See the *Appendix, Table A-2, Surface Speed Table & Lathe Cutting Tool Selection Chart*, pages 485–486.

What is meant by setting the *feed* when performing turning?
The feed is the rate at which the tool is moved parallel to the centerline of the work in inches/revolution or mm/revolution of the work. As with recommended speeds in Table 7–3, the maximum feed you can achieve will depend on the condition of the lathe, its horsepower, tool holder rigidity and the shape and condition of the tool. Table 7–4 lists typical feeds used in production settings for common metals. Note that the feed is listed in inches/revolution (or mm/revolution) of the work.

Material	Roughing Cuts		Finishing Cuts	
	in/rev	mm/rev	in/rev	mm/rev
Low-carbon Steel	0.010–0.030	0.22–0.75	0.006–0.012	0.15–0.30
Medium- & High-carbon Steel	0.010–0.030	0.25–0.75	0.006–0.012	0.15–0.30
HSS	0.010–0.030	0.25–0.75	0.006–0.012	0.15–0.30
Stainless Steel	0.006–0.012	0.15–0.30	0.003–0.015	0.08–0.40
Aluminum	0.008–0.025	0.20–0.62	0.003–0.016	0.08–0.40
Brass & Copper	0.008–0.030	0.20–0.75	0.006–0.014	0.15–0.35
Cast Iron	0.010–0.030	0.25–0.75	0.004–0.015	0.10–0.38

Table 7–4. Cutting feeds for straight turning with HSS and carbide tools.

Section VI – Preparing for Lathe Operation

Cleaning Mating Surfaces

Why is cleaning dirt and chips from threads and mating surfaces important?
Most lathe tools and accessories are secured to the lathe against flat surfaces, threads or tapers. Cleaning these of metal chips, dirt and grease is essential to achieving good alignment and to reduce wear. This cleaning may be done with a brush or rag. Compressed air may also be used, but it presents eye hazards and many shops prohibit using it for cleaning. Compressed air must never be used to clean lathe ways since it drives grit *under* the saddle, leading to rapid wear. *Careful cleaning of mating surfaces is always the first step in installing a tool or accessory.*

Also, inspect the external Morse tapers for burrs and remove them with a file or stone before installing them in the lathe. To remove debris from the female threads on a chuck, use the thread-cleaning tool in Figure 7–88. Use a stiff bristle brush to remove debris from the spindle threads.

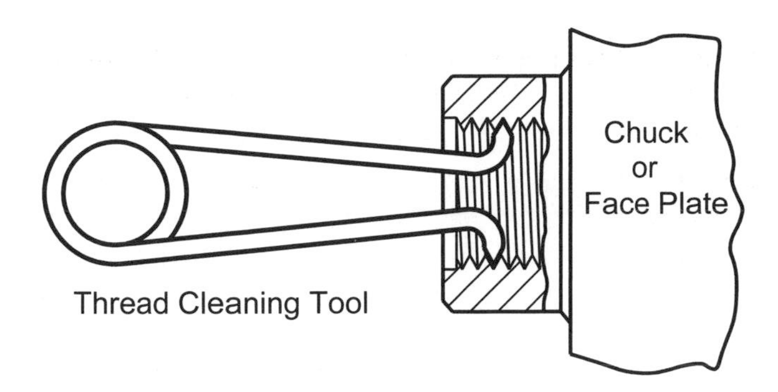

Figure 7–88. Cleaning chuck threads.

Tailstock Alignment

Why is tailstock alignment necessary?

Most lathes have the ability to shift the position of the tailstock back and forth across the ways to cut long tapers. If the lathe tailstock barrel is not concentric with the headstock, the lathe will cut cones instead of cylinders.

Before changing tailstock alignment, check that the lathe is level and has not moved since its last leveling. Check both across and along the ways with a precision level. See Section X of this chapter.

What are four ways to check tailstock alignment?

In order of increasing accuracy:

- The quickest method is to use the tailstock witness marks to indicate alignment, Figure 7–89.

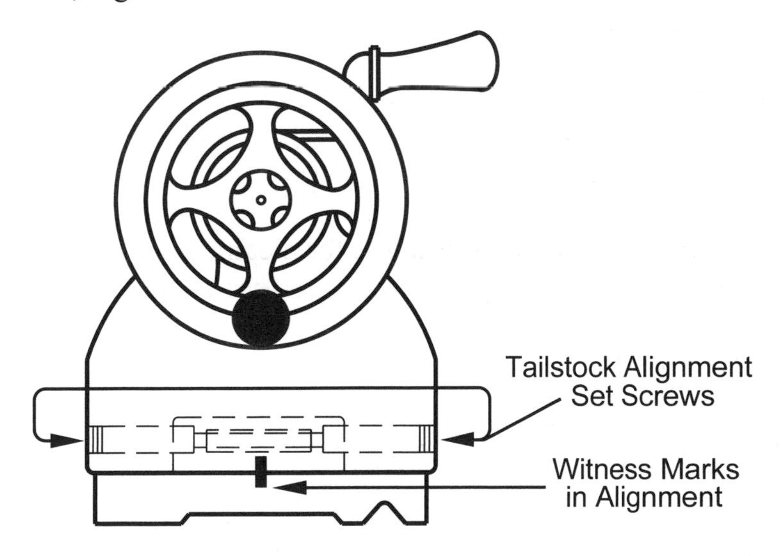

Figure 7–89. Checking tailstock alignment by examining its witness marks.

- Put lathe centers in both headstock and tailstock tapers and push the tailstock so its center just touches the center in the headstock. When adjusted to look like Figure 7–90, the tailstock is aligned.

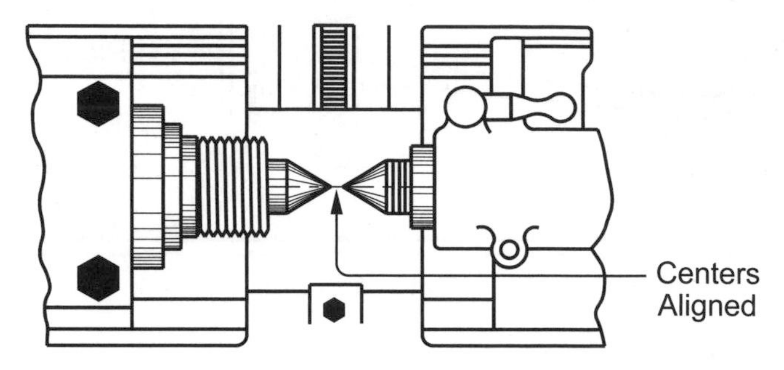

Figure 7–90. Checking tailstock alignment by comparing center tips.

- Another method to quickly check tailstock alignment is to insert a previously machined and accurate test bar between centers and using a dial indicator, verify that there are no dimensional changes as it is moved along the work on the carriage, as in Figure 7–91.

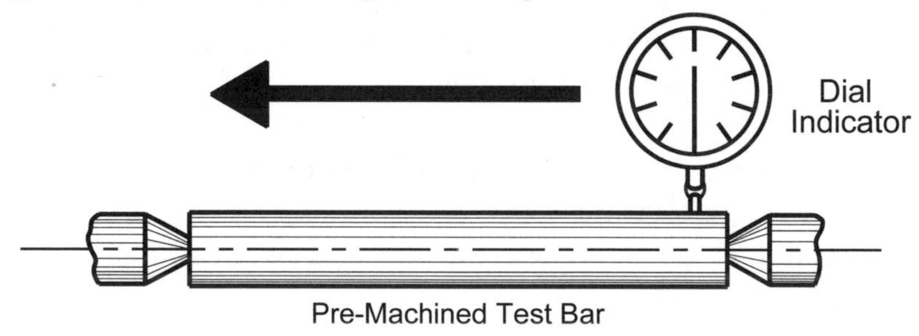

Figure 7–91. Checking tailstock alignment with a pre-machined test bar.

- The most accurate check is to turn a long cylindrical test workpiece or by machining shoulders on a test workpiece. If the diameters at each end are the same, the headstock and tailstock are aligned, Figure 7–92.

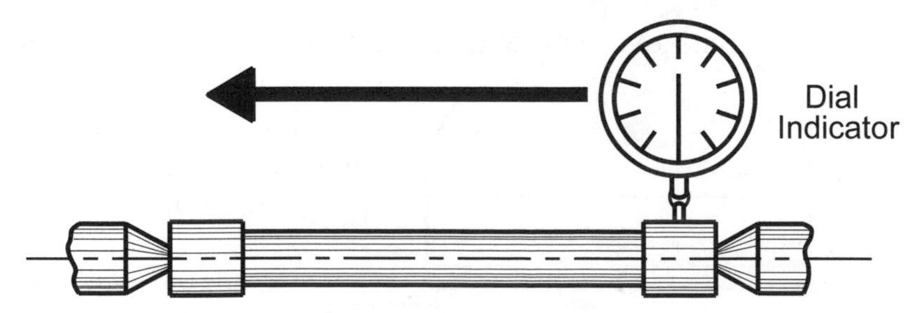

Figure 7–92. Checking tailstock alignment by machining shoulders on a test bar and checking shoulder diameters with a dial indicator. Alignment may also be checked by measuring the shoulder diameters with a micrometer.

Section VII – Lathe Operations

Turning

What is *turning*?

Turning is the most basic machining process performed in the lathe. The lathe rotates the work while the tool moves parallel to the ways and the axis of the work, turning cylinders or cones. Dimensional tolerances in the 0.0005-inch range are possible. The most basic turning set up is shown in Figure 7–93.

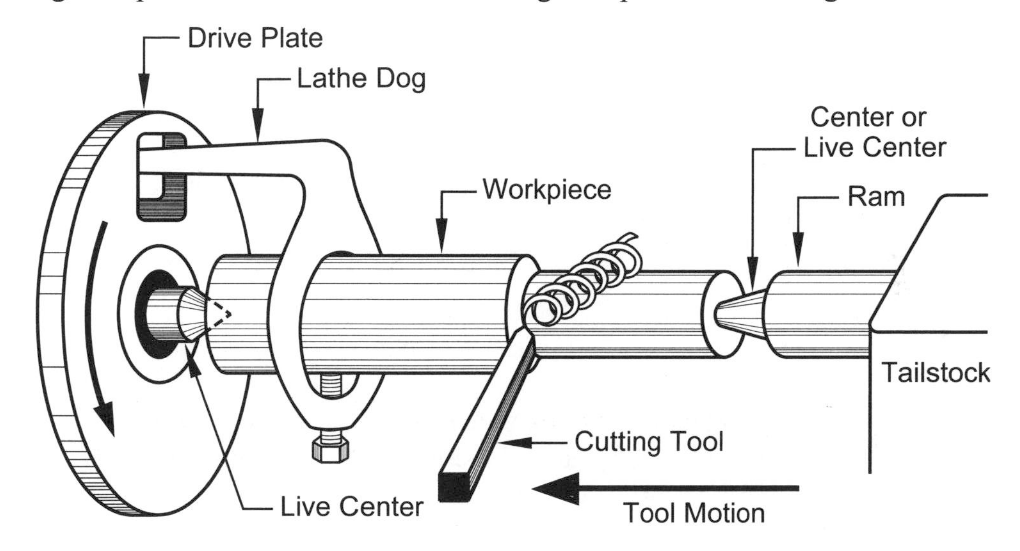

Figure 7–93. Turning between centers with a drive plate and a lathe dog.

The headstock center is sometimes called a *live center* because it turns with the spindle and does not rotate against the work. This live center should not be confused with the *live center* sometimes used in the tailstock which contains ball or roller bearings.

The entire workpiece may be turned to the same diameter by removing it from the lathe, turning it end over end and replacing it between centers to access the section under the lathe dog. Use two pieces of copper or aluminum, one between the lathe dog and the work, and the other between the lathe dog screw and the work, to prevent marking the workpiece finish.

The leg of the lathe dog must fit tightly into the drive plate hole. Some lathe dogs have tapered legs so they can fit snugly into the drive plate. Otherwise, use a wooden wedge between the lathe dog leg and the drive plate hole to take up any slack. Wire the wedge in place. Failure to do this will result in chatter and will be reflected in surface finish on the work.

Are there other ways to hold work for turning?
Yes, there are many different ways:

- Short work can be held in a collet.
- Short work can be held in a 3- or 4-jaw chuck.
- Long work can be held with a 3- or 4-jaw chuck and a center in the tailstock.
- Long work can be held with a 3- or 4-jaw chuck and steady rest.
- Long work can be held with a 3- or 4-jaw chuck, a steady rest and a follower rest.

What are the basic steps to perform turning on centers?

1. Clean debris from internal and external Morse taper surfaces of the headstock, tailstock and centers, and install the centers.
2. Install the tool at proper height using the center in the headstock as a height reference:
   - *For rough cuts*, mount the cutting tool slightly *above center.*
   - *For finish cuts*, mount the cutting tool *exactly on center.*
3. Mount the work holder in the same manner as a chuck, or centers and dogs.
4. If using a dead center, lubricate the center hole at the tailstock end of the workpiece. No lubricant is needed if using a live center which is preferred.
5. Slip the dog over the work if one is used.
6. Place the workpiece between the centers and insure that the tailstock ram is tight enough against the work to prevent end-to-end movement and yet loose enough to allow workpiece rotation.
7. Adjust and tighten the dog, if used. Use a piece of copper or aluminum under the lathe dog's set screw to prevent damage to the work.
8. Position the carriage so the tool clears the work and is located about where turning will begin.
9. Rotate the work/spindle by hand to insure work is secure and free to rotate.
10. Determine and set the proper spindle speed and feed.
11. Start the machine and begin cutting, taking small cuts initially and working up to larger ones. Use power feed if desired.
12. If the tool chatters, increase the feed and reduce the speed.
13. Continue successive cuts until desired diameter is reached as determined by calipers or a micrometer. Remember that the depth of cut indicated by cross slide calibrations on most American-made lathes is taken off the radius of the work, or doubled, since the same amount of material is taken from *both sides* of the work when turning. Conversely, on imported lathes cross slide calibrations read in thousandths off the diameter.

Facing

What is *facing*?

Facing is similar to turning except that the tool moves *across* the face, or end, of the work perpendicular to the ways, Figure 7–94. Facing is one method to adjust the length of the workpiece and square its ends.

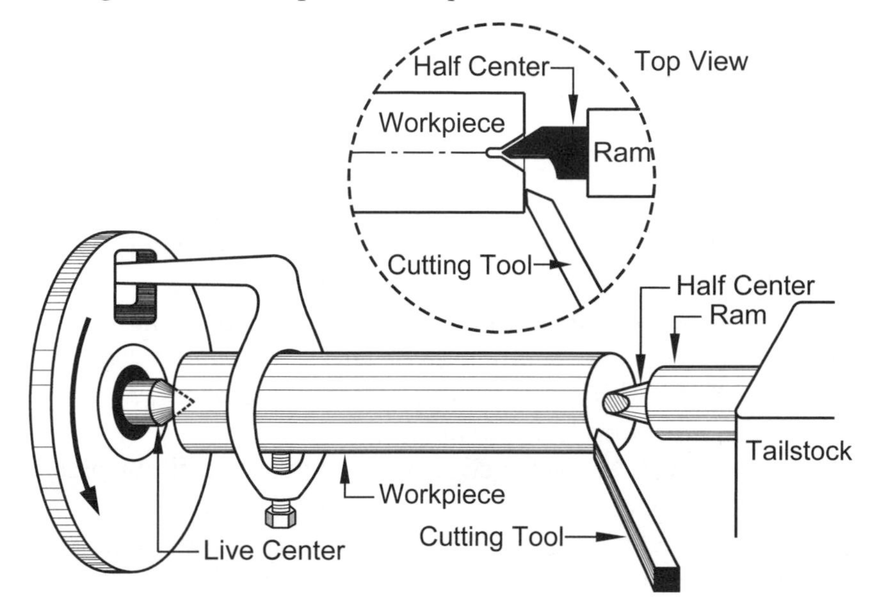

Figure 7–94. Facing operation with a live center and a half center.

If a chuck or collet holds the work, this operation is called *full facing*, Figure 7–95. In order to get a square face, the unsupported work protruding from the collet or chuck must be short and stubby to insure the work does not bend under cutting tool force, or use a steady rest.

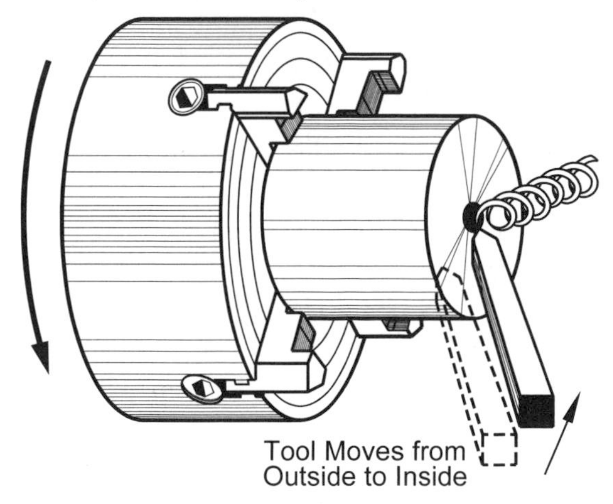

Figure 7–95. Full facing operation with workpiece in a chuck.

What are the basic steps to perform facing?

1. Clean all mating surfaces of debris.
2. Install the tool bit at center height and adjust its angle with the tool holder and compound slide so the tool can access the workpiece end.
3. Mount a work holder such as a chuck or collet.
4. Secure the work in a work holder and center it, if required.
5. Position the carriage so the tool clears the work and is located about where facing will begin, and lock the carriage to the ways.
6. Rotate the work/spindle by hand to insure it is secure and free to rotate.
7. Determine and set the proper speed and cross feed.
8. Start the lathe and begin cutting taking small cuts initially and working up to larger ones. Use a power feed if desired. Making the facing cuts from the center out prevents running the tool into the half-center.
9. If the tool chatters, increase feed and reduce the speed.
10. Continue successive cuts until the desired workpiece length is reached as determined by calipers, micrometer or DRO. *Remember that the depth of cut indicated by cross slide collar calibrations is accurate only when the cross slide is positioned at right angles to the work face.*
11. Use a file to break the outer edge of the work. See filing precautions in *Chapter 3 – Filing & Sawing.*

If the workpiece length is to be cut to a specific length, use a hermaphradite caliper or digital caliper to mark the final length on the workpiece surface. Using a black felt tip marker or layout dye to darken the workpiece surface helps the cut marks show up. It is a good practice to make a very light cut with the lathe tool to mark the final length of the workpiece.

Parting Off

What is *parting off* and why is it useful?

Parting off, also called *parting* or *cutting off,* is a process of cutting off part of a workpiece. Machining a part often starts with a piece of stock longer than the finished part length. This extra stock is held inside a chuck or collet and all of the finished part's length is exposed and available for machining work. Then, when machining is finished, the final part is cut off from the stub end sitting in the collet or chuck, Figure 7–96. Parting off tool bits are usually much narrower than other tools to minimize the material to be removed. Because these tools have a much different cross sectional shape than regular lathe bits, they cannot be held directly in a tool post and require a special tool holder. A medium-sized lathe (typically using a 3/8-inch square tool bit) would normally use a parting tool 3/32 inches in width and 5/8 inches in height.

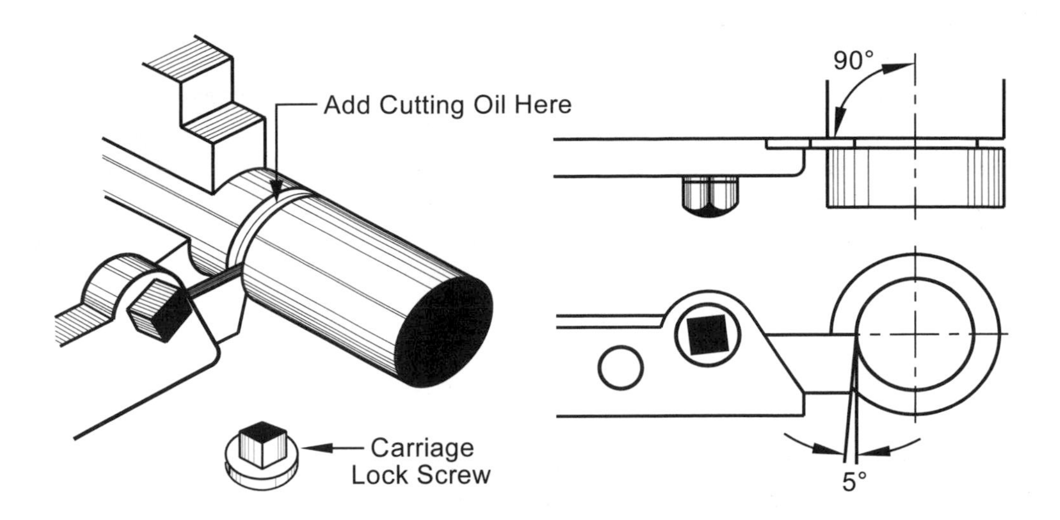

Figure 7–96. Parting off in a chuck.

What do parting tools look like?
Parting tools are long and narrow, and like other lathe tools, they come in HSS, brazed carbide and carbide insert designs. See Figures 7–24 (top left), 7–97, 7–98 and 7–99.

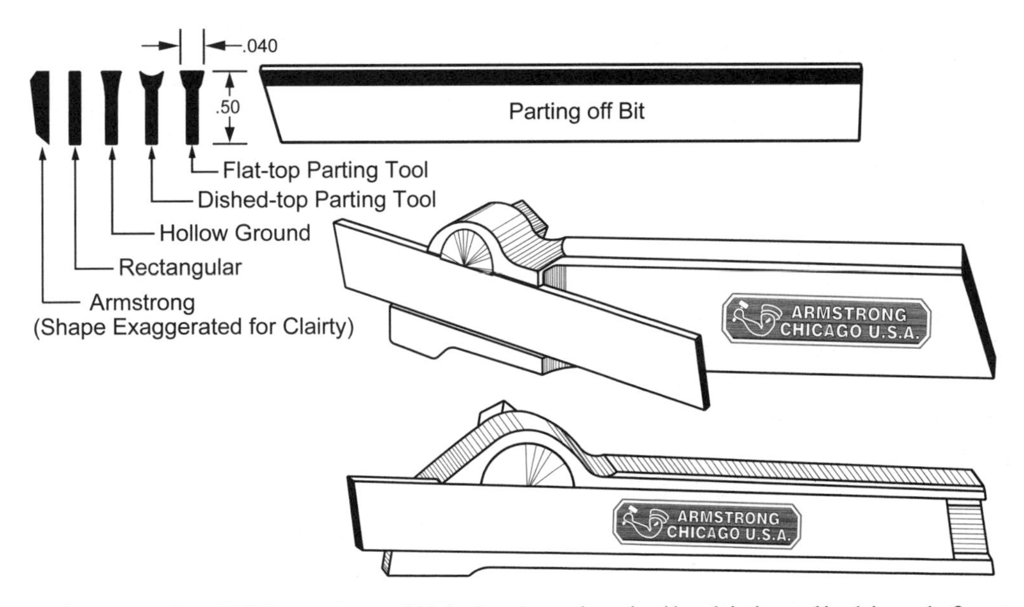

Figure 7–97. HSS parting off bit (top) and a similar bit installed in a left-handed tool holder to part close to the spindle (middle) and installed in a straight tool holder (bottom). A dished-top cutter surface puts a curl on the chips, narrowing them so they can leave the cutting groove more easily and without binding.

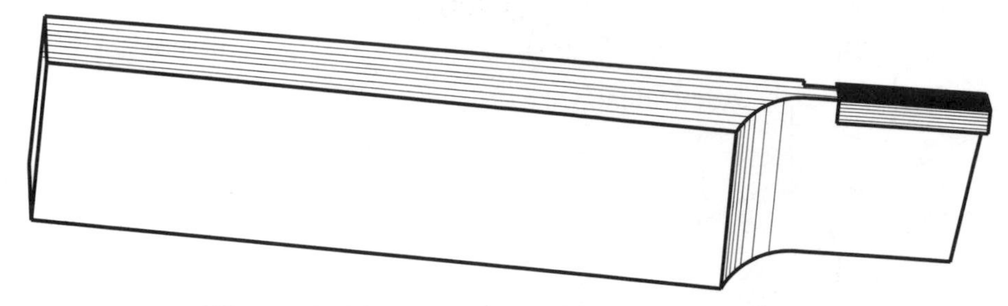

Figure 7–98. Brazed carbide parting off bit.

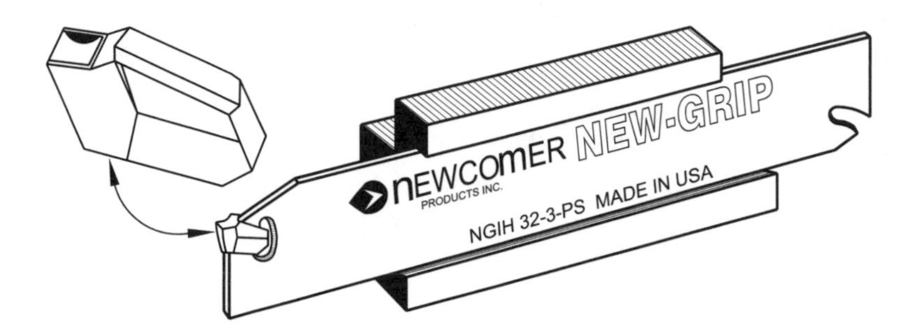

Figure 7–99. Carbide insert parting off bit.

What are the basic steps to perform parting or cutting off?

1. Clean all mating surfaces.
2. Install a work holder and secure the work in the work holder and center it, if required.
3. Set the cutting tool bit exactly on the center height and at a 90° angle to the workpiece, Figure 7–96 (right).
4. Lock the carriage to the ways.
5. Position the tool to clear the work.
6. Rotate the work/spindle by hand to insure it is secure and free to rotate.
7. Set the spindle speed at about one-third of that used for turning.
8. Begin by taking a small initial cut and work up to larger ones. If the tool chatters, reduce the speed and increase the feed.
9. Keep plenty of lubrication in the cutting groove and on the tool.
10. Use a board to protect the ways from damage when the cut is finished and the unsupported end of the work falls away.

Parting cannot be performed on work supported between centers as the two parts will pinch the tool and bind when the cut is complete, Figure 7–100. It would be fine to make a *partial cut* with the parting tool, remove the work from the lathe, put it in a vise and finish cutting the work with a hacksaw.

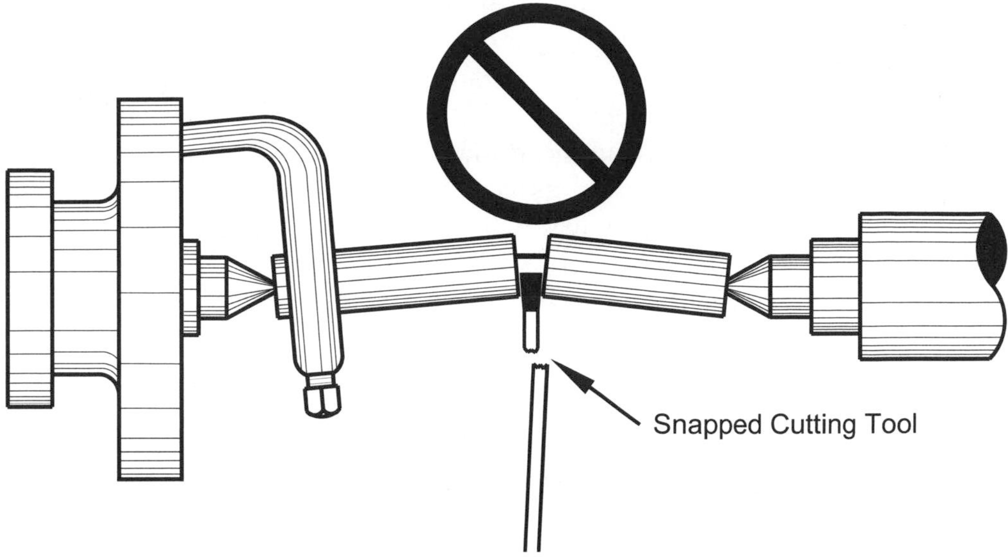

Figure 7–100. Parting off between centers is likely to break the tool bit.

Drilling

What types of drills are used in lathe work and how are they held?

There are two types of drills and holding methods:

- *Standard twist drills* fit into a drill chuck with the chuck secured in the Morse taper of the tailstock.
- *Drills with a Morse taper end* fit directly into the Morse taper of the tailstock. When the drill Morse taper is larger than the tailstock taper, an adapter is used.

See Figure 7–53.

How is *drilling* performed in the lathe?

Drilling begins by first using a center drill to locate the center of the workpiece and drilling a starter hole, then a regular drill is used to complete the hole. The steps are:

1. Clean all mating surfaces of debris, particularly the Morse taper internal and external surfaces.
2. Install the drill chuck, install the center drill in the chuck and then mount the work holder, such as a chuck or collet.
3. Secure the work in the work holder and center it, if required.
4. Move the carriage out of the way.
5. Rotate the work/spindle by hand to insure it is secure and free to rotate.
6. Feed the center drill slowly until it *finds* the center and gets its start, then drill the hole deep enough to give the twist drill a solid starting point.
7. Withdraw the center drill, mount the twist drill and drill the final hole.

8. After the drill hole is two diameters of the hole in depth, withdraw the drill and clear the metal chips from the drill flutes. On subsequent drilling, withdraw the drill after every additional diameter in depth for chip clearing and lubrication. Drills under ⅛-inch diameter must be cleared more frequently or they may shear off.
9. Use a lubricant on steel, stainless steel, but not on cast iron.
10. If the drill chatters, increase feed and reduce the speed.

Is it a good practice to use a series of increasingly larger drills to drill a large hole?
Although this may be necessary if the lathe lacks the power to drill the full-size hole in a single pass, this method is likely to lead to problems. Each successive drill is likely to find a new center and defeat the purpose of using a center drill to locate the actual center.

When drilling a cored hole in a casting, the drill follows the cored hole and runs off center. How can this be avoided?
This can be avoided by machining a bevel on the cored hole to give the drill a proper start, Figure 7–101.

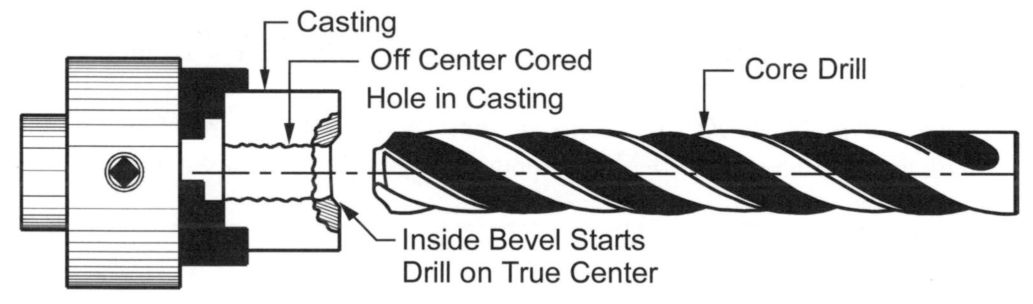

Figure 7–101. Machining a bevel on the edge of a cored hole allows the core drill to run true. Core drills have 3 or 4 flutes and a blunt tip.

What are the common work holding methods for drilling?
- In a chuck.
- In a collet.
- In a chuck with a steady rest.

Boring

How is *boring* performed in the lathe and how does it differ from *drilling*?
Boring is done with a single-point cutting tool on a boring bar. See Figure 7–24 (bottom) and 7–102.

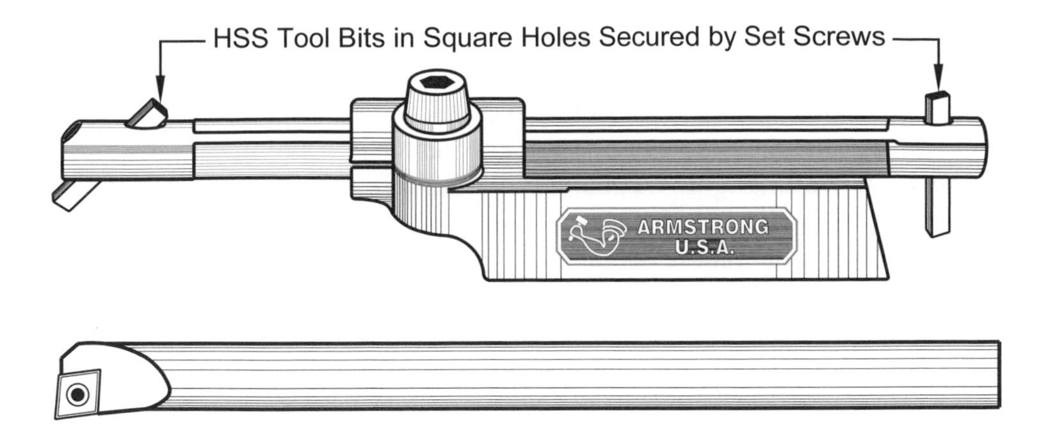

Figure 7–102. Boring bar with conventional HSS tool bits (top) and boring bar with carbide insert (bottom).

Boring is an internal turning operation as seen in Figure 7–103. Drilling is done with a drill and does not necessarily require the use of a lathe. One advantage of boring is that *any* size hole may be bored within the capacity of the lathe. The hole size is not limited to the fixed increments of available drills. On smaller holes, drilling is faster, so boring is usually used for holes too large to drill or where the right diameter drill is not available.

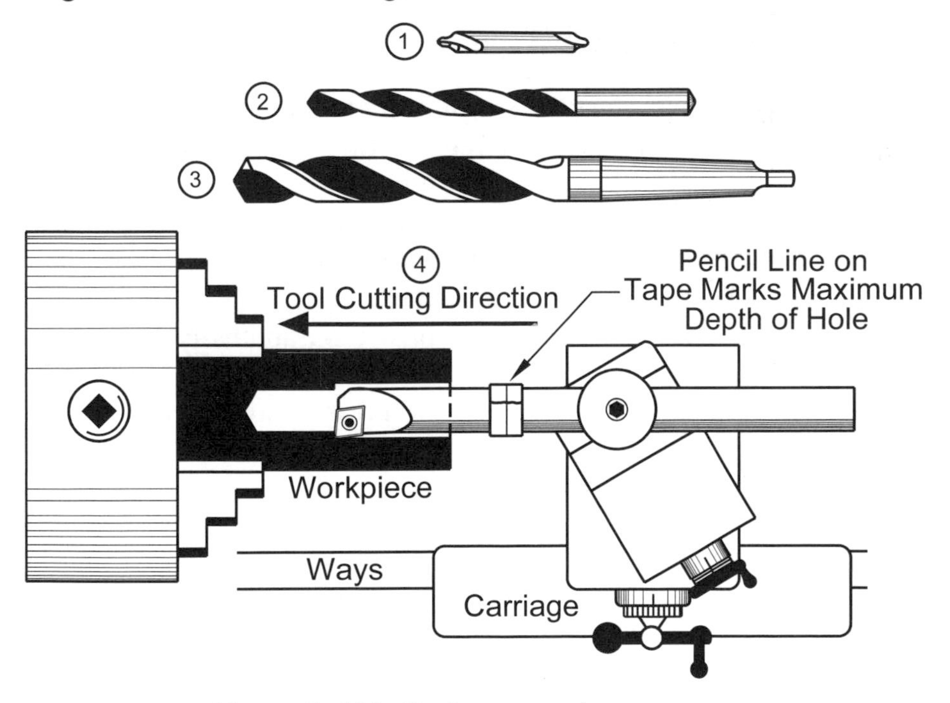

Figure 7–103. Boring operation sequence.

What are the basic steps to perform boring?

1. Clean all mating surfaces.
2. Install the work holder (for example chuck or center and dog).
3. Secure the work in the work holder and center it, if required.
4. Install the cutting tool and position it at the center height.
5. Position the tool to clear the work.
6. Rotate the work/spindle by hand to insure it is secure and free to rotate.
7. Apply a mark or piece of tape, to the boring bar indicating its maximum depth into the work or use a carriage stop. This will prevent making too deep a hole or running the bar into the end of the starter hole. *Do not snug up the clamp on the carriage stop so it will not move if struck by the carriage on a power feed.*
8. Determine and set the proper speed and feed.
9. Set the tool to make an initial light cut.
10. At the end of each cut, stop the machine, set the cross slide micrometer collar to zero, back the tool away from the work and withdraw the boring bar from the hole.
11. Bring the tool back to the zero on the cross feed and increment the cutting depth for another cutting pass.
12. Repeat steps 10 and 11 until final bore diameter is reached.

Here are some other hints on boring:

- Movement of the cross slide screw is reversed from that used to withdraw the tool when turning. The tool is cranked *in* to withdraw it from the work.
- Boring bars are not as rigid as regular tool bit holders; they deflect. Because of this, use the largest, thickest bar that will fit in the starter hole. Several light cuts may be necessary to cope with their deflection instead of one heavy one. Keep the length of the boring bar as short as possible.
- Lubricant may be needed on ferrous metals.
- Plan your work so the final cut is 3 or 4 thousandths of an inch, since boring bar deflections will make smaller increments impossible. The bar will just deflect and not exert enough tool pressure to remove metal.

Shaping & Grooving

How are *shaping* and *grooving* performed in the lathe?

There are several ways to perform these operations depending on the shapes desired:

- For a simple corner rounding, a single cutting operation with a formed tool, like those shown in Figure 7–104, will do the job. The tools in this figure are commercial products. Using a grinding wheel, within a few minutes tool steel lathe bits can be made into almost any desired shape.

The limiting factor is the size of the tool and the cut your lathe can take. Giant form tools will not work as more lathe torque and rigidity is required than the lathe can deliver and large cutting surfaces tend to chatter.

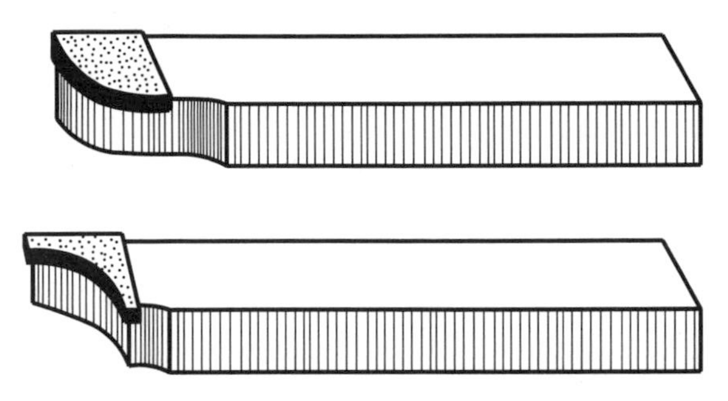

Figure 7–104. Brazed carbide form tools.

- If the shape is larger than a shaped tool can cut, make a template from cardboard or sheet metal and make a series of successive cuts with a pointed tool to get the shape you want. Complete the shape by smoothing with a file and abrasive cloth.
- For a very precise shape, determine a series of dimensions needed to outline the shape and then use the cross feed and compound collar calibrations to make a series of cuts to develop that shape. Finish smoothing with a file and finally an abrasive cloth.
- O-ring grooves or precise ridges can also be readily cut with brazed carbide grooving bits such as those in Figure 7–105.

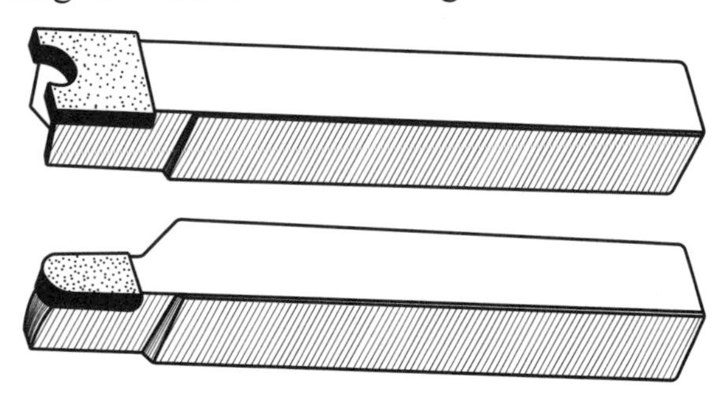

Figure 7–105. Brazed carbide grooving bits.

- External, internal or undercut grooves, as well as ridges are cut with tools such as those in Figure 7–106. Light cuts must be made using undercutting and internal groove tools because of their long, thin unsupported cutting tips.

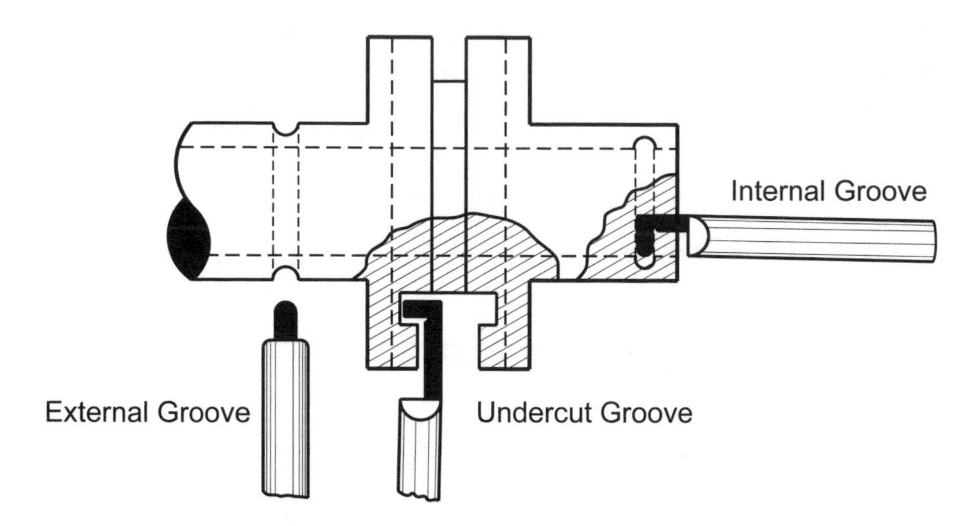

Figure 7–106. Cutting external, internal and undercut grooves.

What are typical shapes put on the ends of cylindrical work?

Usually machining begins with a squared off end. The next step is to either chamfer or round the ends. Chamfering, Figure 7–107 (top), is easily done using a tool bit in the compound. Although rounding can be done with a file, this approach tends to develop into an eccentric rounding. The results depend more on operator skill than tooling. The best rounding is done with a shaped tool bit like that in Figure 7–107 (bottom).

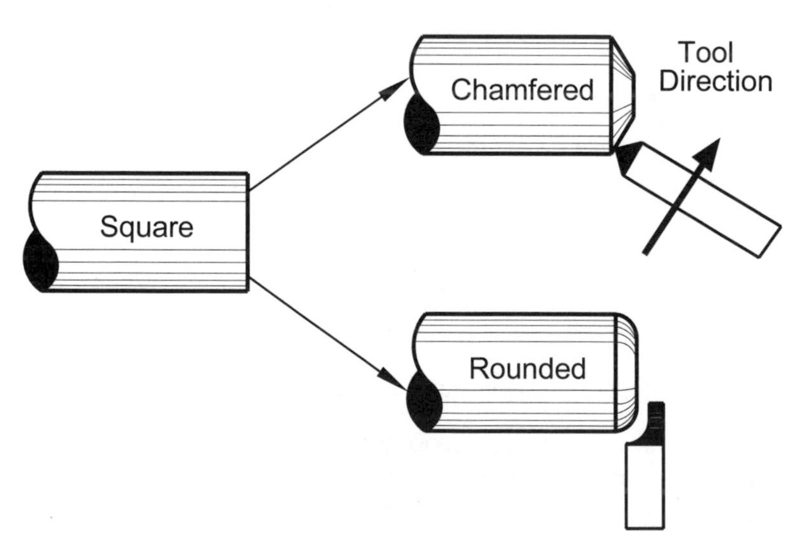

Figure 7–107. Finishing cylinder ends.

How are shoulders cut on cylindrical work?

The most common method used to cut shoulders is:

1. Measure to locate the shoulder on the work and mark the edge with a thin groove.
2. Turn the smaller diameter to size as in Figure 7–108 (left) and face the shoulder. If the base of the shoulder is rounded, use a curved tool, as in Figure 7–108 (right).

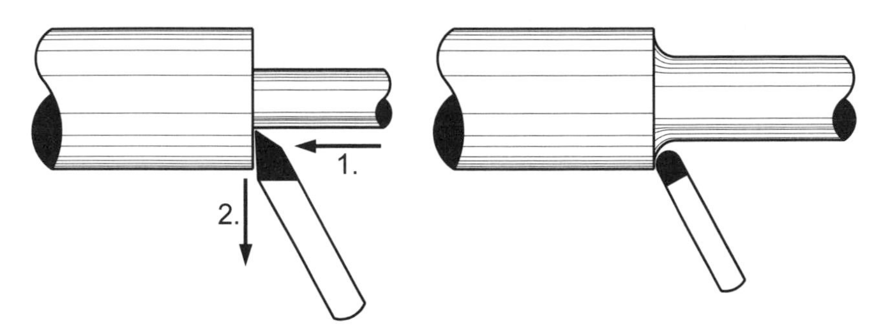

Figure 7–108. Cutting shoulders. Rounded shoulders are usually preferred as they are less likely to generate a stress point leading to cracking.

How can the depth of grooves be readily measured using a regular micrometer?

In order to measure to the bottom of external grooves, use two pieces of round stock of the same diameter and subtract their diameters from the micrometer reading, Figure 7–109 (top). A radius gage may be used to check groove radius and also the cutting radius. See Figure 7–109 (bottom).

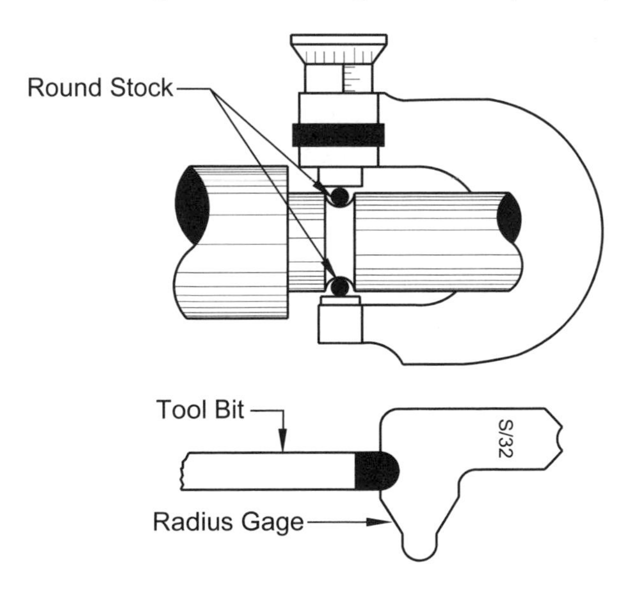

Figure 7–109. Measuring groove depth with round stock (such as dowel pins) and measuring tool bit diameter with a radius gage.

Taper Turning

What is a *taper* and what is its purpose?

A taper is that part of a workpiece which has a uniform change in diameter along its length. Tapers may be external (male) or internal (female). Fitting matched internal and external tapers together provides a rapid and accurate means of concentrically aligning machine parts and is commonly used on twist drills, lathe centers, lathe spindle noses and reamers. Sometimes though, a machine part must be tapered for other reasons and is not made to fit into a mating part.

How are tapers specified?

Tapers in inch units are specified by:

- Inches per inch (tpi).
- Inches per foot (tpf).
- Degrees.

Tapers in metric units are specified:

- By length of the taper in which the diameter changes by 1 mm expressed as a ratio. For example, a 1:20 taper changes 1 mm in diameter over a length of 20 mm.
- In degrees.

Are there other ways to specify tapers?

Yes, there are widely used tapers whose dimensions are industry standard. The most common of these are:

- *Morse Tapers* which are available in 8 standard sizes, designated #0MT through #7MT. These self-holding tapers remain seated due to the wedging angle of their taper. In some applications, *tangs* (flat end projections) are added to the small end of tapers to prevent rotation from torsional forces. Morse tapers have a tpf of *about* 0.6 inches This is the most widely used taper.
- *Brown & Sharp Tapers* are still used on dividing heads and older mills.
- *Steep Tapers*, formerly called the *Standard Milling Machine Tapers*, are self-releasing and have a 3½-inch taper per foot. These have a drive key (a longitudinal fin) to prevent rotation and are mainly used on milling machine arbors and accessories.
- *Standard Taper Pins,* which are used for locking parts in position, come in sizes #6/0 through #10 and have a ground ¼-inch tpf. They are steel.

How can the machinist verify that his taper is accurate?
Morse taper gages, like the Morse taper plug and ring gage seen in Figure 7–110, are used to check fit by lightly coating the male taper with Prussian blue paste dye. The male and female tapers are put together and one rotated with respect to the other, then separated. The male taper is examined for evenness of distribution of the dye. Testing can also be done with chalk stripes applied on the side of the taper plug in a similar test. In the shop it is also possible to use on-hand, known tapers for testing.

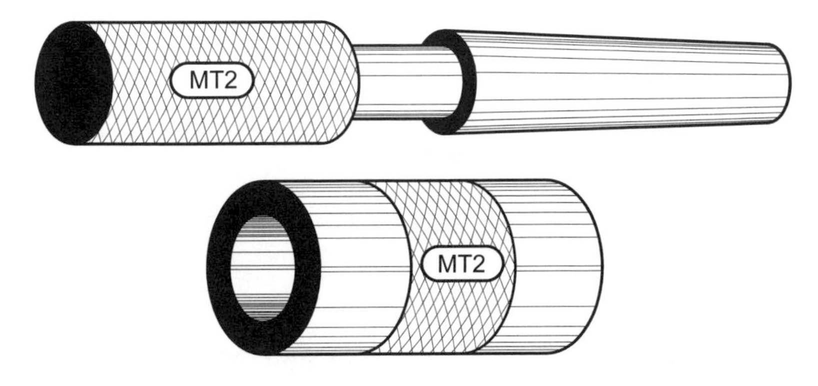

Figure 7–110. Morse taper plug (top) and ring gage (bottom).

How can tpi (taper per inch) be calculated from tpf (taper per foot)?
The formula is:

$$\text{Taper per inch (tpi)} = \frac{\text{taper per foot (tpf)}}{12}$$

Given the dimensions of a part, how can its *tpf* be calculated?
See Figure 7–111 and use this formula:

$$\text{Taper per foot (tpf)} = \frac{(D - d) \times 12}{L}$$

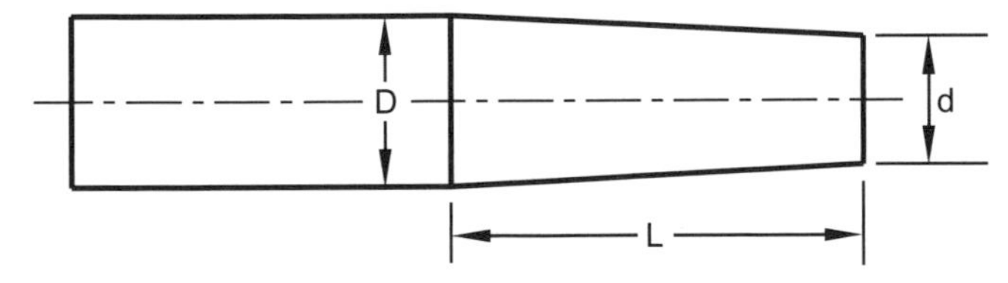

Figure 7–111. Calculating tpf from a part.

Given the dimensions of a part, how can a metric taper be determined?
Use Figure 7–112 and the formula:

$$k \text{ (metric taper)} = \frac{L}{(D - d)}$$

Where D = large diameter
d = small diameter
L = length of taper

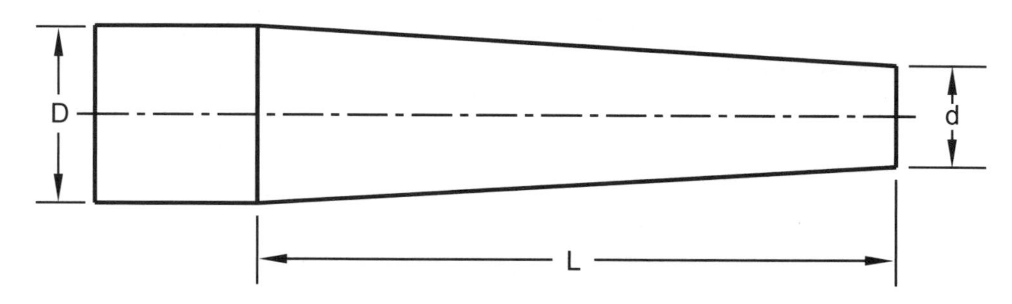

Figure 7–112. Calculating metric taper from a part.

What are five methods for making tapers and what data is required to use them?
See Table 7–5.

Method of Making Taper	Advantages & Disadvantages	Data Needed
Offset Tailstock	Only between centers and only for external tapers.	Taper per inch (tpi), or taper per foot (tpf).
Taper Attachment	Best method.	Angle, taper per inch, or taper per foot.
Compound Slide	Taper length limited, internal and external.	Taper angle.
Reamer	Internal tapers only.	Taper number, limited selection of tapers.
Formed Tool Bit	For short tapers only.	Taper angle.

Table 7–5. Five methods for making tapers with the lathe.

Tailstock Offset Method

How does the *tailstock offset method*, also called the *tailstock setover method*, work?
The tailstock is moved or set over to make the lathe cut a taper on the work. See Figure 7–113. Note that the same tailstock setover on a long workpiece makes a smaller taper than on a short workpiece, Figure 7–114.

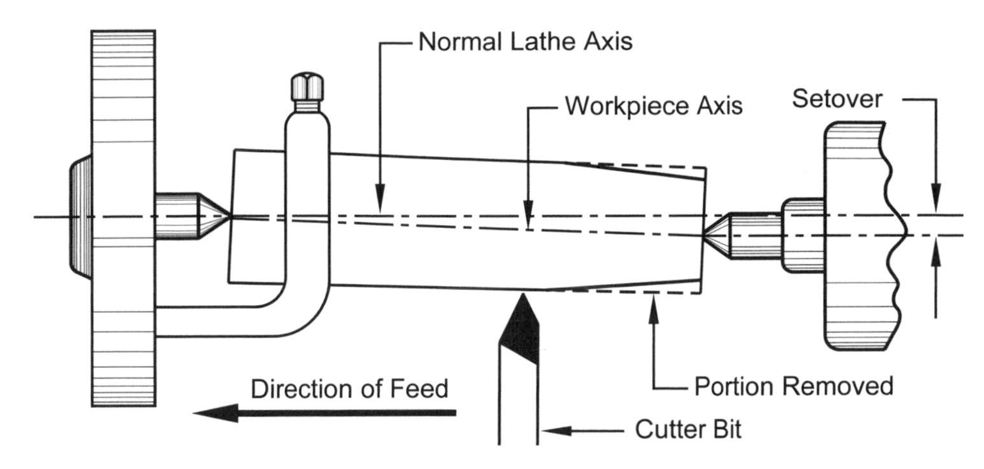

Figure 7–113. How tailstock setover method works.

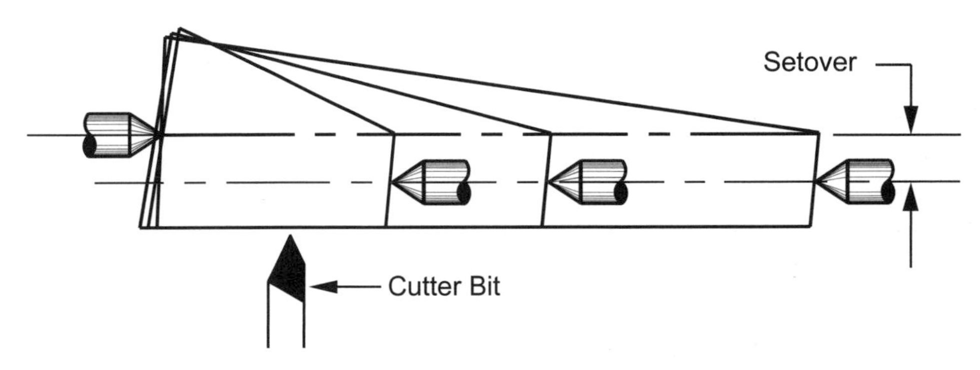

Figure 7–114. How workpiece length affects taper.

How is the tailstock offset calculated to achieve a specific taper?
Use this formula:

$$\text{Tailstock offset (inches)} = \frac{\text{tpf} \times \text{length of work}}{24}$$

If the taper is specified in metric terms, as 1:30, meaning a change in diameter of 1 mm in a 30-mm length, see Figure 7–115 and use this formula:

$$\text{Tailstock offset (mm)} = \frac{(D-d)}{2 \times L} \times \text{OAL}$$

Where D = large diameter
d = small diameter
L = length of taper
OAL = overall length of work

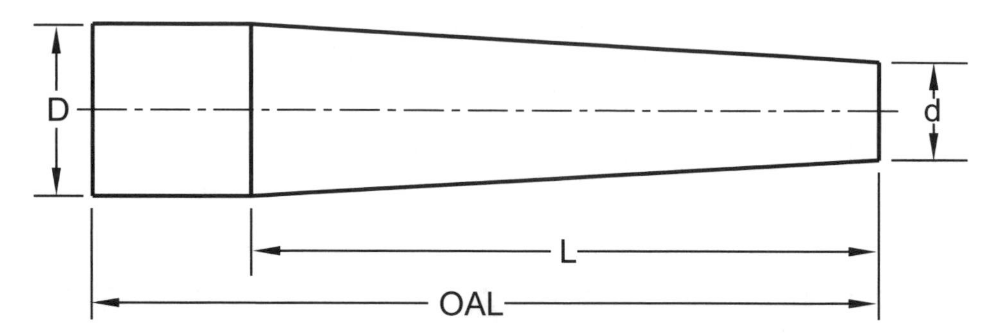

Figure 7–115. Calculating tailstock offset for a metric taper.

How is the tailstock offset to turn tapers?
There are four methods:

- The tailstock is off set by using the graduations below the witness marks on the back of the tailstock. The offset represented by each graduation must be known in advance and may be found in the lathe's manual. See Figure 7–116.

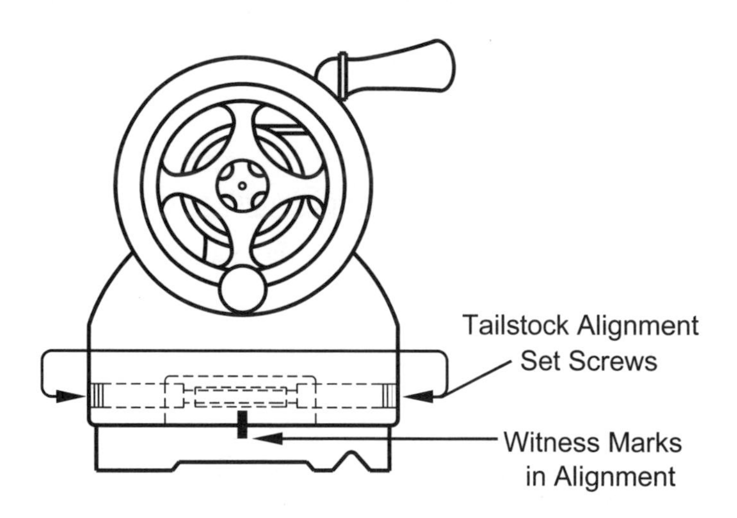

Figure 7–116. Measuring tailstock offset.

- An alternative is to use a ruler between the center points, as in Figure 7–117. The best solution is to install a taper pin when the tailstock is centered. This makes re-centering it simple, just push the taper pin in place with the tailstock loose.

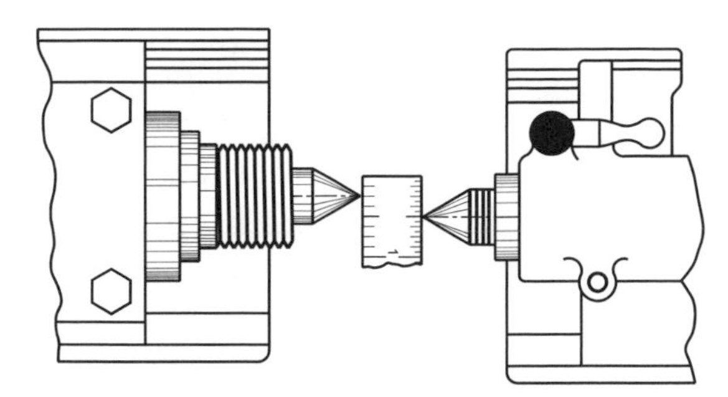

Figure 7–117. Measuring offset between the lathe center points.

- The tailstock can also be off set by using the graduated collar on the cross slide and a piece of thin paper as a feeler gage. Here's how to do it. First, make sure that the tailstock is centered, then position the carriage so the blunt end of a tool just touches the edge of the tailstock ram. Use a thin piece of paper pulled between the blunt end of the tool and the tailstock ram to determine when they just touch. Set the collar on the cross slide to zero and apply its clamp so it will not move. Use the cross slide crank to move the cross slide away from the ram to the desired offset. Next, loosen the tailstock bolts and move the tailstock so the ram just pinches a piece of paper when it touches the back of the tool. Tighten the tailstock clamp and recheck the offset. See Figure 7–118.

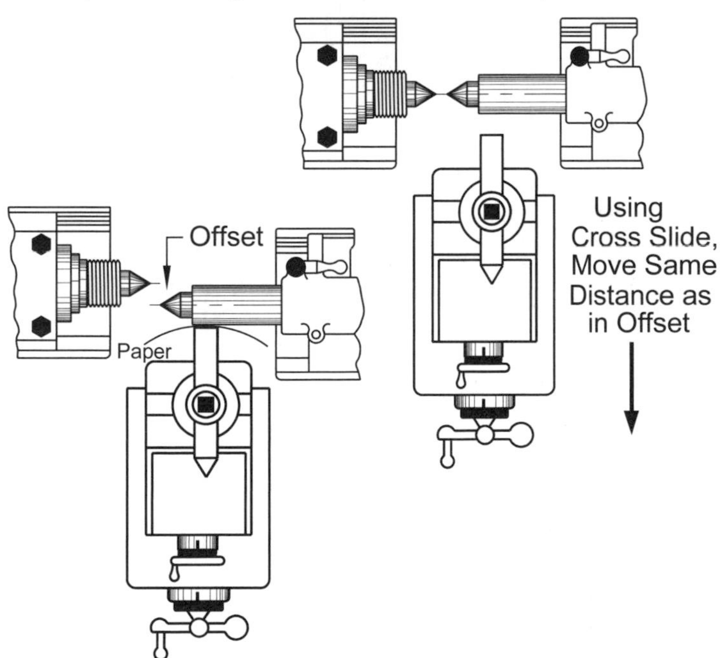

Figure 7–118. Setting the tailstock offset using the cross slide and a piece of paper as a feeler gage.

- Another way to off set the tailstock is by means of a dial indicator attached to the cross feed. This method is quick and accurate. Attach a dial indicator so it will move with the cross slide, Figure 7–119. One way is to put a magnetic based indicator on the cross slide, another is to put the rod holding the dial indicator in the tool holder. To make sure the tailstock is in alignment, set the dial indicator to zero, then, unclamp the tailstock and move it toward the user.

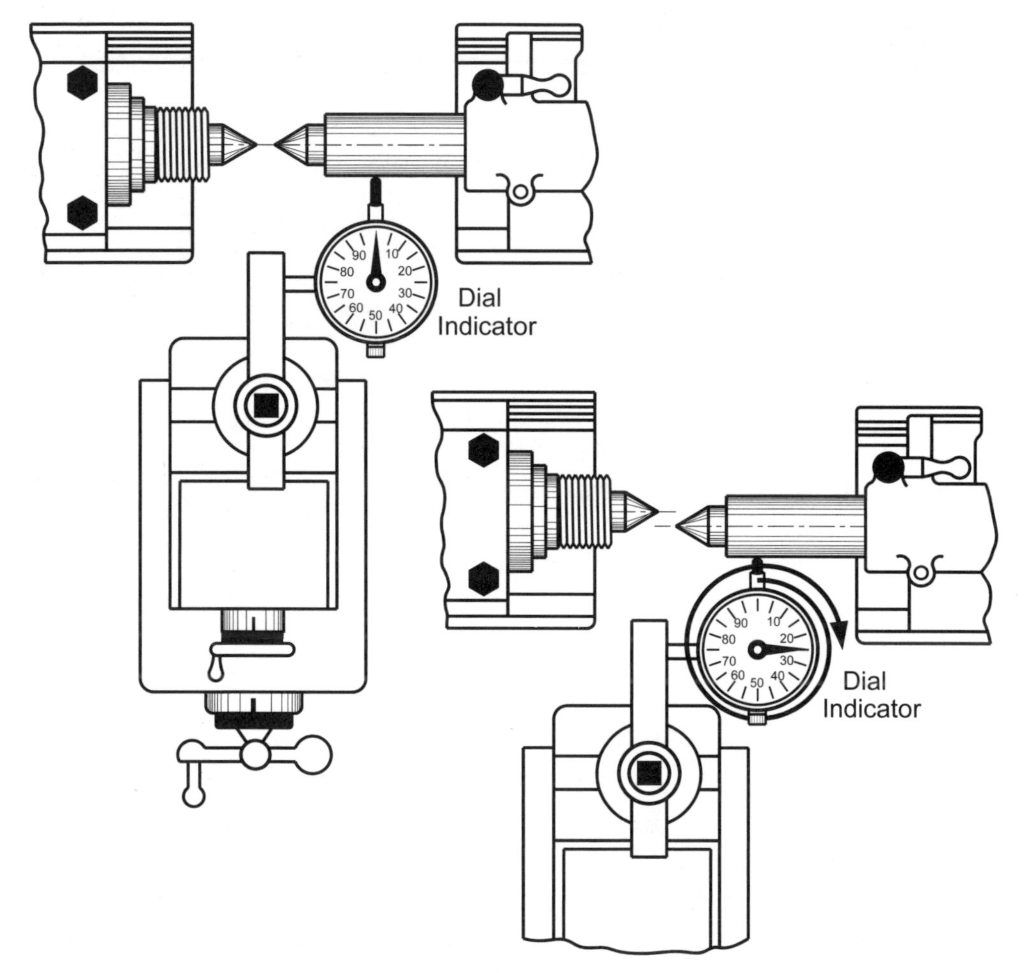

Figure 7–119. Using a dial indicator to set tailstock offset.

What are the disadvantages to the offset method of taper turning?

- On steep tapers the points of the lathe centers are forced to run out of alignment with their center holes, Figure 7–120 (A). This arrangement tends to heat up and burn out the centers if not repeatedly lubricated. To avoid this problem, some machinists use the adjustable center seen in Figure 7–120 (B) or a ball center as in Figure 7–120 (C). Another good choice is drilling the center hole with a radius-type combination drill that will leave a hole that better fits a center during off-axis work than a plain-type center hole, Figure 7–120 (D).

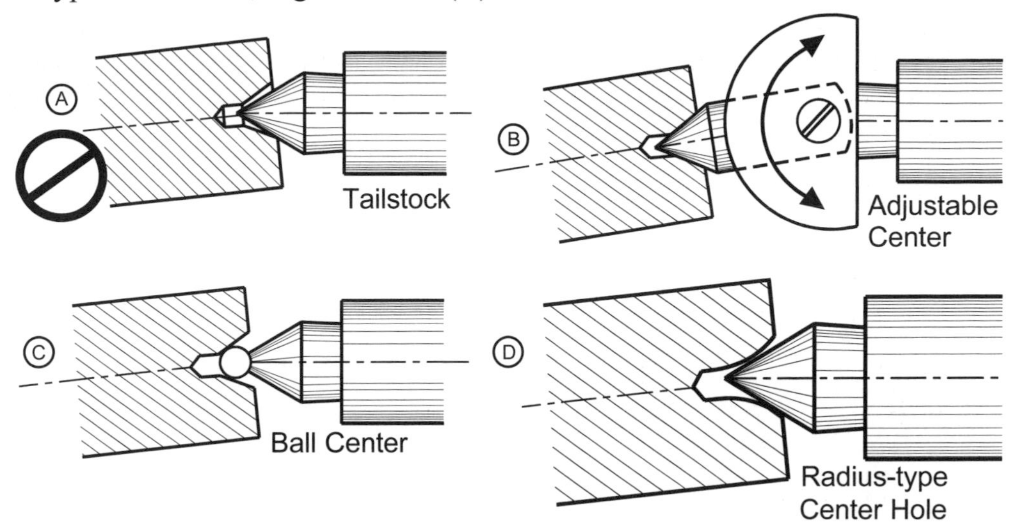

Figure 7–120. Lathe center detail when using the tailstock offset method for turning tapers. The walls of the continuously curved center hole made by the radius-type center drill provides much more contact between the lathe center and the workpiece than a standard center drill would, (D).

- Another disadvantage is that to achieve the same taper, all parts in the same production run must be the same length. Otherwise, the offset must be individually calculated and readjusted for each part.
- All turning must be done between centers. Collets and chucks cannot be used.

What are the basic steps to perform taper turning by the offset method?

1. Calculate and set the tailstock setover.
2. Clean all mating surfaces.
3. Install and position the tool and set its height exactly at center height.
4. Lubricate the center holes and install the work in the center snugly so there is no back-and-forth play in the work, but the work still turns freely.
5. Position the tool to clear the work.
6. Rotate the work/spindle by hand to insure it is secure and free to rotate.
7. Determine and set the proper speed and feed.
8. Begin with a small cut and work up to larger ones.
9. Make repeated caliper measurements to determine when the final diameter at one end is reached.

Taper Attachments

What are the advantages of using the taper attachment?

There are many:

- Lathe centers remain in alignment and do not have to be re-aligned for the next job. In addition, there is less stress and wear on the centers because they remain in alignment.
- The taper remains the same regardless of the length of the workpiece unlike the offset method. This makes it easy to turn duplicate tapers on workpieces of *different* lengths.
- Tapers may be turned using centers, a collet or a chuck.
- A wide range of tapers is available.
- Set up is easy since inch attachments are calibrated in inches of the tpf and degrees. Metric attachments are calibrated in millimeters and degrees. See Figure 7–121.

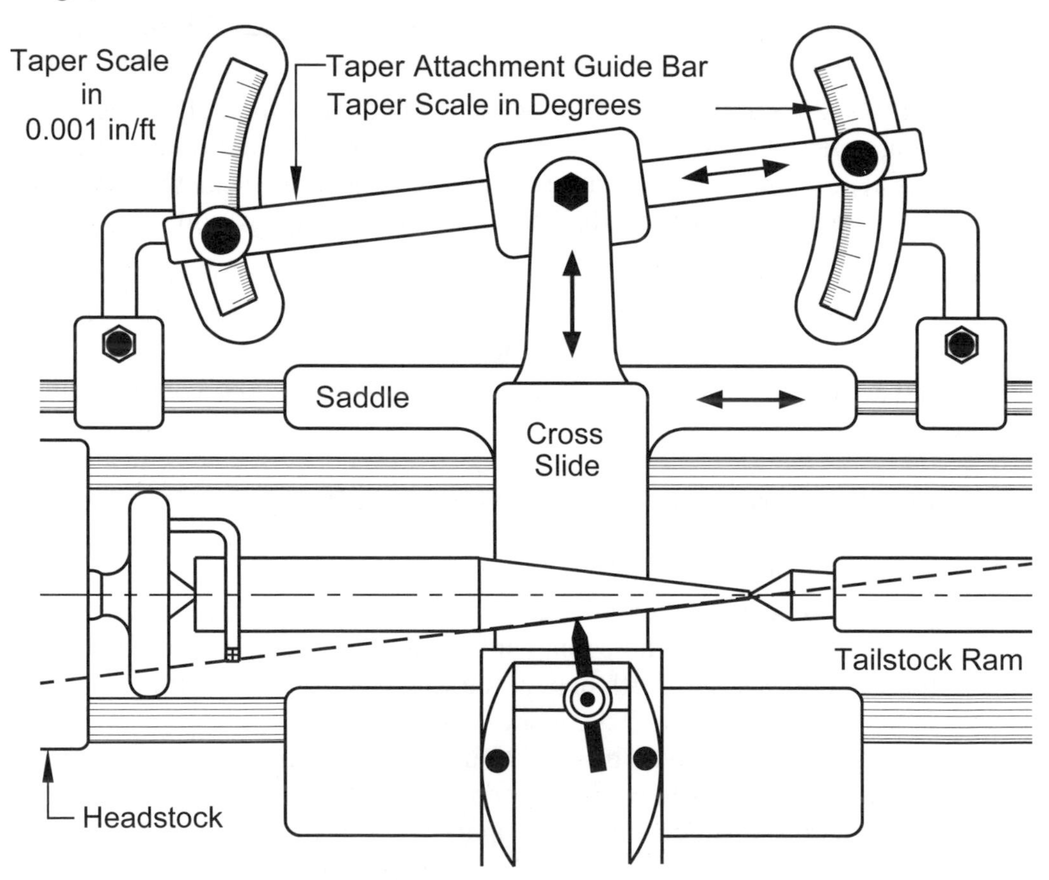

Figure 7–121. Using the lathe taper attachment.

How is the taper attachment used?

1. Clamp the attachment to the ways of the lathe.
2. Clean and lubricate the attachment guide bar.
3. Disconnect the cross slide leadscrew, if necessary. *Warning: Failure to disconnect the cross slide so it is free to be moved by the taper attachment will cause damage.*
4. Place the workpiece in a chuck, collet or between centers and verify that it is held firmly.
5. Install the cutting tool and place it exactly on center height.
6. Mark the end or ends of the taper.
7. Set the taper attachment to the desired taper and lock it into position.
8. Position the cutter bit near the right-hand end of the work, but not touching it.
9. Rotate the work manually to verify that it rotates freely.
10. Set the proper feed and speed.
11. Begin making a light initial cut.
12. Finish turning the taper and check for fit.

Compound Tapering

How can the compound slide be used for tapering?

The compound slide can be used for tapering by setting the angle taper to half the angle desired and cutting the taper from the inside of the work to its outside. See Figure 7–122. This convenient and accurate method is limited mainly by the length of compound movement.

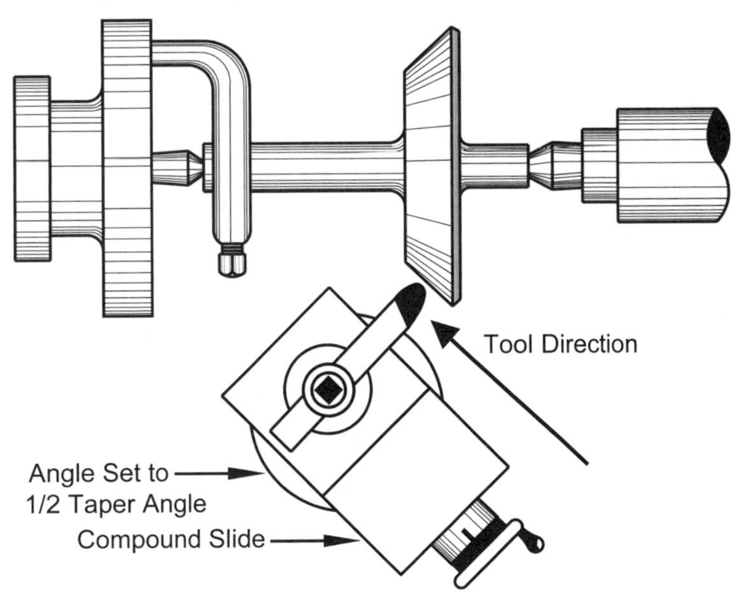

Figure 7–122. Using the compound for taper cutting.

Tool Bit Tapers

How can tapers be applied with a tool bit?
See the method shown in Figure 7–123. This method is limited by the width of the tool bit.

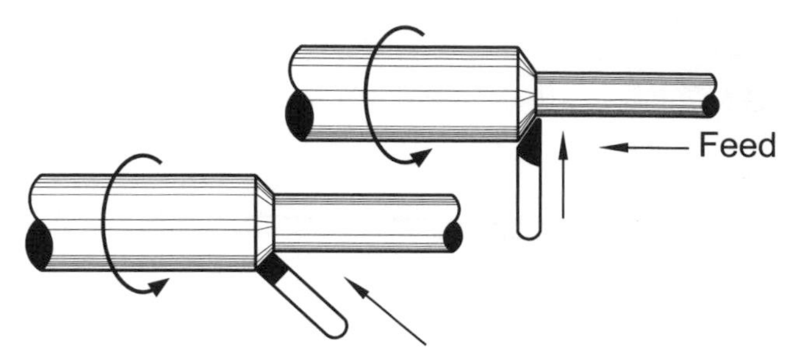

Figure 7–123. Tapering with a tool bit.

Reamer Tapering

How are reamers used to apply internal tapers?
The lathe tailstock holds the finishing Morse taper (MT) reamer to cut a new taper or restore a damaged one. All Morse taper sizes are commercially available. They are used only for the final dimensioning or finishing of the taper by removing several thousandths of an inch, not to hog out the bore. See Figure 7–124. When making a new MT, there are two ways to make the starting cavity for the finishing reamer: bore a slightly undersized taper cavity with the lathe taper attachment or use a roughing MT reamer in a drilled hole.

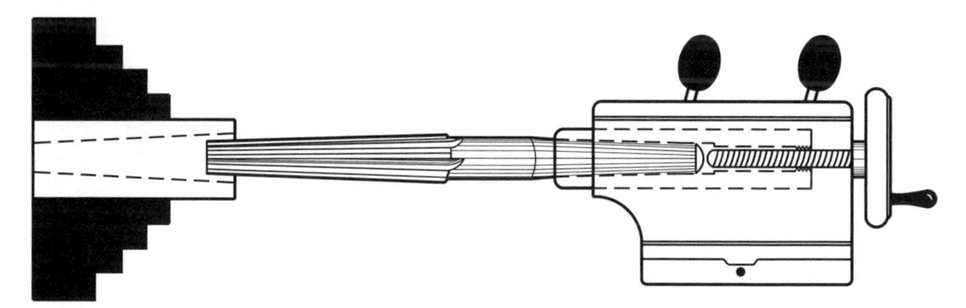

Figure 7–124. Morse taper finishing reamer held in the tailstock by a Morse taper removes the final 12–15 thousandths of an inch.

What are the steps to make a taper using a reamer?
When making a taper using a reamer, the rough boring is done prior to reaming. Be sure to leave enough workpiece material so the reamer actually cuts the work and does not burnish it. See Table 7–6 to determine how undersized the starting bore should be.

Finished Hole Diameter		Undersize Starting Bore by	
(inches)	(mm)	(inches)	(mm)
< 0.25	6.4	0.010	0.25
0.25–0.50	6.4–12.7	0.015	0.38
0.50–1.00	12.7–25.4	0.020	0.50
1.00–1.50	25.4–3.8	0.025	0.064

Table 7–6. Starting diameter for reaming to a given final diameter.

1. Make the initial taper the reamer will bring to final size. This may be done with the taper attachment, an MT roughing reamer or by step-drilling.
2. Clean all mating internal and external surfaces.
3. Install the reamer and the work holder (for example chuck or collet).
4. Secure the work in the work holder and center it, if required.
5. Position the saddle and compound to clear the work.
6. Rotate the work/spindle by hand to insure it is secure and free to rotate.
7. Set the spindle speed to about two-thirds of that used for the same size twist drill.
8. Flood the bore and the reamer with cutting fluid.
9. Feed the reamer in slowly.
10. Frequently withdraw the reamer to clear chips and apply more cutting fluid.
11. Never reverse or stop the spindle with the reamer in the bore.

Always store reamers in separate compartments so they will not strike each other and chip or dull their cutting edges.

Trepanning

What is *trepanning* and why is it used?

The process of trepanning cuts a round hole by removing only a thin annular (donut-shaped) band of material around the outer edge of a hole. The remainder of the workpiece material inside the hole comes out as a single, round piece. Because the material removed is not cut, except for the thin annular band, cutting tool forces on the work are much lower resulting in less energy consumption, less cutting time and less tool wear. Trepanning has other cost advantages over drilling: it requires only a lathe tool bit to make a wide variety of hole sizes and the piece of circular material removed from the hole may be used elsewhere. In Figure 7–125, four clamps space the workpiece away from the face plate so that the face plate is not damaged. Trepanning is much faster than boring because only a small portion of the workpiece is reduced to chips. To cut a hole to a tight tolerance, trepanning is often used to open up the hole, then an additional boring operation brings it to final size.

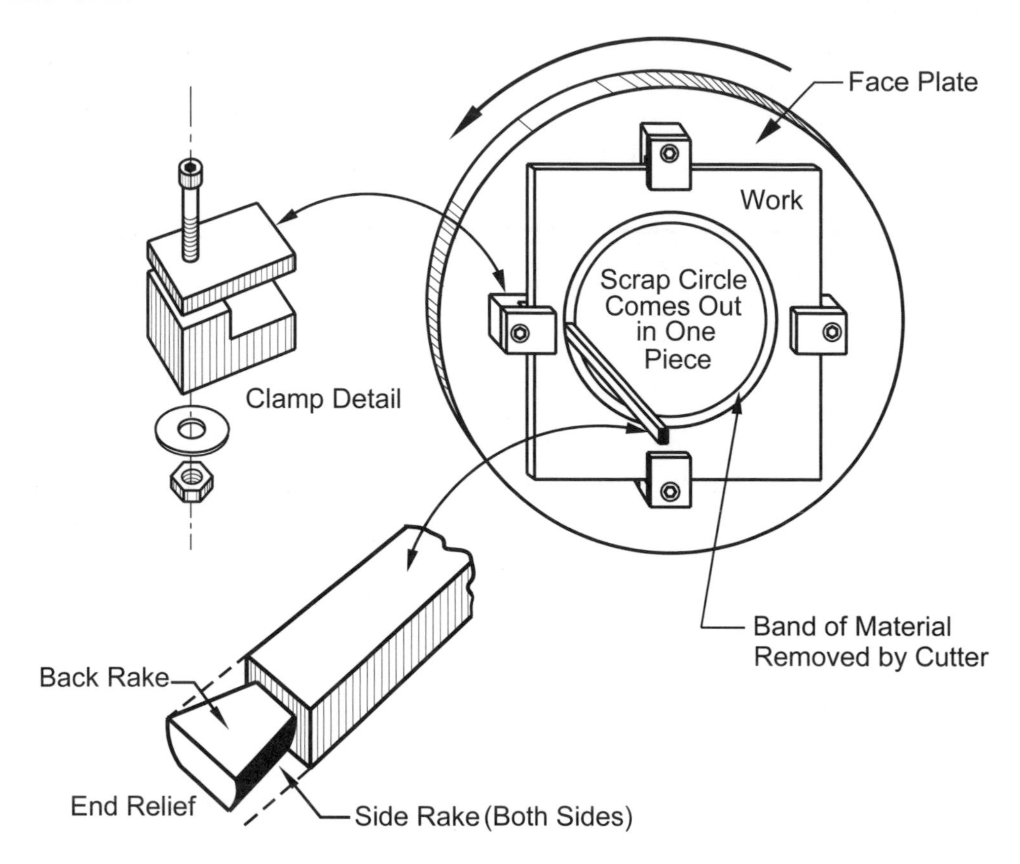

Figure 7–125. Trepanning a flat workpiece mounted on a face plate.

What are the steps to perform trepanning?

1. Clean all mating surfaces.
2. Install the work holder, usually a face plate and hold-down clamps.
3. Secure the work on the work holder, positioning it so that the center of the hole needed coincides with the point of the lathe center in the tailstock.
4. Position the trepanning tool at right angles to the work, use the cross feed to set the tool to the hole radius and then position the tool so it clears the work.
5. Rotate the work and spindle by hand to insure the work is secure and free to rotate.
6. Determine and set the proper speed and feed.
7. Turn on the lathe and begin cutting.
8. Lubricate as needed.
9. Use a board to protect the lathe ways from the falling scrap circle.

Machining Eccentrics

How can an eccentric projection be machined onto a rectangular workpiece?

Clamp the work in a 4-jaw chuck as shown in Figure 7–126. This arrangement permits the eccentric to be positioned accurately by aligning a mark representing the center of the eccentric with a center in the tailstock. The two V-blocks provide more gripping power than using just the four chuck jaws.

Clamping work, as in Figure 7–126, is also suitable for squaring up and facing off the ends of square stock.

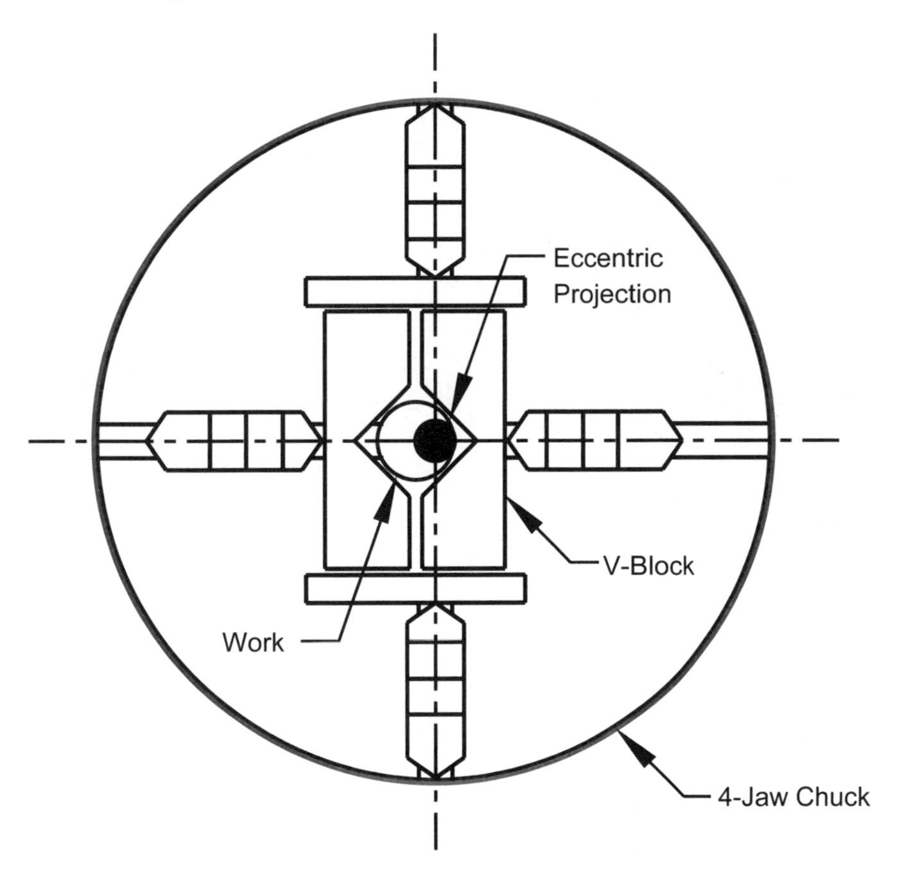

Figure 7–126. Holding round work for machining an eccentric projection.

Machining Crankshafts

What are the most common methods for machining crankshafts?

See Figure 7–127. These methods are for one-offs. Commercial production uses other, faster methods.

Eccentric turning when the throw is too small to allow properly drilled center holes. (Add 3/4" length for each center hole)

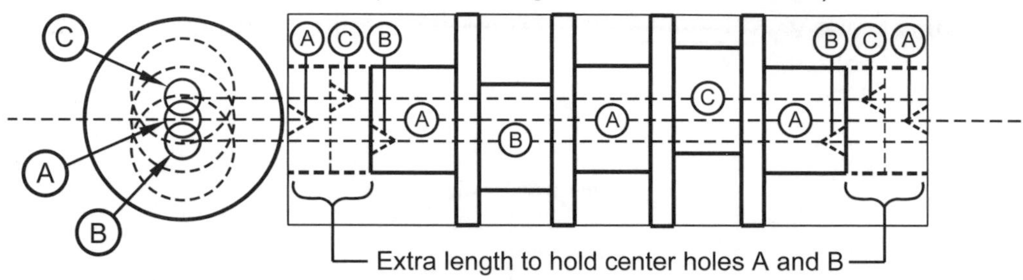

Eccentric turning when the throw is large enough to allow all centers to be located on the work at the same time. in this case, the eccentric throws must be turned before the main center.

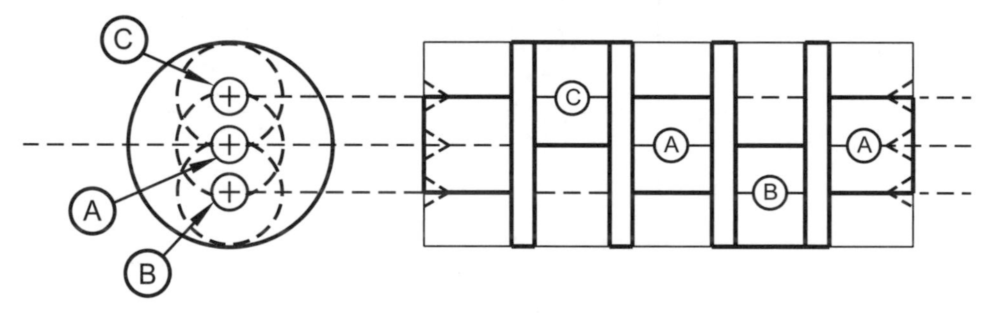

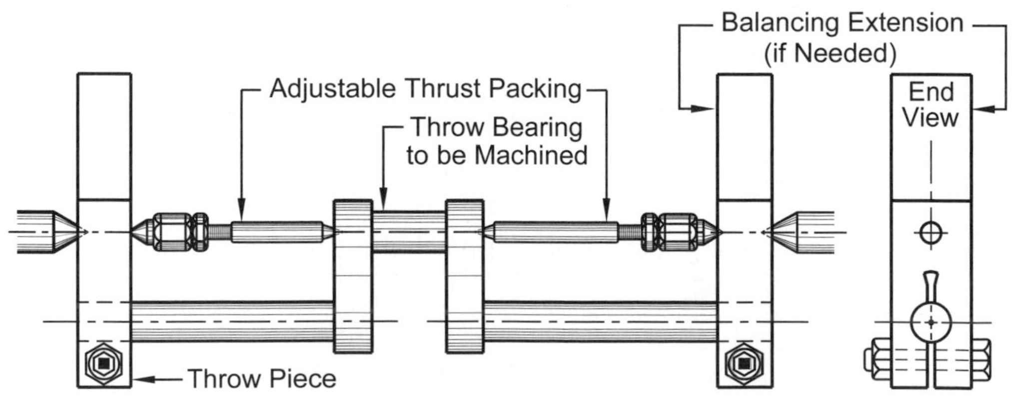

Figure 7–127. Methods of machining crankshafts.

Threading with Taps & Dies

What are the basic steps to perform threading with taps?

1. Clean all mating surfaces.
2. Install the tool bit and work holder (for example chuck or collet).
3. Secure the work in the work holder and center it, if required.
4. Position the tool to clear the work.
5. Rotate the work/spindle by hand to insure it has proper clearance.
6. Lubricate the work and tap.
7. Insert the threaded end of the tap into the hole in the work and secure the other end of the tap in the lathe center in the tailstock.
8. Rotate the work manually to perform threading. Use your right hand to turn the tailstock handwheel to keep the tap aligned as it enters the work and your left hand to turn the tap handle, Figure 7–128.
9. Depending on the workpiece material and the tap, it may be necessary to back out the tap to clear it of chips and add fresh lubricant.
10. *Always back out the tap manually on lathes with a threaded spindle nose or the chuck is likely to unscrew from the spindle and fall onto the ways.* On lathes with other types of spindle nose designs, use a backgeared reverse to back out the tap. Use caution when backing out under lathe power as the lathe has considerable torque and can pinch your fingers.

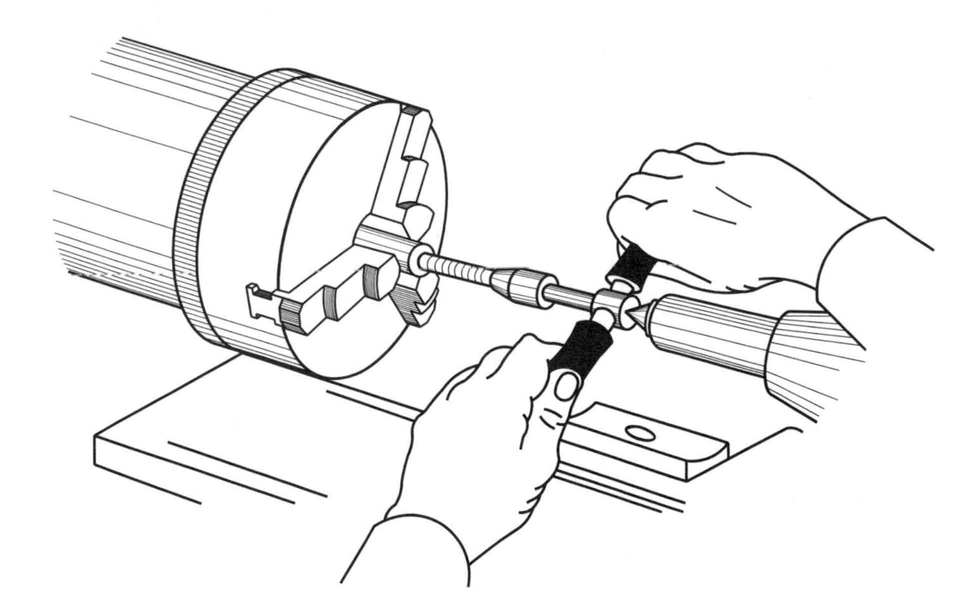

Figure 7–128. Using a tap in the lathe.

How does using a die to apply external threads to a workpiece differ from using a tap?

A die holder, which fits in the Morse taper of the tailstock, holds the die to get it started squarely on the work as in Figure 7–129. Usually the die holder has a hole through it to provide clearance for the work as the die progresses over it. After the threads are started, a conventional manual die holder may be used to finish the remaining threads, if necessary. Threading is usually done by rotating the spindle manually. If the lathe has a slow reverse, it may be possible to use power to remove the work from the die. However, lathes with a threaded spindle are liable to have their chuck unscrew from their spindle.

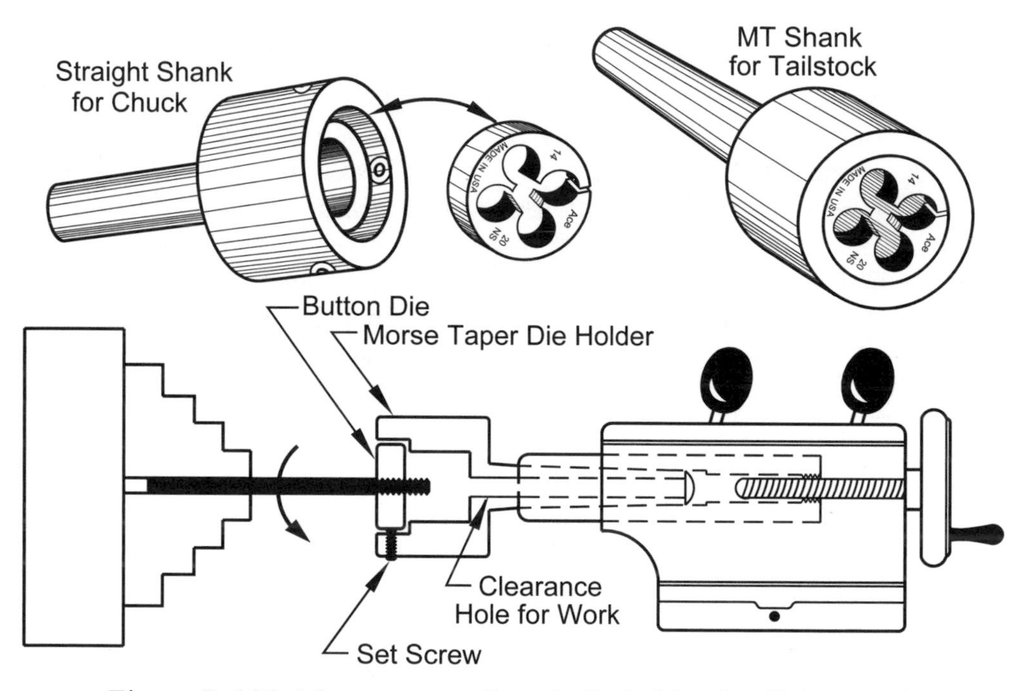

Figure 7–129. Morse taper tailstock die holder detail (top row) and mounted in the tailstock threading work (bottom).

Lathe Filing

Why is filing in the lathe performed?

Filing removes burrs from threads, rounds off the ends of work, blends the edges of formed outlines, breaks the corners on sharp edges and reduces shaft diameter to make a press fit. While a file can be used to shape a workpiece, removing large amounts of materials causes eccentricity.

What are the basic steps to perform filing?

1. Move the carriage out of the way and remove its tool post for safety.

2. Use a 10- to 12-inch long file. A standard *mill file* will work, but a *long-angle lathe* file would be even better. Be sure the file handle is secure.
3. Run the spindle at one-half to one-third of the speed for turning.
4. Take long even strokes, releasing the pressure during the return stroke.
5. Move the file across the work so each stroke covers about one-half the file width.
6. Keep the file moving across the work or the file teeth will load with the workpiece material and score the work surface.
7. Do not use heavy file pressure because it will introduce eccentricity.
8. Check the file teeth frequently and clean with a file card when needed.
9. Some machinists rub blackboard chalk into the file to prevent its teeth from loading with metal and to make it easier to clean. Others use oil.

Also see the material on long-angle files in *Chapter 3 – Filing & Sawing* and particularly Figure 3–8.

Section VIII – Leadscrew Threading

Thread Cutting

What are the advantages of cutting threads with a single-point tool, also called *thread chasing*, with a leadscrew?

Leadscrew threading enables the machinist to cut threads:

- Concentric and symmetrical with the axis of the work.
- For which no tap or die is at hand.
- That are too large in diameter or that have too great a side clearance (like square threads) to be cut with taps or dies.
- Of uncommon or obsolete thread forms.
- Inside very short bores as seen on camera lens adapters.
- Which run right up to an abrupt shoulder or internal wall which cannot be threaded with taps or dies.
- On tapered workpieces.
- That have multiple starts; these have two, three or four separate threads intertwined.

When is threading with taps and dies in the lathe preferable to leadscrew threading?

Taps and dies are usually faster than leadscrew threading for threads less than 0.75-inch (19-mm) diameter and when precise concentricity is not important.

What are the most common methods to handle the start and finish of threads and what are the advantages of each?

Of the three ways of treating the start of threads, Figure 7–130 (A), requires only the breaking of the edge with a file to remove burrs, but starting a nut on these threads is more difficult than when the end has a formed shape, Figure 7–130 (B) or a 45° chamfer, Figure 7–130 (C). Applying the chamfer takes less operator time and care than shaping an end with a form tool.

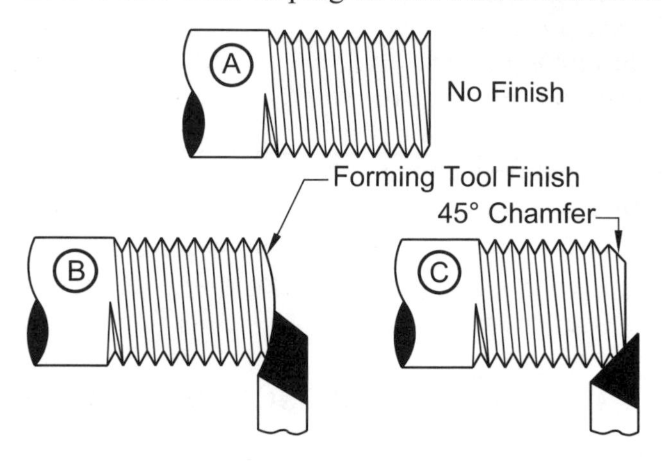

Figure 7–130. Thread starting methods.

Using an *end, necking* or *undercut groove* on the body termination of a thread, Figure 7–131 (A), takes the least threading skill since the machinist has the most time to withdraw the tool bit at the end of the cutting cycle. It should be the first choice when the application permits it. When the part design does not permit the use of an end groove, the next choice is using a drilled hole, Figure 7–131 (B), which still provides some leeway to withdraw the tool at the end of the cut. Withdrawing the tool at the end of the cut takes the most machinist skill, Figure 7–131 (C). Failing to withdraw the tool properly will break the tool bit and ruin the part.

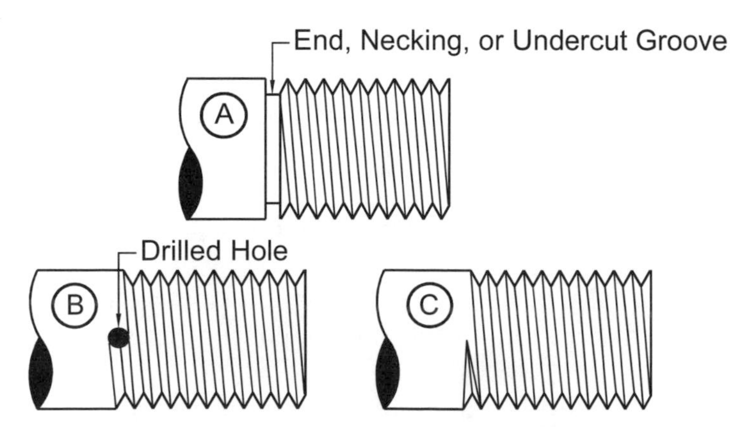

Figure 7–131. Thread terminating methods.

How does *leadscrew thread cutting* work?
Leadscrew threading in the lathe, also called *single-point threading*, is based on linking cutting bit movement with spindle revolution. In Figure 7–132 gears between the spindle and the leadscrew insure that one spindle revolution causes the leadscrew to make the appropriate linear desired tool movement. The turning leadscrew moves the cutting tool along the work. This figure shows how a leadscrew turning at the same speed as the spindle puts the *same* number of threads per unit length on the workpiece as the leadscrew itself has.

If the leadscrew turns faster than the spindle (from having more gear teeth on the spindle gear than the leadscrew), the tool will move more rapidly and the threads will be spaced closer together. Conversely, if the leadscrew turns more slowly than the spindle (from having fewer gear teeth on the spindle gear than on the leadscrew), the threads will be spaced further apart.

Neither the diameter of the work, nor the diameter of the leadscrew are factors in determining how many threads are cut on the work. Only the number of threads/inch (or threads/mm) on the leadscrew, and how many turns the leadscrew makes when the spindle makes one turn, determine the number of threads cut. How fast the leadscrew turns with respect to the spindle is determined by the gears connecting the spindle and leadscrew.

Threads are applied in a series of passes. Three to fourteen cuts are typical. After a cutting pass is complete, the tool is pulled back from the work, the half-nuts, seen in Figure 7–132 holding the leadscrew, are opened and the carriage can be quickly moved back to the starting point. The tool is set into the work and the half-nuts, also called the *split-nut* or *clasp-nut*, are re-closed on the leadscrew to begin the next pass. The controls for closing the half-nuts and leadscrew motion are conveniently located on the lathe apron. See Figure 7–133 for a detail of the half-nuts.

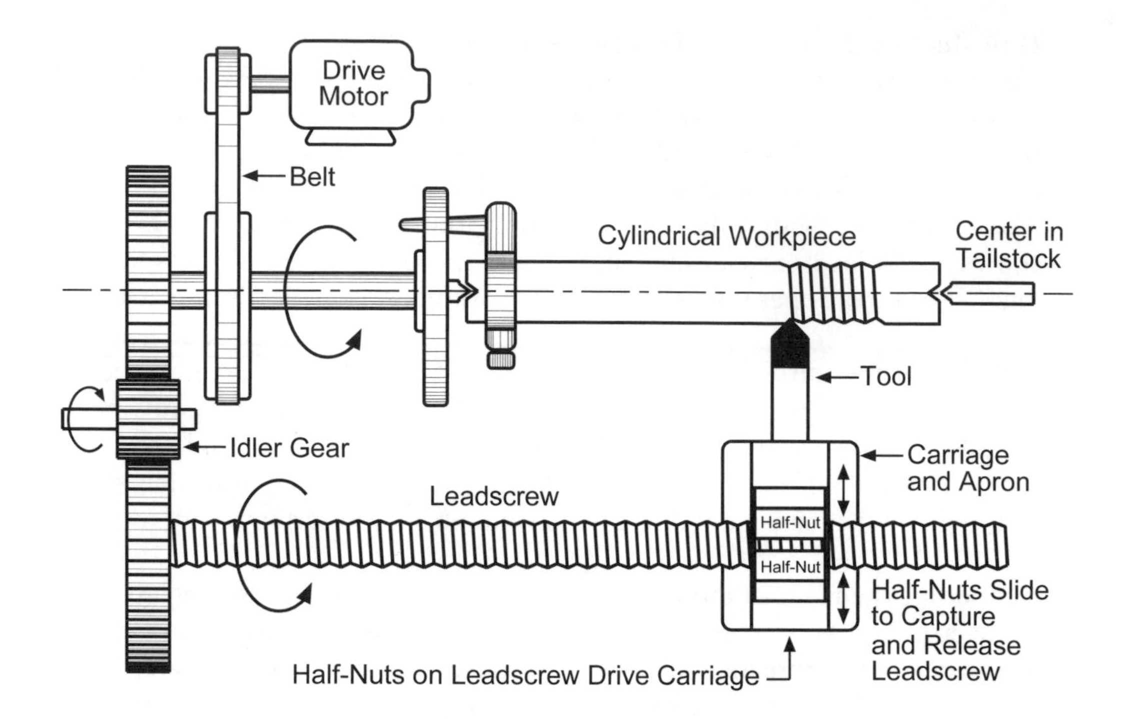

Figure 7–132. Principle of leadscrew thread cutting.

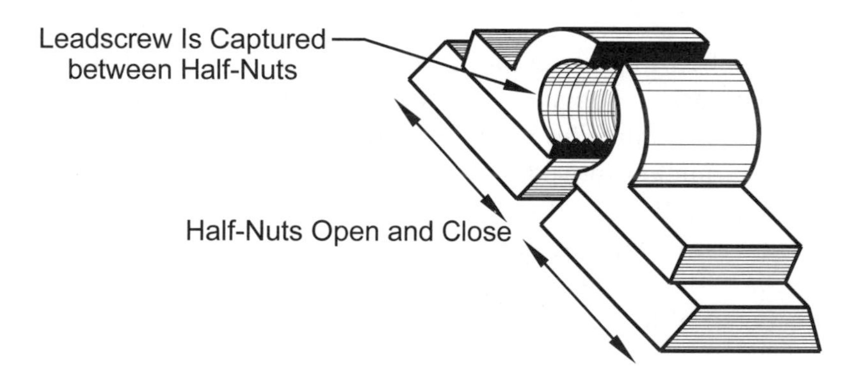

Figure 7–133. Detail of half-nuts.

Thread Cutting Tools

What type tool bits are used for *thread cutting* in a lathe?

Thread cutting, also know as *thread chasing,* is done with a single point tool bit with no back or side rake. The most common thread forms use either a pointed, 60° tool, or a 60° tool with the tip blunted. HSS and alloy steel tool bits are supplied as both unsharpened blanks and sharpened ready-to-use tools. Tungsten carbide tool bits and carbide inserts are supplied sharpened.

How can you tell if the tool bit has the proper shape?
There are several gages commonly used for checking tool bit shape and aligning the tool bit with the work. The most common, seen in Figure 7–134, is the *thread center gage* or *fishtail.* This is suitable for checking all tools with a 60° angle. A similar tool, shown in Figure 7–135, is used for Acme threading tools which have a 29° angle.

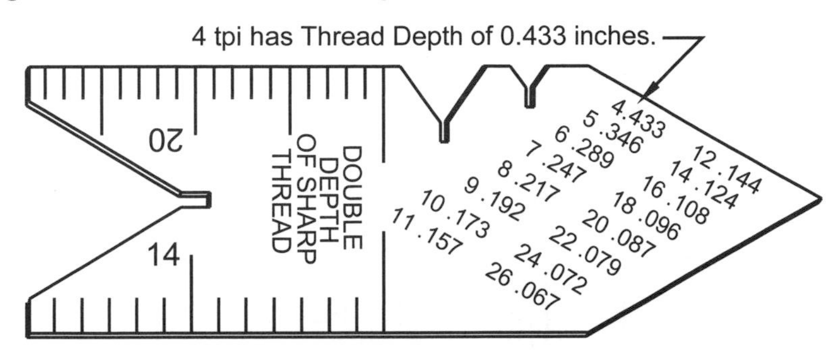

Figure 7–134. Fishtail or thread center gage for 60° threads.

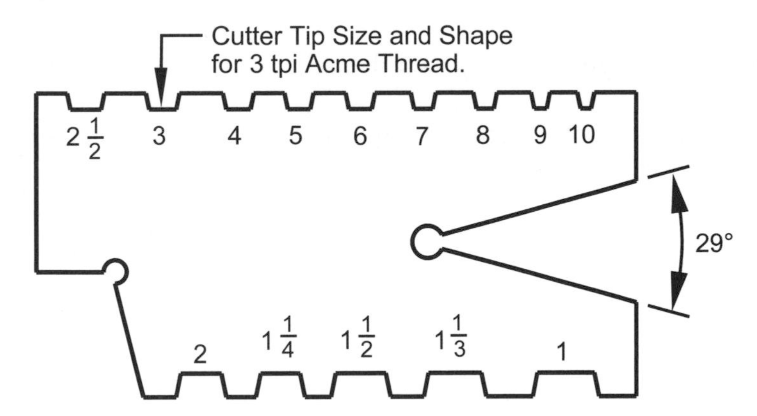

Figure 7–135. Acme center gage for 29° threads; numbers identify tip width for respective tpi thread.

Thread-Chasing Dials

What is the purpose of the thread-chasing dial and how is it used?
The thread chasing dial is used when cutting threads with a leadscrew.

Using a thread-chasing dial enables the user to withdraw the tool at the end of each threading pass, open the half-nuts and manually move the carriage back to the starting end of the workpiece. This is a big time-saver since threading requires between three and fourteen cutting passes depending on the thread. When the half-nuts are opened, the problem is to reclose them so the threading tool enters the existing, partially cut thread at the proper point. The thread-chasing dial, seen in Figure 7–136, performs this function by indicating on its dial when to close the half-nuts. Position the threading tool ahead of the start of the threads, not touching the work and close the half-nuts according to the engagement rules in Table 7–7.

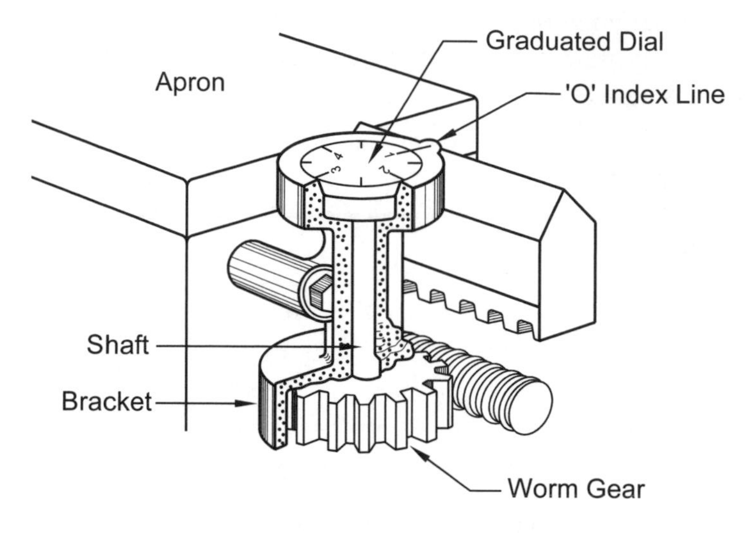

Figure 7–136. Inch thread-chasing dial.

Threads/inch	When to Engage the Half-Nuts	Dial reading
Even number.	Engage at any marked point on the threading dial.	
Odd number.	Engage at any numbered division.	
Any thread with ½ numbers.	Close at 1 & 3, or at 2 & 4 division marks.	
Any thread with ¼ numbers.	Return to original starting mark.	——
Any multiple of the leadscrew tpi.	Any time that half-nuts mesh.	Thread-chasing dial not needed.

Table 7–7. Half-nuts engagement rules.

One way to perform these repeated passes without a thread-chasing dial is to leave the half-nuts engaged during the entire threading process, withdraw the tool at the end of each pass and reverse the lathe to reposition the tool at the beginning of the threads for the next pass. This approach insures that the tool will enter the existing thread cut at the proper spot on the next pass, but it is very slow. For lathes without a thread-chasing dial, the operator must reposition the carriage by reversing the leadscrew or spindle.

How do thread-chasing dials work for metric threads?
Unlike inch thread-chasing dials, metric lathes' thread chasing dials have several gears that can engage the leadscrew. *The proper gear must be selected to match the thread being applied.* See the instruction plate on the lathe.

Does the thread-chasing dial function properly when an inch leadscrew lathe cuts metric threads or vice versa?
The thread-chasing dial works properly only when used to cut threads of the same measurement units as its leadscrew. See Table 7–8.

Leadscrew Units	Thread Units	Thread-Chasing Dial Operation
inch	inch	Yes, this works well.
	mm	No, works only if half-nuts remain closed.
mm	mm	Yes, but use proper gear on thread chasing dial.
	inch	No, works only if half-nuts remain closed.

Table 7–8. Operation of thread-chasing dials on inch and metric lathes.

Change Gear Ratios

How can you determine what gears are required to cut a specific thread?
There are two ways to cut inch threads with an inch-based leadscrew lathe:

- For lathes *with a quick-change gearbox* consult the *index plate* on the lathe as shown in, Figure 7–137. This plate is usually found on or just above the quick-change gearbox. Find the number of threads/inch you want to cut in the larger numbers on the index plate and set the sliding gear and lever corresponding to that number.
- For lathes *without a quick-change gearbox* use individual change gears, also called *change wheels*. They must be installed in the gearbox on the left end of the lathe. Instructions on how to do this (including the gear combinations for various threads/inch) are found on an engraved plate inside the gearbox. For lathes without this information, refer to *Machinery's Handbook,* which has a section for determining the proper gears to use.

16-INCH SOUTH BEND QUICK CHANGE GEAR LATHE

SLIDING GEAR	TOP LEVER	THREADS PER INCH—FEEDS IN THOUSANDTHS								
IN	LEFT	4 .0841	4½ .0748	5 .0673	5½ .0612	5¾ .0585	6 .0561	6½ .0518	7 .0481	AUTOMATIC CROSS FEED EQUALS .375 TIMES LONGITUDINAL FEED
	CENTER	8 .0421	9 .0374	10 .0337	11 .0306	11½ .0293	12 .0280	13 .0259	14 .0240	
	RIGHT	16 .0210	18 .0187	20 .0168	22 .0153	23 .0146	24 .0140	26 .0129	28 .0120	
OUT	LEFT	32 .0105	36 .0093	40 .0084	44 .0076	46 .0073	48 .0070	52 .0065	56 .0060	
	CENTER	64 .0053	72 .0047	80 .0042	88 .0038	92 .0037	96 .0035	104 .0032	112 .0030	
	RIGHT	128 .0026	144 .0023	160 .0021	176 .0019	184 .0018	192 .0017	208 .0016	224 .0015	

Figure 7–137. Lathe index plate.

How are inch leadscrew lathes able to cut metric threads and vice versa? The solution is clever, but simple. The relationship between centimeters and inches is:

$$2.54 \text{ cm} = 1 \text{ inch}$$

If we make a ratio of the two units to represent the relationship we need to convert an inch leadscrew and quick-change gearbox to threads/cm from threads/inch, we get:

$$\frac{2.54}{1.00}$$

This says that to make the conversion we must have the leadscrew turn 2.54 times faster for metric threads than for inch threads. Since we cannot have half-teeth on a gear, the solution is to increase the total number of teeth on the gears, but in the same ratio. If we multiply both the numerator and the denominator by 50 the ratio remains the same, but the result is that a 127-tooth gear running against a 50-tooth gear will produce the 2.54:1 ratio with the smallest number of whole gear teeth:

$$\frac{2.54}{1.00} = \frac{50}{50} \times \frac{2.54}{1.00} = \frac{127}{50}$$

The conclusions are that:

- Inch-to-metric-conversions usually have a 127-tooth gear driving a 50-tooth gear.
- Metric-to-inch conversions usually have a 50-tooth gear driving a 127-tooth gear.

Even on lathes with quick-change gearboxes, the gears linking the spindle and the quick-change transmission must be changed to convert from cutting inch to cutting metric threads and vice versa.

Setting the Compound Rest

How should the compound rest be positioned to infeed the tool bit?

The tool bit itself is set at right angles to the work using the thread center gage, but the compound rest must be set to feed the tool at an angle so only one cutting edge, the left-hand one in this case, forms a chip. If the compound is set at 90° with respect to the ways, both tool edges cut at the same time and the emerging chips collide at the point of the tool causing the tool bit to tear the work metal instead of cutting. This creates a rough surface finish. For threads having a 60° included angle, the compound is set to feed the tool into the threads at a 29° angle. Some machinists prefer a 30°; take your choice. See Figures 7–138 and 7–139.

For Acme threads, seen in Figure 6–11 (center), which have a 29° included angle, use an Acme thread gage to set the tool at right angles to the work. Set the compound rest at 14.5° with respect to the work. Set the tool bit height exactly on center using a lathe center for a center height reference.

Placing a sheet of white paper underneath the thread center gage on the carriage makes it easier to check the cutting tool bit edges against the thread gage by providing a light background against which to silhouette the gage and work. Each edge of the tool bit should be checked against the center gage.

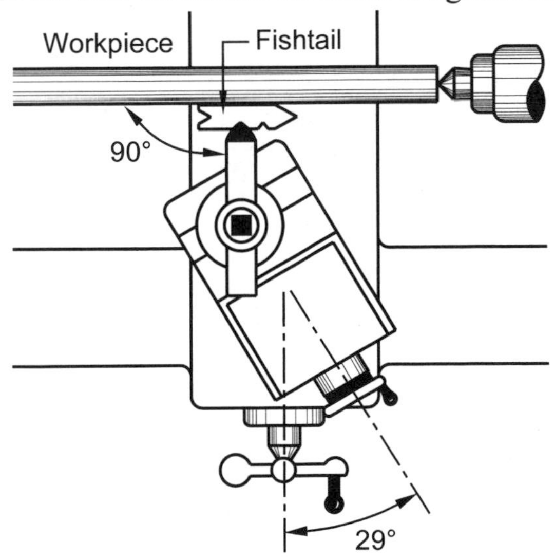

Figure 7–138. Threading tool set at 90° to the work and the compound rest set at 29° to the work for threads having a 60° included angle.

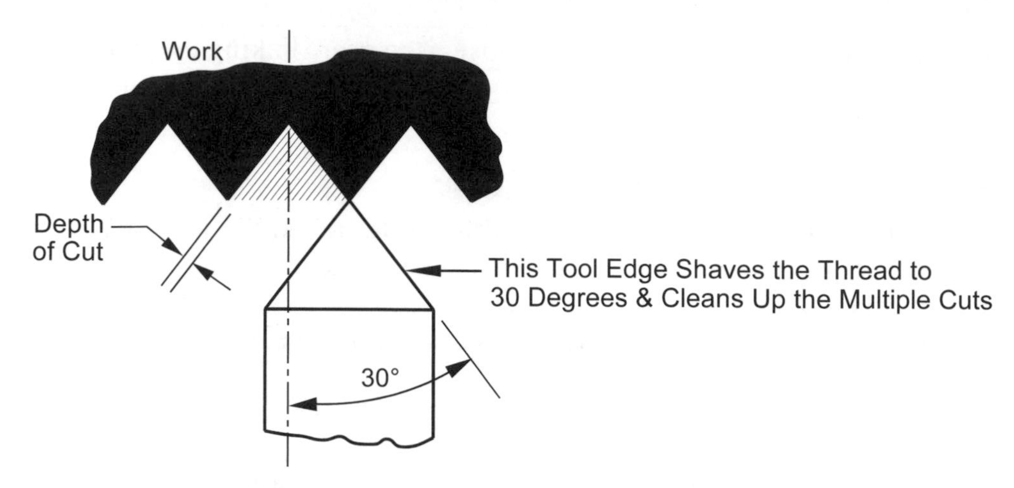

Figure 7–139. The compound rest is set at 29° and the tool cutting edges at 30° making most of the tool cut on only one edge.

Cutting Speed for Threads

At what speed should the lathe be set for thread cutting?

Theoretically, threads can be cut at the same speed as for turning. However, the ability of the operator to stop the lathe at the end of the cut and to pick up the thread at its beginning requires slower speeds. Use one-quarter normal turning speed to begin with. With practice and better operator coordination, higher cutting speeds can be used.

Thread Cutting Lubrication

Lubrication is usually a good idea. It extends tool life, reduces energy costs, reduces forces on the tool and the work and produces a better surface finish. However, lubrication is messy and in some cases unnecessary. Most metals can be threaded dry. The only operation that definitely needs lubricant is tap threading. This is because multiple tool surfaces in contact with the work put large forces on the tap and it may break. Using kerosene, light machine oil or a proprietary cutting fluid, like Rapid Tap® from the Relton Corporation, helps prevent tap breakage.

Depth of Thread Cuts

What depth cuts should be made when cutting a thread?

Making an initial cut 0.001-inches (0.025-mm) deep allows checking of the thread pitch the gearbox setting produces. This light trace on the workpiece is small enough so that if the gearbox needs adjustment, the workpiece is not ruined. The idea is to use larger cuts after the gearbox setting is verified and use progressively smaller cuts as the final depth is approached. Table 7–9 shows a typical sequence of threading cuts.

Thread Cut Number	Cut Depth	
	Inches	mm
1st	0.001	0.025
2nd	0.005	0.127
3rd	0.005	0.127
4th	0.005	0.127
5th	0.004	0.101
6th	0.004	0.101
7th	0.004	0.101
8th	0.003	0.076
9th	0.003	0.076
10th	0.003	0.076
11th	0.002	0.051
12th	0.002	0.051
13th	0.002	0.051
14th and remaining	0.001	0.025

Table 7–9. Sequence of cut depths for threading. The exact number of cuts depends on the metal being threaded.

Cutting 60° External Threads with a Thread-Chasing Dial

What are the steps to apply external right-hand threads?

1. Set the spindle speed at one-quarter the normal turning speed.
2. Set the quick-change gearbox or change gears for the desired pitch using the index plate.
3. Set the leadscrew to turn in the forward direction. When engaged, the carriage will move from right to left.
4. Set the tool height on center and the compound rest angle to 29°.
5. Mount the work between centers, in a chuck, in a chuck plus a center, in a collet or in a collet plus a center and insure it is secure and there is no end-to-end play. Lubricate the lathe centers, if they are used.
6. Set the tool at right angles to the work using the thread center gage (fishtail), Figure 7–134 and Figure 7–138.
7. Measure the diameter of the work. Good practice is to have the work diameter 0.002 inch (0.05 mm) *under* the thread major diameter.
8. Start the lathe and chamfer the right end of the work to just under the thread final minor diameter using the left edge of the cutting tool.
9. Measure and mark the work where the threads will end by cutting a light ring with the threading tool bit. A stopping groove can also be used as in

Figure 7–140. Crank the tool bit away from the work so the carriage can be moved. The spindle will still be turning.

10. Manually position the carriage toward the right end of the workpiece and, using the compound rest, move the tool bit towards the work so it just lightly marks it. *Set both the cross slide and compound micrometer collars to zero*. See Figure 7–140.
11. Back the tool bit away from the work using the cross slide and position the carriage so the tool bit tip is ¼ inch or more to the right end of the work, the starting point.
12. Using the compound rest, move the tool bit into the work between 0.001 and 0.003 inches with your left hand on the compound crank. Engage the half-nuts lever with your right hand when the thread dial reaches the right graduation. This will make the first, or scratch pass.
13. When the tool reaches the left end of the cut, quickly crank the tool away from the workpiece using the cross slide crank and simultaneously disengage the half-nuts.
14. Move the carriage manually back to the starting point and set the cross slide back to the zero position.
15. Stop the spindle, and using either a thread pitch gage or a ruler, check that the cut is the desired pitch. See Figure 7–141. Correct the gearbox setting and try again if the thread count is wrong.
16. Restart the spindle.

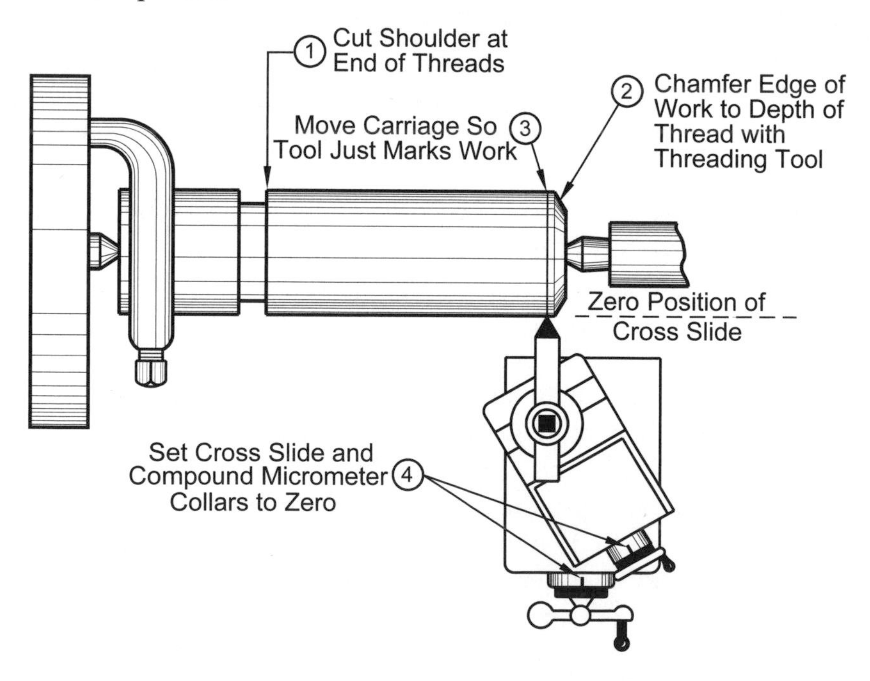

Figure 7–140. Preparation for threading.

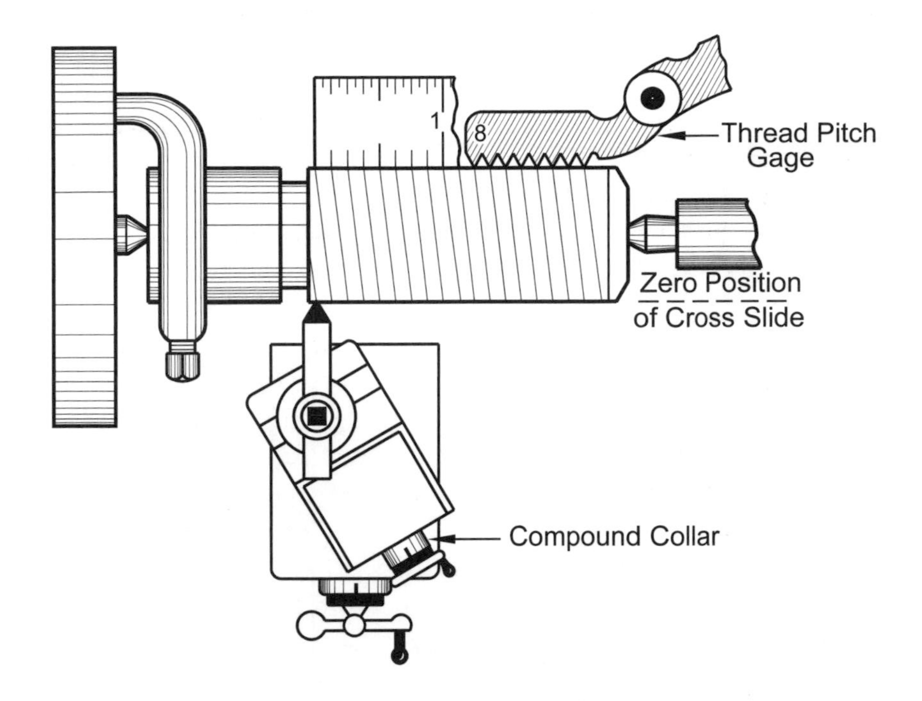

Figure 7–141. Checking thread pitch after initial threading pass.

17. Manually move the carriage to Position 1 as in Figure 7–142. Then, using the collar calibrations, move the cross slide to the zero position, Position 2. With the compound, advance the tool bit in preparation for the next thread cut, Position 3. Use Table 7–9 as a rough guide for the depth of each pass. The idea is to use cross slide motion to pull the tool away from the workpiece at the end of the thread and move the tool back to the zero position for the next pass. Once the compound collar has been set to zero, the compound is incremented for each pass and is never used to pull the tool away from the work.
18. Engage the half-nuts lever with your right hand when the threading dial reaches the correct graduation and the tool will move from Position 3, through Position 4, and to the end of the threads at Position 5. When the tool reaches Position 5, use the cross slide to quickly withdraw the tool to Position 6. Then manually move the carriage back to Position 1. You are now ready to begin the next cutting pass.

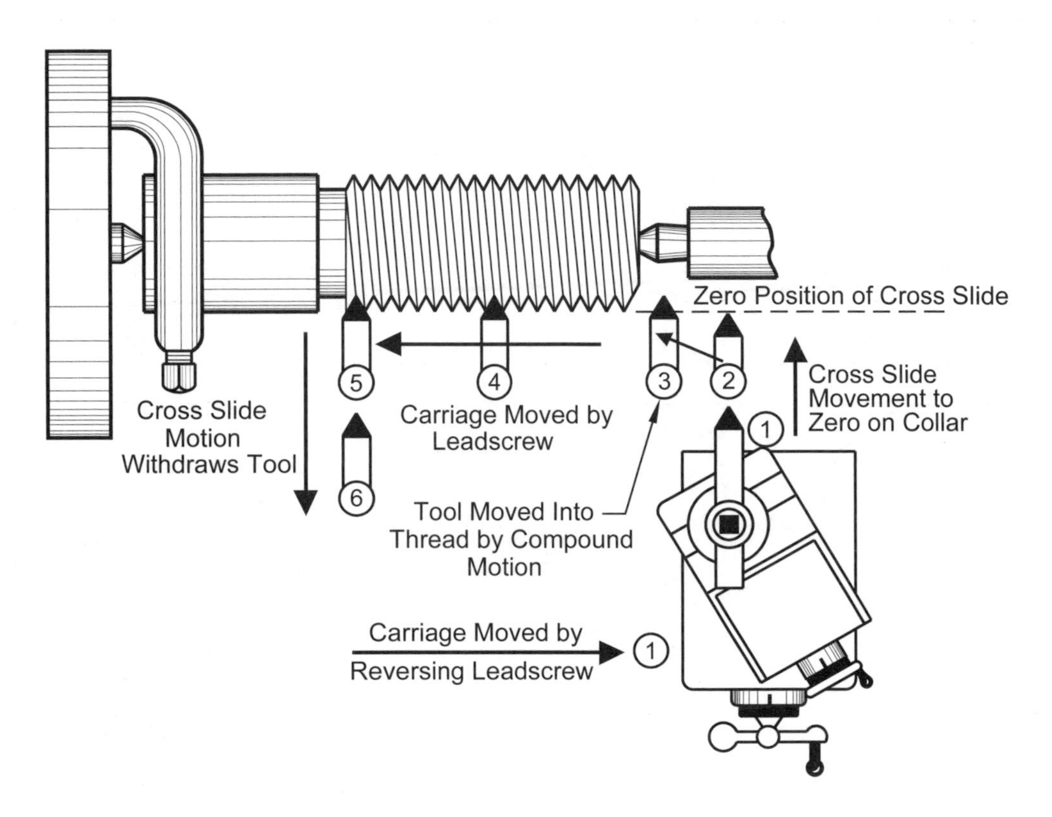

Figure 7–142. Completing the threading process.

19. Tables in *Machinery's Handbook* show the final thread depth for all common thread forms. This depth can be read off the compound collar and can be checked using a threading micrometer. Also, inch thread center gages have the thread depth for various tpi engraved on them. See Figure 7–135. Not having this information at hand, the operator can watch the thread develop until the width at the thread bottom (root) just about equals the width of the thread top (crest). Then take additional small cuts until the mating part or master nut fits properly. Figure 7–143 (A through D) shows the thread development. Repeat steps 17 and 18 until the threading reaches full depth and is completed, as shown in Figure 7–143 (D).
20. Use a file to remove burrs on the top thread edge.

Cutting 60° external threads on a lathe equipped with a thread-chasing dial is the most common and least complicated threading operation.

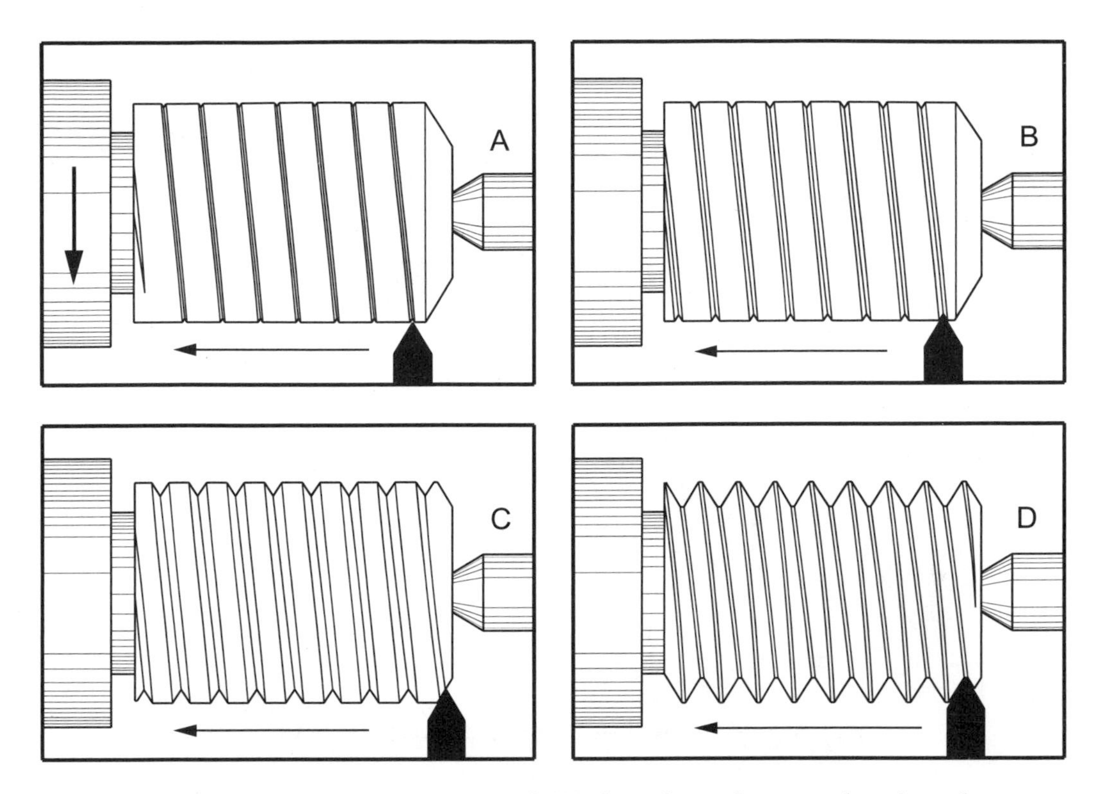

Figure 7–143. Appearance of 60° threads as they are developed.

Threading without a Thread-Chasing Dial

How is threading performed on lathes without a thread-chasing dial?
Whenever a lathe lacks a thread-chasing dial, the half-nuts must not be opened and the carriage must be moved back to the start of the threads by reversing the spindle. *When using a chuck on lathes with a threaded nose spindle, use slow reverse speeds and observe that the chuck does not unscrew during reverse operation.*

Left-Hand Threads

What are the steps to apply external left-hand threads?
1. Install the workpiece in the lathe.
2. Set the compound swung 29° to the left of the cross slide and the tool bit at right angles to the work. This is similar to Figure 7–144 with the exception of the direction of the compound with respect to the cross feed.
3. Set the tool bit on center.
4. Set both the collar on the cross slide and on the compound to zero as for right-hand threads.

5. It is preferable to have a starting groove, or gap, on the left side of the threads in which to begin the threading cuts, Figure 7–144.
6. Follow as for right-hand threads except work from left to right.

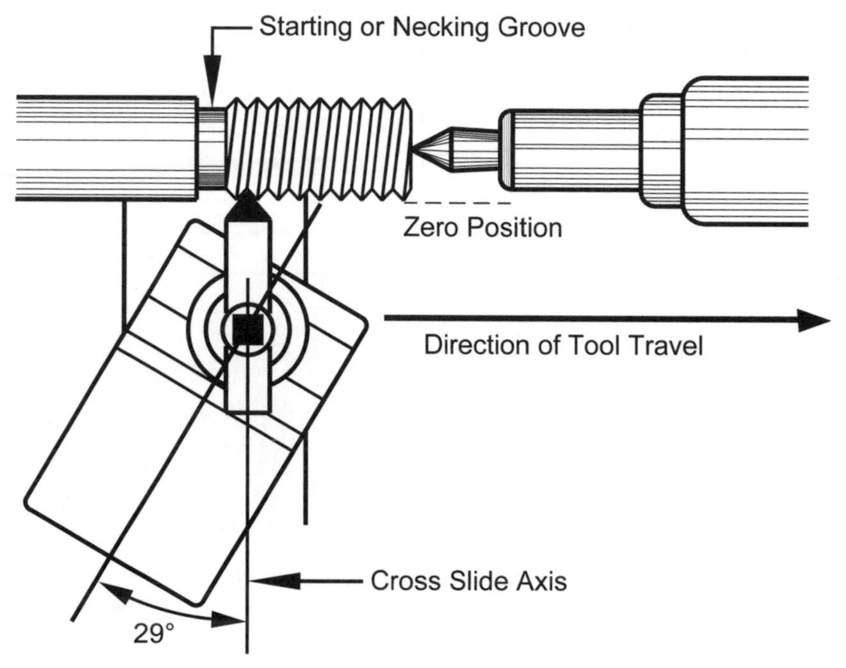

Figure 7–144. Setup for cutting left-hand threads.

Metric Threads with an Inch Leadscrew

What problems are encountered when cutting metric threads on a lathe with an inch leadscrew?

Because the thread-chasing dial will not work for metric threads, the half-nuts must remain closed during the entire threading process.

Here is how to apply 4-mm metric threads on an inch lathe with a quick-change gearbox:

1. Install the 127-tooth gear on the leadscrew.
2. Install the 50-tooth gear on the spindle. Together these two gears convert the gearbox settings from tpi to threads/cm.
3. Convert the 4 mm pitch to threads/cm:

$$10 \text{ mm} = 1 \text{ cm}$$

$$\text{Pitch} = 10/4 = 2.5 \text{ threads/cm}$$

4. Set the quick-change gearbox to 2.5 tpi and, with the changed gearing, it will cut 2.5 threads/cm.
5. Set up the lathe and the work as for cutting a 60° thread detailed above.
6. Make a trial scratch cut as if 60° thread cutting and, using the cross slide, withdraw the tool at the end of the cut. Do not open the half-nuts at the end of the cut. Stop the spindle, then check that the required thread pitch has been cut. Adjust the gearbox if necessary. See Figure 7–145.
7. Reverse the spindle, which will reverse the direction of the leadscrew.

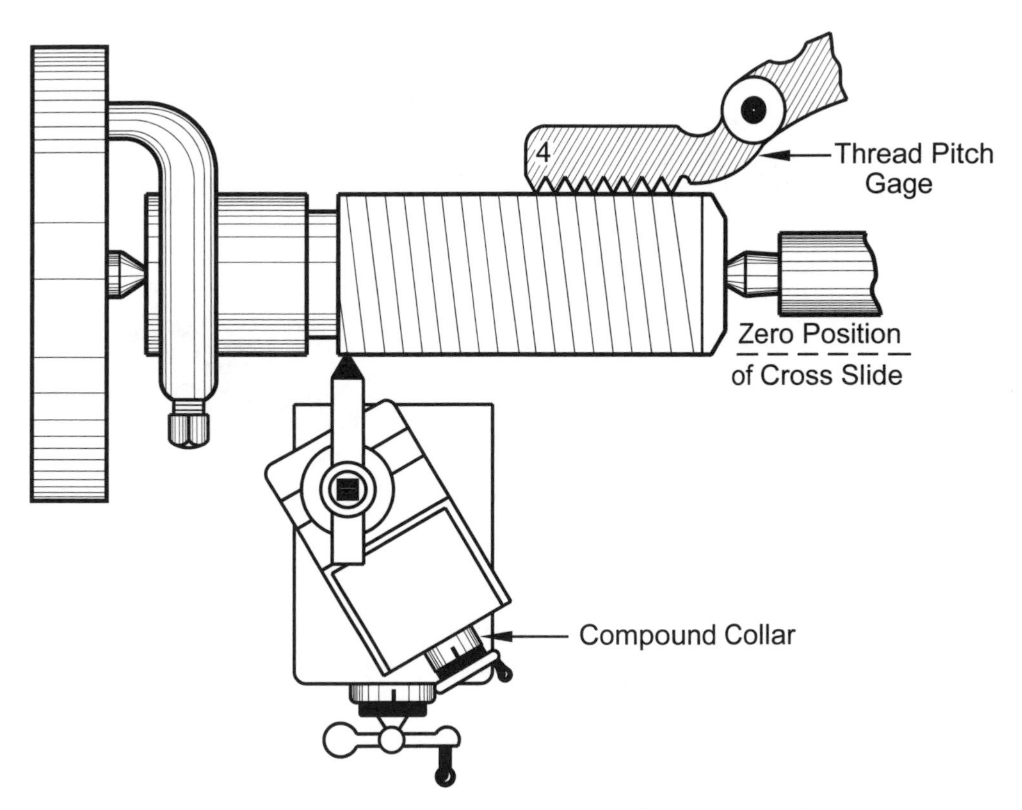

Figure 7–145. Making the scratch cut for a metric thread.

8. With the cross slide still withdrawn, Position 6 in Figure 7–146, start the spindle and move the carriage back to the starting point. Stop the spindle, and when it has stopped turning, set it to turn forward and turn it on.
9. Bring the thread to the required depth with additional passes. Do not open the half-nuts until the thread is completed and has been checked because you will lose the ability to pick up the thread again. The metric thread cutting sequence is shown in Figure 7–146.

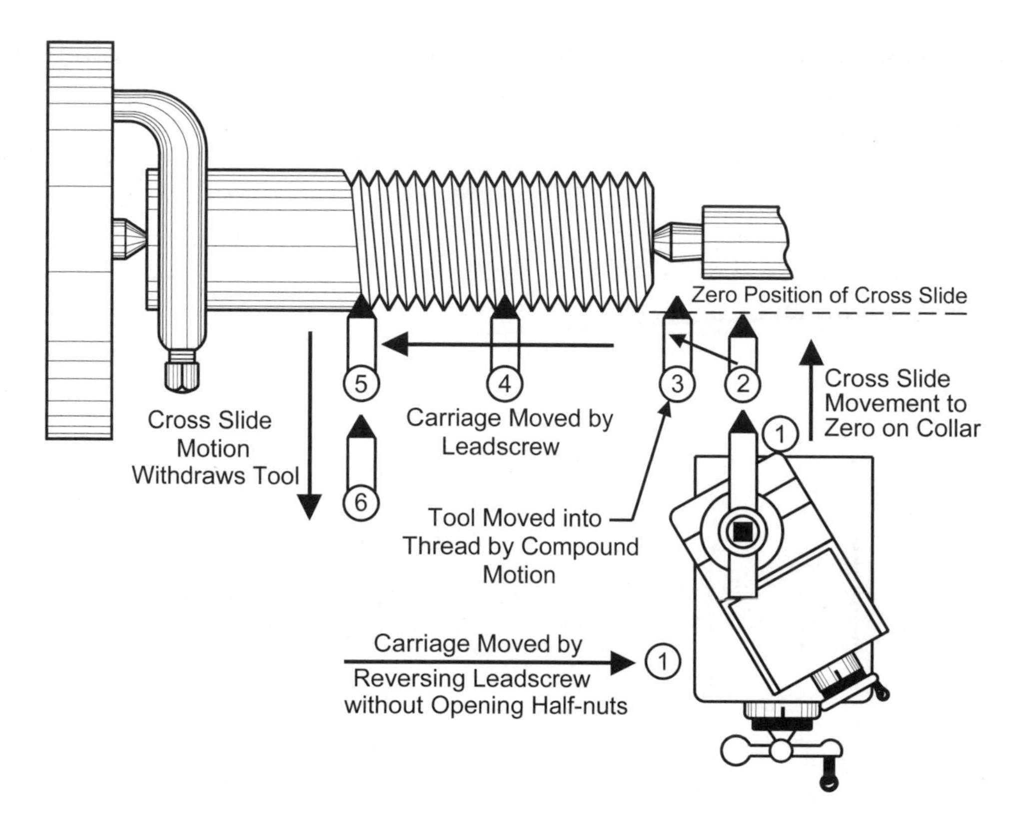

Figure 7–146. Metric thread cutting sequence.

Tapered Threads

What steps must be taken to cut tapered threads?

The threading tool must be set at right angles to the axis of the work, not the face of the section to be threaded. See Figure 7–147.

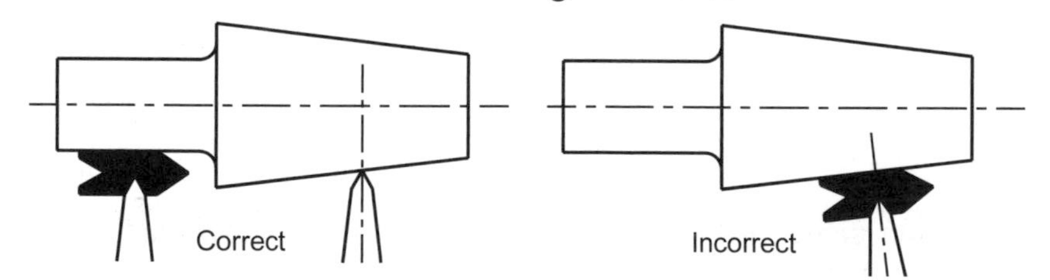

Figure 7–147. Threading tool setup for tapered threads.

There are three other issues when cutting tapered threads:

- Tapered threads are best cut using a taper attachment, not by offsetting the tailstock, because the thread will not advance at a uniform rate and form a true helix. The greater the taper, the more this inaccuracy is a problem.

- Offsetting the tailstock puts additional wear on lathe centers since they do not bear symmetrically against their center holes. One solution to this problem is to use ball centers or adjustable centers, Figure 7–120.
- Relatively mild tapers, like those on NPT (National Pipe Threads) for liquid joint seals, may be cut without problems by offsetting the tailstock.

60° Internal Threads

What tools are used to cut internal threads?

There are two common types of tools:

- Commercial boring bars with either HSS, alloy steel or tungsten carbide tool bits are the most common.
- For holes too small for a boring bar, an HSS tool bit blank may be ground into a small, single-piece threading tool. These are also commercially available. See Figure 7–148.

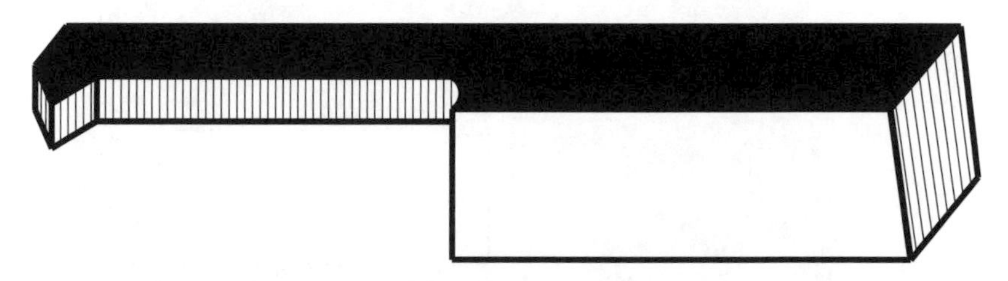

Figure 7–148. Single-piece threading tool ground from a square HSS tool bit blank.

What are two methods for setting the tool bit axis perpendicular to the work for internal threads?

See Figure 7–149.

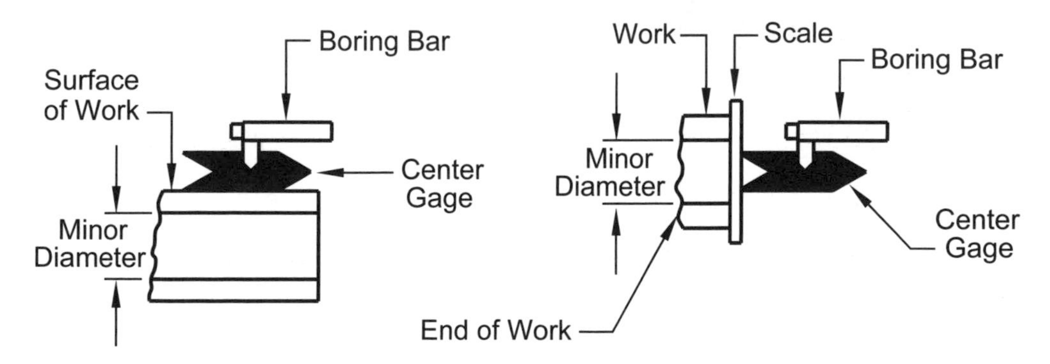

Figure 7–149. Two ways for setting alignment of a tool bit for internal threads.

What are the steps to apply internal 60° right-hand threads?

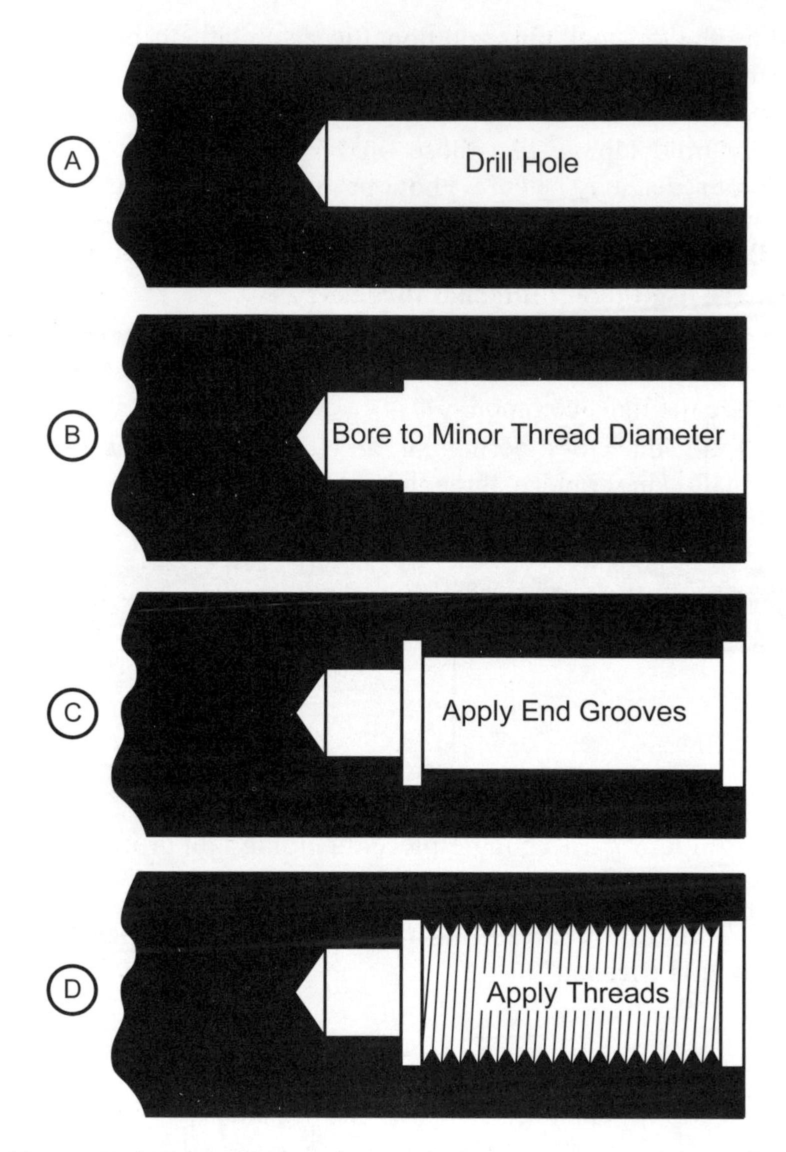

Figure 7–150. Preparing the workpiece for internal threading.

1. Mount the work in a chuck, collet or face plate and drill a starting hole for the boring tool, Figure 150 (A).
2. In the workpiece, drill or bore a hole of the threads' minor diameter, Figure 7–150 (B). This diameter may be obtained from *Machinery's Handbook.*
3. Where the thread does not run completely through the work, use a square-nose tool inside the workpiece to cut a groove where the threads end. This makes it easier to start or stop the threading process because there is no abrupt end of the thread. See Figure 7–150 (C). To avoid leaving a helical

scratch as the tool is withdrawn from the inside of the hole, move the tool away from the sides of the hole before it is withdrawn.

4. Cut a recess in the open end of the workpiece equal to the threads' major diameter so the thread depth can be observed during threading. See Figure 7–150 (C).
5. Threading may be done either starting from the outside of the work, Figure 7–151, or from the inside, Figure 7–152. Starting threading from the inside of the work eliminates the problem of not being able to see the internal end of the threads so you know when to open the half-nuts. This method is only suitable on lathes with a long-taper or cam-lock spindle nose which will not unscrew from the spindle when run in reverse under load. Select one of these methods, then install the boring bar (or single-piece tool bit) with the properly shaped cutting bit in the tool holder and set the compound at an angle of 29° to the work axis.

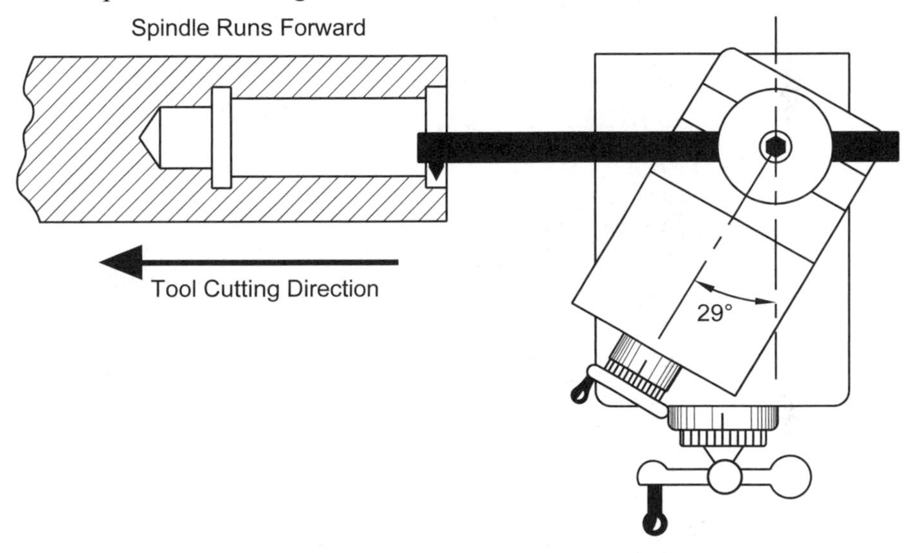

Figure 7–151. Setup for cutting internal threads starting from the outside of the work.

6. Adjust cutter bit height exactly on center, using the tailstock center as a reference. Make sure that there is adequate end clearance on the lower face of the cutter bit and that there is enough room behind the boring bar and cutter bit so that they will not bump into the interior of the workpiece when the tool bit is withdrawn from the threads.
7. Set the spindle and the leadscrew to turn forward, that is, to move from right to left. This will make the tool run into the work.
8. Apply the threads as in external threading operations, except that the cross slide is moved *forward* to clear the tool and return the carriage to the starting point. It is helpful to place a mark or piece of tape on the boring

tool bit arm or on the ways to indicate the stopping point at the end of the internal threads.

9. Observe the depth of the threads in the recessed starting area to determine when the threads are complete, Figure 7–150 (D). Check with a plug gage or the mating part.

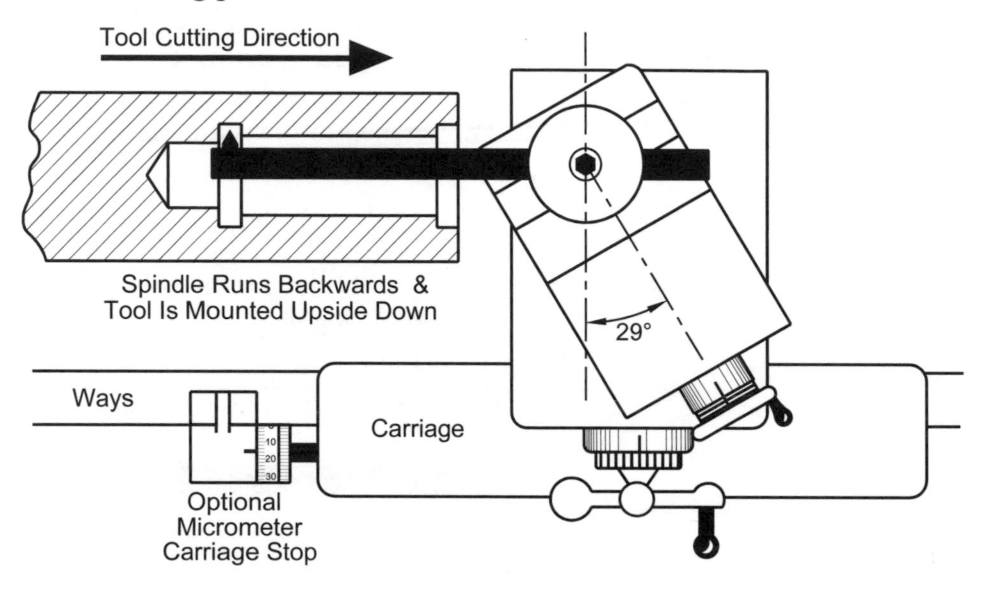

Figure 7–152. Setup for cutting internal threads starting from inside the work. Remember to mount the cutting tool upside down.

Minimize the length of the unsupported boring bar because it will deflect and make taking smaller cuts impossible. Plan to make the final pass more than 0.001 inch depth as the spring in the bar will make such a small cut impossible. Because you cannot see the starting point of the threads, use a *micrometer stop,* which clamps to the ways and provides an accurate starting point to position the carriage. See Figure 7–153.

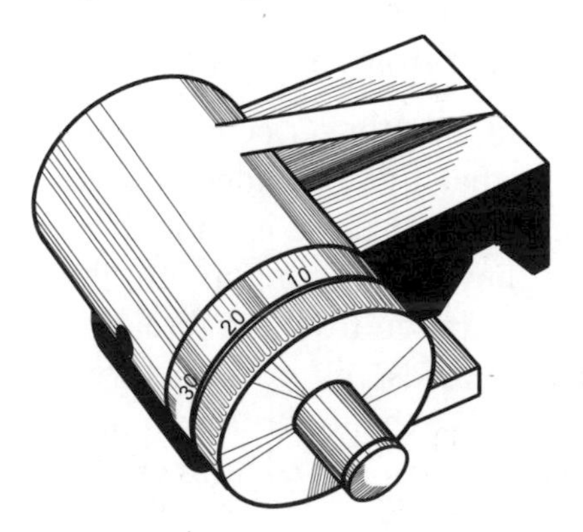

Figure 7–153. Micrometer stop.

Multiple-Start Threads

What are *multiple-start threads*, also called *multiple threads*, and why are they used?

Any thread with two or more individual helices is known as a multiple-start thread. The most common multiple-start threads consist of two, three, four, five or six threads. Figure 7–154 shows a regular, two-start and three-start thread of the same pitch for comparison. They are used when an increase in lead is necessary and a deep, coarse thread cannot be used. These threads are frequently applied on screw-type actuators and camera lens focus threads where a large axial movement or lead is needed. Multiple-start threads are also used on caps, covers and container lids where minimum rotation before catching a thread is required. In order to make room for the additional threads, multiple-start threads are not as deep as normal threads. For example, a two-start thread is one-half as deep as a single-start; a three-start thread is just one-third as deep and so on. These threads are also used where a coarse, large-lead thread is desired but the base metal is too thin to support the related thread depth as on microscope tubes.

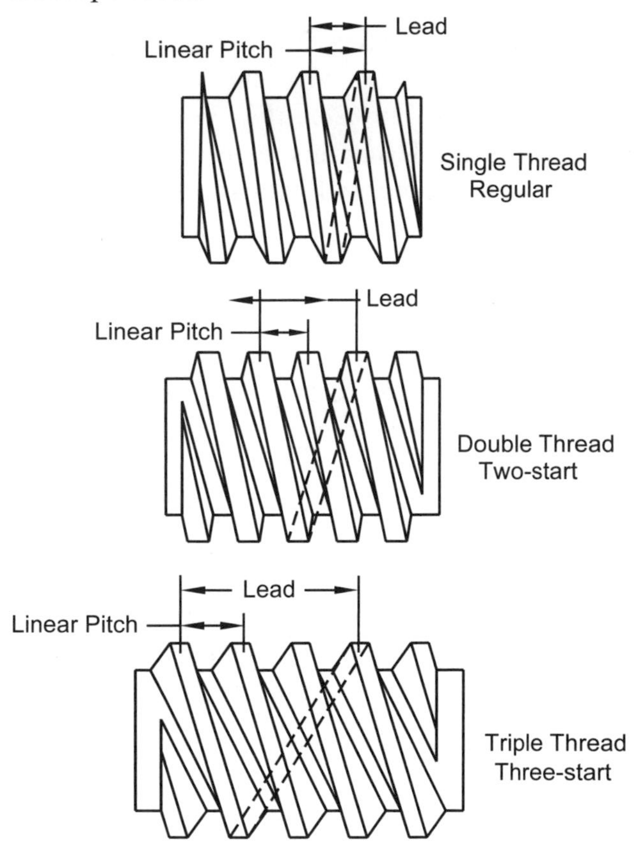

Figure 7–154. Multiple-start threads.

What are the four key points to remember in setting up for multiple-start threads?

On a single-start thread the pitch and the lead are equal, but on a multiple-start thread:

1. Equivalent threads per inch (tpi) = 1/Lead.
2. The quick-change gearbox (or change gears) must be set to move the tool to cut this equivalent tpi.
3. For an N-start thread, each one of the multiple threads is cut 1/N as deep as a normal, single-start thread of the same pitch.
4. When a thread has been applied, the tool (and carriage) must be start-indexed to begin cutting the next thread at the proper position between the existing threads. For a double-start thread, the second thread is placed exactly halfway between the first set of threads.

What are some methods used for *start-indexing*?

The methods are:

- Use an accurately slotted drive plate, catch plate or index plate to rotate the workpiece the proper amount: 180° for a double-start thread, 120° for a triple-start thread and so on. After each thread is complete, stop the lathe, leave the half-nuts engaged, remove the work from its lathe centers, *without removing the lathe dog from the work,* and reinsert the lathe dog in the next drive plate slot. Using this method the work is indexed, not the tool, but the result is the same. In Figure 7–155 the dots on the work and catch plate show their relative positions. This method works well, but only for external threads.

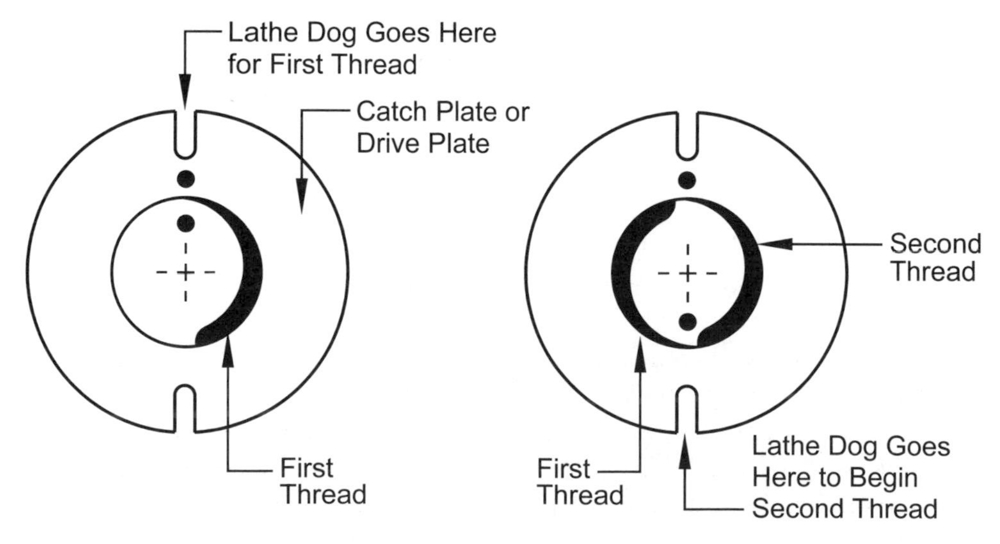

Figure 7–155. Using a drive plate to index work for a two-start thread.

These methods work for both internal and external threads.

- Start-indexing may also be done using a chuck which is rotated with respect to its back plate, the adapter between the back of the chuck and the spindle nose. This is accomplished with a series of screws between the chuck and the back plate, or chuck to spindle adapter, so in precise increments one may be rotated with respect to the other. A variation of this may be made for production so the chuck need not be removed from the lathe to perform indexing.
- The compound rest is set parallel with the work axis and is used to increment the tool bit one pitch distance using the micrometer collar dial. The compound slide must be advanced in small increments during the threading pass so only the leading edge of the tool bit cuts. If this is not done, both edges of the tool bit will cut, the emerging chips will not properly clear the tool edge, and tearing, instead of cutting will result. These small increments must be identical for each thread.
- Multiple tool bits spaced one pitch apart cut two or more threads simultaneously and achieve the same result as individually start-indexing each thread.
- Or use an intermediate gear in the gear train as an index.

What are the steps to apply external, right-hand threads?

1. Set up the lathe and tool bit as for cutting a single-start thread.
2. Adjust the quick-change gearbox or change gears to cut the equivalent number of threads per inch (tpi) required.
3. Cut the initial thread just as you would a normal or single-start thread, but do not open the half-nuts when this threading is complete.
4. Perform start-indexing for the second and subsequent threads.
5. Cut this thread to reduced depth. Half-depth for two-starts, third-depth for three-starts and so on.
6. Repeat Step 5 to complete all thread grooves.

Acme Threads

What is the procedure to cut an Acme thread?

1. Grind a tool bit to fit the Acme thread gage shown in Figure 7–135. Include adequate side clearance so that the tool does not rub on the work.
2. Set the quick-change gearbox or change wheels for the required tpi.
3. Set the compound to half the included thread angle, or 14.5°.
4. At the start of the work, cut a 0.0625-inch long section to the minor diameter. This will give a visual indication for when the full thread depth is reached.
5. Cut the thread as described in the section for cutting 60° threads.

Square Threads

What makes a tool bit for square threads different from other bits?

Because square threads have much larger leads than other threads, their tool bits must be ground with the proper *thread helix angle* to provide clearance so the tool sides clear the thread sides, Figure 7–156 (top).

How is the *thread helix angle* determined?

Construct the right triangle shown in Figure 7–156 (middle). The long leg is the circumference ($\pi \times$ Major Diameter) and the short vertical leg is the thread lead. The smaller of the two angles is the screw thread helix angle, which may be measured with a protractor. The tool width is one-half the thread lead less 0.001 to 0.003 inches to provide for a free fit on the nut.

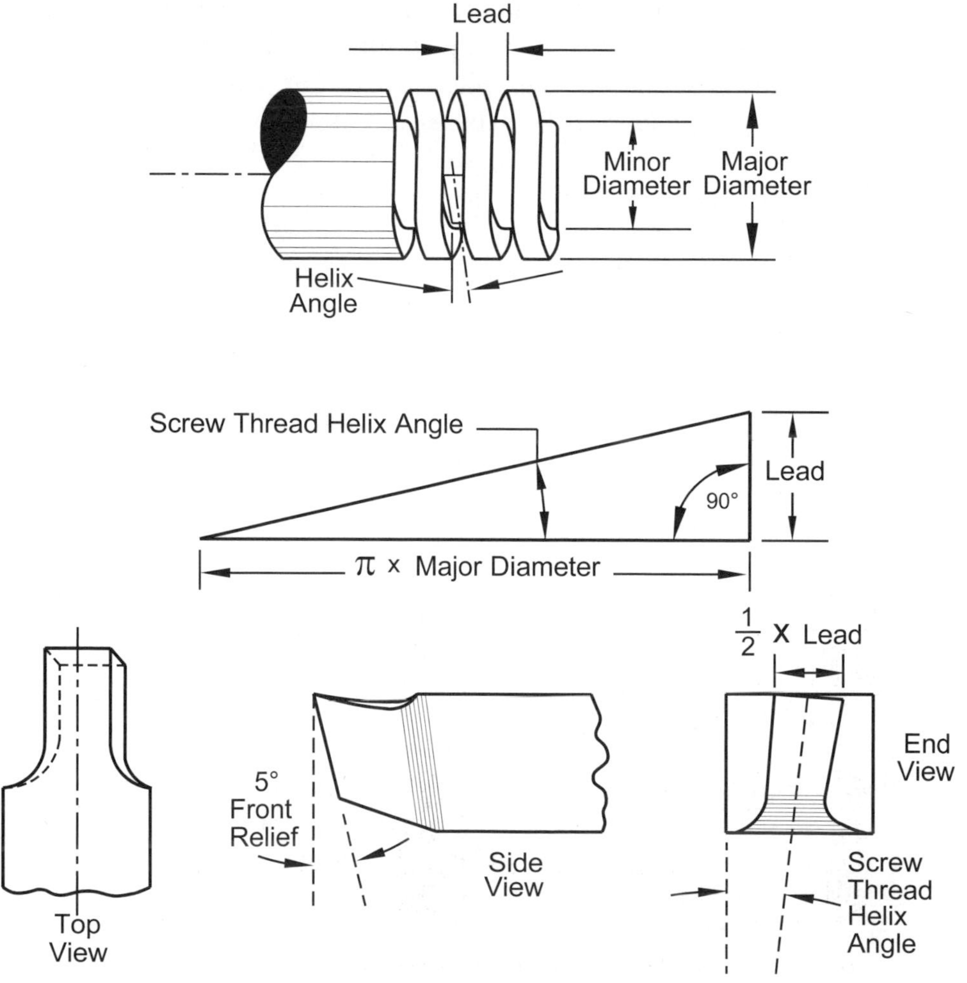

Figure 7–156. Square threads.

What is the procedure for cutting square threads?

1. Grind a tool bit with the proper helix angle as pictured in Figure 7–156 (bottom).
2. Install this tool bit in the tool holder. Set the compound rest 30° to the right, install the tool in the tool holder. Adjust the body of the tool perpendicular to the ways so the tool bit height is exactly on center.
3. Mount the work.
4. Calculate the minor diameter as 0.5/N inches or mm.
5. Turn the starting end of the work down to the minor diameter for about 1/16 inch (1.6 mm) to indicate full thread depth.
6. If possible, cut a stopping groove at the end of the thread to provide a place for the tool to run out. This groove will be the depth of the minor diameter.
7. Start the lathe, set the tool bit to just touch the full diameter of the work, not the minor diameter. Then set the cross feed collar to zero.
8. Make a trial scratch cut from 0.002 to 0.004 inches deep.
9. Check the cut with a thread gage.
10. Use cutting fluid and begin cutting the thread in depth increments of 0.002 to 0.010 inches depending on the size of the work and its material.
11. Remember that the thread sides are square, so all depth adjustments are made with the cross slide.
12. Use any convenient method for thread pickup as used on 60° threads and complete the thread.
13. Some machinists prefer to begin cutting a square thread as a 60° thread and then use the square tool bit to convert the groove to a square during the last few passes, Figure 7–157.

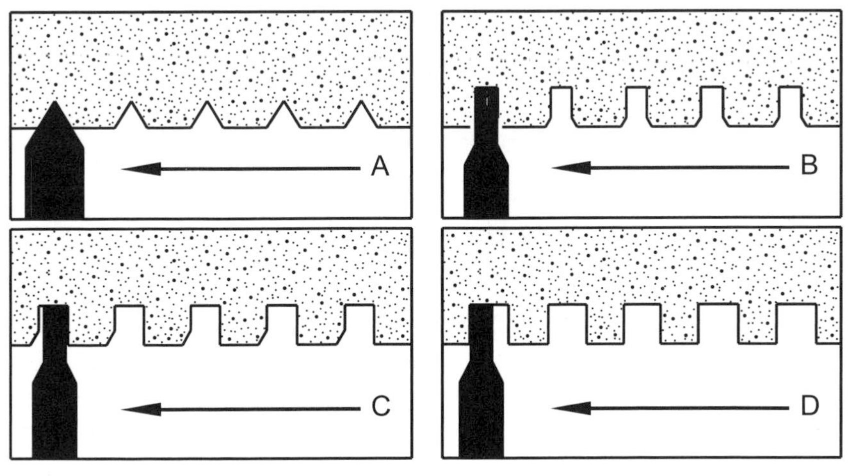

Figure 7–157. Cutting square threads by beginning with a 60° V-groove and widening them out to a square form.

Because this thread is relatively difficult to grind a tool bit for and to cut, it has been replaced with the American National Acme Thread, which is much easier to cut. A lot of equipment remains in service with square threads so spare parts with this thread will still be needed.

Knurling

What is *knurling* and why is it done?

Knurling is the addition of a series of ridges to the workpiece surface. Usually the knurls are parallel to the axis of the part or a diamond pattern. Most often knurling is done on the surface of a cylindrical part, but tools are available to apply knurls to rounded surfaces such as handwheels or the inside of large holes. There are two reasons for knurling:

- To provide an anti-slip hand-gripping surface.
- To add ridges to a pin or axle so it can be pressed into a hole and remain there.

How is knurling performed?

Knurling forces the movement of the workpiece surface metal into ridges. It does not cut or remove material. This is done using a pair of hardened steel rollers called *knurls*. Ridges on the knurl faces are pressed against the work and transferred to it. Knurls with ridges at an angle to the axis of the knurl, Figure 7–158 (left half), produce a diamond-shaped pattern. Those with ridges parallel to the axis, Figure 7–158 (right half), make ridges parallel to the work axis.

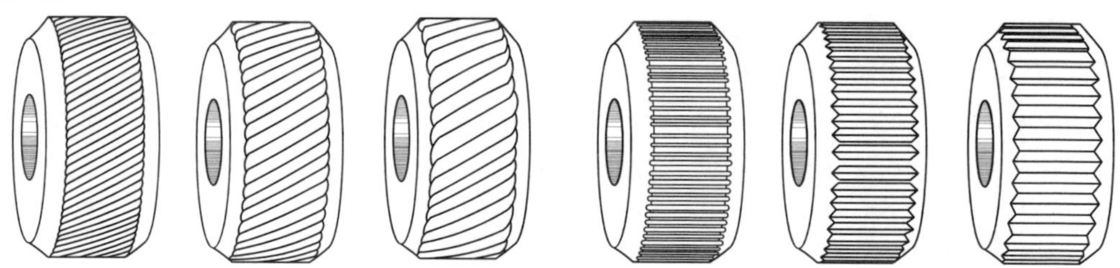

Figure 7–158. Knurls.

What kind of knurling tool holders are available and what are their advantages?

Traditional knurling tools hold a pair of knurls, Figure 7–159 (top left), and fit in the tool post. All knurling forces are transmitted through the tool post, compound slide, cross slide and carriage to the ways. Because large forces are needed to perform knurling, this design works best in large, rigid lathes. The Sherline carriage-mounted opposed-roller pinch-design knurling tool, Figure 7–159 (right), greatly reduces forces on the carriage because the largest forces, those pressing the knurls against the work, balance each other and are not transferred to the carriage. These forces are taken up by the cap-head screws which squeeze the knurls together. Another design eliminates any force on the carriage, Figure 7–159 (bottom left). This design clamps onto the work much like a pipe cutter and forces one roller against two others. The lathe operator prevents the tool from rotating and guides it along the work.

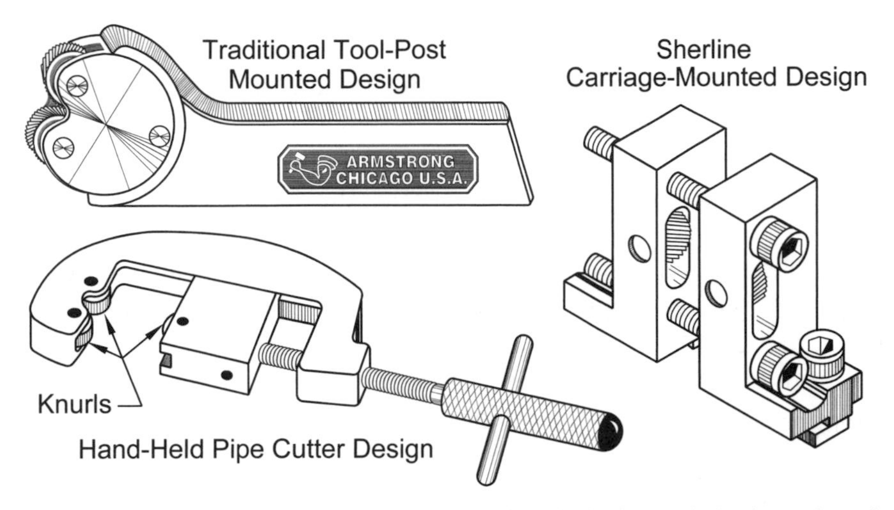

Figure 7–159. Knurling tools. Because of their balanced design, the pipe cutter and Sherline knurlers put much less stress on the lathe than the tool-post-mounted knurler.

What are the basic steps to perform knurling?

1. Plan to perform the knurling before other sections are turned down in diameter to increase the stiffness of the part during knurling. Knurling can be done between centers or in a chuck, but supporting the work with a live center will usually produce better results. Use large center holes to support the large forces of knurling. Set up the job so tool travel is toward the headstock.
2. Clean all mating surfaces and the knurls themselves.
3. Install the tool and work holder (for example chuck or centers and dogs) at right angles to the ways and on center height.
4. Secure the work in the work holder and center, if required.
5. Position the tool to clear the work.
6. Rotate the work by hand to insure it is secure and free to rotate.
7. Set a slow, back-gear speed, typically about one-quarter of the speed for turning the same diameter, and a feed of 0.015 to 0.030 inches/revolution.
8. Mark the section to receive the knurl.
9. Bring the tool up to the work so about half the width of the knurl wheels touches the work, start the lathe and force the knurls into the work until the diamond pattern comes to a point.

10. Engage the automatic feed and run the knurling tool the length of the workpiece.
11. When the tool reaches the end of the first pass, increase the knurling tool pressure, reverse the feed and run to tool back over the work. Repeat until knurling is completed.
12. Use plenty of lubricant to continuously flush away the flakes of loose metal.
13. Do not stop the automatic feed until the job is complete.

My knurls don't look right. Why?
Stops and starts or repeatedly going over the knurled section make ugly knurls. Stopping the knurling tool travel with the spindle turning will create rings. Double impressions are usually caused by one knurling tool being dull. See Figure 7–160. Try raising or lowering the knurling tool to increase the pressure on the dull wheel. Pivoting the tool slightly to the right may be helpful. Sometimes the knurls per inch needs to match the work diameter just as gear teeth must match. Try changing the diameter of the work slightly.

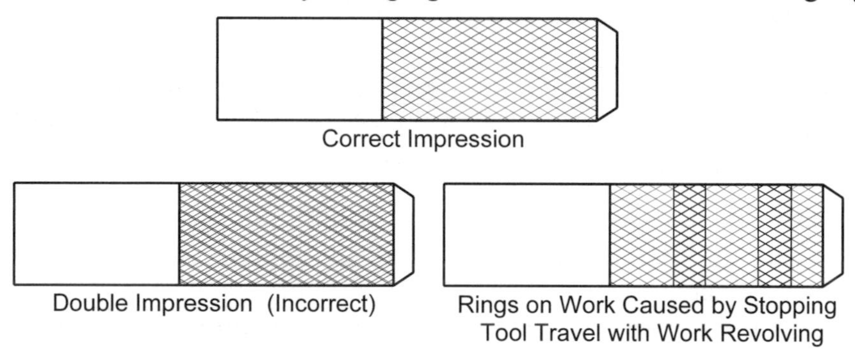

Figure 7–160. Knurling problems and their causes.

Polishing with Abrasive Cloth

What can be accomplished by polishing?
Polishing can remove tool and file marks from the workpiece. Filing removes the larger marks and abrasive cloth the smaller ones. For best results, polish with increasingly finer grades of abrasive cloth with a metal backing plate. See Figure 7–161 (top row) and 7–162.

How is polishing safely performed?
1. Clean all mating surfaces.
2. Move the carriage out of the way and remove the tool post.
3. Install the work holder (for example, chuck or centers and dogs). If a dead center is used, lubricate it. A live center is preferred.
4. Disengage the feed rod and leadscrew.

5. Secure the work in the work holder and center it, if required.
6. Rotate the work by hand to insure it is secure and free to rotate.
7. Protect the ways from the abrasive material that will drop from the polishing material with aluminum foil or paper. *Do not use cloth; it may be picked up by the work or chuck and draw you into the moving parts.*
8. Set the spindle speed to high.
9. Use an aluminum oxide abrasive cloth for ferrous metals. Use a silicon carbide abrasive cloth on nonferrous ones. Cut or tear a strip of cloth about 1-inch wide and 8-inches long from the sheet. Abrasive cloth is also available in tape form 1- or 2-inches wide on a spool like adhesive tape.
10. *The method of holding the abrasive cloth, shown in Figure 7–161 (lower left), is commonly used, but can lead to a dislocated thumb, torn fingernail or a broken finger. Do not use it.* There are two safer ways shown in Figure 7–161 (top row) and Figure 7–162.

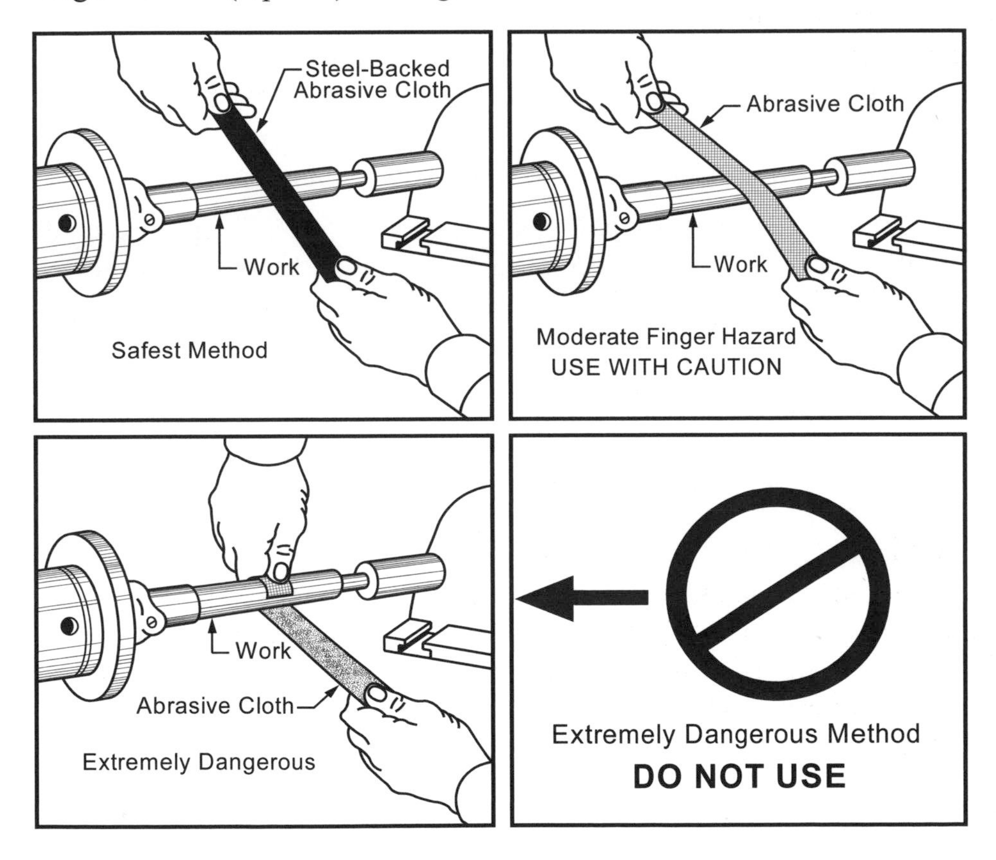

Figure 7–161. Two safe ways of polishing with an abrasive cloth and another to avoid.

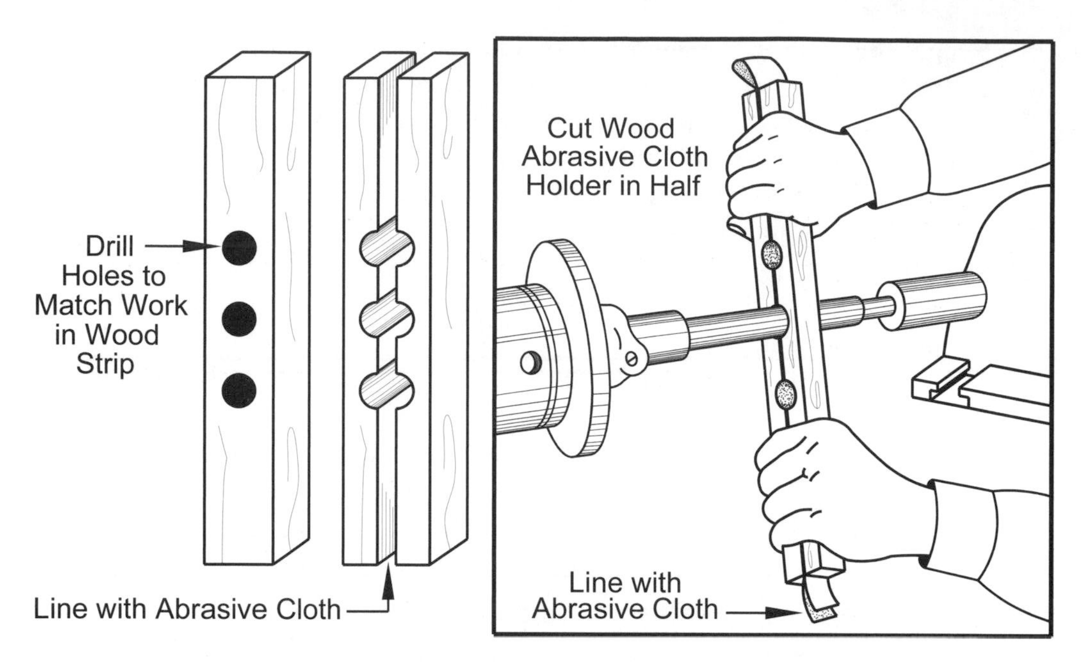

Figure 7–162. Safe and effective way to polish work with abrasive cloth.

11. Hold the cloth against the work using either method and repeat when the cloth is loaded with workpiece metal. As the work becomes polished, switch to finer grades of abrasive cloth.

Radii Cutters

What are the steps for using a *radii cutter*?

Refer to the pictures of radii cutters in Figure 7–84 and 7–85.

1. Clean all mating surfaces.
2. Install the radii cutting tool on the lathe carriage.
3. Secure the tool bit in the radii cutter tool bit holder, set the tool exactly at center height and extended to cut the desired radius. The radii cutter instructions contain details on calculations needed to cut a specific radius.
4. Place the workpiece in a chuck or mandrel. Center the work, if required.
5. Position the carriage so the tool bit clears the work.
6. Rotate the work by hand to insure it is secure and free to rotate.
7. Determine and set the proper spindle speed.
8. Turn on the lathe and slowly swing the cutting arm to form the desired shape.

These are the basic steps needed to use this tool. The exact procedure depends on the workpiece and the design of the radii cutter.

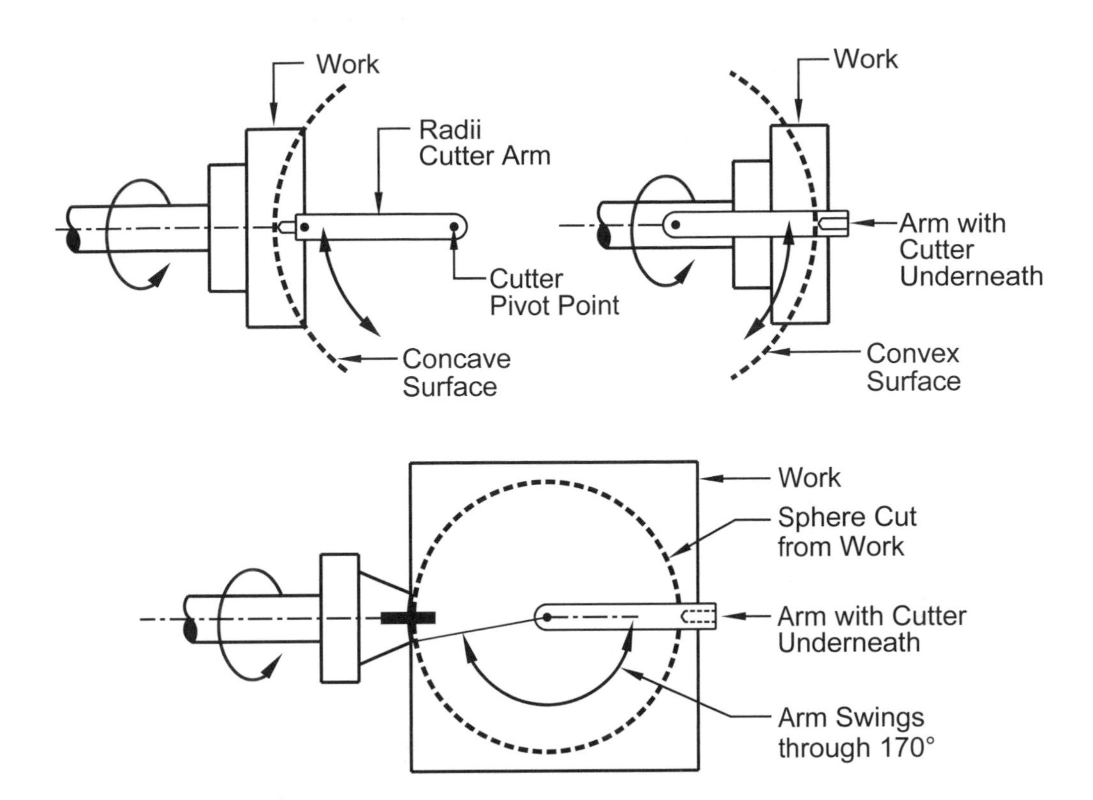

Figure 7–163. Top view of a radii cutter turning a concave surface (top left), a convex surface (top right) and a sphere (bottom).

The radii cutter cannot cut a complete sphere because the chuck or mandrel gets in the way, Figure 7–163. How is this base stub eliminated and a complete sphere made?

To make a complete sphere:

1. Machine a partial sphere.
2. Remove the partial sphere from the lathe.
3. Use a saw to remove the bulk of the base stem. This could be done in a lathe, but using a saw is faster. Do not remove material needed for a complete sphere.
4. Drill and tap the finished end of the sphere to form a support stem, mount the support stem in a chuck or mandrel and trim off the base stem to make a complete sphere. If a sphere without a base hole is needed, epoxy the partial sphere in a mandrel, and taking light cuts, trim off the base stem with the radii cutter. Alternatively, the sphere can be turned between two cup chucks for finishing.

Lathe Milling

How is milling done in the lathe?

Milling can be performed by installing a milling attachment, Figure 7–164, on the cross slide of the lathe, and a milling cutter bit in a collet in the lathe spindle. The milling vise, in combination with the cross and compound slides, provides three-axis positioning for the part. The major drawbacks to milling in the lathe are:

- The lathe milling attachment holds only small parts.
- The small range of three-axis movement this arrangement provides.
- Lack of rigidity limits milling to light cuts.

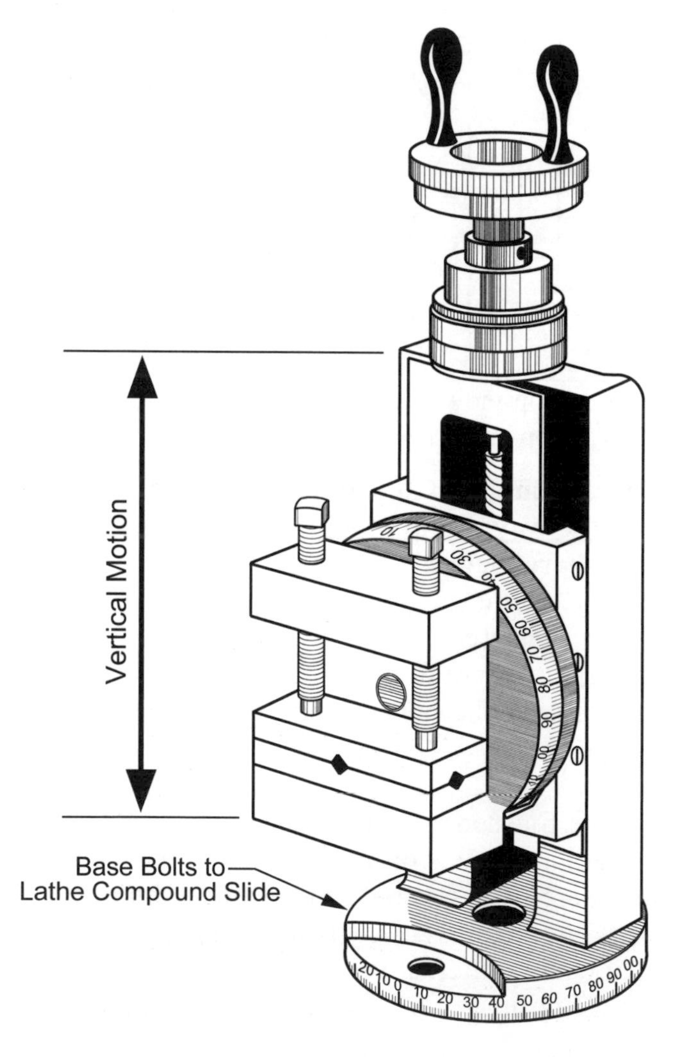

Figure 7–164. Lathe milling attachment.

What are the basic steps to perform milling in the lathe?

1. Clean all mating surfaces.
2. Install the milling cutter in its collet and the milling attachment on the cross slide.
3. Secure the work in the attachment and position it to clear the cutting tool.
4. Position the work where milling is to begin. Use an edge finder or center finder, see *Chapter 8 – Milling Operations* for details.

Tool Post Grinding

Why is *tool post grinding* performed?

Most often tool post grinding is used to rejuvenate hardened lathe centers. There are many other applications for grinding, but there are so many drawbacks to performing tool post grinding in the lathe that it is relatively uncommon. In an industrial setting, tool post grinding for purposes other than center grinding is not usually cost effective. Grinding in the lathe also exposes the ways and other moving parts to the abrasive dust coming off the grinding wheel, a very undesirable situation. See Figure 7–165.

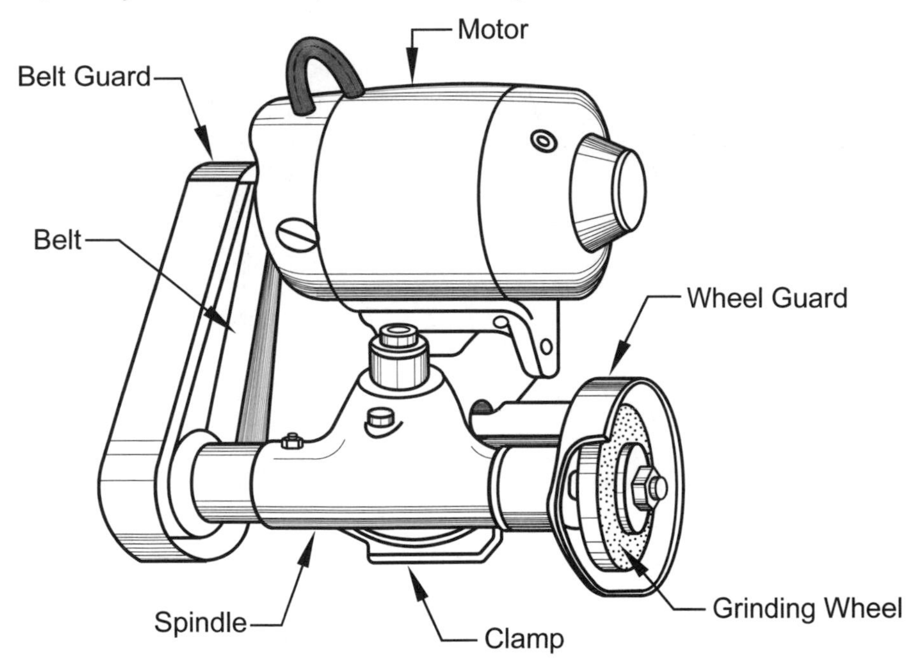

Figure 7–165. Tool post grinder.

What are the basic steps to perform tool post grinding?

1. Clean all mating surfaces.
2. Tape paper over the lathe ways to protect them from the abrasive dust this process will generate. Using cloth will work too, but presents a safety hazard since the rotating chuck, spindle or work may grab the cloth.

3. Install the grinder on the cross slide and the work holder (for example collet, chuck or centers and dogs).
4. Use a diamond tool to true the grinding wheel.
5. Secure the work in the work holder and center it, if required.
6. Position the tool to clear the work.
7. Rotate the work/spindle by hand to insure it is secure and free to rotate.
8. Determine and set the proper speed and feed and turn on the lathe.
9. Take repeated light cuts of 0.001 inch or less.

Section IX – More Work Holding Techniques

Cat Heads

How can a hexagonal-shaped workpiece be supported by a steady rest?
Use a *cat head,* Figure 7–166, over the workpiece as a cylindrical bearing. Use lubricant where the steady rest bears on it. This tool is readily shop-made.

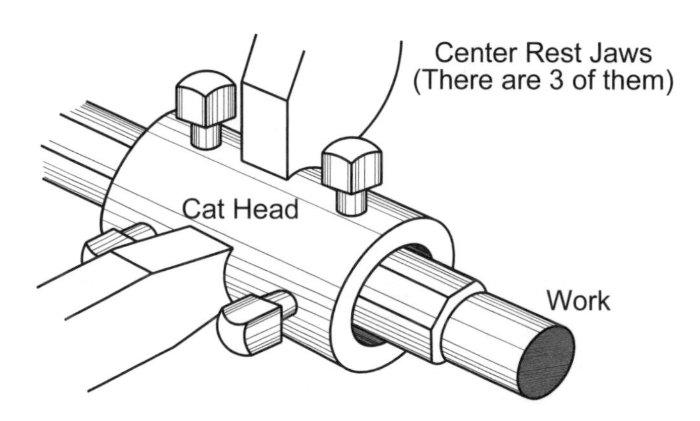

Figure 7–166. Cat head supports non-cylindrical work.

Work Secured in Low-melting Alloys or Pitch

How can a fragile or irregularly shaped workpiece be held for turning?
One method uses a hollow cylinder chucked in the lathe, Figure 7–167. Pitch or a low-temperature melting alloy secures the work within the cylinder. Cerro Metal Products Company, Bellefonte, PA 16823, supplies a family of low-temperature melting alloys containing bismuth, lead and tin suitable for securing irregularly-shaped work during machining. The alloy Cerrobend® with a melting point of 174°F is often used. Although expensive, it may be remelted and reused indefinitely. Other more expensive alloys with lower melting points are also available.

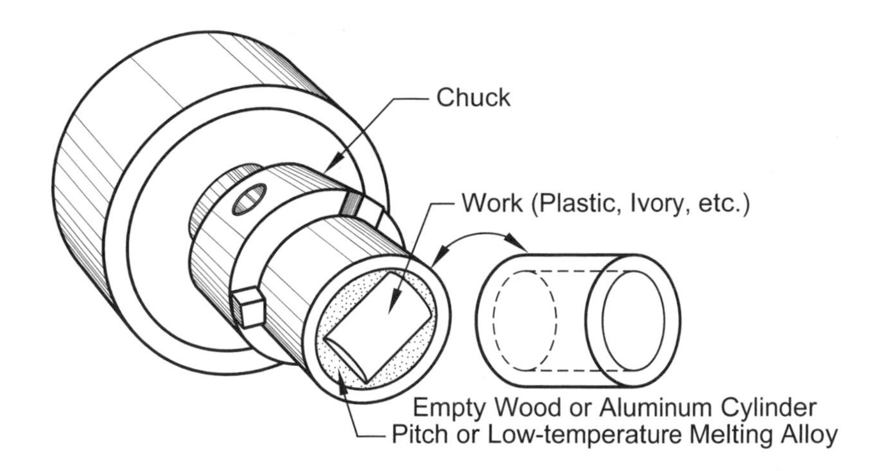

Figure 7–167. Irregularly shaped object embedded in a low-temperature melting alloy or in pitch.

Watchmakers' Technique

How can a small cylindrical part be accurately centered without a chuck or collet?

Here is a traditional approach:

1. Make a cone-shaped depression in one end of a brass, aluminum or steel rod deep enough to hold half the length of the part, Figure 7–168.
2. Fill the depression of the part holder with a thermoplastic material by warming the rod and the material with an alcohol lamp. Remember to warm, not melt. Alcohol lamps are just right for this task; they do not emit soot and their flames are exactly the right size.
3. Insert the part into the softened thermoplastic material and approximately center it.
4. Warm the part and part holder enough to soften the thermoplastic material and place them in the lathe, usually with a chuck.
5. Set the lathe to run about 200 rpm and turn it on.
6. Using a stick, slowly apply gentle pressure to the side of the part near its unsupported end.
7. As the assembly rotates, the part will snap into a centered position just as an edge finder in a milling machine. The thermoplastic material must be hot enough during this centering process to allow the part to find its own center, so more heat may be needed after the lathe is running.
8. Allow the thermoplastic material to cool.
9. Machine the part.
10. Remove the part by heating it again.

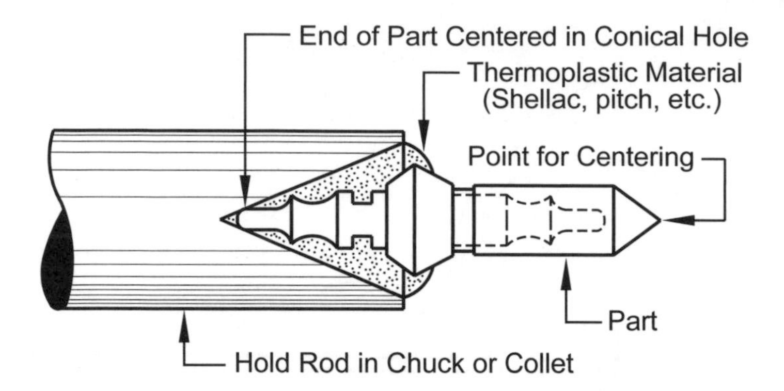

Figure 7–168. Traditional watchmakers' method for centering a small part.

Shop-Made Follower Rest for Threading

How can a long, thin rod be easily threaded?

Make the follower rest, shown in Figure 7–169, from a hardwood like oak or maple, or from aluminum, brass or steel and mount it on the carriage. It will provide rigid support to the rod so it will not bend under threading tool pressure. Be sure to use lubricant.

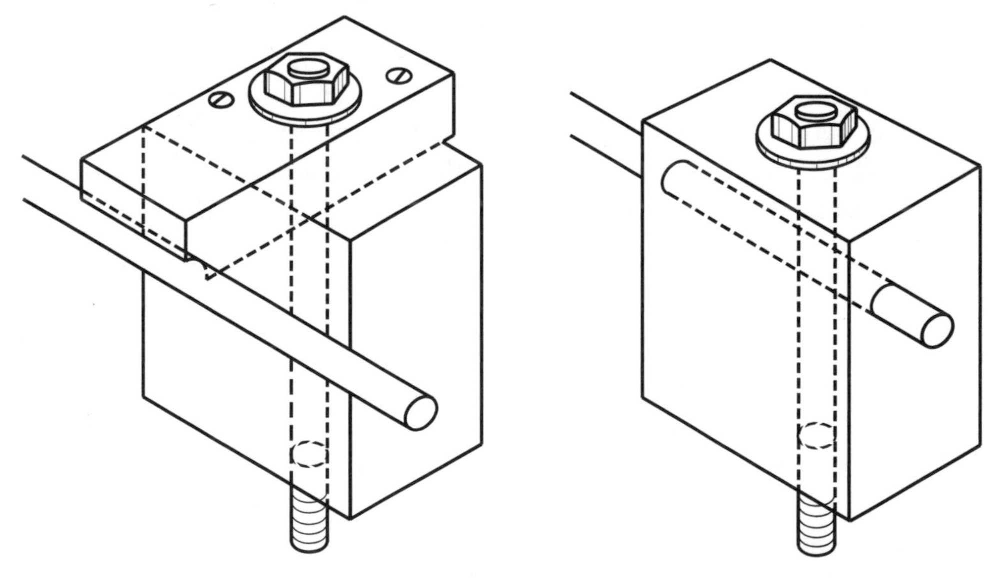

Figure 7–169. Shop-made follower rest for threading thin rod.

Section X - Lathe Installation, Maintenance & Repair

What are the key steps in installing and maintaining a lathe?

They are:

- *Leveling,* which is done to insure the ways are parallel to each other and have no twist. A machinists' level with a sensitivity of 0.005 inches per foot (0.42 mm per meter or 80–90 seconds of arc), like one of Starrett's No. 98 Series, is essential for this task. See Figure 7–170. A master precision level, like Starrett No. 199 with a sensitivity of 0.0005 inches per foot (0.04 mm per meter or 10 seconds of arc) would be even better, especially on a large lathe. A masons' or carpenters' level has a sensitivity of 0.062 inches per foot or less and is of no use. Wood or metal shims are used to level the lathe. Using the level, adjust the shims so the ways, both front and back, and at each end of the lathe are level. Also, level the machine so the ways are level in the transverse direction at both ends. Leveling should be checked periodically as buildings and their foundations shift and settle over time.

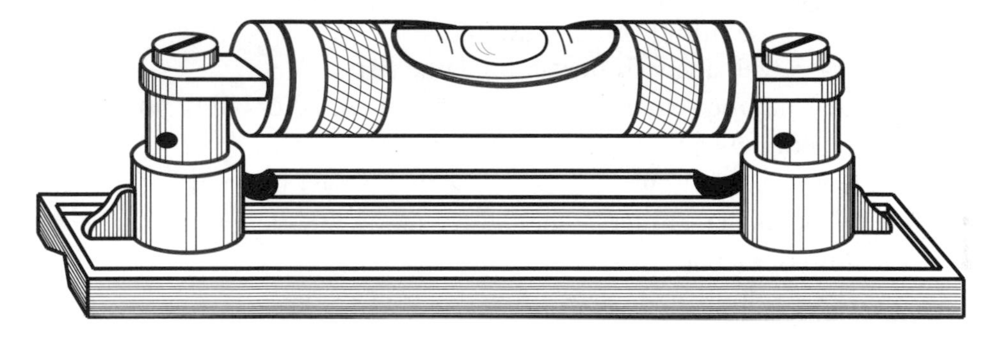

Figure 7–170. Starrett No. 98 Machinists' level.

- *Securing the lathe* to the floor with anchor bolts into concrete or lag bolts into wood prevents lathe movement from vibration and holds the shims in place. Recheck the leveling after tightening the anchor or lag bolts. Small lathes like the Levin and Sherline should be mounted on a plate, board or worktable to reduce vibration and movement. The bases of these small lathes are inherently rigid and do not need leveling.
- *Testing lathe alignment* after installation is done by turning a cylinder between centers without changing the cross or compound slides and checking its diameter, as shown in Figures 7–91 or 7–92. The workpiece will have a uniform diameter along its entire length if the ways are in alignment, level and the tailstock centered.

- *Lubrication* must be applied regularly according to the lathe manufacturer's instructions. A medium-sized lathe will have over twenty points to oil. Old lathes with babbit bearings must have their oilers or reservoirs checked regularly. Large lathes will have oil baths for their apron gears and must be checked and filled regularly.
- *Moving* lathes must be done cautiously because floor-mounted lathes are top-heavy and tip over easily. Use a fork lift or get plenty of manpower.

How can damaged spindle or tailstock Morse tapers be repaired?
Obtain a commercial reamer of the matching taper and re-ream the taper. Usually very little metal needs to be removed.

Section XI – Lathe Safety

What safety precautions should be observed when running a lathe?

- Wear eye protection.
- Because rotating parts of the lathe may draw you into it:
  - Do not wear rings, watches, necklaces or other jewelry.
  - Roll up sleeves, remove ties and put hair up out of the way.
  - Remove loose fitting shirts or sweaters.
  - Keep rags well away from rotating lathe parts.
  - Beware of apron strings catching in the lathe and drawing you into it.
- Keep your fingers away from rotating machinery.
- Understand exactly how the machine works *before* operating it.
- Do not operate machinery when tired or under the influence of drugs, medication or alcohol.
- Clamp work solidly and check often that it is secure.
- Always turn the chuck or face plate manually before starting the lathe to make sure it turns freely. Use the JOG button on larger lathes.
- Stop the lathe, and using pliers, remove chips or turnings. Metal turnings are sharp and can draw your hands into the machinery.
- Do not run lathes without all guards in place.
- Remove burrs and sharp edges from the work with a file before removing the work from the lathe.
- *Never* leave the machine running unattended.
- Do not allow yourself to be distracted while operating the lathe.
- *Never* take your hand off a chuck key while it is in the chuck. This way it cannot remain in the chuck if the machine is turned on accidentally.
- Do not let excessive amounts of long chips accumulate in the chip pan. They can be drawn into the lathe *en mass* and slung all over the shop.

Chapter 8

Milling Operations

Stupidity is the basic building block of the universe.
—Frank Zappa

Introduction

Unlike lathes, which have been known for thousands of years, milling machines are less than two hundred years old. Because they require much more power than hand-driven lathes, their introduction had to wait for the invention of industrial water and steam power. Also, all their mechanical components had to first be made available, such as accurately fitted slides, large castings to resist cutting forces, calibrated leadscrews and hardened steel cutting tools.

Eli Whitney is credited with inventing the first milling machine about 1818, but the knee-and-column support arrangement of the *universal* milling machine of Joseph A. Brown (later of Brown and Sharpe) dates from 1862 and marks an important step in the machine's development. During the last half of the nineteenth century, milling machines gradually replaced shapers and planers which have lathe-type, single-point tool bits that move over the work in a straight line and scrape off metal one stroke at a time. Milling machines, with their continuous cutting action, not only remove metal faster than shapers and planers, they perform additional operations such as cutting helices for gears and twist drills. Today, milling machines greatly outnumber shaping and planing machines. Americans in New England and later the mid-west continuously added features leading to the modern milling machine.

Another important development came in the 1930s when Rudolph Bannow and Magnus Wahlstrom brought out the Bridgeport-style vertical milling machine. This design offers versatility and economy in place of the higher metal removal rates of traditional horizontal milling machines. Because of this versatility, there are more Bridgeport-style mills in existence today than any other milling machine design. Horizontal mills are now usually reserved for production applications where high metal removal rates on identical parts are needed, not prototyping and short runs. Bridgeport-style machines are also called *knee-and-column machines* and *turret mills*. The key features of these machines include:

- Knee-and-column support for the milling table, which provides vertical motion of the work with respect to the tool.
- Saddle which supports the table to provide in-and-out motion from the vertical column.
- One-piece tool head which holds the motor, drive pulleys and spindle.
- Sliding *overarms* or *rams* were eventually added to allow the tool head to be moved in or out with respect to the vertical column. Some machines have provisions for the tool head to be tilted side to side or back to front.

The Bridgeport-style machine offers many advantages over the older horizontal milling machine design:

- The biggest advantage is the quill's ability to advance and retract the cutter easily *without* cranking to raise and lower the milling table. This speeds production and reduces operator fatigue. The retractable quill lets the operator quickly withdraw the tool to clear chips from a hole or check its progress. Tactile feedback through the quill feed handle or handwheel also tells the machinist how the tool is cutting and lets him optimize hand feed with less danger of tool breakage. Vertical table movement is still available for high-accuracy depth adjustment or when more force on the tool is required.
- The second largest advantage is the Bridgeport-style machine's ability to make angle cuts. With the horizontal milling machine, either the milling cutter is made on an angle or the work must be positioned at an angle to the spindle axis. With the Bridgeport-style machine the operator merely needs to tilt the spindle to make an angle cut. Of course, the Bridgeport can also use an angled cutter or mount the work on an angle.
- Vertical milling machines must use smaller cutting tools than horizontal mills because they have less rigid, less massive castings and lower horsepower motors. Still, they can accomplish the same end results as the horizontal mill, just more slowly.
- Vertical milling machines are less complex than horizontal machines because the one-piece tool head eliminates the need for complicated gearing *inside* the vertical column.
- Bridgeport-type milling machines usually have 1- to 5-horsepower motors and smaller castings than most horizontal mills. Because of this they generally cost less.
- Knee-and-column mills offer versatility and economy in place of the high metal removal rates of traditional horizontal milling machines.
- Bridgeport-style mills provide better visibility of the end mill cutter.

There are between 15 and 36 milling machine designs or styles, depending on who is counting, but the focus of this study is the Bridgeport-style vertical knee mill because it is most often used in shops doing prototyping and R&D work. They outnumber all other designs combined. This design has so much to offer that it has been copied in every industrialized country. At one time there were no less than thirteen separate Spanish companies building Bridgeport-style mills. A working knowledge of a Bridgeport-style vertical milling machine also provides a good start for operating any other style milling machine.

Lathes and mills are complementary machines. While lathes rotate the workpiece and produce a cylindrical cut, milling machines move work into a rotating cutter and make a straight line cut. Lathes and mills are both capable of boring large-diameter holes, but mills are better at placing holes anywhere on the surface of the work. Although one can sometimes make do with just a lathe or mill, a well-equipped shop must have both machines.

In lathe cutting, the tool is in continuous contact with the work and makes a continuous cut. Milling machines are just the opposite. They use multi-tooth cutting tools and their cutting action is intermittent as each tooth takes a bite. Metal is removed in small individual chips. Unlike lathe cutting tools, end mills, the most common cutting tool for Bridgeport-style mills, cannot be sharpened freehand because they must be perfectly symmetrical. Sharpening them requires special fixtures and shaped grinding wheels. Smaller shops send their cutters out for sharpening.

Adding a digital readout (DRO) is a great convenience to any milling machine. It reduces the need to repeatedly stop the mill to make measurements and lowers the chance of errors. When reset to zero, the DRO displays the exact displacement from a reference point on the workpiece making it possible for the operator to work directly with the dimensions on his working drawing.

For production applications, there are large, expensive milling machines with three or more axes under computer control. Some machines perform all operations including automatic tool changing. However, today there is an intermediate step between a manual mill and a fully automated one. Adding a computer, digital readouts and actuators to the X- and Y-axes of a Bridgeport-style mill does this. Not only can this enhanced machine tirelessly perform all its existing repetitive functions, it also has added new capabilities. Now the mill can engrave (drive the tool to cut numbers and letters in various sizes and fonts), cut radii and angles without a rotary table, make islands, pockets, and cut ellipses and frames. Entering the position, diameter and number of holes, automates cutting a bolt-hole pattern; the system does the math. The computer can also automatically compensate for the reduced diameter of resharpened milling cutters, saving time and money. The system can be manually programmed through its control panel, use stored programs, "learn" new tasks by memorizing a series of manual operations as the operator makes the first part, or accept files from CAD programs.

In this chapter we will study two mills, the Sherline miniature milling machine and the classic, full-sized Bridgeport® by Hardinge®. First, we will look at each machine, then examine its major components and study each of its adjustments and controls. Next, we will look at milling machine cutting tools and accessories, and learn its step-by-step operation. Finally, we will review milling machine safety issues.

Section I – Modern Milling Machines

Here are two typical quality machines from small to medium size:

- *Sherline Products Incorporated Model 5400-DRO Miniature Vertical Milling Machine*, Figure 8–1, has a 2.75- by 13-inch worktable, which offers two T-slots and rides on precision-machined dovetailed slides with adjustable gibs. The spindle has permanently lubricated bearings with an adjustable preload driven by a one-third horsepower variable-speed SCR drive DC motor. The motor accepts incoming current from 100 to 240 VAC, 50 or 60 Hz and automatically outputs 90 VDC to the motor so it can be operated anywhere in the world without the need for a separate transformer or rewiring. Speed is continuously variable from 70 to 2800 rpm without gear or belt changes. A second pulley position is available for providing additional torque at low rpm. Its three laser-engraved handwheels are calibrated in 0.001-inch increments or 0.01 mm in metric versions. Both the X- and Y-axes have anti-backlash feed screws. Like Sherline lathes, their milling machines have a #1MT spindle and a ¾-16 spindle nose thread. This model's overall dimensions (H × W × D) are: 21 × 15 × 12 inches; travel X-axis: 9 inches, Y-axis: 5 inches and Z-axis: 6.25 inches. The headstock can be rotated up to 90° right or left for angled or horizontal milling. The machine pictured has a DRO for all three axes and also displays spindle rpm. Like larger mills, the DRO can be set to zero at the push of a button. This machine, including DRO, sells for about $1000 and has a shipping weight of 37 lb. Additional optional Sherline column upgrade kits allow headstock column rotation (90° left/right), column pivot (front/back), column swing (90° left/right) and 5.5 inches of column travel (in/out) much like full-size Bridgeport-style mills. Sherline offers dozens of accessories including end mills, milling cutters, flycutters, boring heads, tilting tables, rotary tables, chucks, knurling tools, ball turners and collets. Although these machines are small and relatively inexpensive, they are capable of precision work. Sherline manufactures its products in Vista, California. Sherline's website, www.Sherline.com,, has complete instructions, parts lists and drawings on their tools. There are also dozens of examples of projects made on these mills as well as clever enhancements owners have devised.

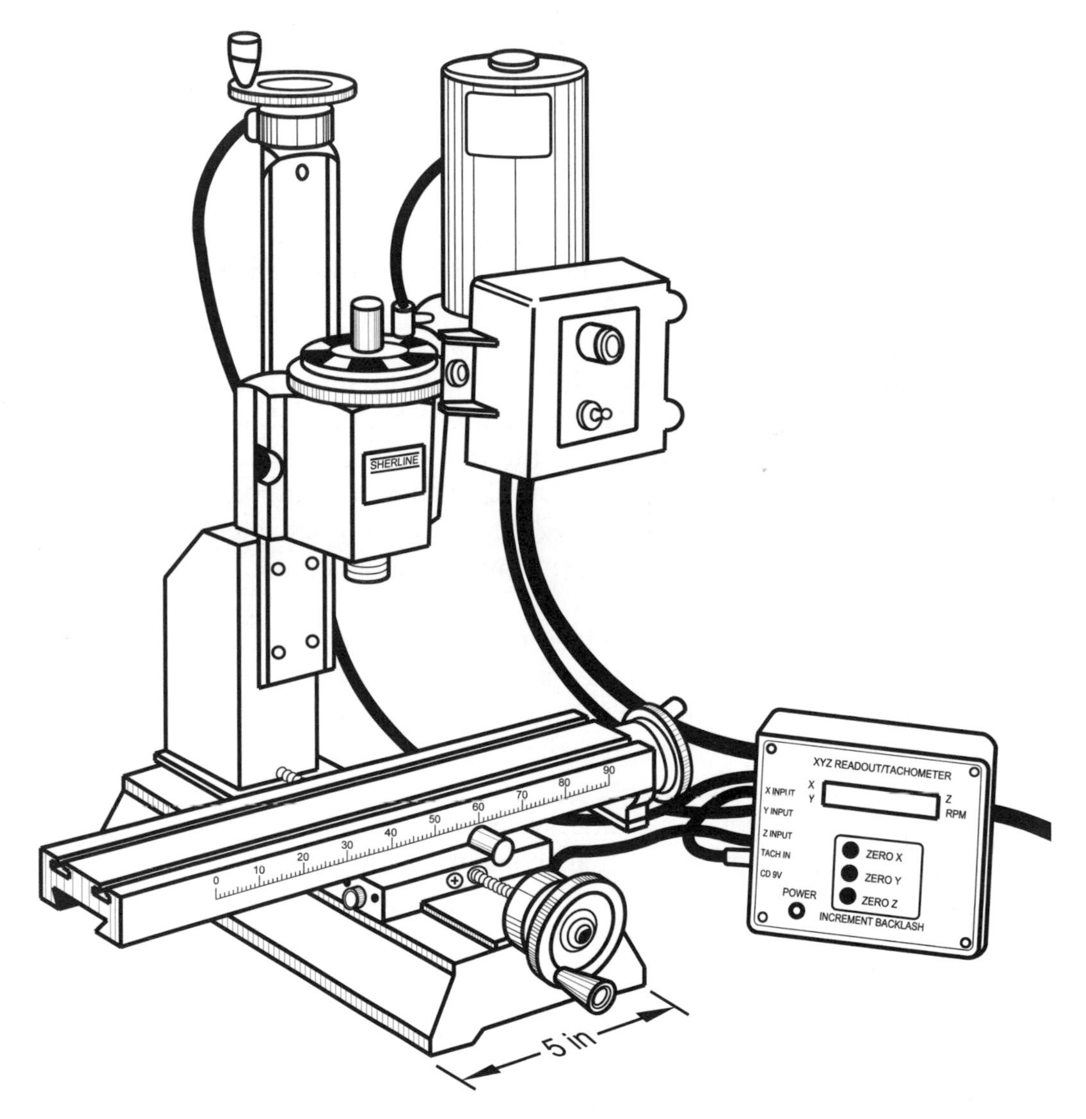

Figure 8–1. Sherline Model 5400-DRO Miniature Vertical Milling Machine.

- *Bridgeport® by Hardinge® Series I Standard Mill,* Figure 8–2, is the latest version of a classic machine tool that has been refined over 75 years. Milling machines based on this design are known around the world as

"Bridgeports." Over the last 60 years the original Bridgeport Company built over 350,000 of these machines. If all competitive look-alike machines in the US were counted, they would number over a million.

This machine uses gray cast iron construction for rigidity. All alignment ways and gibs are hand-scraped to within a tenth of a thousandth of an inch, ensuring full bearing support. The saddle and knee gibs are full length and feature a single gib screw for easy adjustment. There are multiple knee and table locks for maximum rigidity.

The ribbed, cast iron milling table is 48 × 9 inches (1219 × 299 mm), has three T-slots and supports a workpiece weight of up to 750 lbs (340 kg). The X-axis table travel is 36 inches (914 mm), Y-axis travel is 12 inches (305 mm) and Z-axis or knee travel is 16 inches (406 mm).

To eliminate runout, the spindle taper is ground after the spindle is fully assembled. It accepts either R8 collets or tooling. The machine has a 3.375-inch (86-mm) diameter hard chrome-plated and hand-lapped quill. Quill power feed is standard with feeds of 0.0015, 0.003 and 0.006 inches/revolution (0.038, 0.076 and 0.152 mm/revolution). In mild steel, the quill power feed can drive a ¾-inch (19-mm) diameter end mill, a ¾-inch twist drill or bore a 6-inch (152-mm) diameter hole.

The cooling fan is integral to the motor and draws cool air up through the head minimizing heating of the quill and spindle bearings, prolonging belt life. Motor speed is infinitely variable from 60 to 500 rpm on low speed and from 500 to 4200 rpm on high speed.

A metered one-shot lubrication system is standard. Chrome-plated ways and gibs are optional as is a flood coolant system.

This mill is powered by a 4 kVA, three-phase, 208/230/460 volt, 50/60 Hz motor, rated at 3 horsepower for a 30-minute duty cycle and 2 horsepower for continuous duty.

The machine has a shipping weight of 2100 lbs (959 kg), a height of 82 inches (2.1 m) and carries a price of about $11,000. These machines are made and supported by Hardinge, Inc. in Elmira, New York 14902.

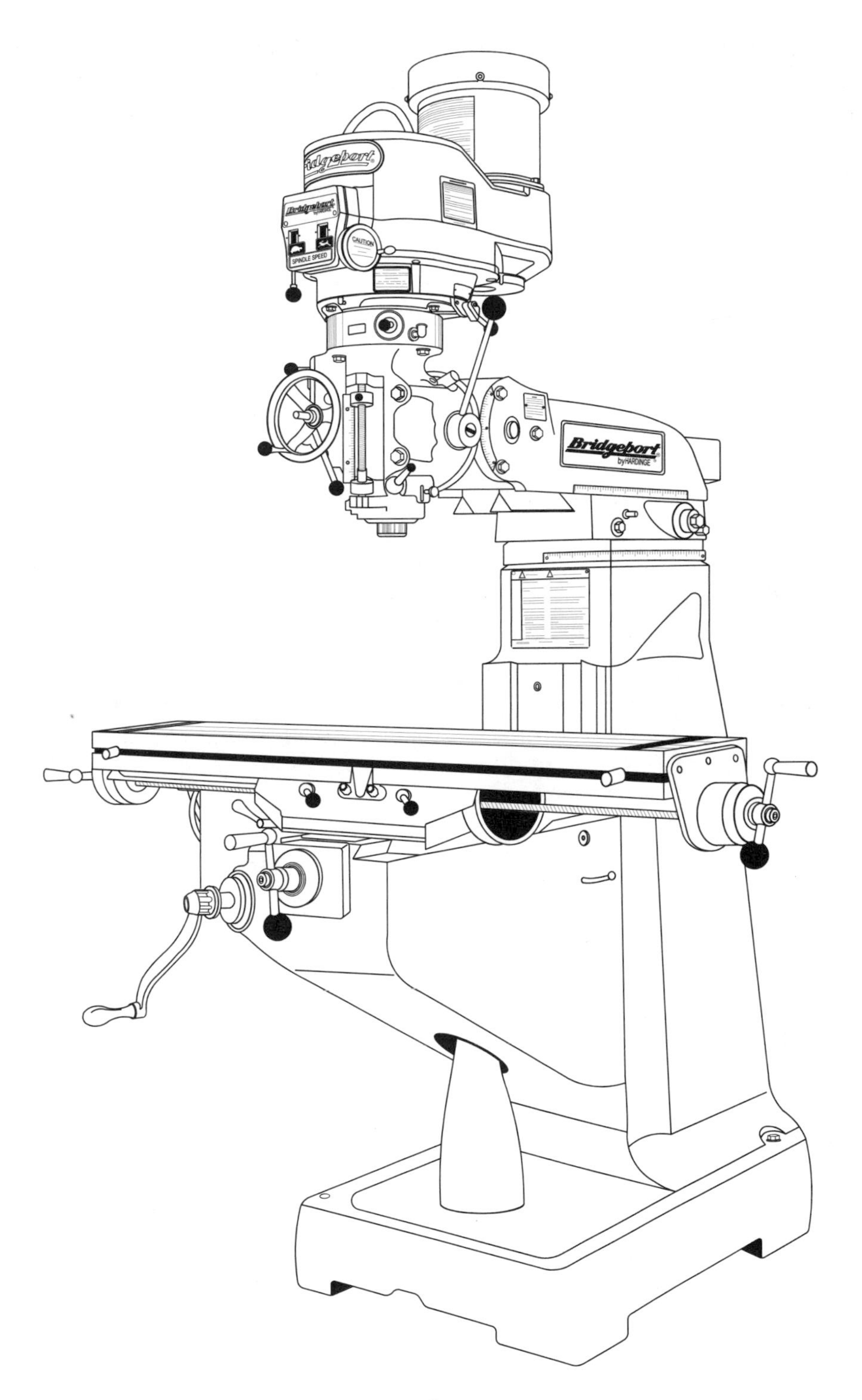

Figure 8–2. Bridgeport® Series I Standard Mill.

Section II – Parts of the Milling Machine

Machine Axes

What are the milling machine's axes?

As the machinist faces the milling machine, the three most commonly referred to table axes are:

- *X-axis*, or *table travel,* left-right table movement.
- *Y-axis*, or *saddle travel*, in-out table movement with reference to the vertical column.
- *Z-axis*, or *knee travel*, or *quill travel*, up-down table or tool movement, respectively.

See Figure 8–3.

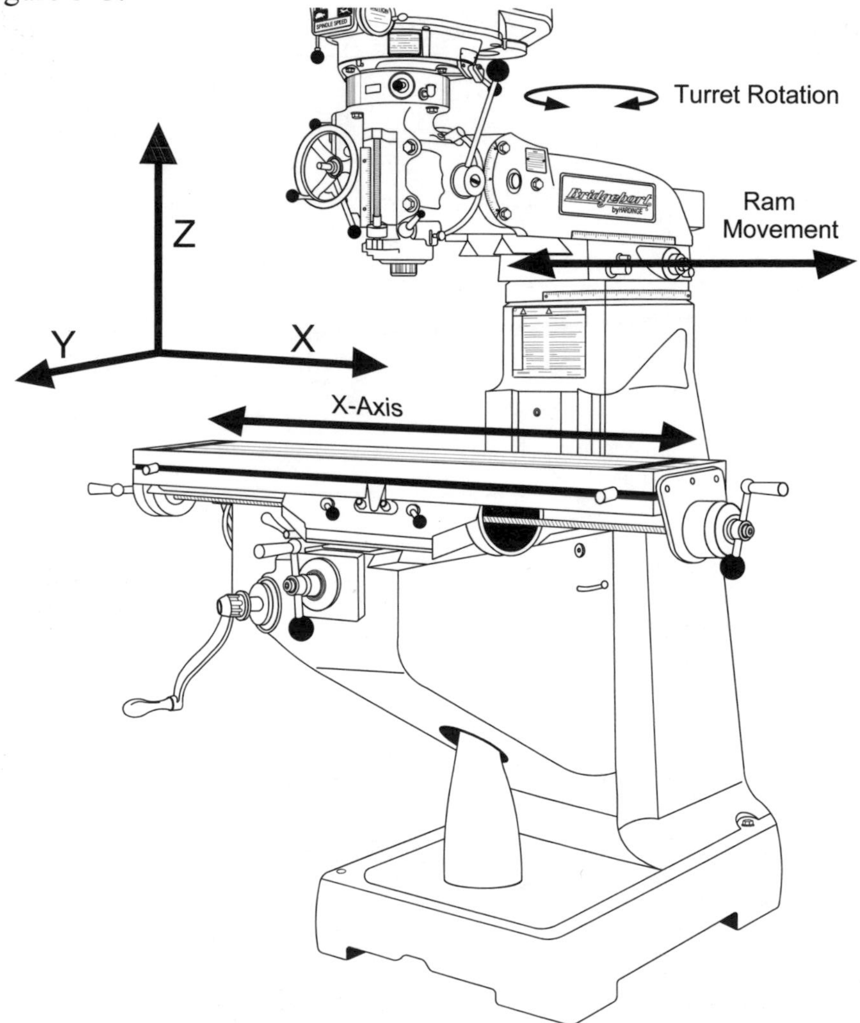

Figure 8–3. Three axes of movement and ram movement.

These are the only directions of movement that are made when the machine is actually cutting metal. There are also other axes of movement that extend the reach or change the angle of the spindle with respect to the horizontal:

- The *ram,* also called the *overarm,* may be slid in-out, away or towards the machinist along the dovetails, extending or shortening the *throat distance.*
- The *turret* can rotate side to side on top of the vertical casting, Figure 8–3.
- The *head* can tilt to the left or right as much as 45°.
- The head can also be tipped toward or away from the operator 45° in each direction.

See Figure 8–4.

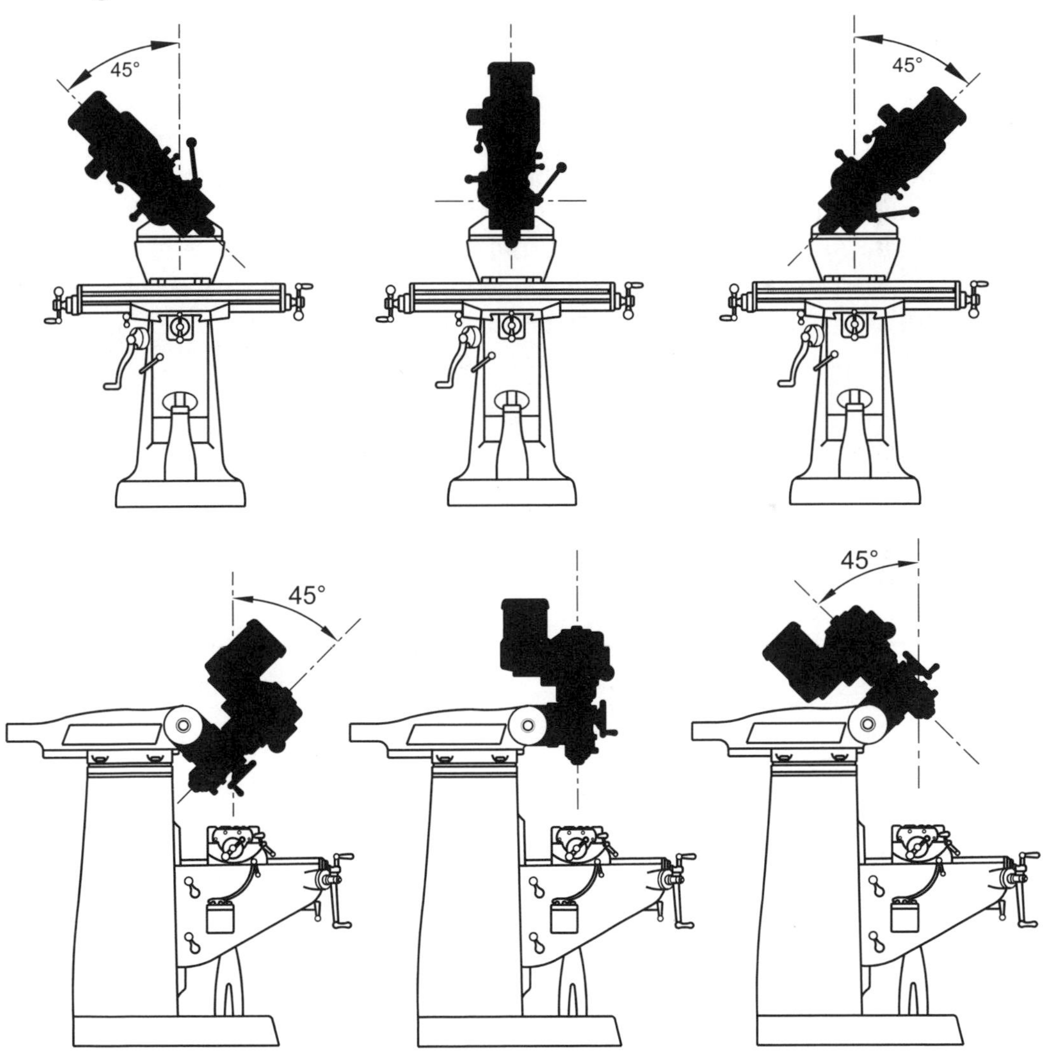

Figure 8–4. Bridgeport head movement 45° side to side (top) and tilting 45° toward front or back (bottom).

Base & Vertical Column

What are the functions of the *milling machine base* and *vertical column*?

The base provides rigid support for the column and knee castings. A vertical dovetail machined into the column guides the knee on the Z-axis. The vertical column also supports the turret, which in turn supports the ram and head. Most bases and columns on Bridgeport-style machines are a single gray iron casting. Just as with lathes, the combined rigidity of the structural castings, the tightness of the dovetails and leadscrew backlash determine the maximum useful horsepower and largest cut that can be made without chatter.

Ram & Turret

What do the *ram* and *turret* do?

The *ram*, also called the *overarm*, enables the cutting head to move in and out in the Y-direction, Figure 8–3. The ram gives the machine greater capacity and flexibility. Minimizing the distance between the vertical column and the cutting head increases overall machine rigidity and reduces the chances of chatter. The *turret* supports the ram and allows it to move in and out on its ways. It also allows the ram to swing side to side as in Figure 8–4. The ram or turret position is not changed while performing cutting operations. Figure 8–5 shows the locking bolts that secure the head.

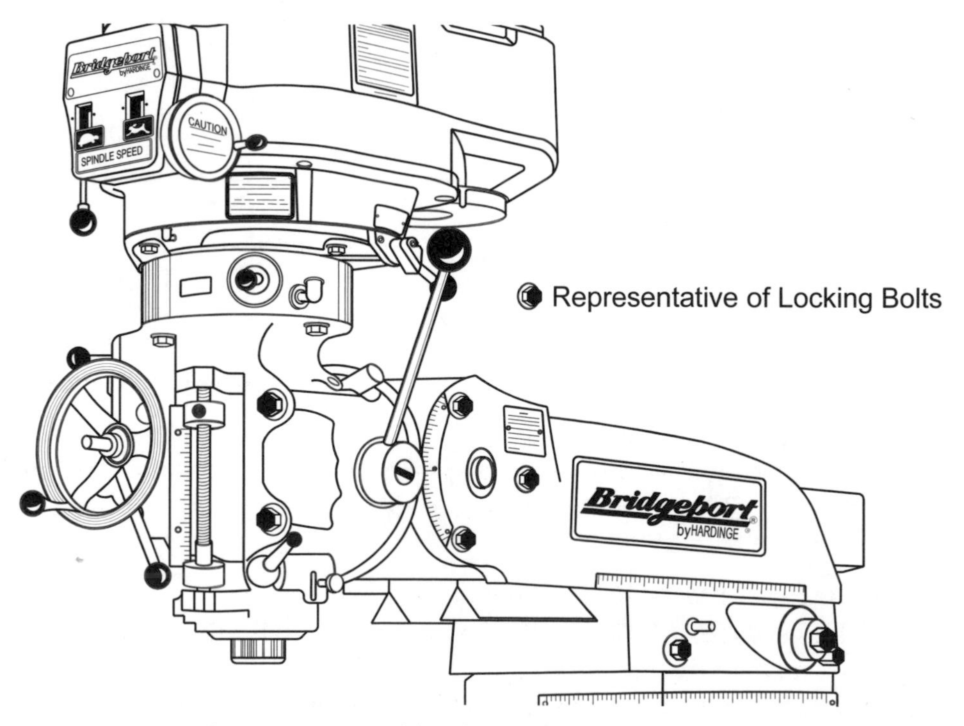

Figure 8–5. Locking bolts for head, ram and turret.

Knee, Saddle, Table & Leadscrews

What is the purpose of the *knee*?

The knee is a casting that supports the saddle and the milling table and prevents their movement under cutting forces on all three machine axes, Figure 8–6 (lower left). After the base and vertical column casting, the knee is the second largest casting. It must support the saddle, the table and the workpiece that alone can weigh as much as 750 lbs.

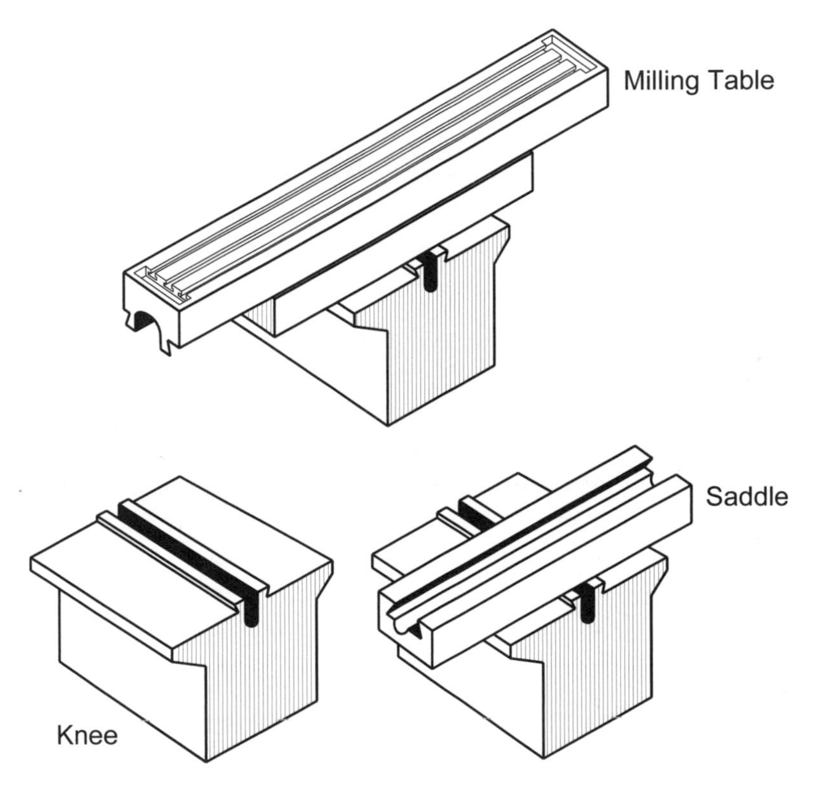

Figure 8–6. The knee casting (lower left), the saddle and knee joined by their dovetails (lower right) and the knee, saddle and milling table together (top).

What is the function of the *saddle*?

The saddle casting has dovetails at right angles to each other that permit milling table movement along both the X- and Y-axes. See Figure 8–6 (lower right).

What is the purpose of the *milling table* and how is it made?

The milling table provides a base on which to clamp the workpiece and hold it rigidly against machining forces, Figures 8–6 and 8–7. Many milling machines provide a stream of coolant, which the milling table traps and recycles.

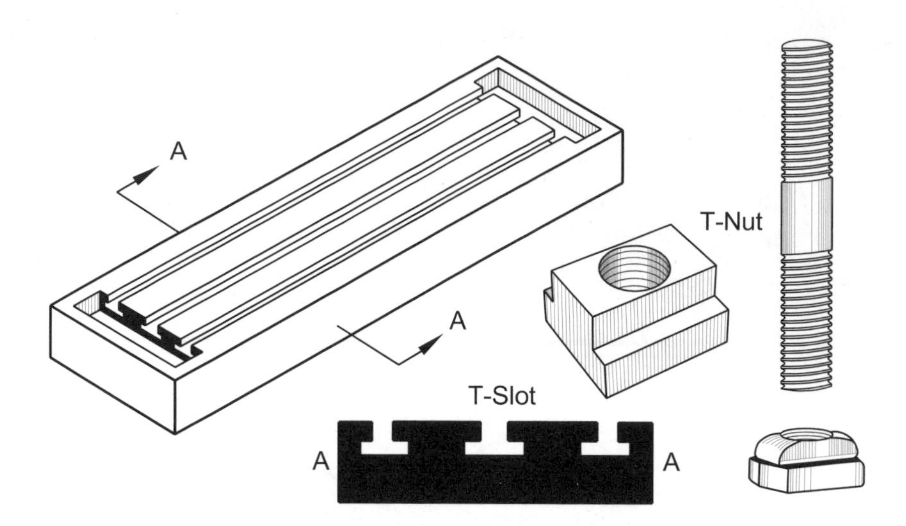

Figure 8–7. The milling table, a stud and two T-nuts that fit in the mating T-slots for securing work to the milling machine table.

Where are the *leadscrews* positioned and what turns them?

The leadscrews, which move the X- and Y-axes, lie between the upper and lower dovetails on the saddle in the channels, or depressions, made for them. Handwheels at the right and left ends of the milling table provide X-axis movement, and the single handwheel on the face of the knee provides Y-axis movement. Another leadscrew below the knee provides vertical knee movement. A crank, to give the mechanical advantage required to raise the knee under heavy table loads, turns this leadscrew. For even more mechanical advantage, the vertical crank usually provides only one-half the table movement the same rotation on the X- or Y-axis handwheels does.

What is *backlash*, what causes it, and what mechanism is used on the leadscrews to minimize it?

Backlash is the lost motion, or dead band, that occurs between a leadscrew and its nut. When a handwheel turns the leadscrew and the milling table fails to move, this is backlash. Usually backlash is only a small part of a handwheel revolution, typically 5 to 10 thousandths of an inch, and is readily compensated for by always taking up the slack in the leadscrew in the same direction.

Lack of a perfect, zero-clearance fit, between the leadscrew threads and its nut threads causes backlash. The nut threads must be slightly smaller than the leadscrew threads to allow a good fit. Normal wear on the leadscrew and its nuts degrades this fit. Backlash causes excessive chatter when milling because of the intermittent cutting action of a multi-toothed milling cutter. Backlash must be reduced to minimize chatter and to produce a smooth, milled surface.

Figure 8–8 shows how a backlash eliminator works. Two threaded nuts fit over the leadscrew and are prevented from revolving by a spline. Compressing these two nuts together has the overall effect of making the pair of nuts fit tightly on the leadscrew, preventing slack motion in both directions and greatly reducing backlash. Leadscrews are steel, and the two nuts in the backlash eliminator are a softer material, usually bronze. These bronze nuts wear faster than the leadscrew and are easily and inexpensively replaced when worn. A backlash eliminator is essential when climb milling.

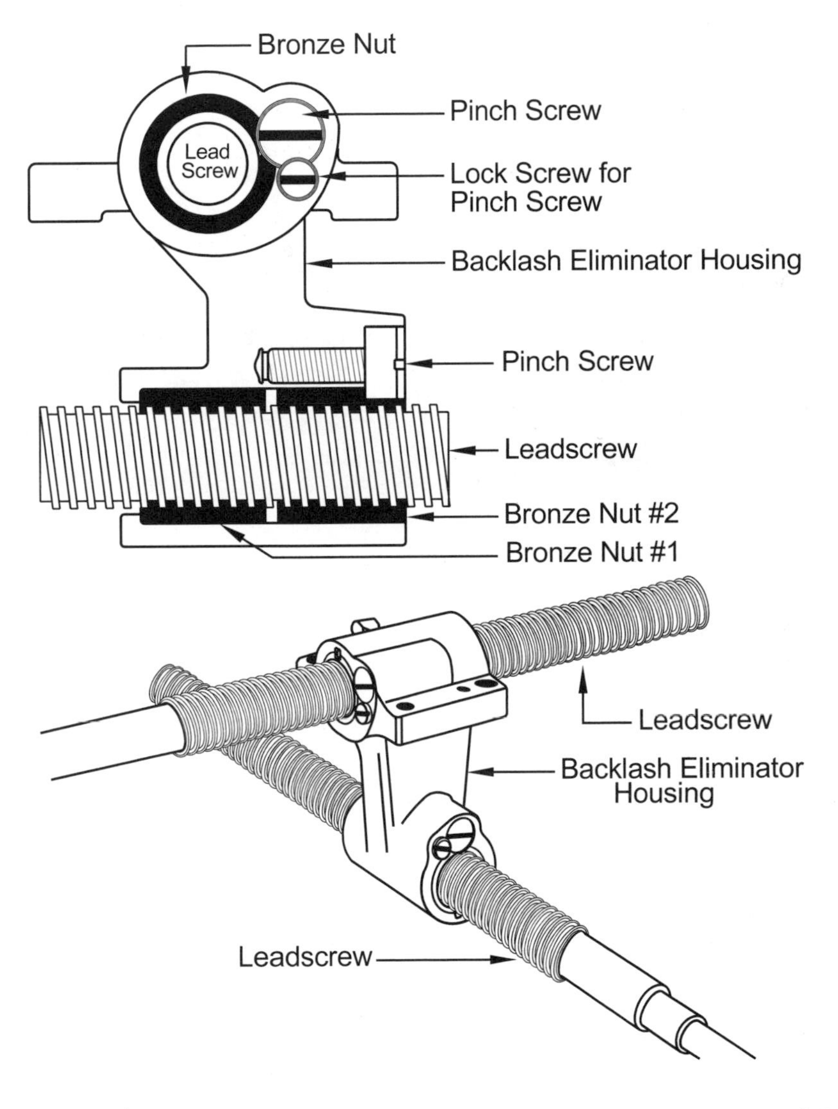

Figure 8–8. Cross sectional view of a backlash eliminator on a Spanish-built Kondia vertical knee mill (top) and the X- and Y-axes leadscrews and their two backlash eliminators in the same casting (bottom).

Quill & Spindle

What is the purpose of the *quill* and *spindle*?

The purpose of the quill is to support the spindle in ball bearings and guide it vertically. The spindle is bored on its lower end to accept an R8 taper collet. The spindle holds a collet and the collet holds the cutting tool. A drawbar, which projects from the top front of the milling machine head housing, holds the collet in place, Figure 8–12.

How does the *R8 collet* compare with the 5*C collet* frequently used in lathes?

They are of similar design, but different enough so that they are not interchangeable. See Figure 8–9.

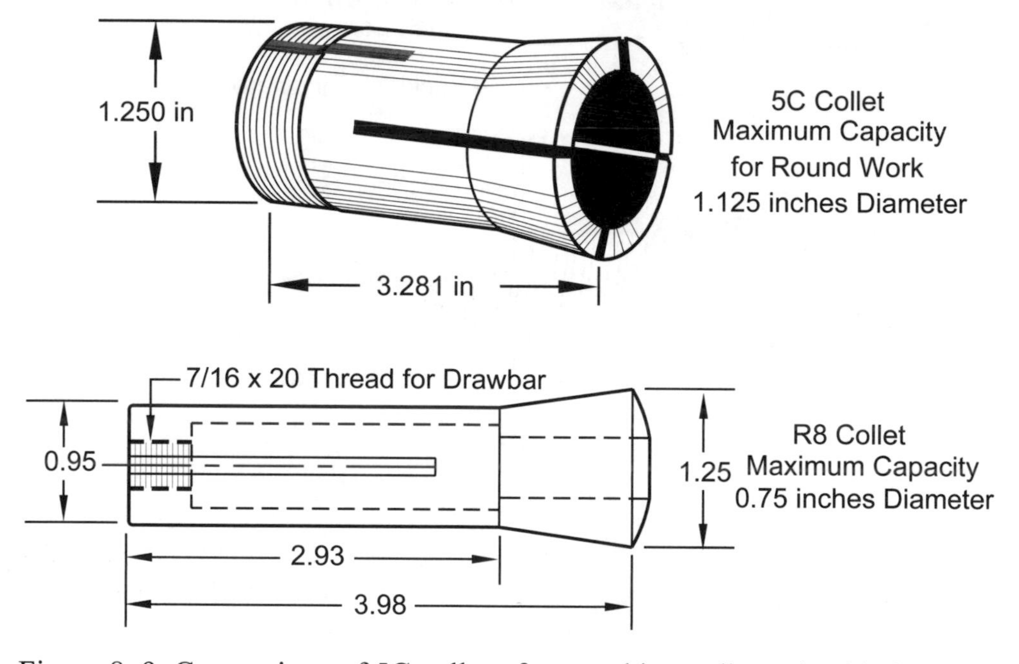

Figure 8–9. Comparison of 5C collet often used in medium-sized lathes and R8 collet used in Bridgeport-style milling machines.

Power Feeds for the Worktable

Where are the power feed units located, how are they used and what advantages do they offer?

Power feeds are variable-speed electric motor drives that move the worktable to:

- Prevent operator fatigue when making long cuts.
- Provide uniform table motion to produce a smooth, even cut.

To install a power feed unit, one handwheel (or Z-axis crank) is removed, the power feed unit slipped over the leadscrew end, and then the handwheel is replaced by fitting it over the power feed. This way both manual and power feeds can be used to move the worktable. Although power feeds are available for all three axes, for economy they are often fitted only to the X-axis because that is the axis on which most long cuts are made. Many milling machines are shipped from the factory with power feeds installed; others have them added later. See Figure 8–10.

Figure 8–10. Milling table power feed.

Digital Readouts

Why are digital readouts (DROs) installed on milling machines?

- DROs show the operator exactly where the table is at all times on all three axes, eliminating the need to repeatedly stop the mill to re-measure. They also eliminate errors due to leadscrew wear and backlash because the DRO reads table position directly without using the leadscrews.
- DROs make it possible to position the worktable to a dimensional reference point on the work and then zero its display. This saves time and prevents errors. However, DROs do not remove the need to take into account leadscrew backlash due to cutting tool pressure when cutting.
- Many DROs can switch between inch and metric measurement units at the push of a button permitting work in either system even though the milling machine calibration is in only one system.

See Figure 8–11.

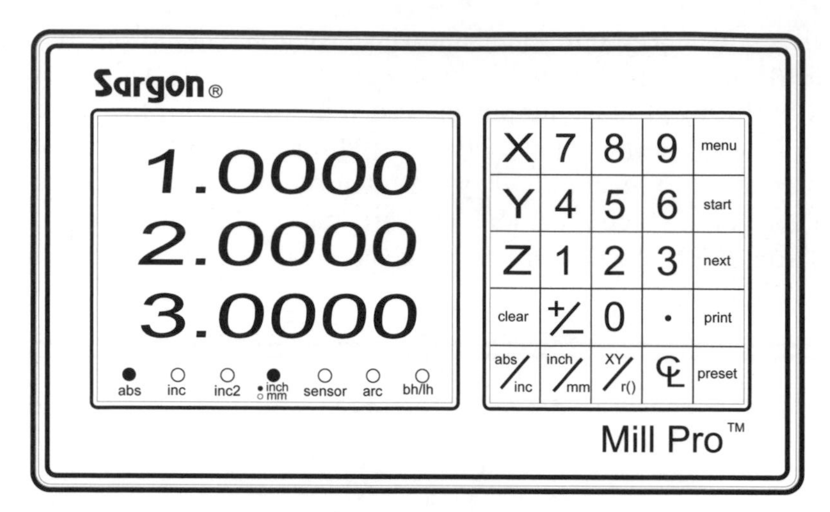

Figure 8–11. DRO for a milling machine.

What kind of accuracy can be expected from a DRO?
DROs for full-size mills provide three- or four-decimal place readouts. They are much more accurate than leadscrew collar readings because DRO electronics measures table position by counting lines etched on a glass ruler which eliminates errors from leadscrew wear and backlash. Typical performance from a quality DRO is an accuracy of 0.0005 inches and a resolution of 0.00025 inches. A cover protects the glass ruler from metal chips and oil, so it is not usually visible.

The Sherline mill's DRO provides a 3½-decimal place readout, but since it does not use an etched glass ruler and determines table position from encoders mounted on the leadscrews, its accuracy may be slightly less than its readouts. Nevertheless, it is a great convenience. DROs are so much help and so cost-effective that few milling machines in daily use are without them.

Section III – Bridgeport-Style Milling Machine Controls & Adjustments

Note: SMALL CAPITAL LETTERS *are used in this section to denote the industry standard names for milling machine controls, settings or functions the first time they appear.*

Motor, Drive & Controls

How does the electric motor drive the Bridgeport-style machine?
On the HIGH-SPEED RANGE SETTING (500–4200 rpm), the motor drives the spindle through a belt that connects pulleys on the motor and the spindle.

Because of this direct drive arrangement, the motor shaft and milling machine spindle turn in the same direction. The spindle turns clockwise when viewed from the top of the machine when running in the FORWARD DIRECTION and rotates right-handed twist drills and end mills correctly.

On the LOW-SPEED RANGE SETTING (60–500 rpm), a backgear provides speed reduction and added torque. Using the backgear causes the spindle to rotate in the opposite direction to the motor shaft or counterclockwise as viewed from the top of the milling machine.

Warning: Failure to match the motor power switch setting (high or low) to the high-neutral-low lever setting (high or low) will cause the spindle to run backwards. Only mismatch these settings when reverse spindle rotation is desired. Reverse spindle rotation is used with left-handed end mills.

How do you change from one speed range to the other?
With the motor turned off, move the HIGH-NEUTRAL-LOW LEVER, Figure 8-12, to the position desired. When switching from high speed to low speed, rotate the spindle by hand to get the teeth on the backgears to mesh. Do not force this lever. The NEUTRAL POSITION permits the spindle to rotate freely and is useful when using a dial test indicator (DTI) in the spindle.

Warning: Do not move the HIGH-NEUTRAL-LOW LEVER *when the motor is on.*

How is power applied to the motor?
First, make sure the SPINDLE BRAKE, Figure 8–12, is not applied, or damage to the drive belt and motor could result. Then turn the power switch setting to match the high-neutral-low lever setting. The power switch is sometimes called the HIGH-LOW RANGE SWITCH.

When is the spindle brake useful?
The spindle brake prevents spindle rotation and is used when the draw bolt, which holds the spindle in place, is tightened or loosened.

Once a speed range has been selected, how is the spindle speed varied?
Spindle speed is varied by the SPEED CHANGE HANDWHEEL on the upper right side of the motor housing, Figure 8–12. There are two windows, one for each speed range, which indicate the approximate spindle speed.

Warning: Do not turn the speed change handwheel with the spindle stopped. Damage to the machine will result.

Note: Many machines have stepped pulleys instead of a variable speed mechanism. This requires that the motor be stopped and the belt be manually shifted from step-to-step to change speed.

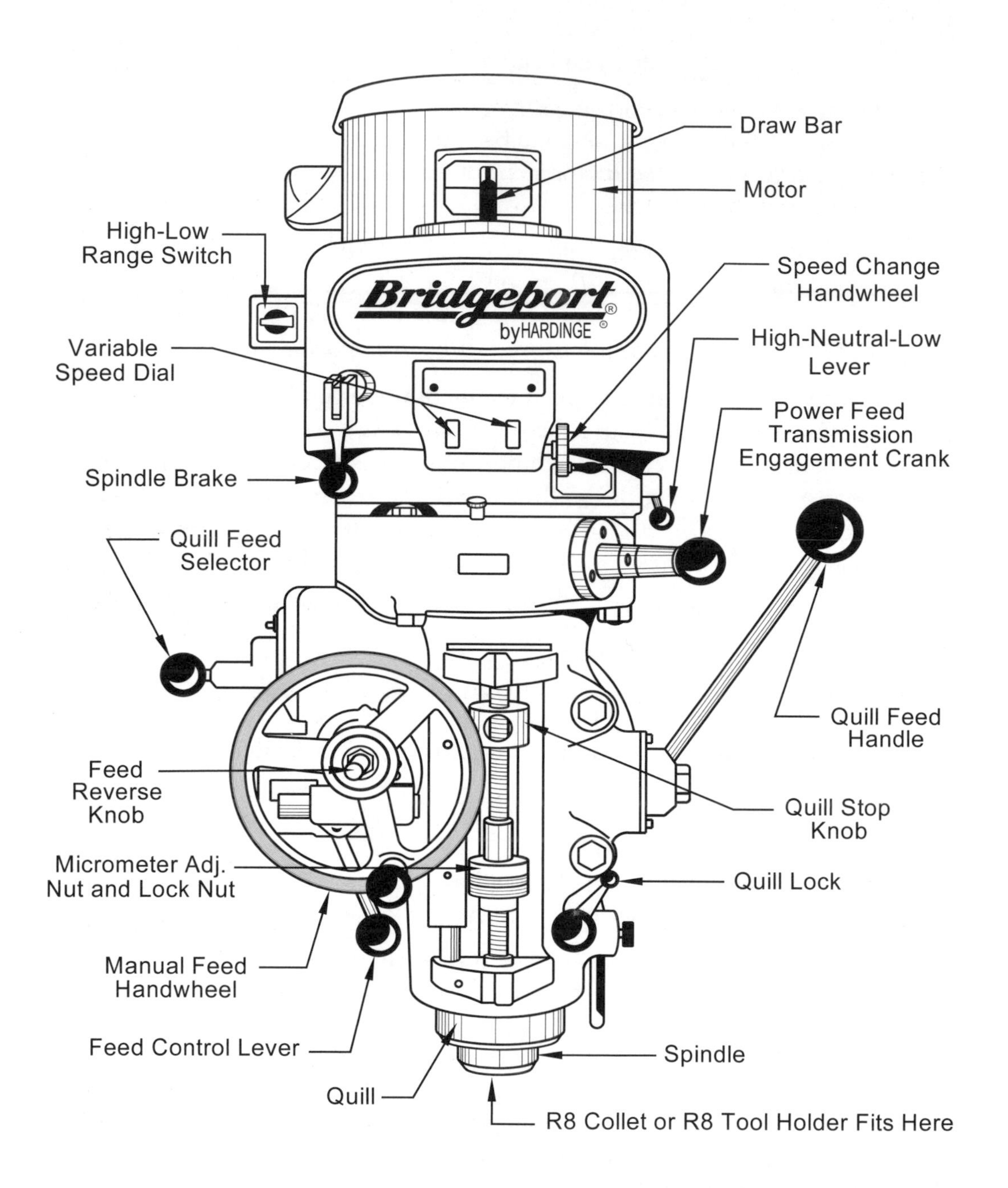

Figure 8–12. Bridgeport milling machine head, motor and drive.

Quill & Spindle

How is the *quill* held rigidly in a fixed position?

The QUILL LOCK, Figure 8–12, on the right-hand side of the quill casting, holds the quill at any depth. The quill lock should be applied whenever the AUTOMATIC QUILL FEED is not being used.

What are the three ways to actuate the quill?

1. Pulling down the QUILL FEED HANDLE lowers the quill; raising the handle raises the quill. A spring return on the quill feed handle, much like that on a drill press, tends to pull the quill back to its upper-most position. This spring tension is adjustable. However, if this spring is broken, the quill will fall to its lowest position.
   Warning: Remove the quill feed handle when using automatic quill feed to eliminate the chance of injury. The handle's automatic movement rotates with considerable torque when the quill moves downward.
2. Turning the MANUAL FEED HANDWHEEL clockwise lowers the quill; turning it counterclockwise raises it. Use the manual feed handwheel or the quill feed handle, whichever is more convenient. The manual feed handwheel gives the operator more mechanical advantage than the quill feed handle. The manual feed handwheel can be removed by pulling it straight off its shaft. For added safety, consider removing the manual feed handwheel whenever using the power feed.
3. The automatic quill feed can move the quill either up or down.

What does the quill stop knob do?

The QUILL STOP KNOB performs two functions:

1. When the quill is actuated by the feed handle or manual feed handwheel, the MICROMETER NUT limits the maximum downward movement of the quill by acting as a stop on the quill stop knob.
2. Automatically disengages the automatic quill feed when the quill stop knob pushes against the micrometer nut. This function operates when the spindle is on automatic feed in either the up or down direction.

How does the quill stop knob set the lower quill stop point?

The quill stop knob is fastened to the quill through a vertical slot in the head casting, Figure 8–12 and 8–13. As the quill moves vertically, the quill stop knob rides along the threaded rod on the face of the milling head. The quill stop knob has no threads, just a center hole. The micrometer nut is threaded onto the vertical threaded rod. Because they are connected, stopping downward motion of the quill stop knob, stops the quill. Positioning the micrometer nut along the rod using the vertical depth scale and 0.001-inch calibrations around the nut, provides accurate quill depth settings. A MICROMETER LOCK NUT below the micrometer nut prevents it from turning with vibration. This is a relative depth setting and must be reset whenever the vertical tool position is changed.

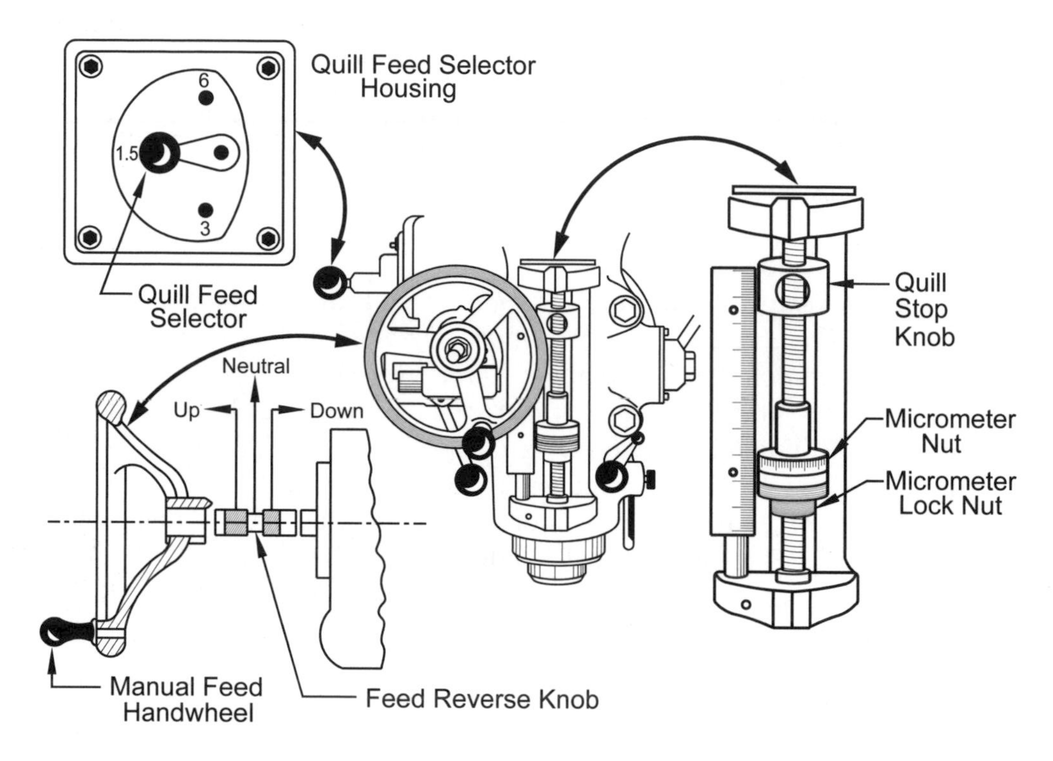

Figure 8–13. Quill feed details of a Bridgeport head.

How is the *automatic quill feed* actuated?

The steps are:

1. There are three spindle feed rates: 0.0015, 0.003 and 0.006 inches/spindle revolution. Set the crank on the QUILL FEED SELECTOR to the rate desired by pulling out the knob and setting it to the proper feed rate. Feed rates are stamped on the quill feed selector housing, Figure 8–13. The selection setting is most easily made when the quill is running.
2. Set the lower limit of quill movement by positioning the micrometer nut. Each graduation on the micrometer nut represents one-thousandth of an inch depth. Larger increments of movement are read directly from the scale on the left of the nut. Lock the micrometer nut in place with its locking nut.
3. The FEED REVERSE KNOB, which protrudes through the middle of the manual feed handwheel, determines the direction of automatic quill feed

motion. There are three positions for the feed reverse knob: PUSHED IN drives the spindle DOWN, MIDDLE POSITION puts the drive in NEUTRAL, and PULLED OUT drives the quill UP. Set this knob for the desired quill direction.

4. To engage the automatic quill feed, pull the FEED CONTROL LEVER to the left or away from the spindle casting. The automatic quill feed will trip out when the quill stop knob pushes against the micrometer nut. Push the feed control lever back toward the spindle to manually stop it. The feed control lever must be engaged to use either the quill feed handle or the manual feed handwheel. These may be removed from the milling machine when not in use.
   Warning: Turn the quill lock off when actuating the automatic quill feed.

What advantages and disadvantages does using the VERTICAL TABLE FEED have over the manual or automatic quill feed when vertical tool movement is wanted?

The vertical table feed, sometimes called the Z-AXIS FEED, is a more accurate way to control tool motion and maximum depth than either the manual or automatic quill feed. The vertical leadscrew and its calibrated collar are inherently more accurate than the quill and its micrometer stop nut. The vertical table feed mechanism is also much stronger, and its hand crank provides greater mechanical advantage, and so, more vertical force against the tool. In addition, larger diameter drills have a tendency to "dive" into soft materials, like brass and aluminum, rather than drill into them. Using the vertical table feed, which moves only when the crank does, avoids this problem. In general, use the quill manual or automatic feed for drills under 0.25-inch diameter and when depth accuracy is not required. Using the quill feed handle or manual feed handwheel for drilling small holes saves a lot of cranking and makes the work go faster. For larger drills and boring tools, use the vertical table crank.

Why must holes larger than 0.75 inch (20 mm) be drilled manually and not with the automatic quill feed?

The FEED CONTROL OVERLOAD CLUTCH is preset at the factory for a maximum downward pressure on the quill of 200 lbs and will trip out if this force is exceeded. A slow manual feed will prevent the clutch from tripping to protect the machine. On larger holes, use the manual feed handwheel or the vertical table crank.

Warning: Do not attempt to adjust this clutch; it is properly set at the factory to prevent overstressing the quill feed mechanism.

Quill Stop

What is a QUILL STOP and how is it used?

This metal device shown in Figure 8–14, resembles a spring clothespin. It clips onto the quill stop micrometer screw and forms a second depth stop in addition to the micrometer depth stop.

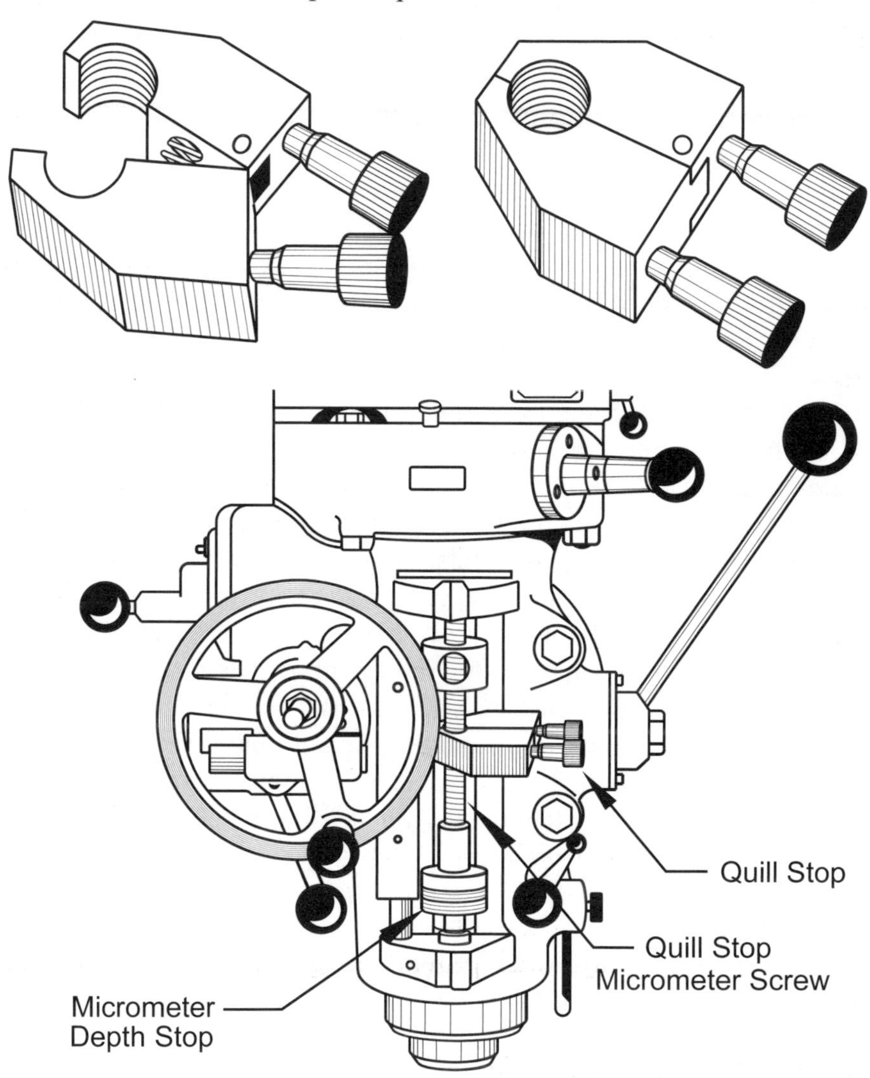

Figure 8–14. Quill stops open and closed (top) and quill stop clipped to quill stop micrometer screw (bottom).

Threads in the quill stop match threads on the depth control. Quill stops are used for tasks where two stops are needed. For example, when a hole must be drilled to accurate depth and then spot faced. The micrometer depth stop is used for drilling depth and the quill stop for spot facing, which must be done against a solid stop. Although less accurate than the micrometer stop, quill stops are often adequate and save a lot of time from readjusting the depth control micrometer nut when two depth stops are necessary.

Locks

Besides the QUILL LOCK and SPINDLE LOCK, what other locks are there on the milling machine?

There are three:

- KNEE LOCKS
- SADDLE LOCKS
- TABLE LOCKS

Keep each lock applied except when moving the machine along its axis. This will reduce unwanted table movement and chatter.

Gibs

When first using an unfamiliar or newly assigned milling machine, what should the machinist do?

Vertical knee mills have a pair of ways for each of their three axes. Like the gibs on lathe ways, there are gibs on the milling machine ways to adjust for tightness as they wear. You will find gib adjustment screws either at the ends of the slide ways or along them depending on the mill design. Go over the three sets of gibs and adjust them so the machine moves smoothly and does not bind. Proper gib adjustment minimizes chatter and increases accuracy. There is a fourth set of ways on the ram, but they are not used during milling operations and their adjustment is not critical.

One-Shot Oiler

When should the ONE-SHOT OILER be used and what does it do?

The one-shot oiler should be actuated by pulling its lever before starting to use the machine each day. This will put a few drops of oil on each of the ways. Properly oiled ways will provide long service and maintain their accuracy.

Flood Coolant System

What is a FLOOD COOLANT SYSTEM on a vertical knee mill?

This system consists of a coolant pump, delivery tubing, coolant filter and reservoir. It delivers a steady stream of coolant to the cutting tool and work eliminating the need to add coolant by hand. Coolant flows into the table and drains into the machine base where the coolant is recycled.

Section IV – Cutting Tools

Cutting Tools Overview

What cutting tools are used in vertical knee mills?

- *End mills*–teeth on the circumference and the end.
- *Face mills*–radial teeth only.
- *Plain mills*–teeth only on the circumference.
- *Plain side mills*–teeth on the circumference and on one or both sides.
- *Staggered tooth mills*–alternate teeth have right- and left-hand helix angles.
- *Shell mills*–teeth on the circumference and the end, usually large diameter.
- *Angle cutters*–teeth are on an angle to the rotational axis.
- *Formed mills*–curved teeth.
- *Other mills*–special purpose designs for T-slots, keyseats and dovetails.
- *Flycutters*–single tooth lathe-type tool bits for holes in thin, flat stock.
- *Boring heads*–single tooth lathe-type tool bits for very accurate holes.
- *Slitting saws*–teeth on circumference, but may also have side teeth.
- *Twist drills*–from 1/32- to 1-inch diameter.
- *Spade drills*–carbide inserts in a holder for rapid drilling of large holes.
- *Reamers*–straight and tapered.

End Mills

What are *end mills* used for?

End mills cut slots, keyways, pockets and islands, Figure 8–15. They are the cutting tools most often used in the vertical knee mill.

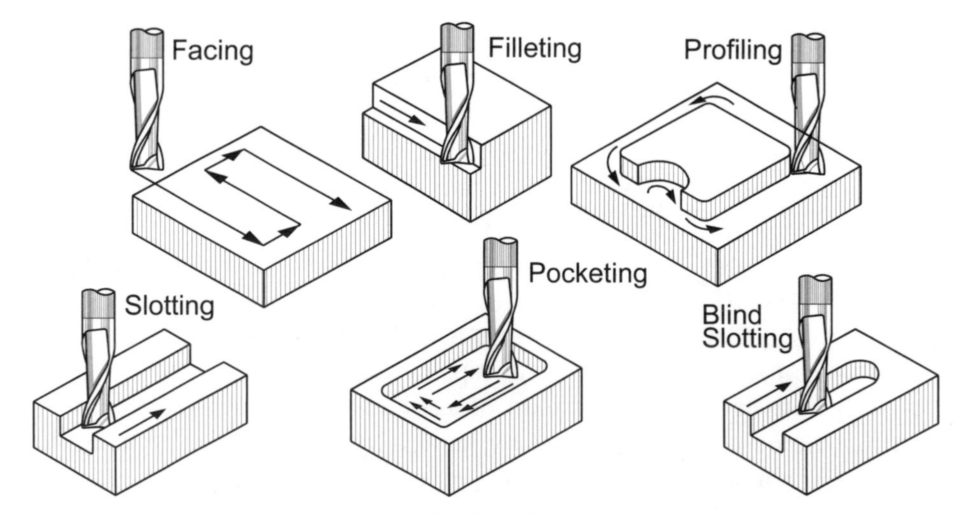

Figure 8–15. Typical end mill applications.

What are the most common end mill designs?

There are several common designs:

- *Two-flute end-cutting end mills* with square ends are used because they can make *plunge cuts* into the work like a drill as well as cut sideways. Their end teeth extend to their centers to allow plunge cuts. Some three-flute end mills can also perform plunge cuts. Both two and three-flute end mills are suitable for non-ferrous metals.
- *Two-flute end mills with a ball end* produce rounded-bottom slots or filleted edges.
- *Four-flute finishing end mills* with a square end make smoother cuts than two-flute end mills and stay sharp longer, but cannot make plunge cuts. These are best for ferrous metals. There are also end-cutting four-flute end mills.
- *Four-flute roughing end mills* remove metal faster than four-flute finishing end mills with less horsepower, cutter deflection and vibration. Their chip breakers produce smaller, more controllable chips.

These four common designs are shown in Figure 8–16. Their shanks and cutters are one piece of steel so they are sometimes called *solid end mills*.

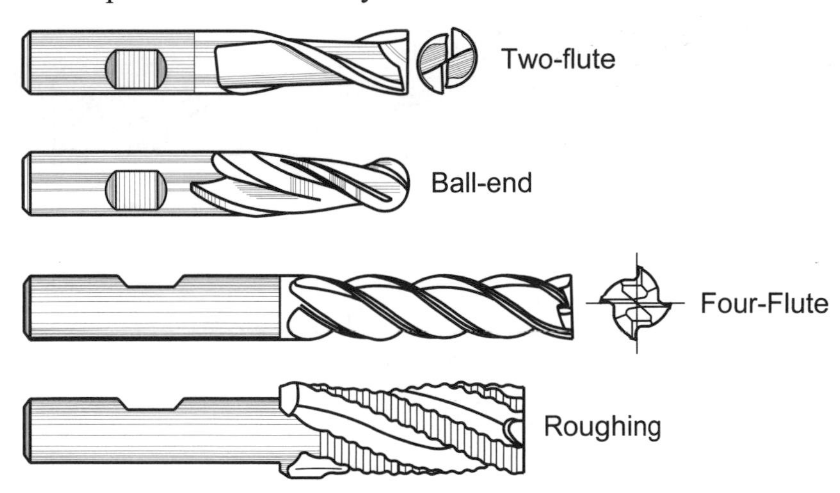

Figure 8–16. Single-ended solid end mills.

Why are *flutes* in the sides of cutting tools helical rather than straight?

The helical shape reduces chatter because one or more points of the flutes remains in contact and cutting metal all the time. This presents a continuous load to the machine instead of a pulsing load, which occurs each time a straight flute begins and ends contact with the work. Helical flutes also help remove chips. Making helical flutes is considerably more complex than straight ones, but reduction in chatter and the smoother finish they produce is worth the extra cost. For this reason, many cutter designs have helical flutes.

What are the options in solid end mills?
Here are the common options:

- *Square end, ball end* or *90° point.*
- *Single-ended*, Figure 8–16, or *double-ended*, Figure 8–17. Double-ended end mills cost only about 20% more than single-ended ones, so buying double-ended cutters is economical.
- *Number of flutes*, or cutting teeth, ranges from two to four. While end mills with more than four flutes are available, they are not satisfactory in vertical knee mills, and so not used. When not plunge cutting, four-flute solid end mills are preferred because they last longer and produce a smoother surface finish.
- *Cutter material*: HSS, HSS Cobalt (5%, 8% or 10% Cobalt), brazed-on tungsten carbide cutting tips and solid tungsten carbide.
- *Cutter coatings*: bright finish or titanium nitride (TiN). Other more exotic coatings are available, but not needed in a non-production work setting.
- *Right-hand cutting* (clockwise as viewed from the top of the tool) or left-hand cutting, Figure 8–17. Left-handed cutters are used when the direction of cutting forces must be reversed to avoid stressing the workpiece.

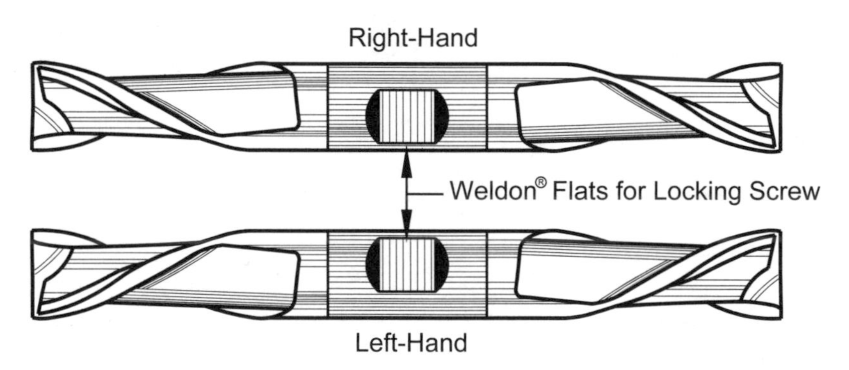

Figure 8–17. Double-ended, two-flute right- and left-hand cutting end mills. Weldon® flats provide a flat bearing surface for set screws in the tool holder that prevent the tool from spinning. There are either one or two flats depending on the cutter size.

- *Length of flutes:* stub, regular, long and extra long.
- *Twist of flutes:* varies from straight flutes (no twist) to 10° to 25°.
- *Tapered end mills:* commonly tapered ½°, 1°, 1½°, 2°, 3°, 5°, 7°, 10°, 15°, 20° and 25° per side. These are especially useful in mold making to provide draft, or taper, to ease the release of the product.

- *Cutter bit diameter tolerance:*
  - Two- and three-flute end mills have a tolerance of +0.000 and − 0.002 inches and are often used to cut slots or keyways.
  - Four-flute end mills havc a tolerance of +0.002 to − 0.000 inches and are used for finishing operations where width is not critical.

While not every combination of every option is available, many are, and industrial tool catalogs require sixty or more pages to list their end mill offerings.

In what sizes are solid end mills available?
See Table 8–1.

	Cutter Diameter Range	Diameter Increments	Shank Diameter
Inch Sizes (inches)	More common: 1/8–1 Less common: 1/32–2	1/8, 1/16, 1/64	3/16, 3/8, 1/2, 5/8
Metric Sizes (mm)	More common: 3–25 Less common: 2–45	1.0, 0.5	For inch collets (inches): 3/16, 3/8, 1/2, 5/8, 3/4, 7/8, 1 For metric collets (mm): 6, 10, 12, 14, 16, 18, 20

Note: Intermediate, non-standard sizes, can be special ordered at added cost.

Table 8–1. Common single- and double-end solid end mill diameters.

Methods for Holding Cutting Tools

How are cutting tools held in milling machines?
Collets and tool holders are used to hold cutting tools. The Bridgeport-style knee mill has a female R8 taper in its spindle. These tapers hold both R8 spring collets and R8 tool holders by pulling them up and holding them in place with the spindle drawbar. Some tools, like drill chucks, have an integral R8 taper that fits directly into the spindle.

R8 collets work like other spring collets: when the drawbar pulls the collet into the taper, the collet is compressed against the tool it holds. Tool holders are also pulled into the taper by the drawbar, but have no collet. They hold tools with a set screw, threads or a special proprietary system made to hold a particular tool. Both collets and tool holders provide excellent concentricity and the rigidity to support lateral loads. Their external portion can be modified to hold many different tools.

How are end mills held by the Bridgeport-style milling machine?
There are three methods:

- Use an *R8 spring collet* to hold end mills with shanks from 0.125- to 0.75-inches diameter, Figure 8–18 (top). Three-quarters of an inch is the R8 collet maximum capacity diameter.
- Use an *R8 end mill tool holder*, Figure 8–18 (bottom). This is a better method because the two set screws in the side of the tool holder bear on the Weldon® flats on the side of the cutter shank. This provides a positive method to prevent cutter spinning. Smaller end mills have one flat, and larger ones have two flats. Under high cutting loads, a cutter will sometimes spin or pull down in an R8 spring collet because the collet has no set screws and depends on friction. R8 end mill tool holders, with a capacity of up to 1.25-inches diameter are available. The R8 end mill tool holder is much less likely to allow the milling cutter to spin inside it because of the two set screws on the Weldon flats. These holders are much better for power feed applications. A tool spinning during a power feed stops cutting but continues to be fed into the work, leading to a cutter snapping off and damaging the work.

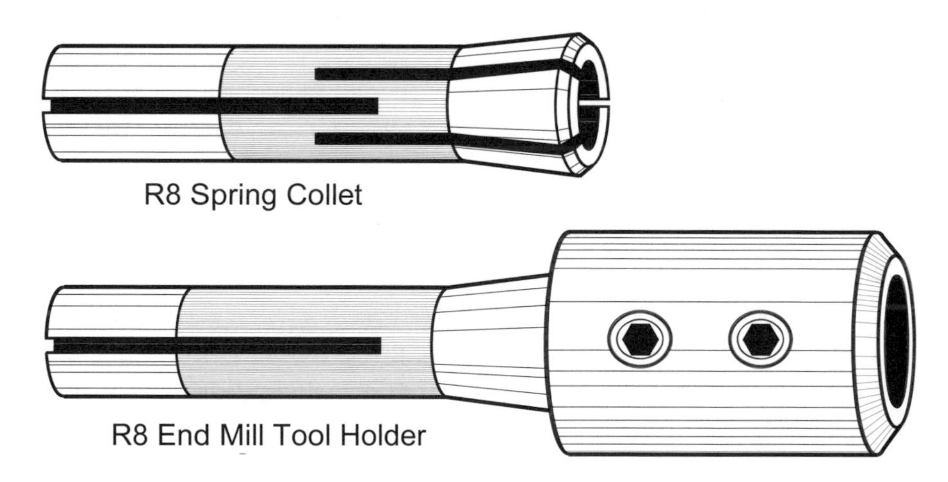

Figure 8–18. Two R8 devices for holding milling cutters.

- Use a *double-angle collet chuck extension,* Figure 8–19. This tool holder fits into the milling machine spindle and is held there by an R8 spring collet. This tool holder extends the reach of the spindle 3 to 6 inches and allows milling in the bottom of a narrow casting or close to the edge of a step, Figure 8–20. A *double-angle collet* in the end of the extension holds the end mill. This collet design is used because it has slits (cuts in the sides of the collet) that extend from both ends allowing the collet to collapse around the milling cutter evenly along its length. This collet design also provides a slightly larger range of tool diameters than a conventional spring collet, which has slits cut from only one end. Center drills and twist drills can also be held in the chuck extension. Lighter milling cuts must be taken because of the extension's flexibility.

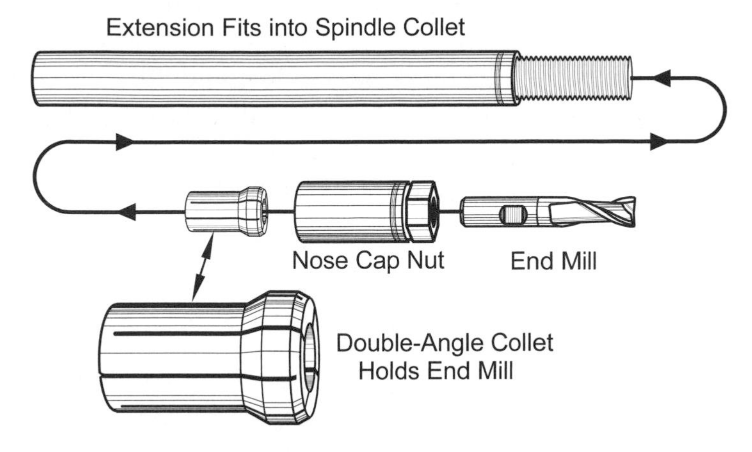

Figure 8–19. A double-angle collet chuck extension holds an end mill.

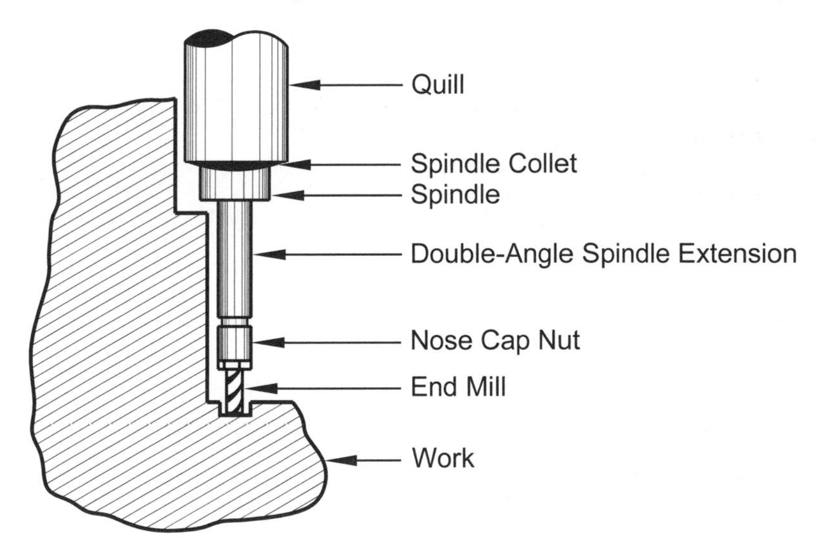

Figure 8–20. Double-angle collet chuck extensions position the tool into tight or narrow locations.

Warning: Do not attempt to perform milling by holding an end mill in a drill chuck. Not only are chucks not designed for lateral loads, they will not grip the end mill firmly enough to keep it from spinning.

How are R8 collets and tool holders installed in the vertical knee mill?
There are four steps:

1. Use a tapered brush or fingers to clean chips and debris out of the inverted interior of the spindle taper. *Never insert fingers in a turning tool holder.*

2. Wipe down the portion of the collet that will be seated inside the mill's quill taper. Clean the taper surfaces to insure concentricity.
3. Insert the collet up and into the quill taper and hand tighten the drawbar into the collet.
4. Finally, insert the milling cutter into the collet and hold the cutter in place as it will tend to fall out of the loose collet. Engage the spindle brake and tighten the drawbar with a box wrench. Do not use an open-end wrench; it will eventually round the corners on the drawbar. The drawbar should be tight.

Warning: Properly securing a milling cutter in its collet is critical when using a table power feed. If the cutter is loose in the collet and spins instead of cutting, the table power feed will snap the cutter off and damage the part. The power feed mechanism may also be damaged. This is why it is always preferable to use an R8 end mill tool holder, Figure 8–18, instead of a collet. The R8 end mill tool holder secures the milling cutter by one or two set screws bearing against Weldon Flats preventing the cutter from spinnng.

How are R8 collets and tool holders removed from the vertical knee milling machine?

To remove collets and tool holders:
1. Move the quill to its highest position and apply the quill lock to hold it vertically in place.
2. Engage the spindle brake.
3. Partially unscrew the drawbar from the collet, but leave a few threads engaged.
4. Gently tap the top of the drawbar nut to break the spindle free from its taper. Holding a rag under the spindle (and cutting tool) prevents the cutting tool and collet from falling on the milling table.
5. Unscrew the drawbar completely and remove the collet from the quill.

How does the Sherline milling machine hold end mills?

Collets that fit directly into the spindle's #1MT hold end mills with ⅛-, 3/16- or ¼-inch diameter shanks, Figure 8–21. These collets are pulled tight with a drawbar. The external tool holder, Figure 8–22, holds cutters with ¼- and ⅜-inch diameter shanks and screws onto the ¾-16 spindle nose threads. These tool holders have two holes, one with a set screw to secure the milling cutter, and the other to accept a tommy bar (rod wrench) to tighten and loosen the holder from the spindle.

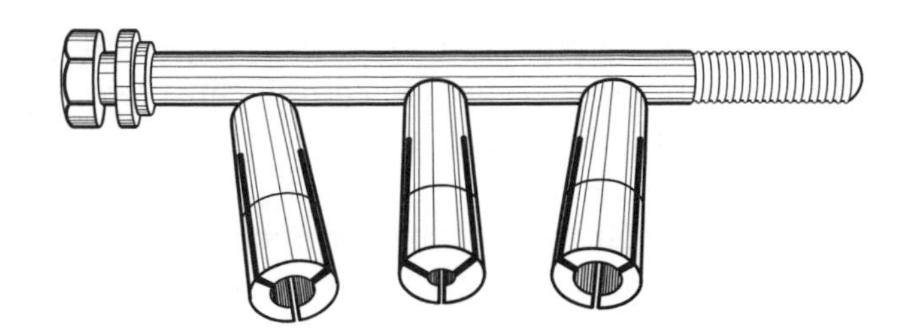

Figure 8–21. Sherline end mill collets and drawbar.

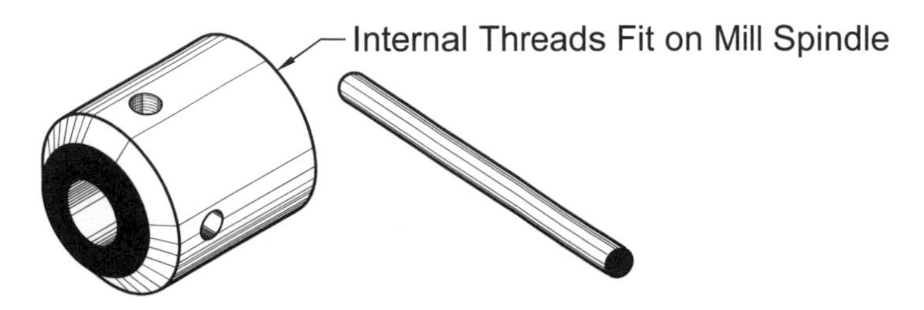

Figure 8–22. Sherline external end mill tool holder with its tommy bar.

Face Mills

What is the purpose of *face mills*?

Face mills, Figure 8–23, are used to smooth large flat surfaces and to remove large volumes of metal rapidly. Face mills with removable inserts are available in diameters from ½ to 3 inches and these will work fine in aluminum, but, lack of mill frame rigidity and the resulting chatter, limit vertical knee mills to smaller diameter face mills on harder metals. Face mills are more suitable for large massive C-frame mills.

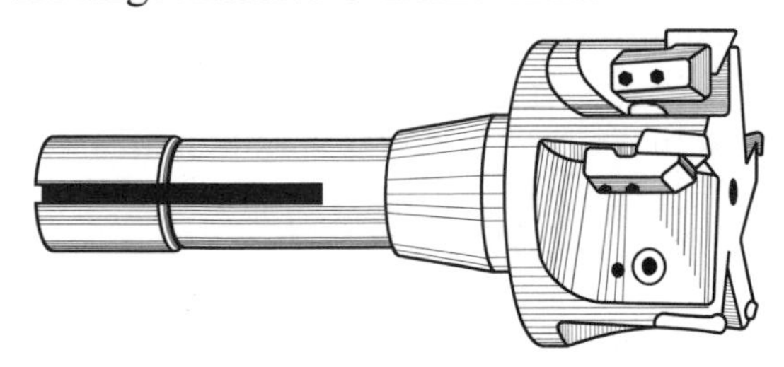

Figure 8–23. Face mill with insert teeth on an R8 spindle.

Plain Mills

What are *plain milling cutters* used for?

Figure 8–24 shows a typical plain milling cutter. Usually they are used on horizontal milling machines on an arbor supported at both ends. Horizontal end milling machines provide the power and rigidity to fully utilize the high metal removal rates possible with plain milling cutters. However, a *horizontal milling attachment,* Figure 8–85, permits the use of plain milling cutters on vertical knee mills. This is a very useful accessory, even though, because of rigidity and chatter limitations, a vertical knee mill must take lighter cuts than a horizontal mill. In a vertical knee mill with a horizontal milling attachment, the plain milling cutter removes more metal faster than other cutters. In some instances, lacking a horizontal milling machine, using several cutters on the same arbor is simply the best way to do the job. Plain mills are usually wider than the work and cut completely across it. Plain milling cutters remove a large volume of metal rapidly, smooth large, flat surfaces and provide a good finish.

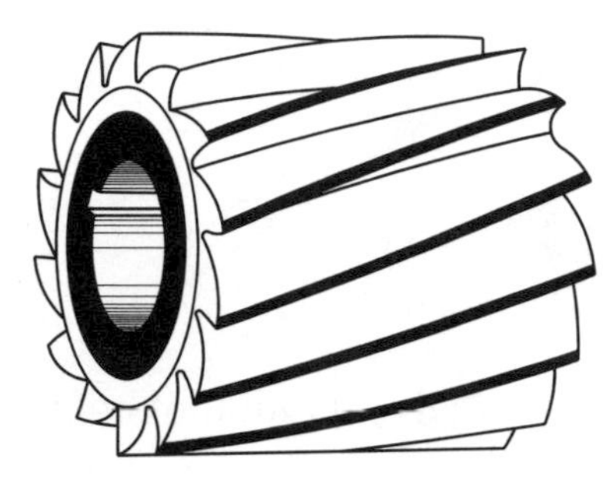

Figure 8–24. Plain milling cutter.

Plain-Side Mills

What are *plain-side milling cutters* and how are they used?

Plain-side milling cutters are similar to plain milling cutters, but they can also cut on their sides, Figure 8–25. They cut slots, cut off heavy metal bars, smooth faces or cut material into strips. They can be mounted on an R8 stub tool holder, but are best employed in a horizontal milling machine or in a horizontal milling attachment on a vertical knee mill, which is more rigid. Like the plain milling cutter, plain-side mills can be mounted with several other milling cutters to make multiple cuts simultaneously.

Figure 8–25. Plain-side milling cutter.

Angle Cutters

What are the most common designs for *angle cutters* and what are their uses?

Angle cutters are either single or double. Single-angle cutters come with either 45° or 60° included angles, Figure 8–26 (left). They are used for cutting dovetails and ratchet wheels. Double-angle cutters are available in 45°, 60° and 90° included angles, and are most often used for milling notches, V-grooves, angles and chamfers, Figure 8–26 (right).

Figure 8–26. Angle cutters: single-angle (left) and double-angle cutters (right) in stub-mounted tool holders.

Stub Mandrels

How are plain mills, plain-side mills and angle cutters mounted in Bridgeport-type milling machine spindles?

Stub arbors that fit in R8 spindles are used, Figure 8–27. The size of the cutter and the location of the cut determine the bushings' size and position.

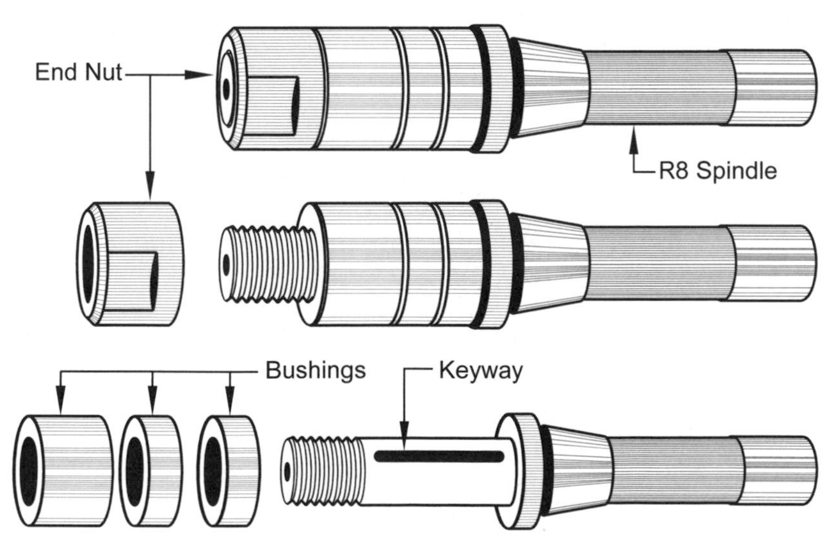

Figure 8–27. R8 stub mandrel holds larger diameter milling cutters in Bridgeport-style mills.

Shell End Mills

What are *shell end mills*, when are they used, and what are their limitations in vertical knee mills?

Shell end mills, Figure 8–28, are cutters that cut on both their face and sides leaving a smooth finish. These cutters are available in diameters from 1¼ to 6 inches, and with flutes cutting from ½ to ¾ inches depth. Because of both horsepower and rigidity limitations, vertical knee mills are limited to the smaller shell mill sizes. Larger diameters may be used effectively on aluminum and plastics.

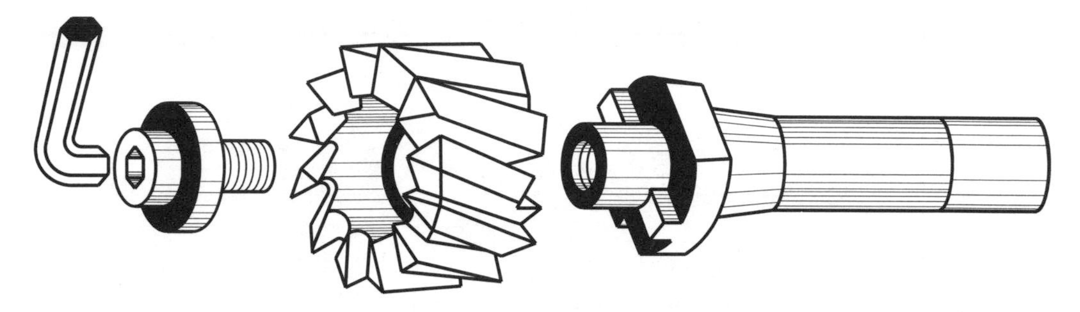

Figure 8–28. Shell end mill and its R8 arbor.

Formed Cutters

What are *formed cutters* and what are their uses?

These specialized cutters form the exact negative shape (mirror image) of their profile on the workpiece. Their most common designs are:

- *Concave cutters.*
- *Convex cutters.*
- *Corner rounding cutters.*
- *Involute form gear cutters.*

Several examples are shown in Figure 8–29. Form cutters are usually arbor-mounted for use in horizontal mills, but smaller shank-mounted cutters are also available.

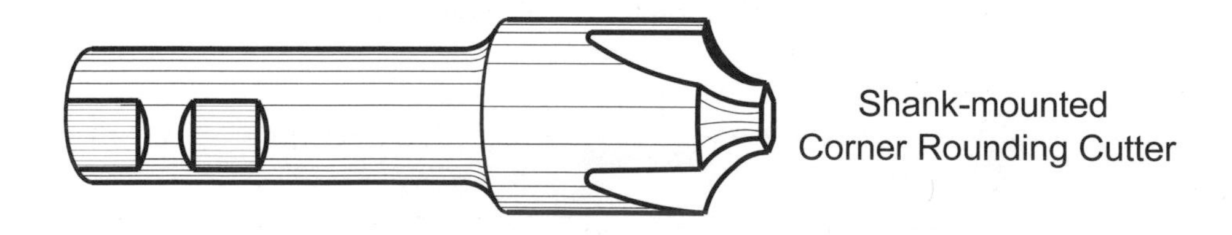

Figure 8–29. Formed cutters.

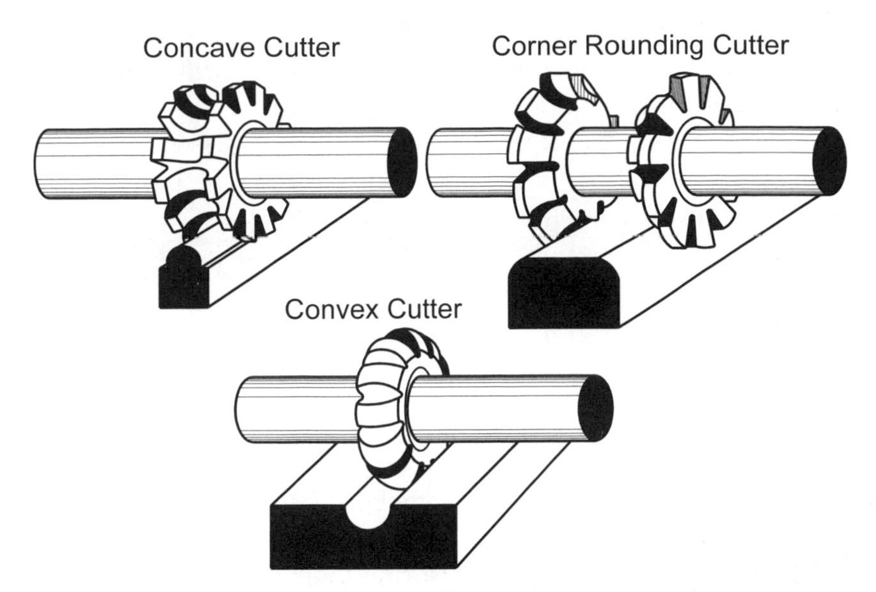

Figure 8–29. Formed cutters (continued).

Other Milling Cutters

What other common cutter designs are used and for what applications?

Other common designs are the dovetail, Woodruff keyseat and T-slot staggered tooth cutters, Figure 8–30. Each is designed for a particular task.

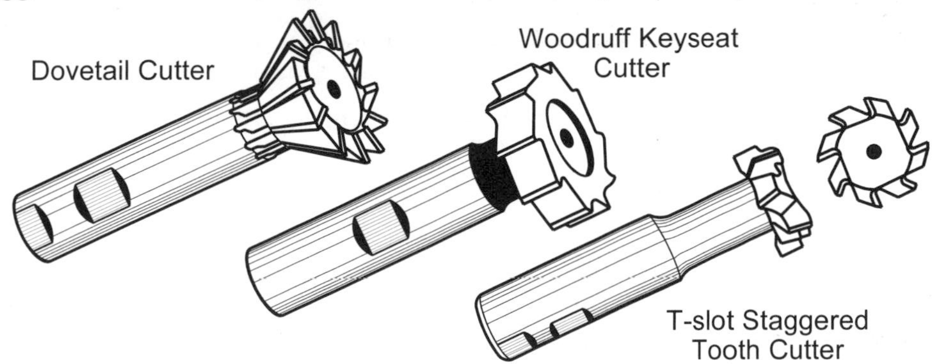

Figure 8–30. Other common milling cutters.

Boring Heads

What are the advantages of a *boring head* over a twist drill in a milling machine?

A boring head provides more accurate hole positioning, more perpendicular hole walls to the work surface, smoother side walls and more accurate final diameter than a twist drill. See Figure 8–31.

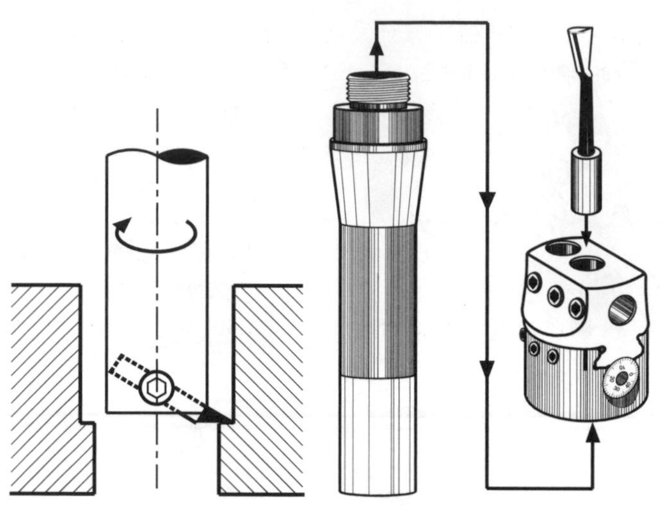

Figure 8–31. Enlarging an existing hole by boring with a lathe-type cutter bit (left) and an R8 spindle tool holder (center), boring head and boring cutter shown inverted for clarity (right). Boring tools may be held in either the end holes or the side holes of the head.

Other advantages of boring heads are that they:

- Cut a larger maximum hole diameter than twist drills.
- Eliminate the need for stocking drills in many diameters.

Flycutters

What are *flycutters* used for and what are their advantages?

Flycutters hold a lathe-type, single-edged cutting tool, and spin it over the work, Figure 8–32. The cutter may be a carbide insert, brazed carbide or HSS, which is easily sharpened.

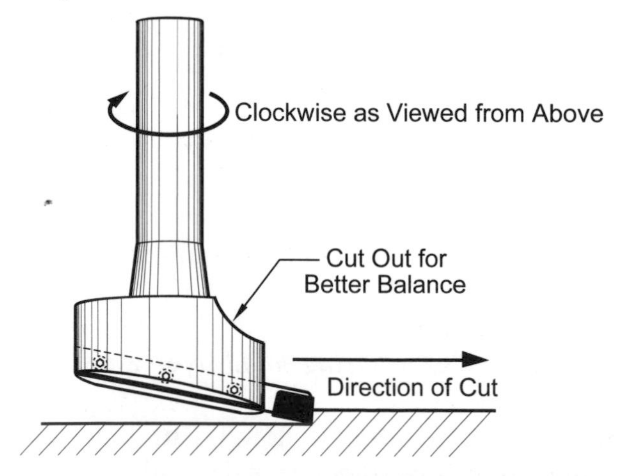

Figure 8–32. A flycutter holding a lathe-type brazed-insert cutter. The flycutter is mounted on an R8 spindle. Note the cutting action.

Figure 8–33 shows two Sherline flycutter designs. The flycutter on the right has the advantage of being able to cut up to a shoulder. Because flycutters have only one cutting edge, they take a much smaller bite with each pass, and remove metal more slowly than an end or face mill. However, they can smooth, or face, a wider surface than end or face mills before exceeding available milling machine torque. In addition, they impose lower cutting forces on the head, and have less chance of chattering. Use flycutters when the work is too big to face on a single pass with other type cutting tools. Be sure the work is securely fastened to the milling table. When a flycutter produces a better surface finish cutting in one direction than in the other, the spindle is not perpendicular to the milling table.

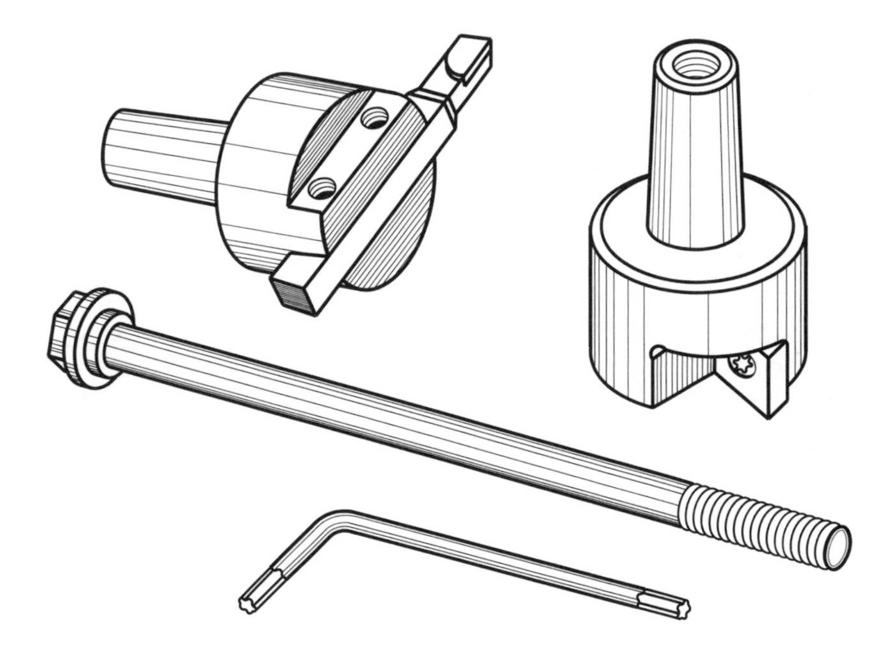

Figure 8–33. Two Sherline flycutter designs: brazed carbide cutting tool (left), carbide insert (right), drawbar and Torx® wrench for carbide insert (bottom).

Slitting Saws

What are *slitting saws* and what is their purpose?

Slitting saws are milling cutters that:

- Are much thinner than other milling cutters, usually between 1/16- and 1/8-inches thick, and 2- to 4-inches diameter.
- Have sides that are often ground to a slight concavity to avoid dragging or binding in the cut.
- Are HSS or solid tungsten carbide.

They are used to:

- Slit or cut entirely through metal.
- Cut slots only part way through the work to make a channel.
- Plunge cut slots in thick stock or make a slot beginning *on* the work.
- Cut-off thin sheet and tubing.
- Apply slots in screw heads.

See Table 8–2 and Figure 8–34.

What are the different types of *slitting* and *slotting saws*?
See Table 8–2 and Figure 8–34.

Saw Type	Common Diameters (inches)	Range of Thickness (inches)	Application
Metal Slitting Saws with Side Chip Clearance	2½, 3, 4	1/16–1/8	Deep sawing, plunging and slotting
Metal Slitting Saws with Staggered Teeth	3, 4	3/32–3/16	Deep, heavy-duty slotting
Plain Metal Slitting Saws	2½, 3, 4	1/32–1/8	Slotting and cutting off
Jewelers' Slotting Saws	1, 1¼, 1½, 1¾, 2, 2½, 3 4, 5, 6	0.008–0.128	Slotting sheet metal or tubing
Screw Slotting Saws	1¾–2¾	0.080–0.102	Slotting screw heads, cutting sheet metal tubing

Table 8–2. Slitting and slotting saw types, dimensions and applications.

The slotting saws in Figure 8–34 have cutting teeth on their sides and may be thought of as very thin, plain side mills. However, they are not used to smooth the sides of work as is a typical plain-side cutter, because they are too thin to support lateral forces.

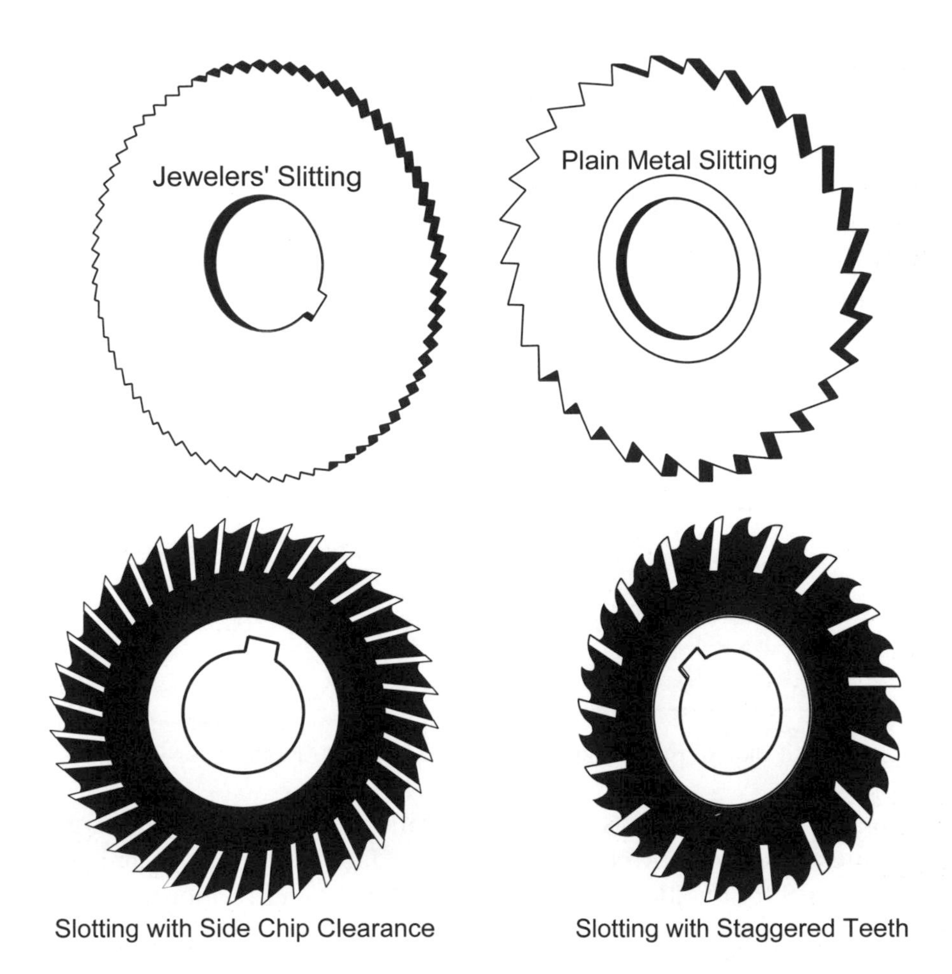

Figure 8–34. Saws.

Figure 8–35 shows arbors used to hold saws in Bridgeport-style (top) and Sherline mills (bottom).

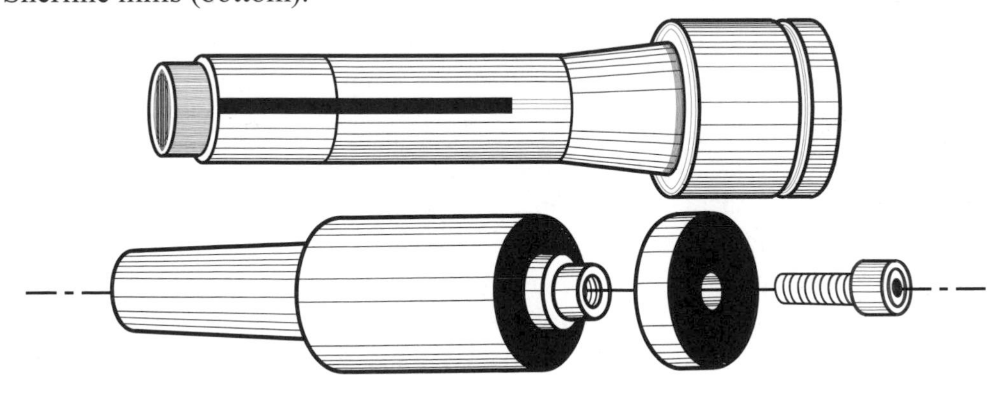

Figure 8–35. R8 slitting saw arbor (top) and Sherline slitting saw arbor on #1MT (bottom).

Drills

What are the two methods for holding twist drills in milling machines?
As on a lathe, smaller diameter twist drills are held in drill chucks. These chucks, which have an internal Jacobs taper (JT), are mounted on R8 tool holders that have Jacob chuck tapers on their lower ends, Figure 8–36 (top). Larger drills usually have Morse tapers, and these are held in an R8 tool holder with an internal Morse taper, Figure 8–36 (bottom).

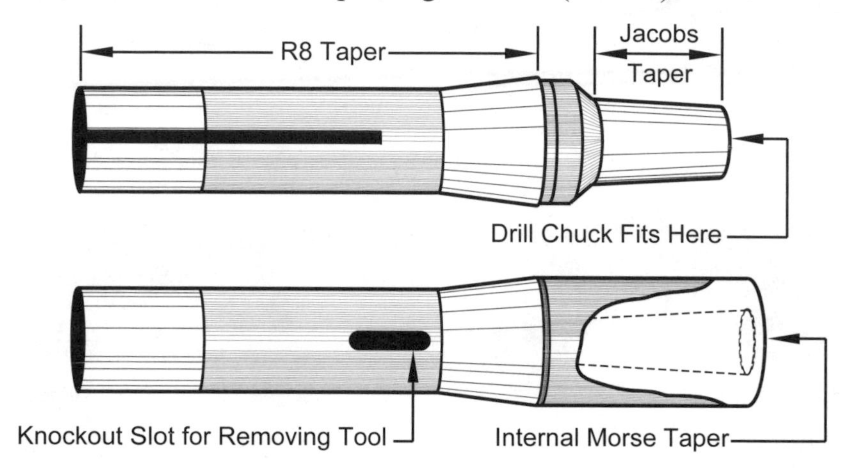

Figure 8–36. Two methods for holding drills in vertical knee mills: a drill chuck fitted to an R8 tool holder with a Jacobs taper (top) and an R8 tool holder with an internal Morse taper (bottom).

To hold twist drills, the Sherline miniature mill uses a #1 Morse taper adapter screwed into a chuck, Figure 8–37. The taper adapter fits in the spindle taper and is held in place by a drawbar.

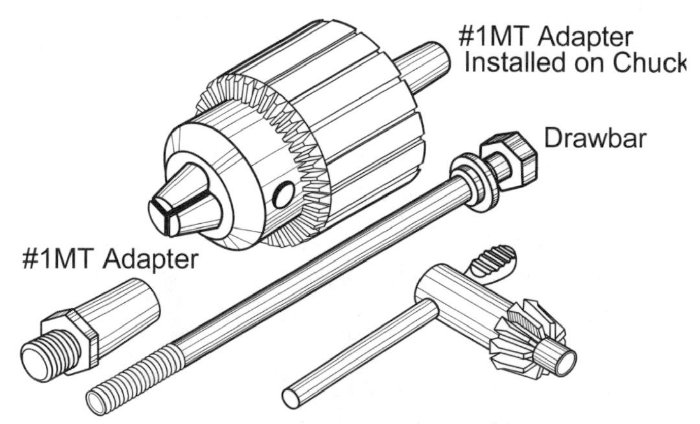

Figure 8–37. Sherline mills mount drill chucks on a #1MT adapter, which fits into the spindle and is held by a drawbar. (All components not to same scale.)

What operating precaution must be observed when using twist drills in a milling machine?
Apply the quill lock to prevent vertical quill motion. Use the vertical feed or Z-axis feed to bring the work *into* the drill. If this approach is not taken and the quill is used to feed the drill into the work, large drills have a tendency to self-feed and screw themselves into the work rather than drilling into it. Using the Z-axis crank also gives the machinist a greater mechanical advantage than the quill feed and puts drilling forces on the strongest components of the milling machine. When using twist drills, always make sure the work is securely clamped to the milling table.

Section V – Cutting Mechanics

Conventional vs. Climb Milling

What are the differences between *conventional* and *climb milling*?
During conventional milling, also called *up-cut milling:*

- The direction of cutter rotation *opposes the advance* or *infeed* of work.
- Chip thickness is smallest at the beginning of the cut and largest at the end, Figure 8–38 (left).

During climb milling, also called *down-cut milling*, the opposite conditions hold:

- The direction of cutter rotation *aids the advance* or *infeed* of work, producing large forces, which tend to pull the work under the cutter in the direction of infeed.
- Cutter forces push the work down against the milling table holding it in place. These same forces press the worktable against the gibs.
- Chip thickness is largest at the start of the cut and smallest at the end, Figure 8–38 (right).

Climb milling has two advantages over conventional milling:

- Produces a smoother surface finish.
- Prolongs cutter life because the cutter teeth initially take a big bite. This prevents the teeth from being dragged through the metal while building up pressure to cut into it.

Conventional milling also has two advantages:

- When milling abrasive materials, such as cast iron, which may have molding sand embedded in its surface, conventional milling protects the cutter by removing metal from below, before the cutter can contact the sand adhered to the metal's surface.

- It is less sensitive to backlash and looseness in the table gibs.

Climb milling is often more desirable, but it requires a rigid machine with backlash eliminators on the table leadscrews. *Climb milling is not recommended for light or older machines.* On vertical knee mills in good condition, climb milling may be used, with caution, while taking small cuts. Serious accidents can result if the table is not tight and the work holding device is not securely mounted. Climb milling is more often used during horizontal milling with larger, more rigid milling machines, and when the cutters are held on an arbor supported at both ends.

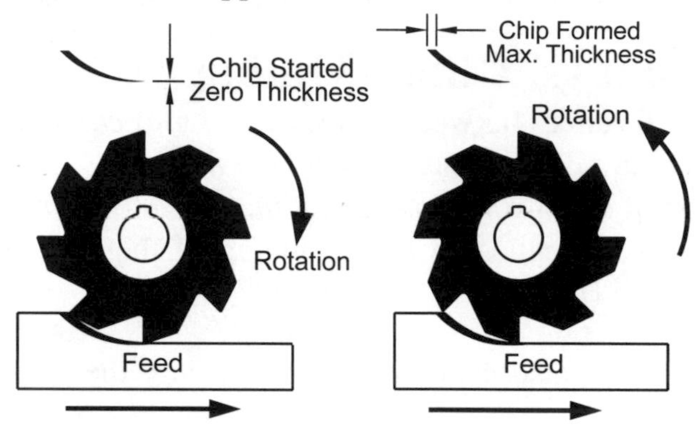

Figure 8–38. Conventional (left) and climb milling (right).

Feeds & Speeds

How are cutting speed and cutting feed defined?

Cutting speed is the speed at which the tool point moves with respect to the work and is measured in feet per minute, also called surface feet per minute (sfm or sfpm), or meters per minute.

Cutting feed is the rate at which the work moves into the cutter and is measured in feed per tooth per revolution (fpt). Cutting feed affects:

- Machining time.
- Cutting tool life.
- Surface finish.
- Milling machine horsepower required.

What factors affect speed and feed?

- Depth and width of cut.
- Type of cutter and its design.

- Cutter sharpness.
- Workpiece material.
- Size and shape of the workpiece, which influence rigidity and its ability to withstand cutting forces.
- Final workpiece surface finish.
- Allowable tolerance of the cut.
- Rigidity and horsepower of the machine and its work holding devices.

How are *cutting speeds* determined and where are they listed?

Just like speed and feed rates for lathes, cutting speeds are determined experimentally. Because of variations in machines, cutters and materials, speeds are not exact and cannot be. Recommended cutting speeds offer a good starting point. Table 8–3 shows some typical cutting speeds. Extensive cutting speed tables appear in *Machinery's Handbook* and in cutter manufacturers' product data sheets.

For a given cutting speed, usually taken from a table (see *Table A–2*), how is the proper spindle speed in revolutions/minute (rpm) calculated?

$$\text{For inch units: Spindle Speed (rpm)} = \frac{\text{Cutting Speed (ft/min)}}{\text{Circumference (in)}}$$

Using 3 for π, and converting inch units to feet, this equation simplifies to:

$$\text{Spindle Speed (rpm)} = \frac{4 \times \text{Cutting Speed (ft/min)}}{\text{Cutter Diameter (in)}}$$

For metric units and after similar simplification:

$$\text{Spindle Speed (rpm)} = \frac{320 \times \text{Cutting Speed (m/min)}}{\text{Cutter Diameter (mm)}}$$

Material	High Speed Steel Cutter		Carbide Cutter	
	ft/min	m/min	ft/min	m/min
Aluminum	600–1000	180–300	1500–2500	450–760
Brass & Bronze	150–300	45–91	375–750	114–227
Cast Iron	50–100	15–30	125–250	38–76
Steel, Cold Rolled	80–120	24–36	200–300	61–91
Steel, Free Machining	80–120	24–36	200–300	53–75
Steel, Hot Rolled	70–100	21–30	175–250	61–75
Steel, Tool	50–75	15–23	125–188	38–57

Table 8–3. Milling machine typical cutting speeds.

How is milling machine *feed* defined?

Milling machine feed is the distance in inches (or millimeters) per minute that work moves into the cutter. The feed rate on milling machines with power feed is adjustable and independent of the spindle speed. For example, on Bridgeport-style vertical knee mills, the X-axis power feed has a range from ¾ to 45 inches per minute.

How is *feed rate* determined?

Use the recommended feed rate per tooth for the workpiece material plus the spindle speed to calculate the feed rate. The recommended feed rate per tooth is determined principally by the workpiece hardness and is available in *Machinery's Handbook,* Tables 10-15, Feeds and Speeds Section. Table 8–4 below shows recommended feed per tooth figures for HSS cutters for a few common metals.

Here is the formula for calculating feed rate:

$$\text{Feed} = \text{Number of teeth in cutter} \times \text{Feed/tooth for metal} \times \text{rpm}$$

Note: This formula is the same for either inch or metric units. Using an inch rate of feed/tooth gives inch/min units and similarly, using a mm rate of feed/tooth, gives mm/min units.

Material	Feed per Tooth per Revolution for HSS Cutters (inches per tooth in thousandths of an inch)				
	End Mills	Face Mill	Slab Mill	Slotting & Side Mills	Saws
Aluminum	9–23	15–40	8–15	10–20	5–10
Brass & Bronze	7–14	11–33	6–12	8–16	4–8
Cast Iron	4–8	7–18	3–7	4–10	1–3
Steel, Cold Rolled	5–10	8–20	4–8	6–12	5–10
Steel, Free Machining	5–10	8–20	4–8	5–11	3–6
Steel, Hot Rolled	5–10	8–20	4–8	6–12	5–10
Steel, Tool	2–7	5–12	1–5	3–6	1–3

Note: Feeds for carbide cutters are 2–3 times higher than the above figures.

Table 8–4. Recommended feed per tooth for HSS cutters.

Recap of Feed & Speed Calculations

How can feed and speed be calculated?

1. Determine the workpiece material.
2. Look up the recommended milling machine cutting speed.
3. Calculate the spindle speed using:

   For inch units: $\text{Spindle Speed (rpm)} = \dfrac{4 \times \text{Cutting Speed (ft/min)}}{\text{Cutter Diameter (in)}}$

   For metric units: $\text{Spindle Speed (rpm)} = \dfrac{320 \times \text{Cutting Speed (m/min)}}{\text{Cutter Diameter (mm)}}$

4. Using tables, look up the recommended milling machine feed per tooth for the cutter material (HSS or carbide), cutter type (end mill, face mill, slotting saw, etc.), and the number of cutter teeth. Then use this data, plus the spindle speed, to calculate the feed rate.

   $$\text{Feed} = \text{Number of teeth in cutter} \times \text{Feed/tooth for metal} \times \text{rpm}$$

   Remember, this is just a starting point and the rate may have to be adjusted depending on results.

Speed Guidelines for Sherline Milling Machines

What end mill cutting speeds should be used for common metals in the Sherline miniature milling machine?

See Table 8–5.

Material	Cutting Speed (ft/min)	End Mill Cutting Speeds for Slot & Side Milling Cutters (rpm)		
		⅛-inch	⅜-inch	⅝-inch
Aluminum, 2024	200	2800	2500	2000
Aluminum, 6061	280	2800	2500	2000
Aluminum, 7075	300	2800	2500	2000
Aluminum, Cast	134	2800	2000	1300
Brass	400	2800	2800	2800
Gray Cast Iron	34	1000	500	350
Stainless Steel, 303	40	1200	600	400
Stainless Steel, 304	36	1100	500	350
Stainless Steel, 316	30	900	450	300
Steel, 1018	34	1000	500	350
Steel, 12L14	67	2000	1000	650
Steel, 4130	27	800	400	250

Table 8–5. Sherline recommended end mill cutting speeds for their miniature milling machines.

Note: There are no cutter feed guidelines because the manual Sherline mill does not have a power feed.

Coolants

What is the purpose of *coolants,* also called *cutting fluids*?
The purpose of coolants is to:

- Cool the cutting tool to keep it from overheating and losing its hardness.
- Lubricate the surface of the cutting tool to help cutting chips slide along it and away from the work.
- Flush away chips.
- Improve the surface finish.
- Permit higher feeds and speeds by reducing the necessary horsepower.

What are typical coolants and which ones are used on which metal?
See Table 8–6.

Material	Recommended Coolants for Milling
Aluminum and its alloys	Kerosene, kerosene and fatty oil, soluble oil
Brass	Dry, soluble oil, kerosene
Bronze	Dry, soluble oil, fatty oil, mineral oil
Cast Iron	Dry, air jet, soluble oil
Copper	Dry, soluble oil, kerosene
Magnesium	Dry, with an air blast, mineral oil
Malleable Iron	Dry
Plastics	Dry
Steel, Alloy	Soluble oil
Steel, Forged	Soluble oil
Steel, Low & Medium Carbon	Soluble oil, fatty oil
Steel, Tool	Soluble oil, fatty oil

Table 8–6. Coolants for milling.

Coolants for the Sherline miniature mills are only necessary during slitting saw and cutoff operations, and sometimes helpful when drilling. The advantages of lubrication are usually offset by the messes it creates. This small machine, because of its limited horsepower and frame rigidity, does not generate the cutting forces and heat seen in full-size machines, so coolants are not usually required.

Section VI – Securing & Aligning the Work

Table Clamps

How can work be secured to the milling machine table?

Putting bolts into the milling table T-slots is an excellent way to secure workpieces, Figure 8–39.

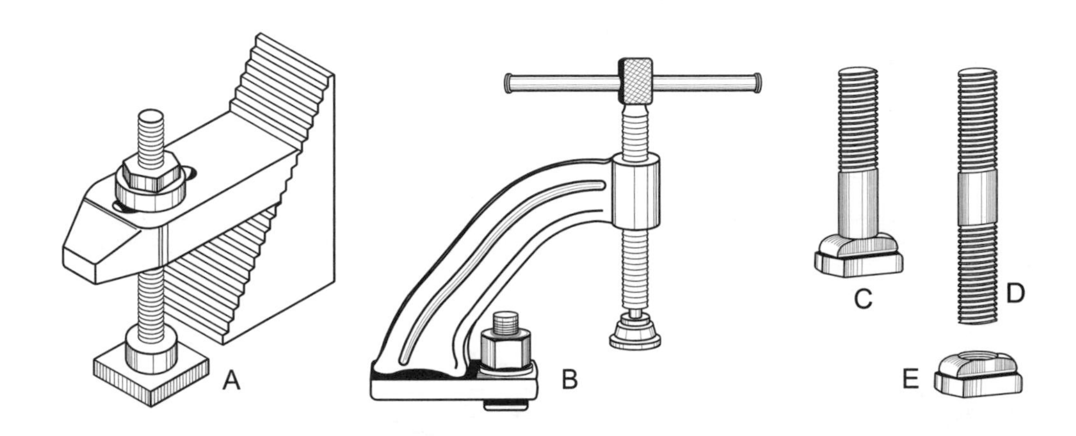

Figure 8–39. (A) Sherline miniature step block and T-nut, (B) table clamp, (C) T-bolt, (D) clamping stud and (E) T-nut.

What precautions should be observed when using table clamps to secure work?

Some standard precautions are:

- Place the clamp bolt so it is close to the work, Figure 8–40 (A), not close to the step block, Figure 8–40 (B). This gives the bolt a mechanical advantage and increases its clamping force.
- If the clamp will mar the work, place an aluminum or brass shim between the clamp and the work, Figure 8–40 (C).
- Place the clamp on the step block so it remains level to provide better clamping action on the work, *not* as in Figure 8–40 (D).
- If the work is to be clamped in a bridge off the milling table, place the spacer blocks so the clamping force acts on them and does not act to bend the work, Figure 8–40 (E & F), respectively.

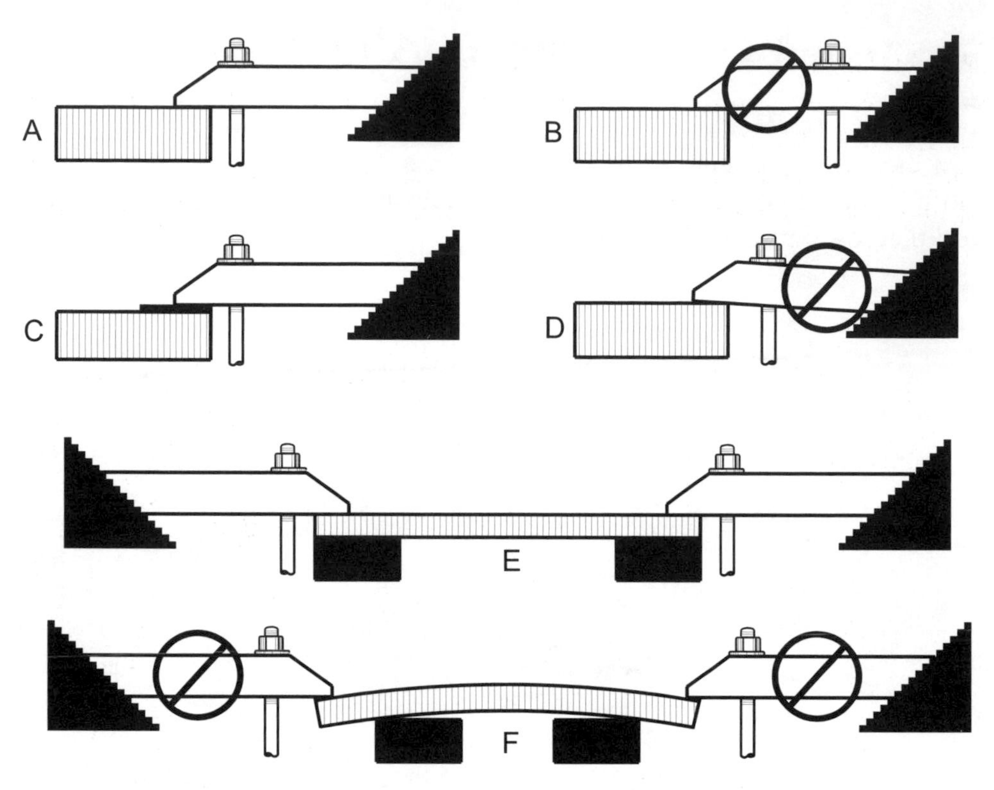

Figure 8–40. Correct (A, C & E) and incorrect ways (B, D & F) to use table clamps.

Jacks

What is the purpose of milling table jacks?

They support the overhanging portion of a part from deflecting under milling tool forces, Figure 8–41.

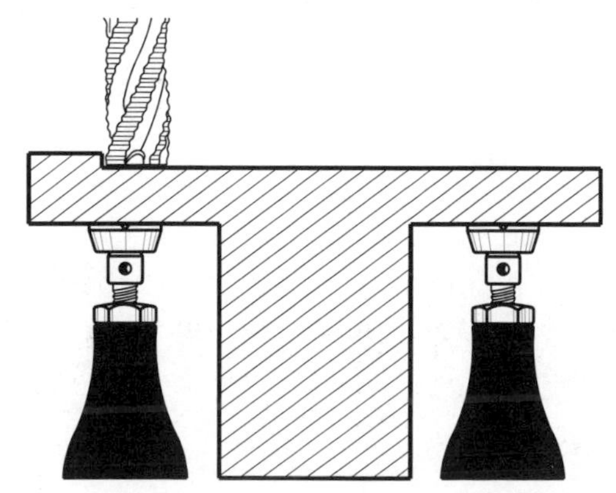

Figure 8–41. Small milling table jacks support overhanging work against tool forces.

Milling Vise

What are the advantages of a milling vise to hold work over using table clamps in T-slots?

Once the fixed jaw of a vise has been set parallel with the X- or Y-axis of the milling machine, parts placed into the vise are automatically aligned with this axis, saving time. See Figure 8–42.

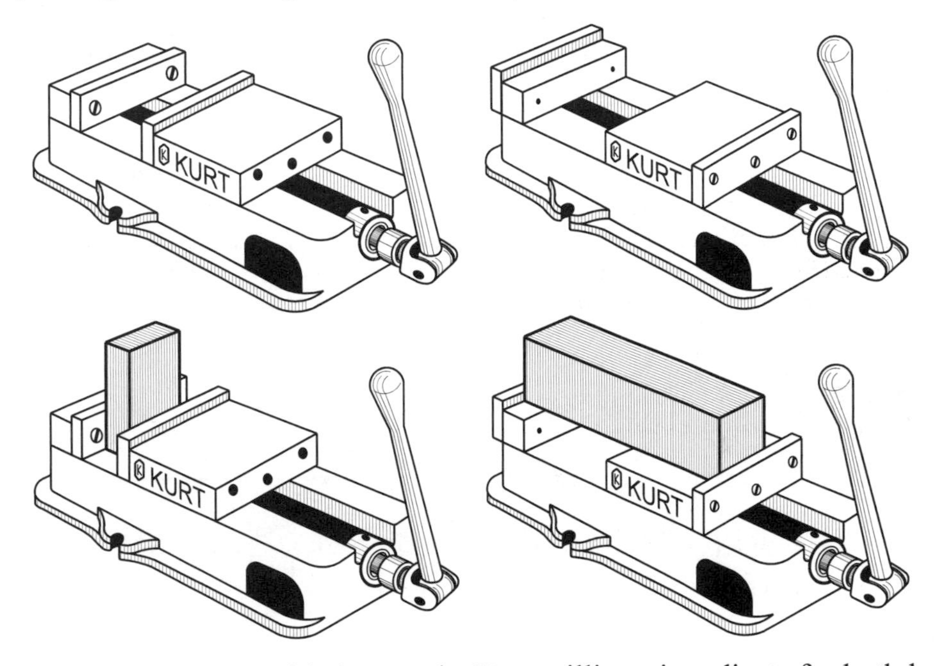

Figure 8–42. The moveable jaw on the Kurt milling vise adjusts for both large and small workpieces, supporting both ends of the work at all times.

How can angled cuts be made without removing work from the milling vise?

Use a swivel base with the milling vise, Figure 8–43.

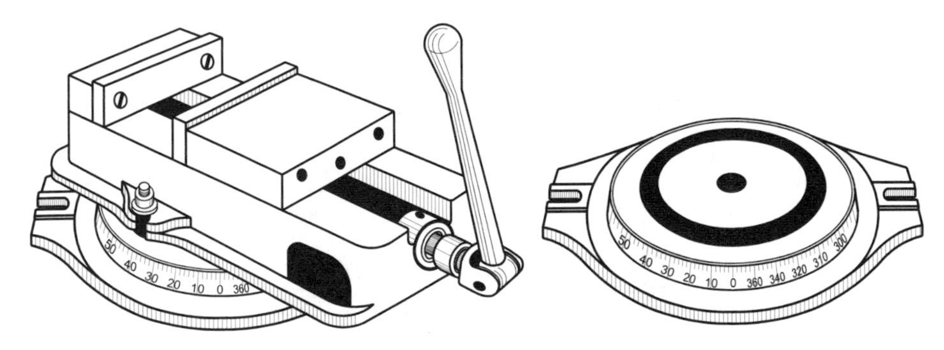

Figure 8–43. Milling vise with swivel base (left) and swivel base casting (right).

What precautions should be observed when clamping work in a milling vise?

- Use a set of parallels to be sure the work is square and flat to the bottom of the vise and using a soft-faced hammer strike the workpiece with moderate blows to insure it is fully seated in the vise or on parallels.
- Try pulling on the parallels to insure they are not loose.
- Never use a hammer to tighten the vise handle.

See Figure 8–44.

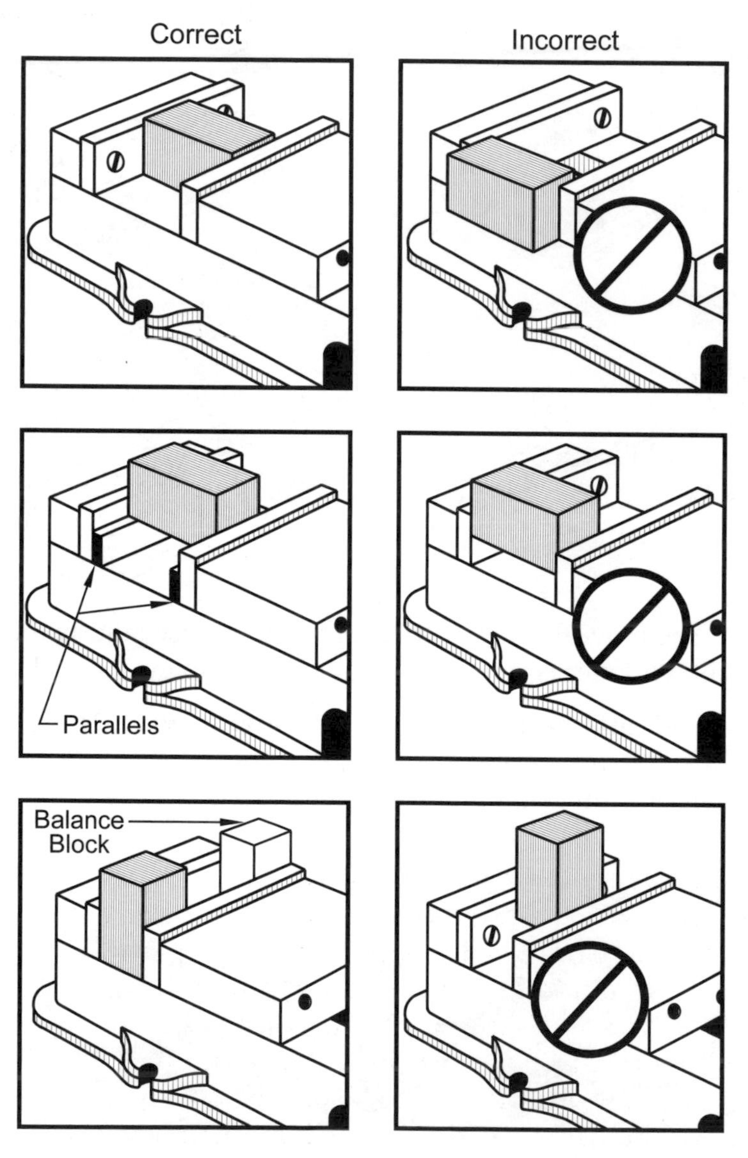

Figure 8–44. Correct and incorrect ways of clamping work in a milling vise.

Toolmakers' Milling Vise

How does a *toolmakers' milling vise* differ from a standard milling vise?

The tightening bolt on the toolmakers' vise is set on an angle and pulls the jaw closed as well as downward, Figure 8–45. In general, these vises provide more accurate alignment of parts than regular milling vises.

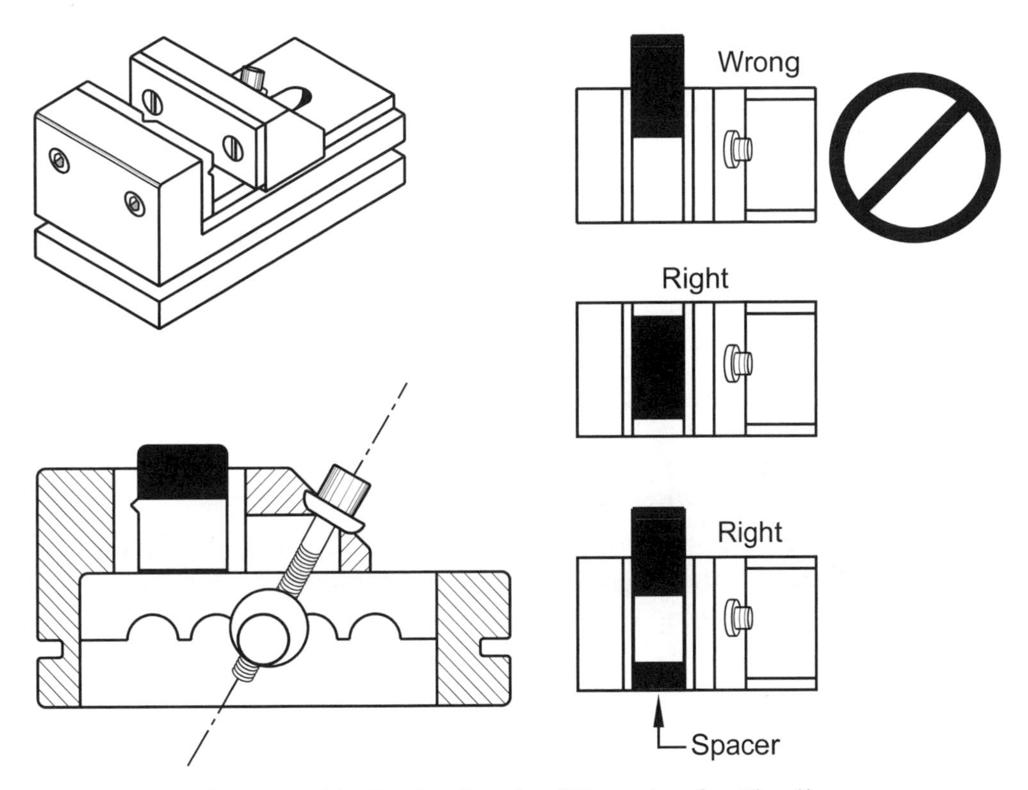

Figure 8–45. Toolmakers' milling vise for Sherline mill (left) and clamping practice (right).

Parallels

What are *parallels* and how are they used?

Parallels are rectangular, hardened steel spacers that come in pairs, Figure 8–46. Simultaneously grinding both edges of a pair of two parallels insures the edges are parallel and widths identical. Usually parallels are supplied in sets of four to eight pairs in ⅛-inch height increments. Parallels for Bridgeport-style knee mills are usually 6-inches long and ⅛-inch thick. Parallels from 3- to 12-inches long are stock items from industrial supply houses. The 3-inch length is most suitable for the Sherline milling machine. The holes in their sides are useful for clamping parallels when extra support is needed.

Typical specifications of high-quality parallels:

- Pairs parallel within 0.0002 inches.
- Pairs differ by not more than 0.0004 inches.
- Rockwell C hardness 52 to 58.

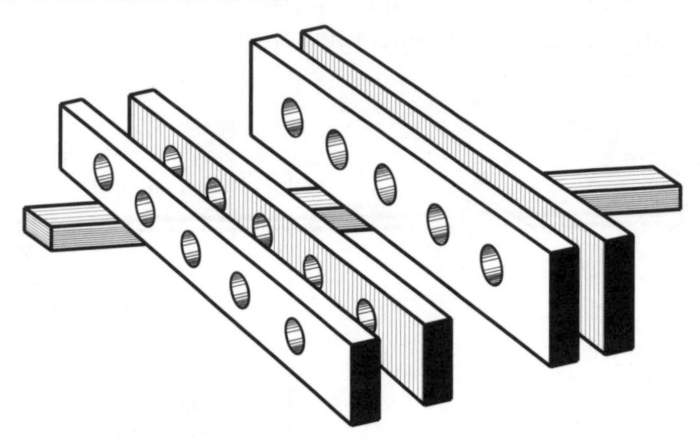

Figure 8–46. Two pairs of parallels.

V-Blocks

How are *V-blocks* used when milling?

V-blocks are used to support round stock and hold it steady. Workpieces may be placed in them and then clamped directly to the milling table, or V-blocks may be used inside a milling vise, Figure 8–47.

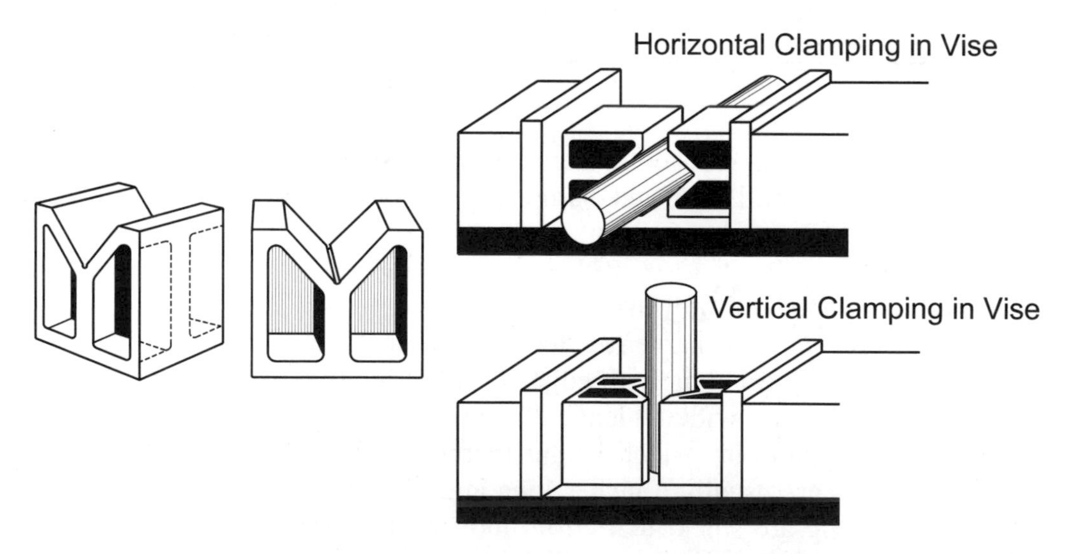

Figure 8–47. V-blocks (left) and V-blocks holding rounds in a milling vise (right).

1-2-3 Blocks

What are *1-2-3 blocks* used for?

These 1 × 2 × 3-inch steel spacer blocks, Figure 8–48, have nearly identical dimensions which makes them ideal for spacing workpieces up off the milling table or supporting a workpiece clamped to an angle block. They are similar to parallels, but more stable, and may be used outside a vise. Quality 1-2-3 blocks are hardened and ground and supplied in matched pairs. Commercial 2 × 4 × 6-inch spacer blocks are also available. Using aluminum, many machinists make their own blocks in much larger sizes.

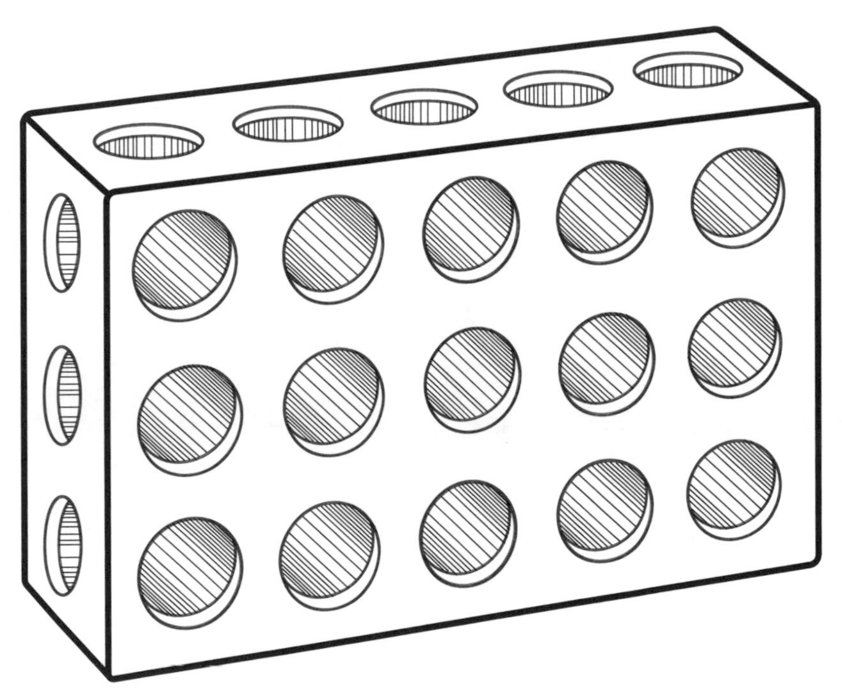

Figure 8–48. A 1-2-3 block used for spacing work off the milling table.

Angle Plates & Angle Tables

When are *slotted angle plates* used on the milling table?

Cast iron slotted angle plates hold work too big or too long to fit in a milling vise. A series of slots on their vertical faces allows work to be bolted to them, Figure 8–49. Clamps may also be used to secure work to the angle plate, Figure 8–50. Less expensive plates for general machine shop work are machined, but plates for precision work are stress-relieved and ground. They are secured to the milling table by T-bolts through slots in their bases. They come in a wide variety of sizes from 2 × 2 inches to 12 × 16 inches. Some plates also have threaded holes for mounting work.

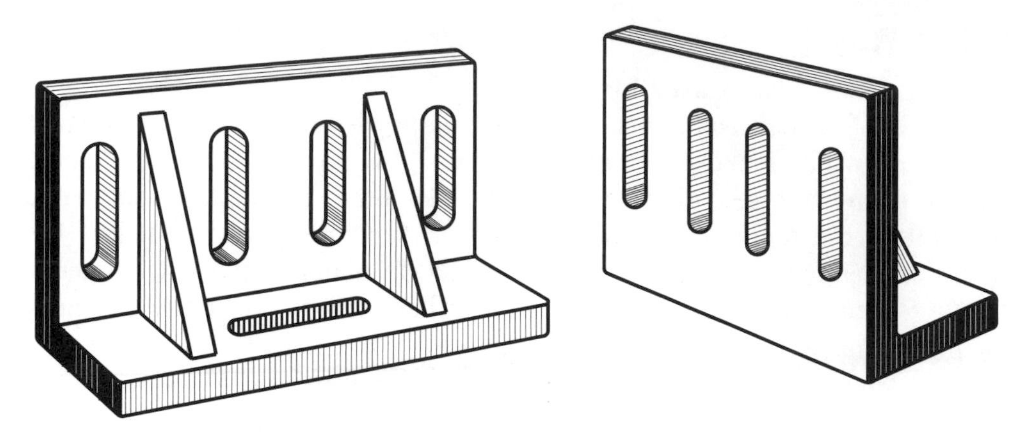

Figure 8–49. Slotted angle plates.

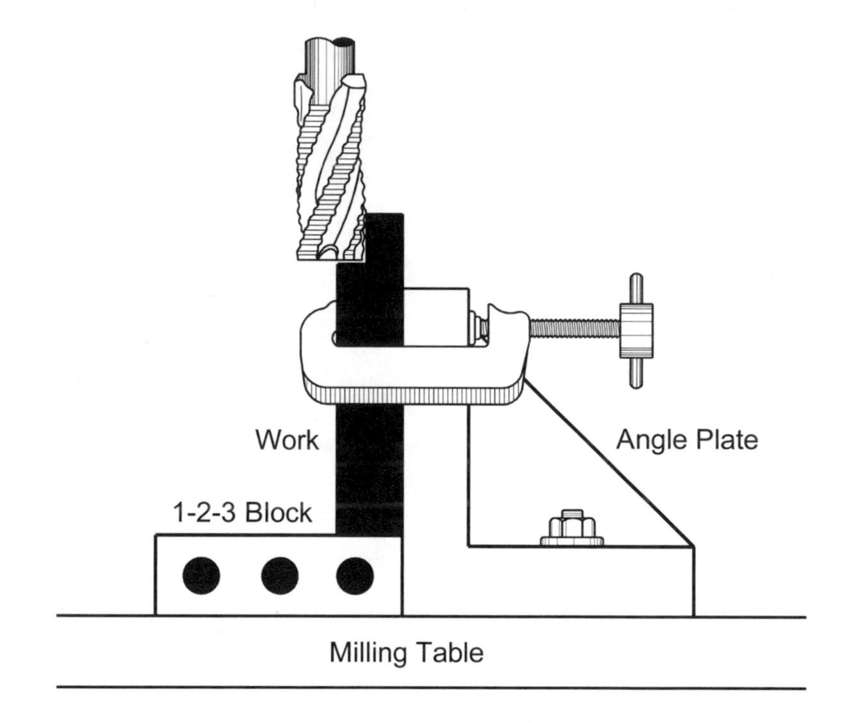

Figure 8–50. Work clamped to an angle plate during milling using a 1-2-3 block to raise the work parallel to the table. Place the clamp as close to the cutting point as possible for stability and chatter reduction. Use two or more clamps whenever possible.

Are there other angle plate designs?

There are many designs that tilt and swivel to position work at an angle to the milling table. Figure 8–51 shows a tilting angle table for the Sherline milling machine. Plain and threaded holes make it easy to mount a vise or a rotary table, and work can also be bolted directly to it.

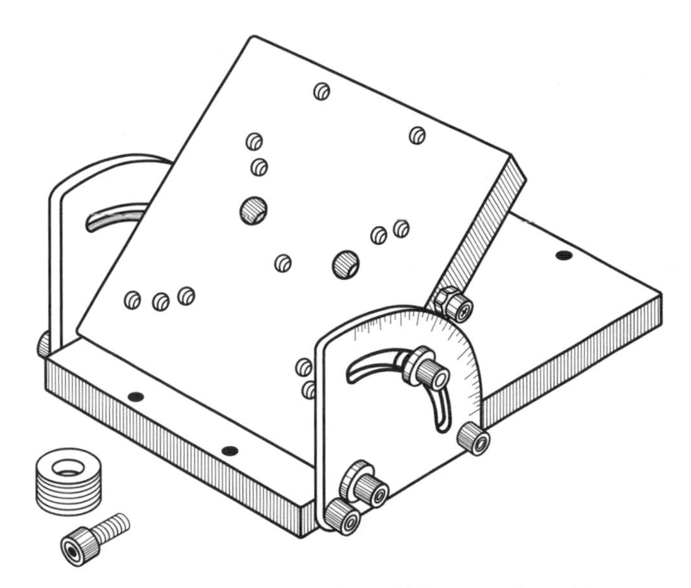

Figure 8–51. Sherline tilting angle table.

Squaring the Vise

What is *squaring the vise* and how is it done?

Squaring aligns the fixed jaw face of the vise parallel with the motion of the milling table, usually along the X-axis. Milling cuts in the work are then parallel with its edge. Here is how to perform this alignment:

- Wipe off the base of the vise and the surface of the worktable to remove chips and debris that prevent the vise from seating fully.
- Mount the milling vise on the worktable with T-bolts. Leave the bolts loose enough so the vise can be adjusted by tapping it with a soft-faced hammer. Tightening one bolt more than the other forces the vise to swivel about the tighter bolt and simplifies alignment.
- Secure a dial test indicator in the spindle using a collet, a Jacobs chuck, or by clamping it around the quill using a dial test indicator holder. See Figure 8–52.

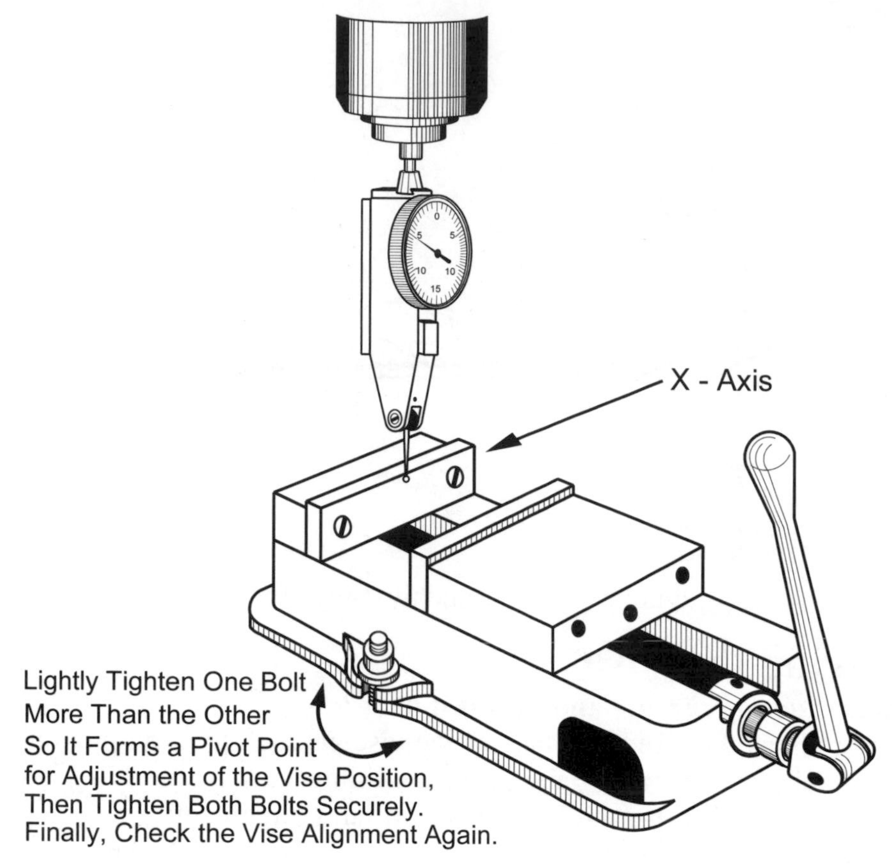

Figure 8–52. Squaring the vise with the milling table motion.

- Lower the spindle and adjust the table so the dial indicator touches the fixed face of the vise about ¼ inch below the top edge of the vise jaw and the dial indicator turns about one-half revolution. Make sure that the dial indicator leg does not get caught in the screw holes in the face of the vise.
- Set the dial indicator bezel to zero.
- Slowly move the table back and forth in the X-direction so the dial indicator leg sweeps from side to side of the vise face.
- Adjust the angular position of the vise by lightly tapping the sides of the vise with a soft-faced hammer so the deflection of the dial indicator remains constant as the dial indicator sweeps from side to side of the vise jaw. This shows that the vise face is parallel to the X-axis.
- Tighten the T-nuts on the vise and re-check its alignment.

The vise face can be aligned with the Y-axis in the same way, but a faster, although less accurate method, is to use a square between the face of the vertical column and the fixed face of the vise, Figure 8–53.

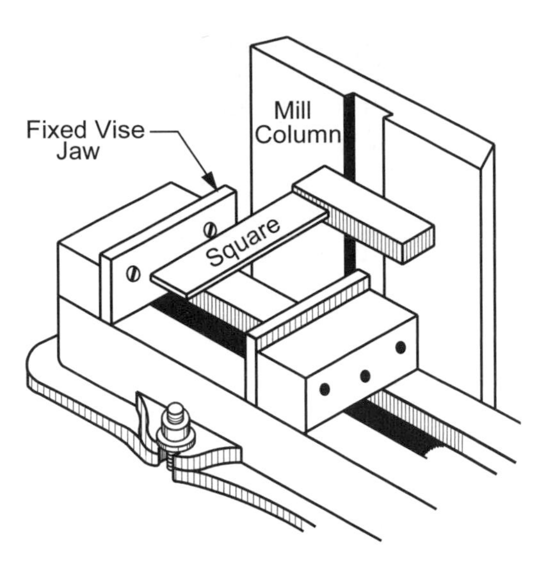

Figure 8–53. Using a square to align the vise jaw with the Y-axis of the mill.

Is there a faster way to align a rectangular piece of work with the X-axis without using a dial indicator?

Yes, a common practice is to insert two hardened steel pins in the milling machine table T-slots. These pins are slightly larger than the T-slot width and remain vertical when inserted into the T-slots. Only light pressure is needed to insert or remove them. The T-slots are parallel to the X-axis, so work placed against them is automatically parallel with the X-axis. See Figure 8–54.

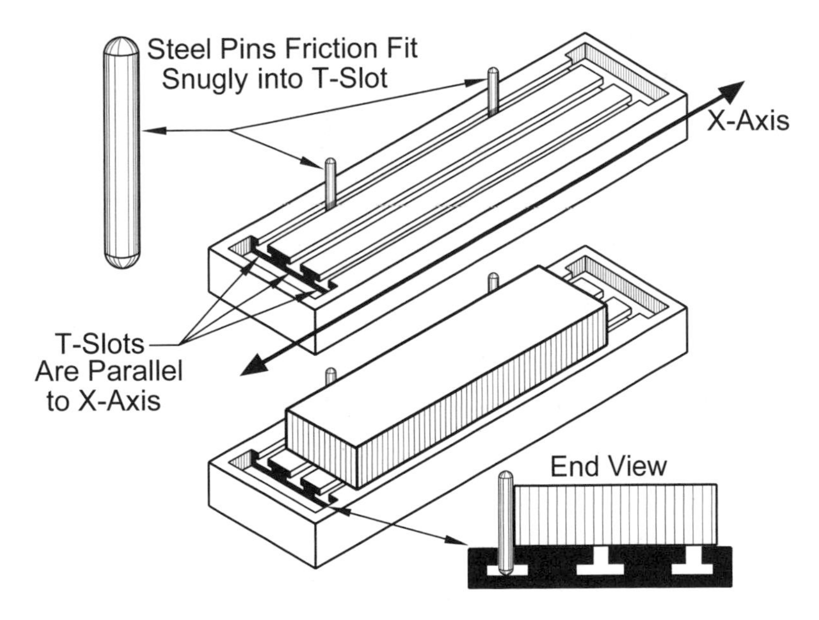

Figure 8–54. Using steel pins in T-slots to align work with the X-axis.

Tramming the Head

What does *tramming the head* mean and how is it done?

Tramming means adjusting the position of the milling head so the axis of the spindle is *perpendicular* to the milling table surface. Some milling heads can be rotated side to side or also tilted front to back. These adjustments add flexibility, but also allow the machine to drift out of vertical alignment. Here is how tramming the head is performed:

1. Wipe the worktable clean of dirt and debris.
2. Install a dial indicator in the spindle so that its leg touches the table at least 6 inches away from the spindle axis. Mount the dial indicator support rods in a collet, Figure 8–55.
3. Raise the table so the dial indicator deflects about one-half a revolution.
4. For side to side (left/right head alignment), put the dial indicator leg at the 3 or 9 o'clock position and slowly rotate the spindle back and forth a half turn and observe its dial. You may have to depress the dial indicator plunger so it does not get caught in the T-slots as it rotates. If so, do not change the vertical position of the DTI.
5. Adjust the head position so the dial indicator shows no deflection when swung through this half circle.
6. Tighten the support bolts and recheck. When the head castings on a milling machine are warped, tightening the bolts may throw the machine out of alignment, so check the alignment again.

To check if it is perpendicular side to side, repeat the process using the 6 and 12 o'clock positions.

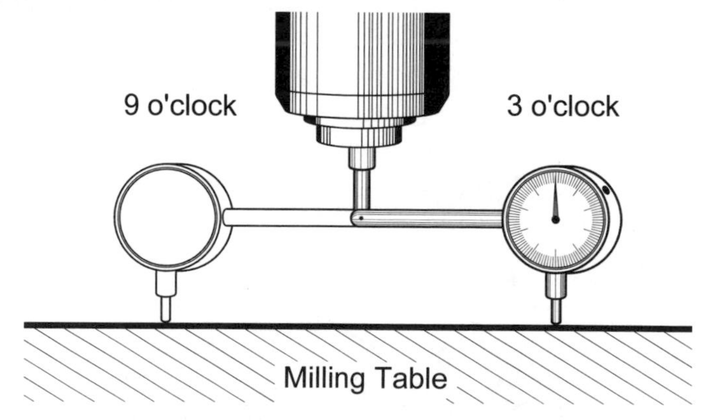

Figure 8–55. Tramming the head so it is perpendicular side-to-side.

Angling the Head

How can a milling machine head be set to cut at an angle to the worktable other than 90°?

This is done very much like tramming the head as described above, except the dial indicator works against a device used to establish the angle instead of the worktable. This device could be a protractor head on a square, a universal bevel gage, an angle block or a machinists' triangle. See Figure 8–56.

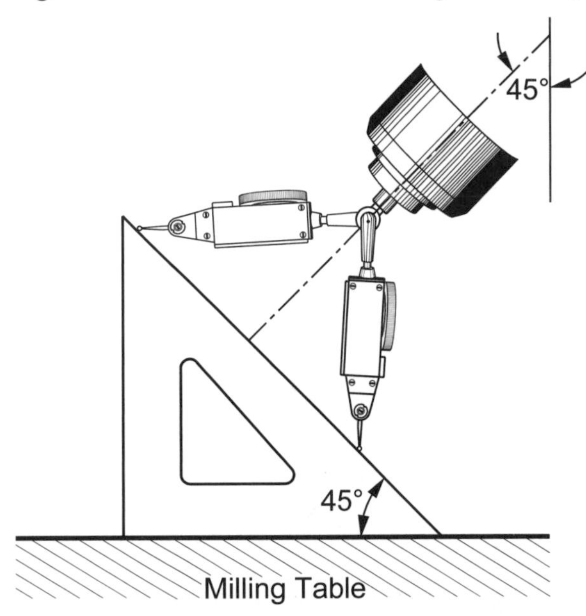

Figure 8–56. Using a triangle to set the milling head to a 45° cutting angle.

Locating Work under the Tool

You have secured a workpiece on the milling table or in a milling vise and the work has a layout mark for a hole. How can the work be precisely positioned under a drill or cutter?

This is a common problem. When layout lines and holes have already been marked, their centers can easily be found using a *wiggler*, sometimes called a *center finder* or *line finder*. Here is how:

1. Install a wiggler, Figure 8–57, in a collet in the spindle. Use a collet for best concentricity, not a drill chuck. Adjust the milling table so the wiggler tip clears the work when the tip is vertical.
2. Adjust the wiggler pointer 15 to 20° off center. This setting is not critical.
3. Set the milling machine spindle to run 500 to 900 rpm; turn on the motor.
4. Use a pencil to apply gentle pressure to the upper end of the wiggler. The wiggler will begin to come into alignment.
5. Slide the pencil about three-quarters of the way down the wiggler and the wiggler will come into position and stay there. It is concentric with the spindle axis. Position the table with X- and Y-axis handwheels so the wiggler pointer lies exactly over the layout mark. Set the micrometer

collars or DRO to zero. The other points on the workpiece may be located using the handwheel collars or the DRO.

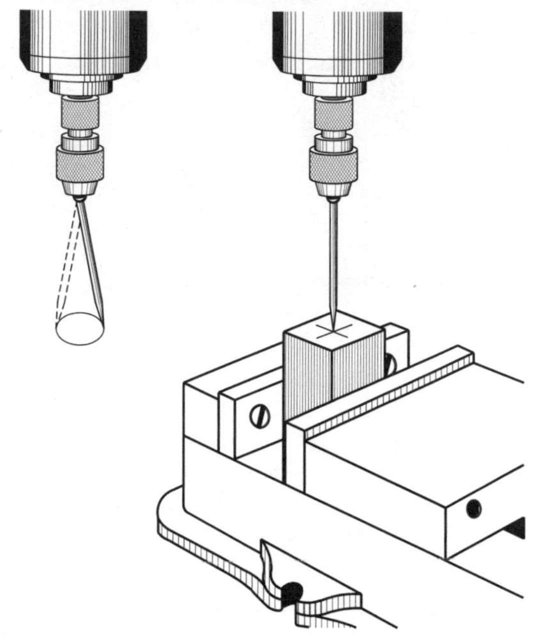

Figure 8–57. Positioning the work exactly under the spindle using a wiggler: Initial setting of the wiggler is 15 to 20º off-axis (left) and after straightening, positioned over work (right).

Using the Edge Finder

You have a workpiece secured to the worktable or in a milling vise and you want to locate the edge of the workpiece with respect to an axis of the spindle. This is done so you can use the handwheel collar graduations or the DRO to position the work. How can you do this?

A wiggler could be used, but an *edge finder,* Figure 8–58 (top left), will be more accurate, usually with less than 0.0005-inch (0.0012-mm) error. Here is the procedure:

1. Install the edge finder in the spindle using a collet.
2. Adjust the height of the worktable so that about one-half the lower segment of the edge finder contacts the edge of the work.
3. Push the lower portion, or contact portion, of the edge finder to offset it from the main, upper portion. There is a spring inside the edge finder so it will remain offset.
4. Set the spindle to turn 500 to 800 rpm.
5. Slowly move the worktable to bring the work edge into contact with the edge finder's spinning, eccentric contact portion, Figure 8–58 (A).

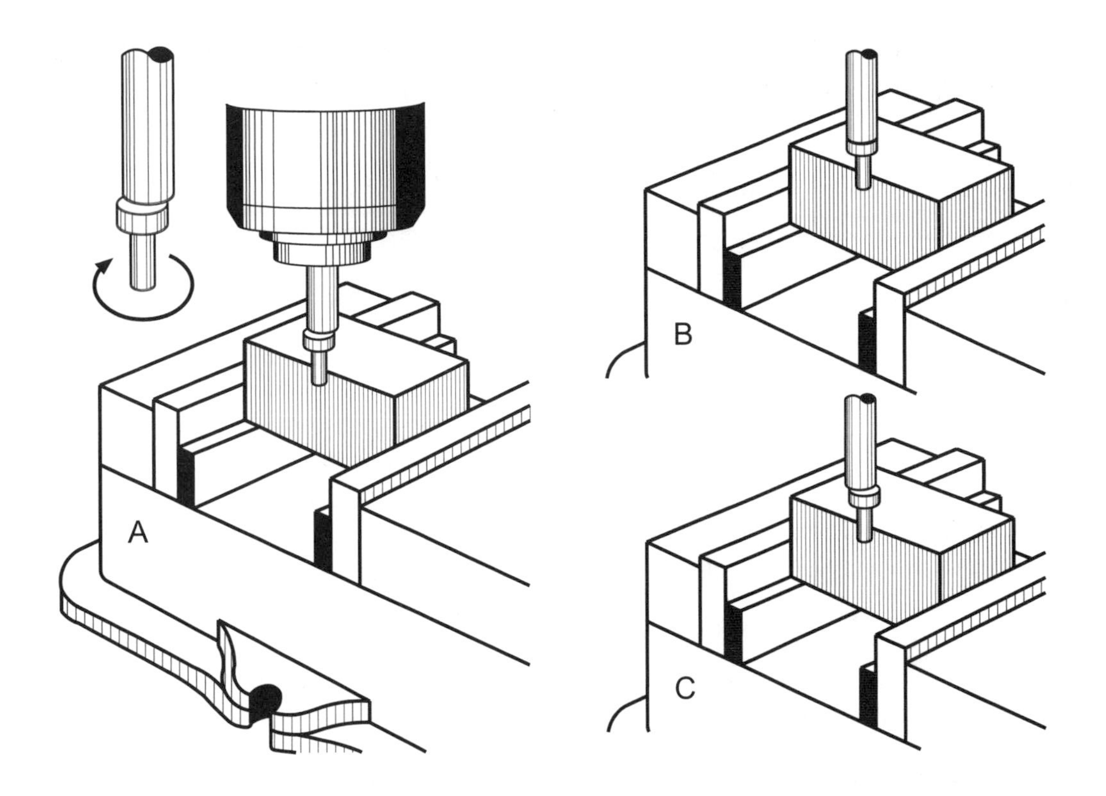

Figure 8–58. Using an edge finder.

6. As the edge finder touches the edge of the work, the eccentric base will come into alignment with its upper shank, Figure 8–58 (B).
7. Continue to move the table very slowly toward the edge. When the edge has been found, the lower end of the edge finder will snap sideways between 1/32 and 1/16 inch because it is dragging completely around its periphery, Figure 8–58 (C). Stop the table motion, the edge has been located and lies half the diameter of the edge finder from the spindle's vertical. Starrett edge finders have a contact diameter of 0.200 inch (6 mm for metric tools). Hence the edge is 0.100 inch (3 mm for metric tools) from the spindle center.
8. Set the handwheel collar or DRO at this breakout point. Repeat at least once to verify the result.

Double-ended edge finders, Figure 8–59 (bottom), have a serious problem. When installed in an R8 collet and spun, the upper end piece moves off center and remains that way. This makes removing the edge-finder from the collet difficult or impossible. Avoid using them. Aside from this removal problem, either end of the double-ended edge-finder works well.

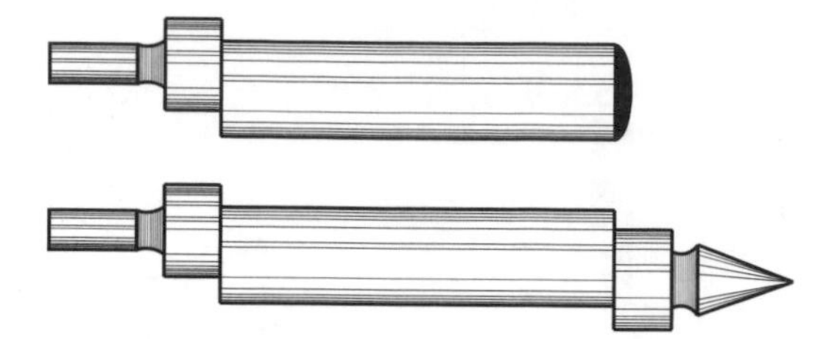

Figure 8–59. Single- and double-ended edge-finders. The double-ended edge finder is likely to hook itself onto the top end of a collet inside the spindle of a milling machine and may be very difficult to remove.

Although edge finders with 0.500-inch diameter contact cylinders are available, they are more difficult to use than edge finders with a 0.200-inch diameter contact cylinder. Half the contact diameter must always be added to the spindle position, and adding 0.250 inches is more difficult and error prone than just adding 0.100 inches.

Using a DTI to Locate an Edge

How can you locate the edge of a workpiece perpendicular to the X-axis without an edge finder?

Using a DTI held by a flexible joint holder in the spindle is a quick way to locate an edge to better than 0.0005-inch accuracy. Unlike an edge finder, the spindle axis is placed directly over the workpiece edge, not offset by half the diameter of the edge finder. Here is the method:

1. Install the DTI in the milling machine spindle.
2. Position the spindle slightly *beyond* the workpiece edge, lower the quill so the DTI leg touches the work, and move the table to deflect the DTI about one-half a revolution. Swing the spindle through a 10–15° arc while observing the DTI and adjust the spindle so the DTI deflection is minimum. This positions the DTI perpendicular to the edge.
3. Set the DTI to zero by turning its bezel, Figure 8–60 (A), and also zero the DRO or calibrated collar of the table axis perpendicular to the edge.
4. Raise the spindle to clear the workpiece, turn the spindle one-half turn, Figure 8–60 (B) and move the milling table to the right .
5. Hold or clamp a flat piece of metal against the workpiece edge, projecting above it to form a stop for the DTI, Figure 8–60 (C). Move the table to deflect the DTI and as in Step 2 above, swing the spindle back and forth to find the minimum DTI deflection.
6. Adjust the table to zero the DTI.
7. Read the DRO or calibrated table handwheel collar which shows how much movement was needed to re-zero the DTI.
8. Move the table back *half* the total distance moved in Step 6 and read the DRO. This positions the spindle axis over the edge, Figure 8–60 (D).

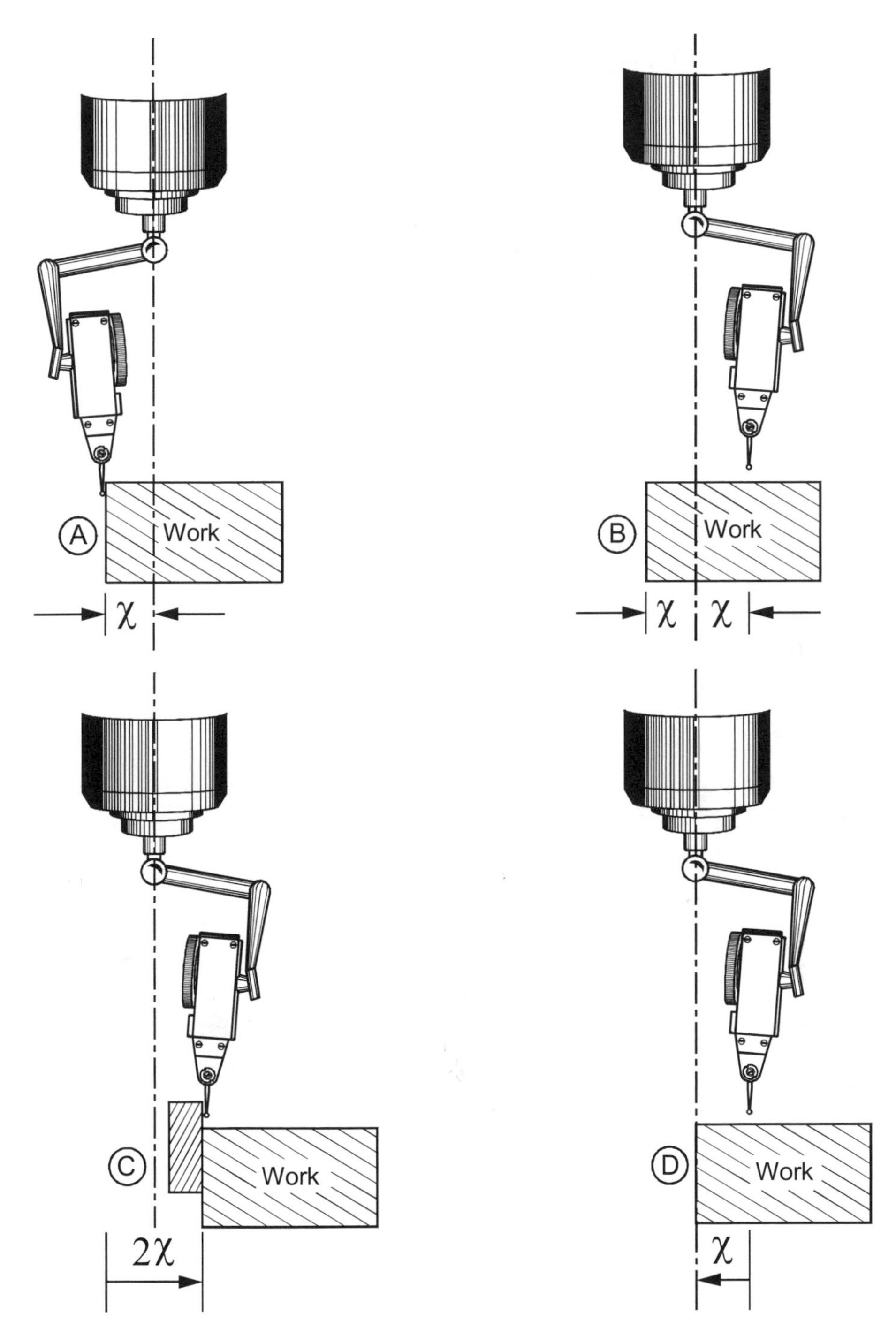

Figure 8–60. Using a DTI to locate an edge.

Locating over Round Stock

How can the center of a round workpiece be located?

The method in Figure 8–61 works for all types of milling cutters.

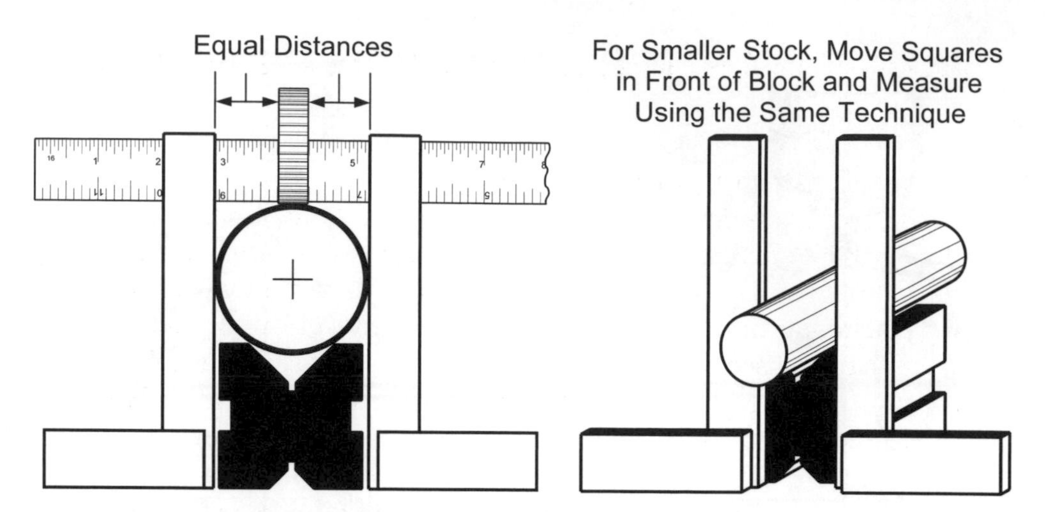

Figure 8–61. Locating the center of a round workpiece.

What other centering method works well with a side cutter when the workpiece is too small in diameter to use the methods shown in Figure 8–61?

1. Secure the work in a milling vise that has its jaw aligned with the milling table movement, usually the X-axis.
2. Set the height of the milling cutter by eye so it will touch the work at its center. Use a slip of paper between the tool and work to sense they touch.

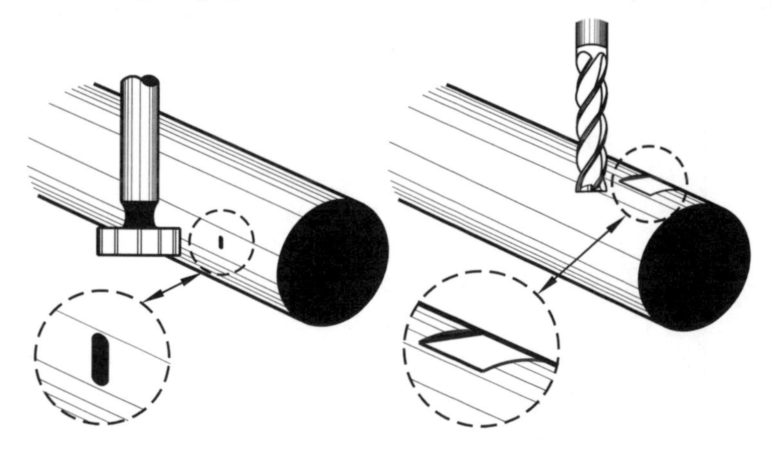

Figure 8–62. Finding the center of a short workpiece by marking the work with the milling cutter.

3. Touch the cutter to the work to make a small reference mark. This mark indicates the center of the work, Figure 8–62 (left).
4. Re-center the cutter over the reference mark.
5. Take the final cuts. Use plenty of coolant.

This method also works for locating the top of shafts, Figure 8–62 (right).

Section VII – Milling Machine Operations

End Mill Selection

What are the basic guidelines for selecting end mills?

- Choose the largest practical diameter end mill to maximize its rigidity, extend tool life and minimize chatter. Where possible, select a cutter large enough to span the entire surface of the part in a single pass. This will save machining time. Usually 0.75-inch diameter (19-mm) end mills are the largest size used in a Bridgeport-style milling machine, and 0.375-inch (9.5-mm) diameter end mills for the Sherline miniature mill. However, there are carbide insert cutters with 0.75-inch spindles for Bridgeport-style mills, but with 1-inch cutting diameters which are suitable for aluminum.
- If a lot of metal must be removed, use a roughing mill to remove most of the metal, and then finish with a regular cutter. Use the same spindle speed as on a regular cutter, but higher feed rates on the roughing cutter.
- If the workpiece is subject to bending from cutting tool forces, springing the work can be avoided by using a small cutter or supporting the work from behind the cut.
- Select two-flute end mills for soft materials (aluminum, brass and plastics) and higher feeds and speeds because they provide greater chip clearance. When plunge cutting is needed, a two- or three-flute end mill must be used. Use four-flute end mills for tougher materials, such as steel and cast iron, and reduce the speed and feed, or use higher feeds at the same spindle speed. Although they have less chip clearance, four-flute end mills are stronger and provide a smoother finish.

End Mill Cut Size

How much of the end mill should engage the work when side flutes of the end mill perform most of the cutting?

Limit the depth of cut to one-tenth the diameter of the cutter, Figure 8–63 (A), and make repeated passes to reach final depth. Deep side cuts put high bending forces on the tool that may result in a curved, rather than a straight cut. Also, too much side stress may snap off the cutter. Use conventional, not climb cutting.

If cuts are under 0.006-inches depth, the cutter may simply bend away from the work and not remove workpiece material. Plan work for a larger final cut.

When the end teeth of an end mill do most of the cutting, sometimes called end face milling, how much of the cutter should enter the work?

Best results are achieved with a cut depth of not more than 90% of the diameter of the end mill, Figure 8–63 (B). Use conventional cutting.

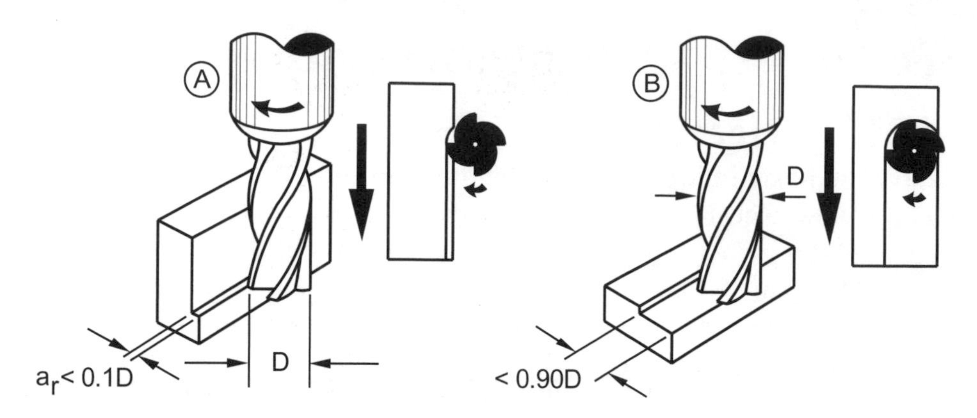

Figure 8–63. Guidelines for end mill use on side and end cuts.

When slotting with an end mill, what is the maximum recommended cut depth on a single pass?
Use a slot cut depth of not more than one-half the cutter diameter. See Figure 8–64.

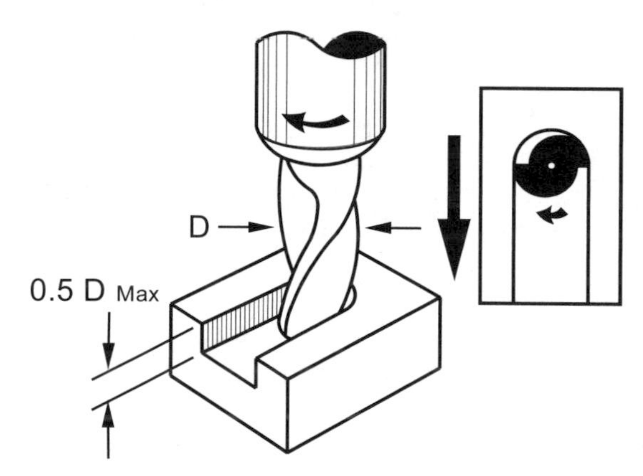

Figure 8–64. Cuts for blind slots, those without an open end, should not exceed a depth equal to one-half the diameter of the end mill.

When using a ball end mill for slotting, how deep should the cut be?
Half the diameter of the flat-bottom end mill is a good, conservative starting point, but about one-third the diameter for a ball-end mill.

To enlarge an existing hole, what is the maximum recommended depth of an end mill?
No more than three-quarters of the flute depth, Figure 8–65 (left).

When using an end mill to make plunge cuts, what is the maximum recommended cut depth?
For soft materials, make plunge cuts no deeper than the end mill diameter, and for hard materials, no more than half the end mill diameter, Figure 8–65 (right).

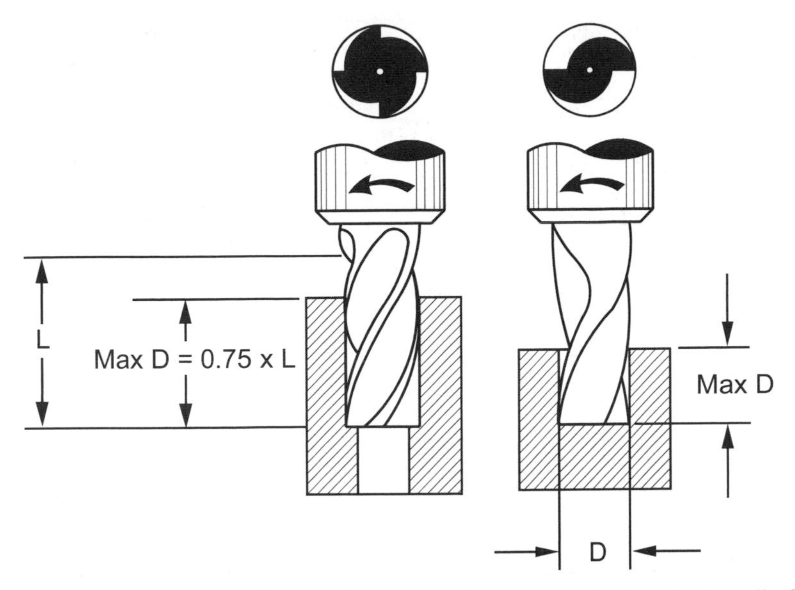

Figure 8–65. End mill practice for enlarging existing holes (left) and plunge drilling a new hole (right).

End Milling

What is the basic procedure for end milling?

1. Calculate the feed and speed for the cutter and workpiece material. Set the spindle speed on the milling machine.
2. Wipe all mating surfaces of debris on the end mill shank, inside the collet or end mill holder and inside the spindle.
3. Install the collet or end mill holder in the spindle and tighten the drawbar.
4. Install the end mill in its tool holder.
5. Secure the workpiece to the table, in a milling vise or in a fixture. Make sure it is firmly held and that the workpiece is aligned to the axes of the milling table. End milling produces large cutting forces on the work.
6. Though not essential, it is a good idea to mark out the center line and the width of the slot(s) needed on the workpiece. This will reduce errors.
7. Move the worktable and position the spindle so the cutter is ready to enter the work but will clear it when the motor is started.
8. If flood coolant is available, turn on the pump.
9. Gradually move the cutter into the work and begin milling.

Keyway Slot Milling

What is the procedure for milling a *keyway slot*?

1. Layout the keyway on the shaft to reduce the chance of errors. Use *keyseat clamps*, Figure 8–66, to keep the steel rule parallel with the shaft axis.

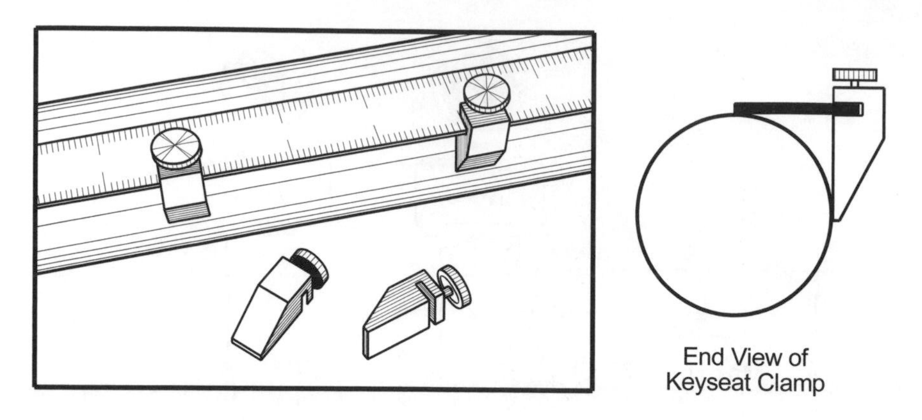

Figure 8–66. Keyseat clamps hold the steel rule parallel to the shaft axis for easy marking.

2. Use a two-flute end mill to permit plunge milling.
3. Position the milling cutter over the center of the shaft, plunge the cutter into the shaft until *both* sides of the cutter just touch the shaft, and set the z-axis collar or DRO to zero.
4. Cut to the final depth of the keyway slot, which equals one-half the height of the key, Figure 8–67.

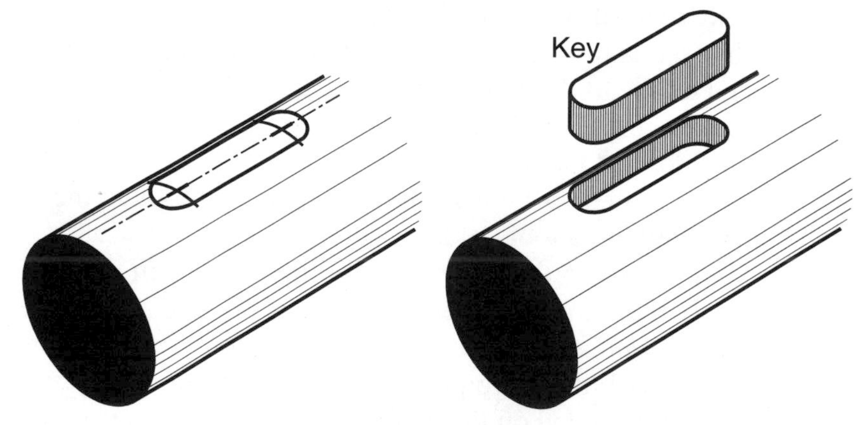

Figure 8–67. Marking out keyway (left) and completed keyway (right).

What are the three methods for locating and positioning a keyway milling cutter over a shaft?

Method 1: With the motor off, lower the spindle and move the milling table so that the side of the milling cutter just touches the side of the shaft. Make sure to do this slowly and gently using a piece of paper to sense touching. Then, raise the spindle so the end mill clears the shaft to be machined, and move the table one-half the diameter of the end mill plus one-half the shaft diameter. The spindle is now exactly over the mid-point of the shaft. See Figure 8–68.

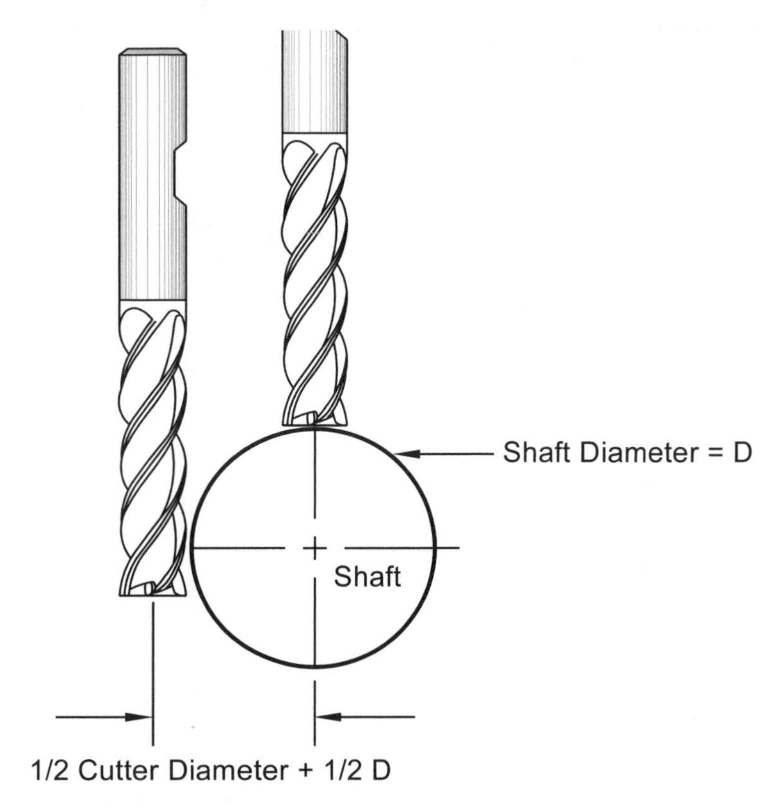

Figure 8–68. Locating the center of a shaft
for milling a keyway by Method 1.

Method 2: Use an edge finder to locate the edge of the shaft, then raise the spindle and move the table toward the shaft one-half the diameter of the edge finder plus one-half the shaft diameter. The spindle is now exactly over the middle of the shaft. This is very similar to Method 1.

Method 3: See Locating over Round Stock in Section VI and Figure 8–61.

Machining a Woodruff Key

What is a *Woodruff key* and what is its purpose?

A Woodruff key is a half-moon shaped piece of metal that fits into a mating depression in a shaft. This keys, or locks, the shaft to a gear, pulley, roller or other component that is to rotate with the shaft, Figure 8–69. Woodruff keys and their matching milling cutters come in standard sizes shown in *Machinery's Handbook*. The size of the cutter is stamped or etched on the cutter shank. These keys may be purchased ready-made or can easily be fabricated from round bar stock. Woodruff keys are popular because cutting slots for them is easier than machining a keyway for a square key.

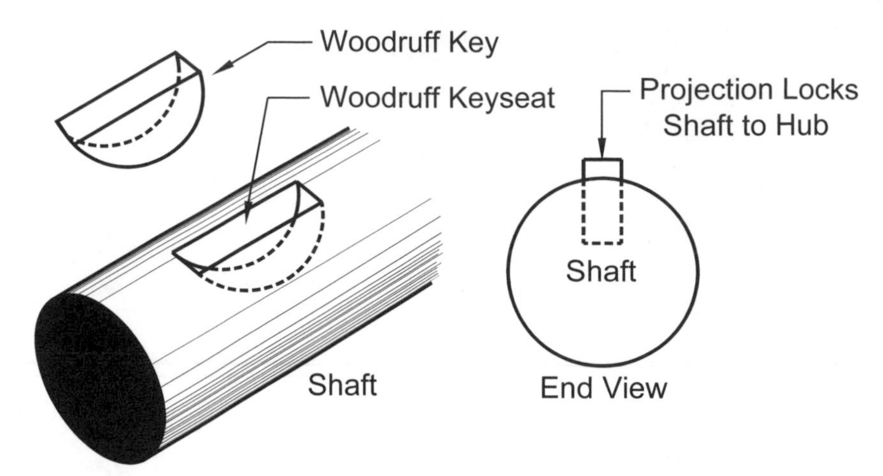

Figure 8–69. Woodruff keyseat.

What is the procedure for cutting a Woodruff key?

Note: These instructions are for a vertical knee mill, so the keyseat, also called keyslot, will be milled *horizontally* into the side of the shaft.

1. Tram the head to confirm that it is perpendicular to the milling table.
2. Mark out the position of the keyslot. Key dimensions are standardized and can be located in *Machinery's Handbook.* Keyslot sizes are often identified by their *key number*, which is listed in tables with the key's dimensions and also etched on the shank of keyseat cutter.
3. Secure the workpiece shaft horizontal and parallel to the milling table using V-blocks or in V-grooves in the faces of a milling vise.
4. Mount the Woodruff key cutter in the spindle and check spindle rotation direction.
5. To locate the center of the shaft, touch the top of the workpiece with the bottom of the cutter, Figure 8–70 (1st Position) and zero the Z-axis crank collar. Use a piece of paper or a 0.002-inch feeler gage to determine when the cutter touches the shaft.
6. Move the table horizontally so the cutter can be positioned alongside the shaft.
7. Raise the milling table one-half the diameter of the shaft plus one-half the width of the cutter to position the cutter in the middle of the shaft height.
8. Turn on the spindle and move the table to bring the cutter into contact with the shaft and zero the Y-axis (cross feed) micrometer collar, Figure 8–70 (2nd Position). As in Step 5, a strip of paper or feeler gage, may be used to determine when the cutter just touches the shaft.
9. Cut the keyseat to depth by using the cross feed micrometer collar. The depth is determined by the size of the key and the diameter of the shaft; see Table on ANSI Standard Woodruff Keys in *Machinery's Handbook.*

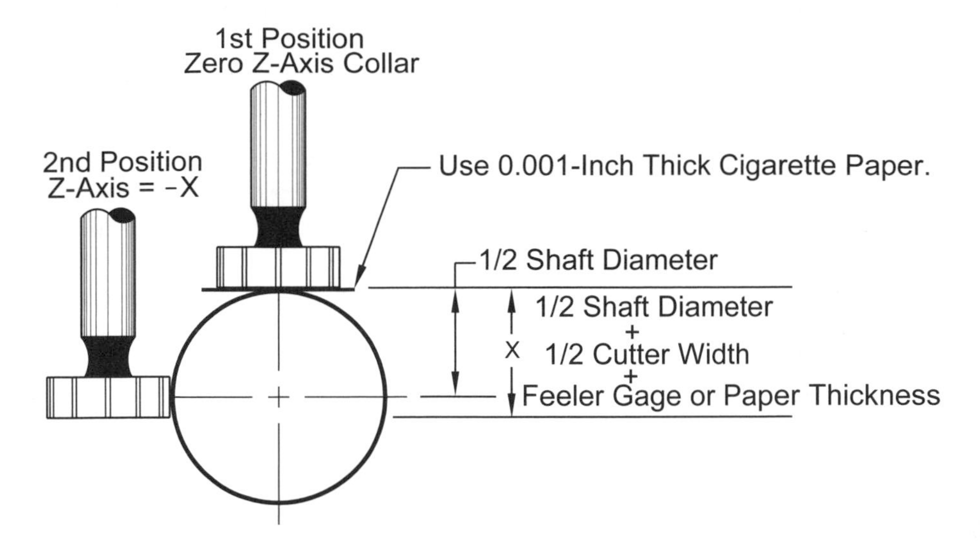

Figure 8–70. Locating cutter vertical position for a Woodruff keyseat.

There is an important and subtle detail regarding the depth of the pocket for the Woodruff key. While the nominal depth of the key pocket is half the height of the key, this "half-height" is measured from that point when the cutter is cutting its full width. Initially the cutter does not cut its full width because the cross section of the shaft is a circle so there is a small extra distance more than the "half height. This distance can be obtained from the tables in *Machinery's Handbook.* Alternatively, the machinist begins the X-axis cut, and when the cutter is *fully engaged across its entire width*, he sets the collar on the X-axis handwheel to zero and then cuts to the nominal depth.

Important: To avoid snapping off Woodruff cutters on their own swarf, direct a compressed air stream across the top of the cutters as they emerge from the cut to remove chips in their teeth. Do not let the cuttings re-enter the cut.

Flycutting

What is the procedure for *flycutting*?

Flycutting in a milling machine and turning in a lathe are similar processes, both of which use single-point cutting tools, so lathe speed and feed tables may be used for flycutting. Here is how to perform flycutting:

1. Wipe debris from mating surfaces on the flycutter and inside the spindle.
2. Install the cutting tool in the body of the flycutter and check that the set screws holding the lathe-type tool to the flycutter are tight.
3. Install the flycutter in the spindle and tighten the spindle drawbar.

4. Secure the workpiece in a milling vise clamped to the milling table or in a tooling fixture.
5. Lock the quill and position the flycutter so the cutter clears the work when the milling machine is turned on.
6. Determine proper spindle speed and set feed rate on table feed motor, if used.
7. Plan the cutter path so the flycutter sweeps over the entire width of the work in a single pass, if possible.
8. Raise the table to begin the initial cut. Rough cuts are in the 0.010- to 0.025-inch (0.3- to 0.6-mm) range, and finish cuts in the 0.003- to 0.010-inch (0.08- to 0.3-mm) range.
9. Engage horizontal table feed, if available, and confirm it is correct by observing the cut.
10. Repeat as needed to reach final dimensions.

See Figures 8-32 and 8-33.

T-Slots

How are *T-slots* cut?

1. Determine T-slots dimensions in *Machinery's Handbook.*
2. Use a conventional milling cutter to cut a vertical slot, Figure 8–71 (left).
3. Mount and use a T-slot cutter to undercut the slot, Figure 8–71 (center & right).
4. Use plenty of coolant and a very slow feed.

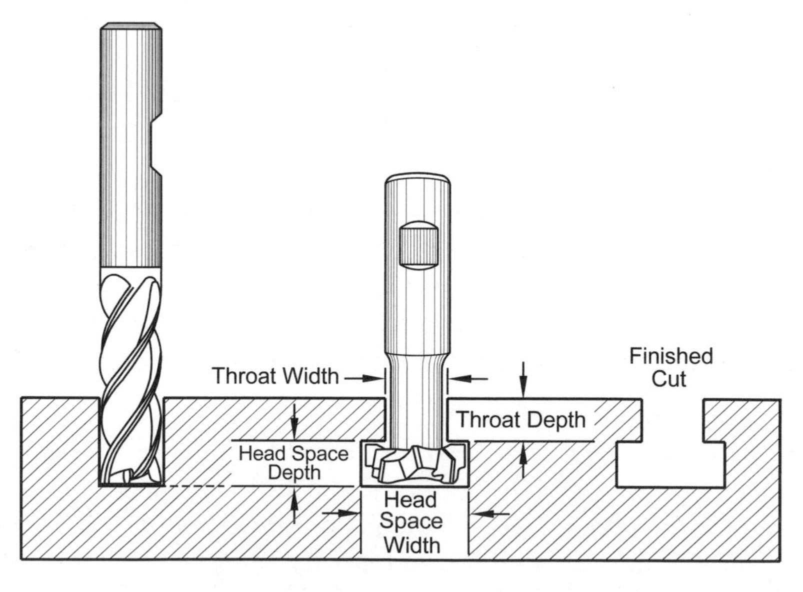

Figure 8–71. Milling a T-slot.

Boring on a Milling Machine

What is the procedure for *boring* on a mill?

Boring tools, Figure 8–31, can maintain bore diameter tolerances to within one-thousandth of an inch. Here is how to use them:

1. Position the spindle over the center of the bore using a wiggler or edge finder.
2. Lock the X- and Y-axes of the worktable.
3. Drill a pilot hole with a center drill.
4. Drill the starting hole centered in the pilot hole. This starting hole should be less than the final bore size, large enough to accommodate the tool in the boring head, and small enough to be easily drilled by the milling machine. Very large starting holes may need to be drilled in two steps.
5. Install the boring tool on the spindle.
6. Raise the table with the table crank to control boring depth. Do not use the quill feed as it is not as accurate.
7. Take a light initial cut of 0.010 to 0.015 inches (0.25 to 0.40 mm) to round up the drilled hole and provide a good start for subsequent cuts.
8. Check the bore size with a *telescopic gage* (unless a more expensive measuring device, like an Intramike® is available, see the Starrett catalog) and determine how much more the bore must be enlarged.
9. Make all remaining cuts, except the final cut, equal to 0.010 to 0.020 inches.
10. Make the final cut of 0.005 to 0.010 inches. Because of springiness of the mill, boring head and cutting tool, taking very small cuts often does not work well as the tool merely flexes and does not cut the expected amount. *You have to plan the size so your final cut is large enough.*

Dovetails

How are *dovetail* dimensions measured using the *two-rod method*?

To measure dovetails, a caliper, a calculator and two pieces of round stock or drill rod of the same diameter are needed. Place the rods against the sides of the dovetails. See Figure 8–72. Best accuracy results when the rods contact the side of the dovetail below and under the top edge as at point *e*. Where:

$$x = D\,(1 + \cot \tfrac{1}{2}A) + a \quad \text{and} \quad y = b - D\,(1 + \cot \tfrac{1}{2}A)$$

Warning: The two-rod method for measuring dovetails works for symmetrical dovetails only: Angle A = Angle B. Some imported machine tools have asymmetrical dovetails. Should you decide to make mating dovetails for them, you will have to devise a template or fixture to measure their dimensions.

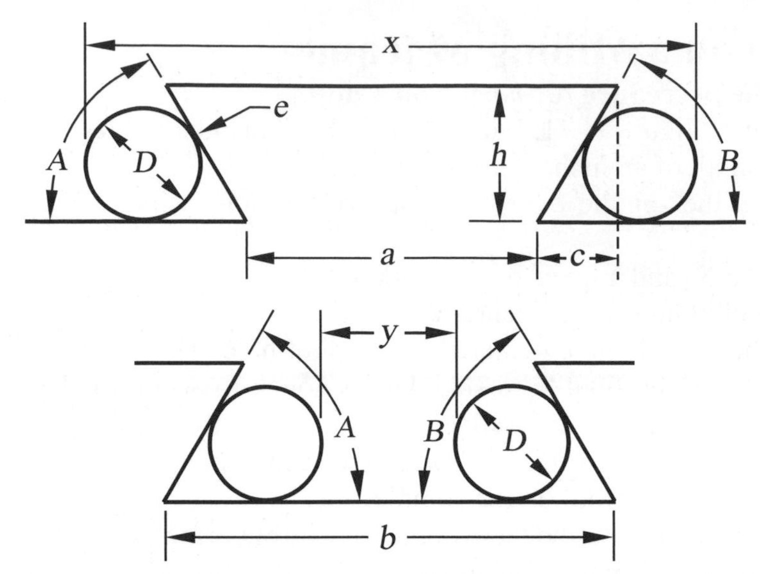

Figure 8–72. Measuring dovetails by the two-rod method.

How are *internal dovetails* milled?

1. Consult your workprint or check *Machinery's Handbook* for dovetail dimensions.
2. Mark out the edges and center of the dovetails as seen looking down on them from above and from the ends of the workpiece, Figure 8–73 (A). While not essential, these markings serve as a check on micrometer collar readings and reduce errors.
3. Secure the work to either a:
   - Milling vise.
   - Work holding fixture.
   - Milling table by clamping directly to it.
   - Rotary table, the preferred mounting method, because it will save measuring steps and reduce chances of error.
4. Align the workpiece so the axis of the dovetail parallels the X-axis, Figure 8–73 (B). The dovetail centerline is usually parallel with an external edge of the work.
5. Install an end mill that is slightly *smaller* than the finished width of the dovetail. Setting the Z-axis collar to zero, position the end mill over the dovetail center and locate the top of the work by gently touching the end mill against the work surface.
6. Move the end mill to clear the work, Figure 8–73 (C). And make repeated cutting passes until the cutting depth reaches 0.030- to 0.050-inch (0.8- to 1.3-mm) *deeper* than the specified dovetail depth. This will provide clearance for the dovetail cutter, and later, chips and debris. Lock the saddle (Z-axis) in position.

7. End mill the slot, Figure 8–73 (D).
8. Move the table along the Y-axis one-half the difference between the finished width of the dovetails and the existing, machined slot width. Make the cut with the end mill, Figure 8–73 (E). Check the width after this cutting pass. *Remember to remove table leadscrew backlash.*
9. Move the table over to the other side of the dovetail and make the finish end mill cut, Figure 8–73 (F).
10. Mount the dovetail milling cutter and adjust the saddle height so the bottom of the dovetail cutter just touches the top of the work and re-zero the Z-axis collar.
11. Move the table to clear the cutter and adjust milling table height to make a cut 0.005- to 0.010-inch (0.12- to 0.25-mm) *less* than the final dovetail depth, Figure 8–73 (G).
12. Determine the width of the finished dovetail and move the cutter to cut one side of the dovetail 0.010-inch (0.25-mm) *less* than the finished width. Make a rough cut on this side with the dovetail cutter, Figure 8–73 (H).
13. Move the same distance to the other side of the center line and make a rough cut for this side, Figure 8–73 (I).
14. Adjust the cutter to final depth and cut both dovetails to final depth.
15. Using the two-rod method, check the dovetail width.
16. Make the final dovetail width cuts equal to one-half the difference between the existing and finished depth, Figure 8–73 (J).
17. Check finished dimensions.

If the work is mounted on a rotary table, all cuts for the "other side" may be made by rotating the work 180° and leaving the table in the same Y-axis position. This reduces measuring steps and the chance of errors. An external dovetail is cut in a similar fashion.

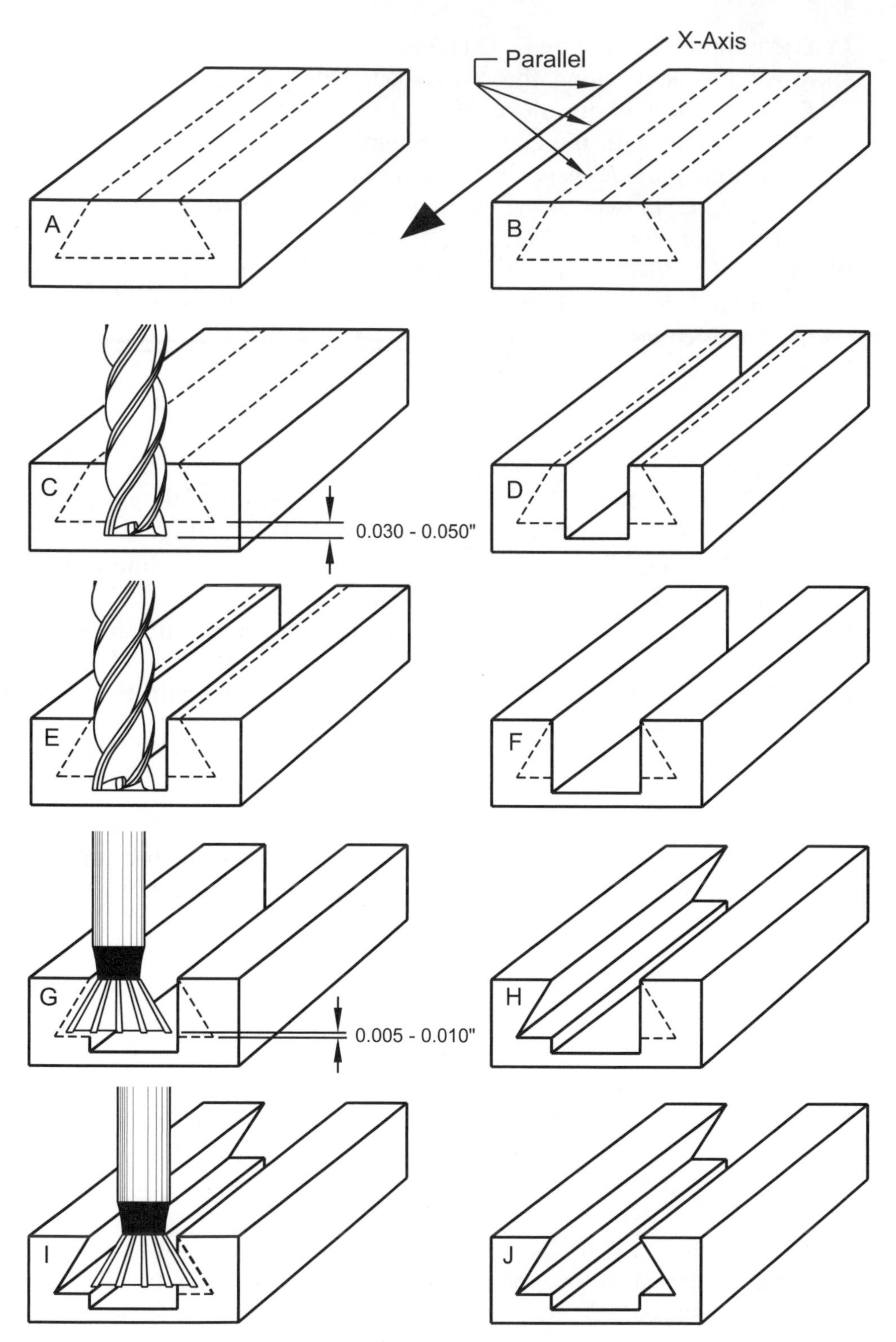

Figure 8–73. Cutting internal dovetails.

Drilling

What are the precautions for using twist drills in a milling machine?

- To prevent a workpiece from spinning on the end of a drill, firmly secure the work to the milling table, in a milling vise or in a fixture. *Warning: Never attempt to hold work by hand.*
- Use the quill lever or handwheel quill feed for small drills and the knee table feed (Z-axis) for large ones. The knee table feed offers more mechanical advantage and provides more force on the drill than can be safely applied with the quill.
- On Bridgeport-style vertical knee mills, the power quill feed may be used on drills up to 0.75-inches (19-mm) diameter. Use the manual vertical table feed for larger drill sizes.
- Remember that large drills can "dive" into soft metals such as brass and aluminum, or when breaking through the back side of thin materials just like a wood screw. Avoid this by locking the quill and using vertical table feed.
- Use the center drill with lubricant and the proper spindle speed. Holes are drilled very much the same as in a drill press. On stainless steel, do not let drills rotate in the bottom of the hole without cutting. This will work harden the work metal and burn out the drill. To avoid this, "bump" the drill against the hole bottom and apply moderate force to the drill to start it cutting again. When the drill does cut, reduce the force slightly.

Slitting & Slotting Saws

What precautions must be observed when using *slitting* or *slotting saws*?

- Make sure the saw is mounted so its teeth are pointing in the correct direction. They will cut only in one direction and will be destroyed in the other.
- Calculate and use the correct cutting speed. Too high a cutting speed will instantly destroy the saw.
- Coolant must be used to cool and to keep the small teeth from loading up with cuttings. While flooding is not needed, never let the saw run dry.
- When cutting into material with internal holes or other openings, residual stress may cause the sides of the slit to clamp down on the saw when the saw breaks through into the opening. Stop the saw, slip in a small wedge to keep the slit open, and continue cutting.
- On a slitting saw, usually the teeth are not perfectly concentric, so some teeth will begin cutting sooner than others, and as these higher teeth wear, the next highest teeth will begin cutting. This is not a problem, just use the proper cutting speed and light cutting pressure. You can confirm the presence of this condition by the uneven noise the slitting saw makes.

Squaring Up Stock

What procedure is used for *squaring up stock* in a mill?

Here is how it is done:

1. Using a file, disk or belt sander, remove any chips, burrs or dirt that might prevent seating the work in the vise.
2. Machine the largest side or surface first. Since the part in our example is rectangular, we place the work in the vise horizontally adding spacer(s) as needed to clamp the part rigidly, Figure 8–74 (A), and take one or more light cuts with a flycutter, face mill or end mill. When using a flycutter, set the tool to approach the workpiece so cutting forces are against the *fixed* jaw of the vise. One side of the work is now flat. Should the work be too small to fit into the vise and project above the vise jaws, use a pair of parallels to raise the work.
3. Remove the work and deburr it. Rotate it 90º counterclockwise so just the flattened face is against the fixed vise jaw, then re-clamp the work and make the cuts needed to flatten Side 2, Figure 8–74 (B).
4. Remove the work, deburr it, rotate it 90º counterclockwise and smooth Side 3, Figure 8–74 (C).
5. Remove the work, deburr it, rotate it 90º counterclockwise and smooth Side 4, Figure 8–74 (D). The four sides of the workpiece are now square.

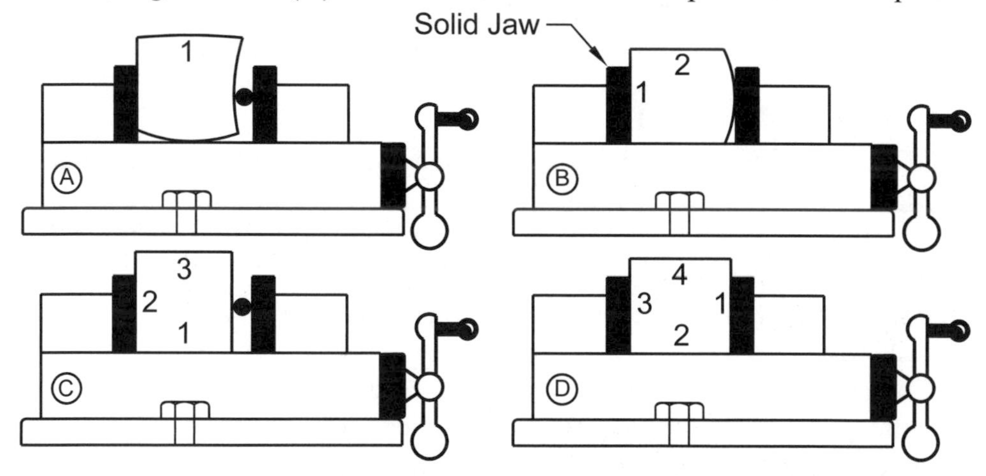

Figure 8–74. Squaring up stock. Always place the squared side of the workpiece against the solid jaw and the curved side against the movable jaw.

6. To square up the ends, deburr the part and secure it in the vise vertically using a solid square, Figure 8–75 (left). If the part is too long to be stable, place it in the vise horizontally and machine its ends square, Figure 8–75 (right). Another alternative is to clamp or bolt the work to a right angle plate, Figure 8–50. Check for squareness and deburr after each of these millings.

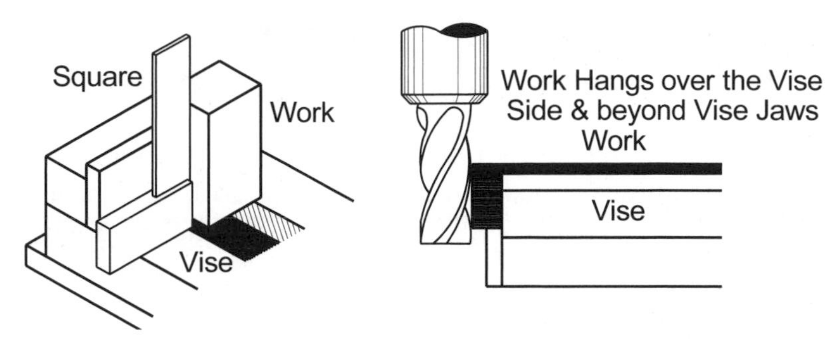

Figure 8–75. Squaring up stock in a milling vise.

Section VIII – Dividing Operations

Collet Blocks

What is the purpose of *collet blocks*?

Collet blocks hold round, square or hexagonal work in collets to position the work for milling or drilling operations. Collet blocks are clamped in a milling vise or other fixture and allow the work to be rotated at 60º, 90º or 120º increments. For these angle increments, collet blocks perform the function of a rotary table or indexing head, but they are much less expensive and offer less chance of operator error. See Figure 8–76.

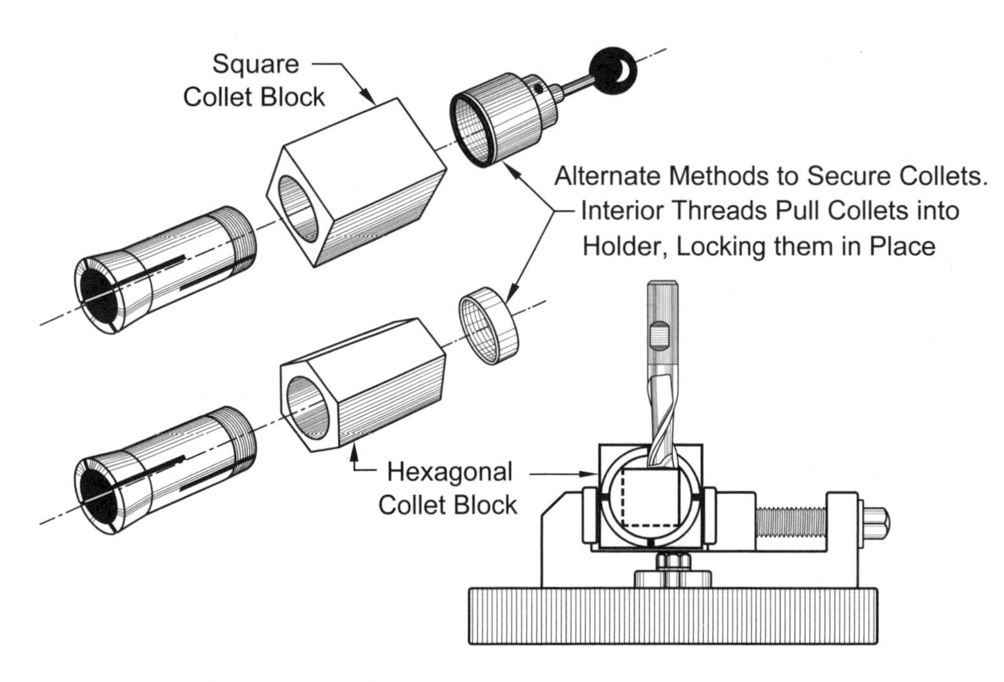

Figure 8–76. 5C collets, collet blocks and pull-up nuts.

Rotary Tables & Dividing Heads

What is the function of *rotary tables* and *dividing heads,* sometimes called *indexing heads*?

Rotary tables and dividing heads rotate the workpiece through one or more angles and hold it securely in position for milling or drilling. A worm or bevel gear takes many turns (typically 40, 72 or 90) to make the main work-holding gear—and the work—turn through 360º. Calibrations on the collar of the driving gear divide a circle into small, accurate increments. For example, the rotary table, Figure 8–77, has a 90:1 gear ratio and micro-collar graduated to one minute with a ten second vernier scale, or better than 200,000 parts. A full-size, high-precision dividing head divides circles into nearly 400,000 parts, Figure 8–78. Today most rotary tables and dividing heads have provisions to be used either with their axis horizontal or vertical. For most applications the resolution of these devices greatly exceeds the precision actually required.

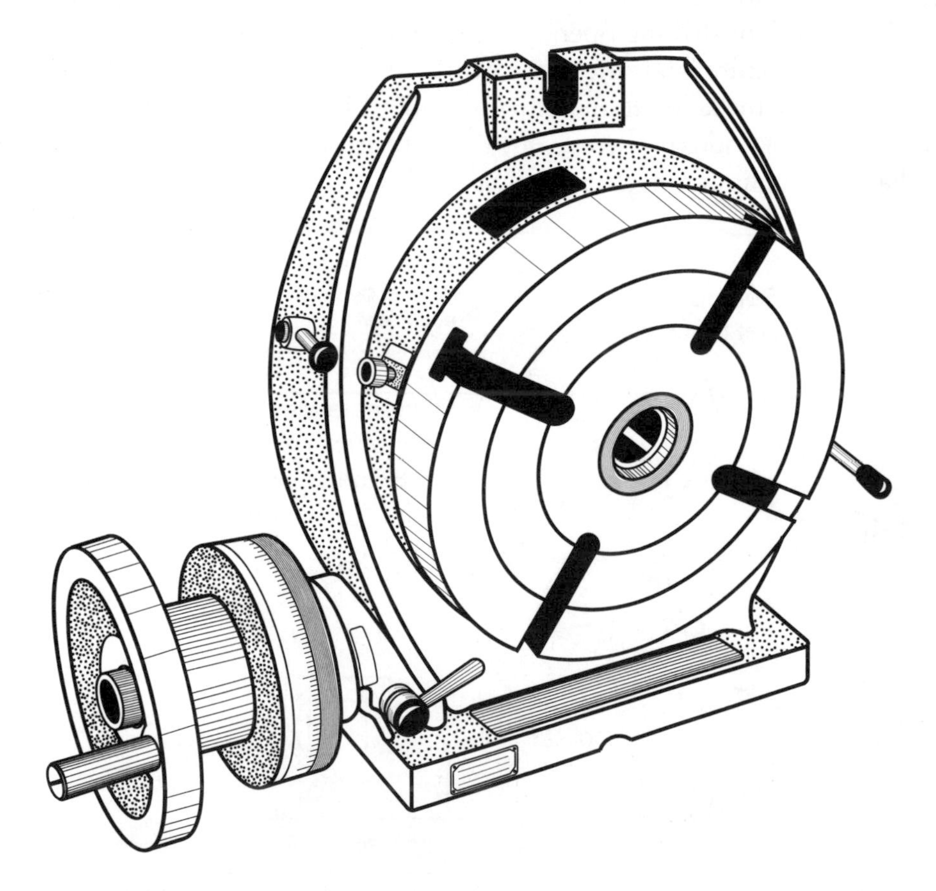

Figure 8–77. Phase II rotary table, available in 4-, 6-, 8-, 10-, 12- and 16-inch table diameters. Ninety handwheel revolutions turn the table one revolution.

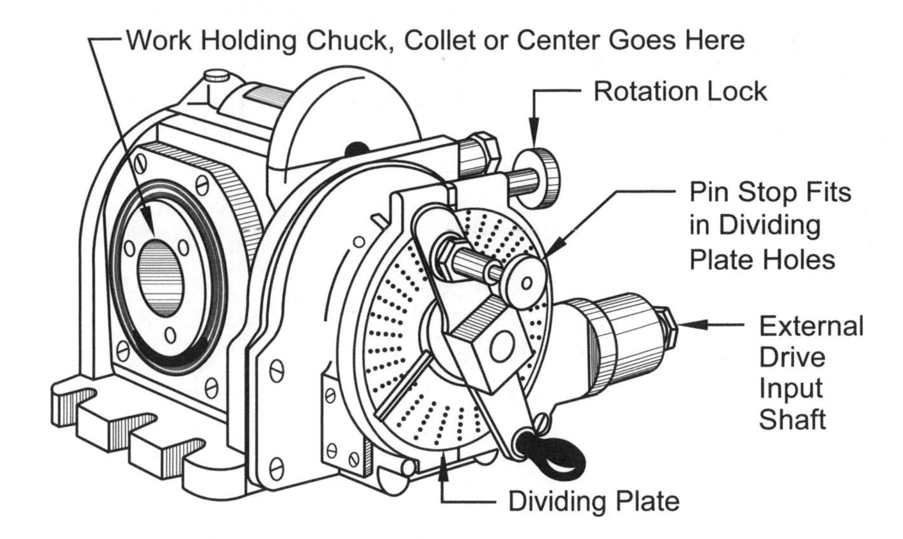

Figure 8–78. Typical dividing head with provision for an external drive from the milling table leadscrew for cutting a helix.

How is the workpiece held by rotary tables and by dividing heads?
Usually there are multiple provisions for holding work:

- Clamped into the T-slots in a rotary table or in T-slots in a faceplate adapted to the dividing head.
- Between centers with one center in the socket of the rotary table or dividing head and the other center held in a tailstock, Figure 8–79.
- Secured in a 3- or 4-jaw chuck adapted to the rotary table or dividing head with an adapter plate, Figure 8–80.

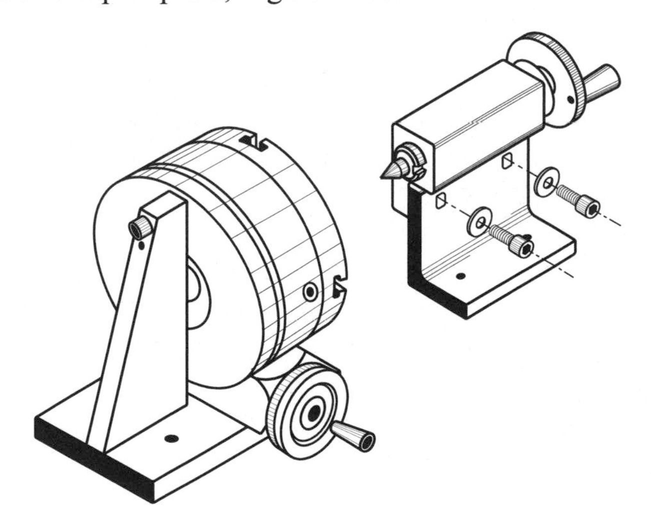

Figure 8–79. Sherline 4-inch rotary table mounted in a vertical bracket (left) and a tailstock (right) for holding work between centers with the rotary table.

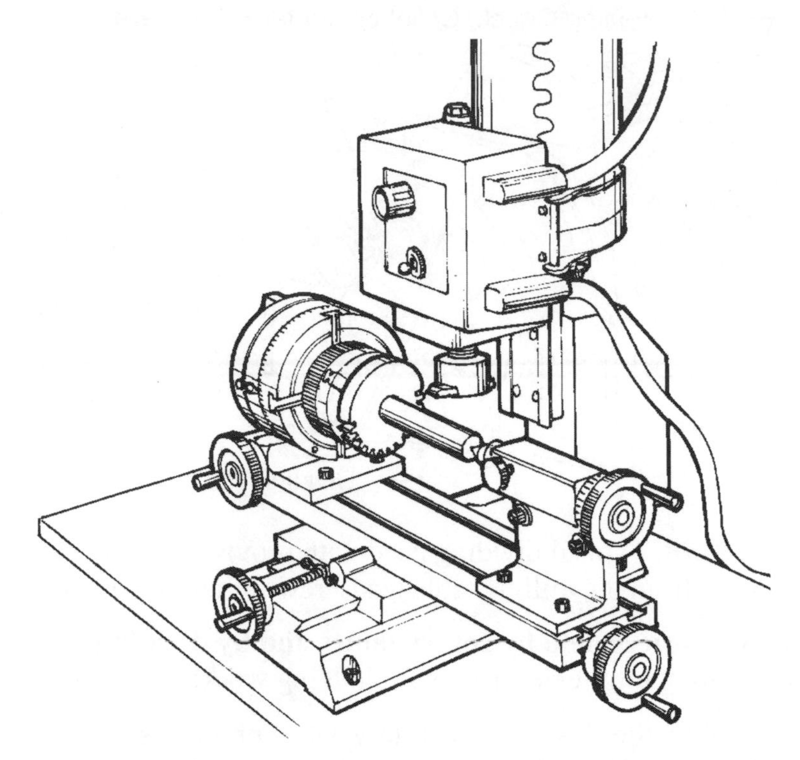

Figure 8–80. Sherline milling machine cutting a gear held between centers in a rotary table and tailstock. Note the 4-jaw chuck.

- Held in a collet that fits into a taper socket in either the rotary table or dividing head. This socket may be a Morse taper or a Brown & Sharpe (B&S) machine tool taper depending on the manufacturer and the time the device was manufactured. Older machines tend to be B&S, newer ones are often MT. Using a collet to secure work insures precise, repeatable positioning with the axis of the table or dividing head.
- Custom fixture to position a part accurately and easily.

Angle Setting Methods

How are angular increments set on a rotary table or dividing head to divide a circle into equal parts such as when drilling a bolt-hole pattern or cutting a gear?

There are several ways:

- *Using the graduated collars on their handwheels* makes calculating the angular increment easy. It is simply 360° divided by the number of divisions. Also, collars usually have vernier markings for added accuracy. The hard part is accurately advancing the collar and handwheel when the increment contains fractional degrees or worse, degrees, minutes and seconds. Getting this right every time is difficult. Imagine incrementing the table 103 times (to cut a 103-tooth gear) at the rate of 3.495° for each step and getting the increment correct every time.

- *Using a dividing,* or *indexing plate,* Figure 8–78, is another method. For example, if the dividing head had a gear ratio of 40:1, one complete turn of the crank handle would revolve the head 360°/40 or 9°. If a 24-hole pattern on the plate were used, pin movement from hole to hole would revolve the head 9°/24 or 0.375°. One complete crank revolution plus an increment of the hole pattern will produce the desired angular increment without the need for using the collar and its vernier. This increases the chance of making the right increment.
- *Using a microprocessor-driven stepper motor* greatly simplifies incrementing the angle settings. A press of the button moves the head or table to its next position. Figure 8–81 shows a Sherline motion controller and stepper motor on a rotary table.

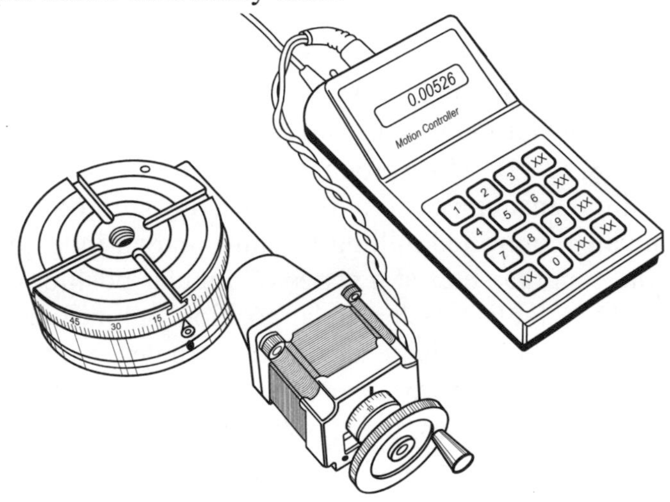

Figure 8–81. Sherline motion controller and stepper motor on a rotary table.

What additional capabilities do dividing heads and rotary tables add to a milling machine?

Whole new classes of parts can be made, many of them essential to modern life. Going from the simple to the complex, here are some of them:

- *Bolt-hole patterns* on flanges and hatches, Figure 8–82 (A) and *weight-reducing hole patterns* on flywheels, gears and pulleys, Figure 8–82 (B). These patterns of symmetrical holes can be applied other ways, but a rotary table is the fastest and most accurate method.
- *Partial circles* or *arcs* in flat stock, Figure 8–82 (C), are made with a rotary table and an end mill.
- *Spokes* in a gear or wheel, Figure 8–82 (D). A milling machine with a rotary table is the fastest way to cut these.

- *Splines* on a shaft, Figure 8–82 (E), lock other components to the shaft so all parts rotate together. Cutting these requires a dividing head to position the shaft.
- *Polygons:*
  - Milling one or more flats on just the end of rounds, as shown in Figure 8–76 (bottom), using collet blocks is also easily done. A dividing head fitted with a collet will also work well.
  - Milling flat stock or plate, Figure 8–82 (F, top), is usually best done with a rotary table.
  - Milling rounds into shapes with an equal number of flat sides such as pentagons, hexes or octagons, Figure 8–82 (F, bottom), is best done with a dividing head and tailstock.
- *Ratchets,* Figure 8–82 (G). These are best cut with a dividing head and tailstock.
- *Spur* and *bevel gears* require a rotary table or indexing head. Spur gears, Figure 8–82 (H), are relatively easy to cut as are bevel gears, Figure 8–82 (I). Setting the angular increments correctly is the biggest problem.
- *Cams,* Figure 8–82 (J). Most production cams are made with computer-controlled milling machines and indexing heads, but prototypes can be cut using a manually controlled milling machine and an indexing head. First, determine the cam diameter at a series of angular positions, then use the indexing head to establish each angular position and mill the part to match these dimensions. The cam is finished with a grinder or a hand file.
- *Helices,* Figure 8–82 (K & M), require an indexing head and a tailstock. However, to make these cuts, the straight-line motion of the milling table must be coupled to the rotary motion of the work. Connecting a leadscrew, usually on the X-axis to the indexing head through a gear train, couples these motions, Figure 8–83. Although this drawing shows a miniature Sherline milling machine, the biggest milling machines work exactly the same way. The illustration of a full-size, industrial dividing head, Figure 8–78, shows the input shaft for leadscrew gearing on the lower right-hand side of the head. Worm gears, twist drills, helical end mills, helical cams and helical taps are cut this way. A large percentage of horizontal milling machine time is devoted to helix cutting operations. *Machinery's Handbook* thoroughly discusses how to calculate change gears for helical milling and includes extensive change gear tables.
- *Spirals*, Figure 8–82 (L), also require coupling between milling table motion and a rotary table. The spiral illustrated is from a 3-jaw universal chuck.

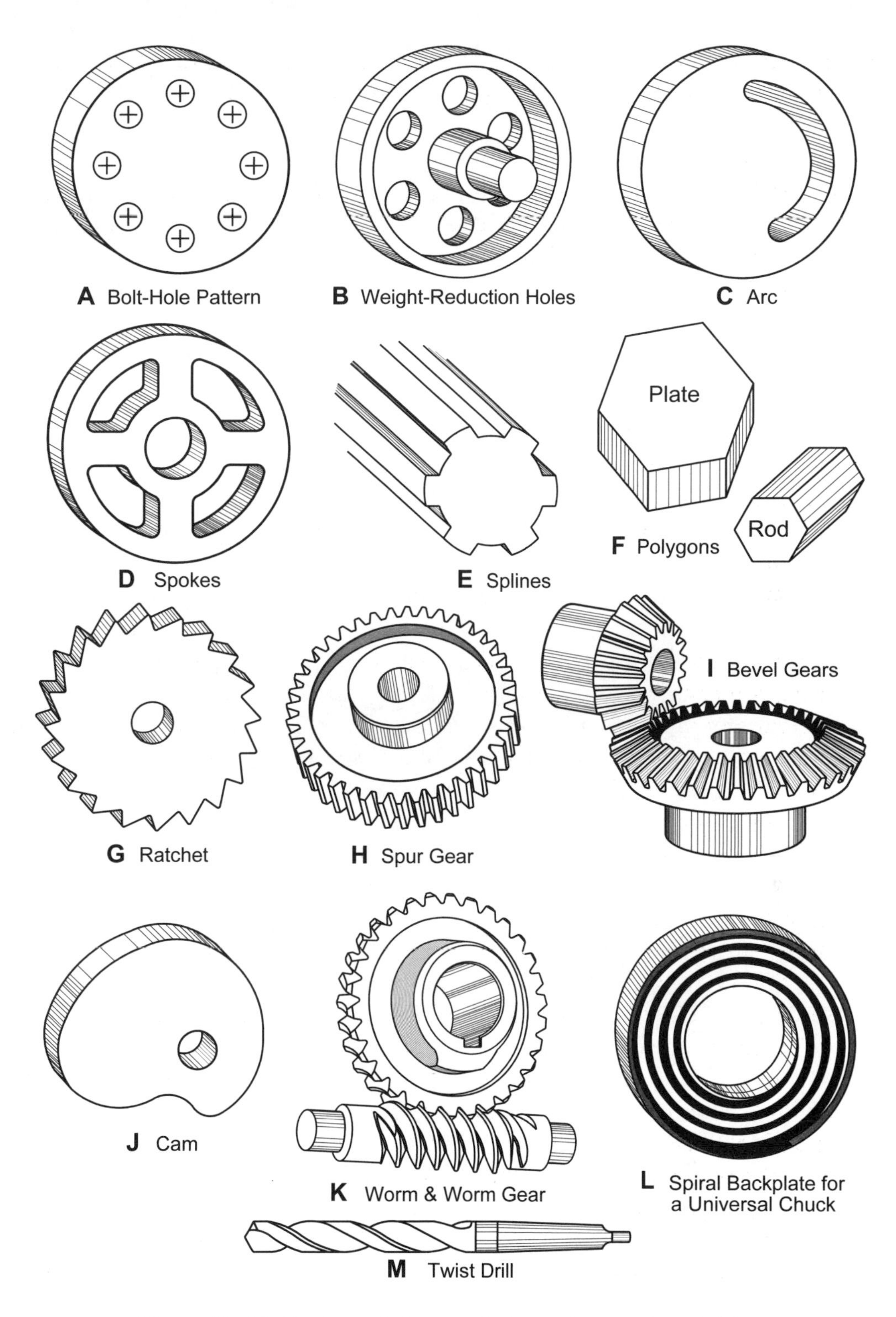

Figure 8–82. Parts and shapes that require a rotary table or dividing head and milling machine.

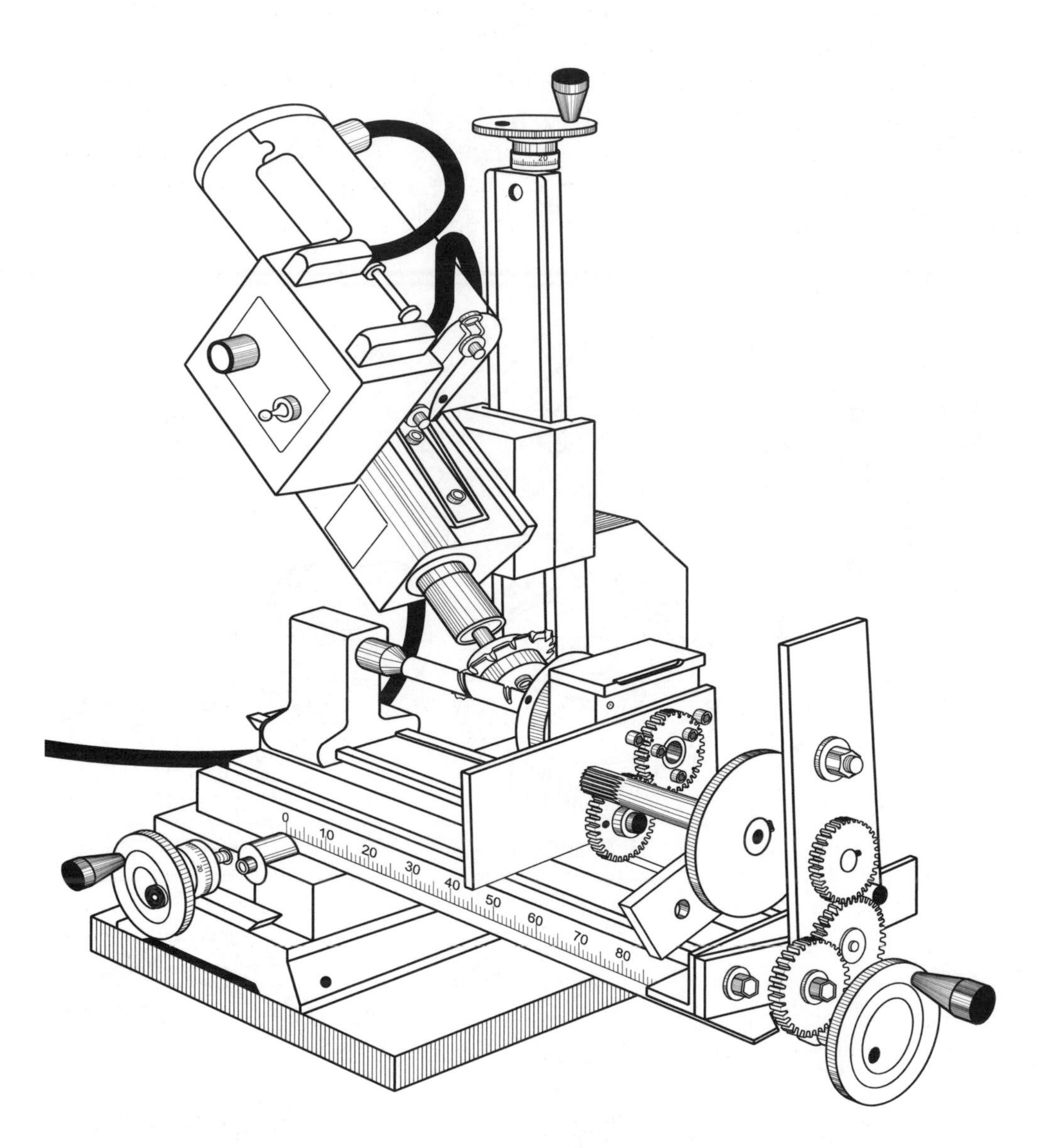

Figure 8–83. Coupling horizontal table motion to rotary table and part rotation through gearing enables this milling machine to cut a helix.

Bolt-Hole Patterns

What are *bolt-hole patterns,* also called *bolt-hole circles,* and how are they located in a milling machine?

Bolt-hole patterns are a series of holes arranged symmetrically around a circle, Figure 8–84, and are used to bolt hatches, covers and ports over openings in vessels, usually to contain pressure. Bolt-hole patterns are also used for joining flanges. The number of holes varies from 3 to more than one hundred for very large work, but most jobs require fewer than a dozen holes.

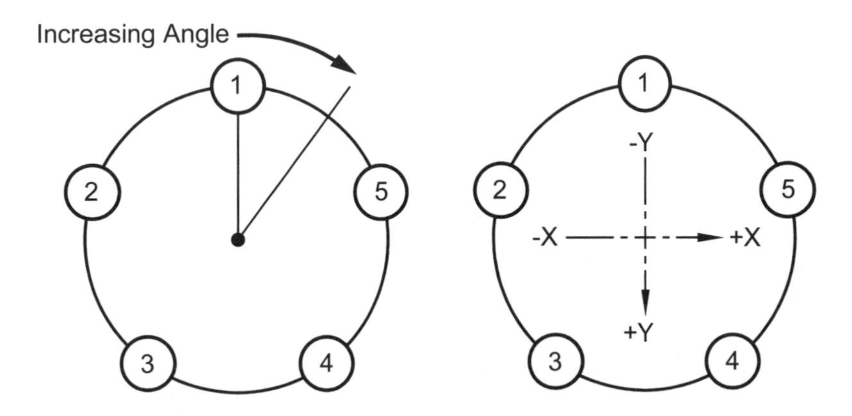

Figure 8–84. Drilling a bolt-hole pattern with a rotary table (left) and with X- and Y-coordinates (right).

There are two methods to locate the position of these holes using a milling machine:

- The easiest method uses a rotary table. For a pattern with N holes, where N is the number of holes, the holes will be 360/N degrees apart. Table 8–8 (columns 1 & 2) shows the angles for a five-hole pattern. Angles in degrees are set directly on the rotary table.
- Without a rotary table, use rectangular coordinates on the X- and Y-axes of the milling table to determine hole position.

The *Jig Boring Section of Machinery's Handbook* has tables for 3- to 28-hole patterns on a circle with a diameter of 1.000 in. Table coordinates need only be multiplied by the diameter of the circle required.

Table 8–8 (columns 1, 3 & 4) shows the rectangular coordinates for a five-hole pattern.

Hole Number	Angle from Hole Number 1 (clockwise in degrees)	X-Coordinate Factor	Y-Coordinate Factor
1	0	0.00000	−0.50000
2	−72 or 288	−0.47553	−0.15451
3	−144 or 216	−0.29389	+0.40451
4	−216 or 144	+0.29389	+0.40451
5	−288 or 72	+0.47553	−0.15451

Table 8–8. Angles and rectangular coordinates for a five-hole pattern.

Section IX – Horizontal Milling

How is horizontal milling performed on a vertical knee mill?

A horizontal milling attachment, Figure 8–85, installs on the milling machine and consists of three main components, the:

- *Right-angle drive* clamps over the quill, couples to the milling machine spindle with its own internal R8 collet, and turns a spiral bevel gear set to drive the horizontal arbor. A slight speed reduction through the drive increases torque. In addition to horizontal milling with the arbor supported at both ends, where size is a factor, such as on jobs where milling must be performed on the *inside* of a rectangular box-shaped workpiece, the right-angle drive can be used with a stub arbor, which is supported at only one end.
- *R8 milling machine arbor* fits into the right-angle drive on one end and into the arbor support on the other. A series of arbor spacers and shims fitted over the arbor provide lateral support to the milling cutters as they are tightly clamped in place. This holds the cutters in the exact location required. A milling cutter replaces an arbor spacer of the same size, Figure 8–86.
- *Arbor support* fits onto the ram's dovetail and with its end bearing provides rigid support to the arbor.

How effective is the horizontal milling attachment?

While the horizontal milling attachment is *not* a replacement for a heavy-duty C-frame milling machine with a 20- to 50-horsepower motor, it is often a time and money saving tool, especially for tasks where cutters on a horizontal arbor are the best way to remove metal quickly. But, in addition to starting with only a 3-horsepower motor on vertical knee mills, power is lost in the horizontal milling attachment's right-angle drive.

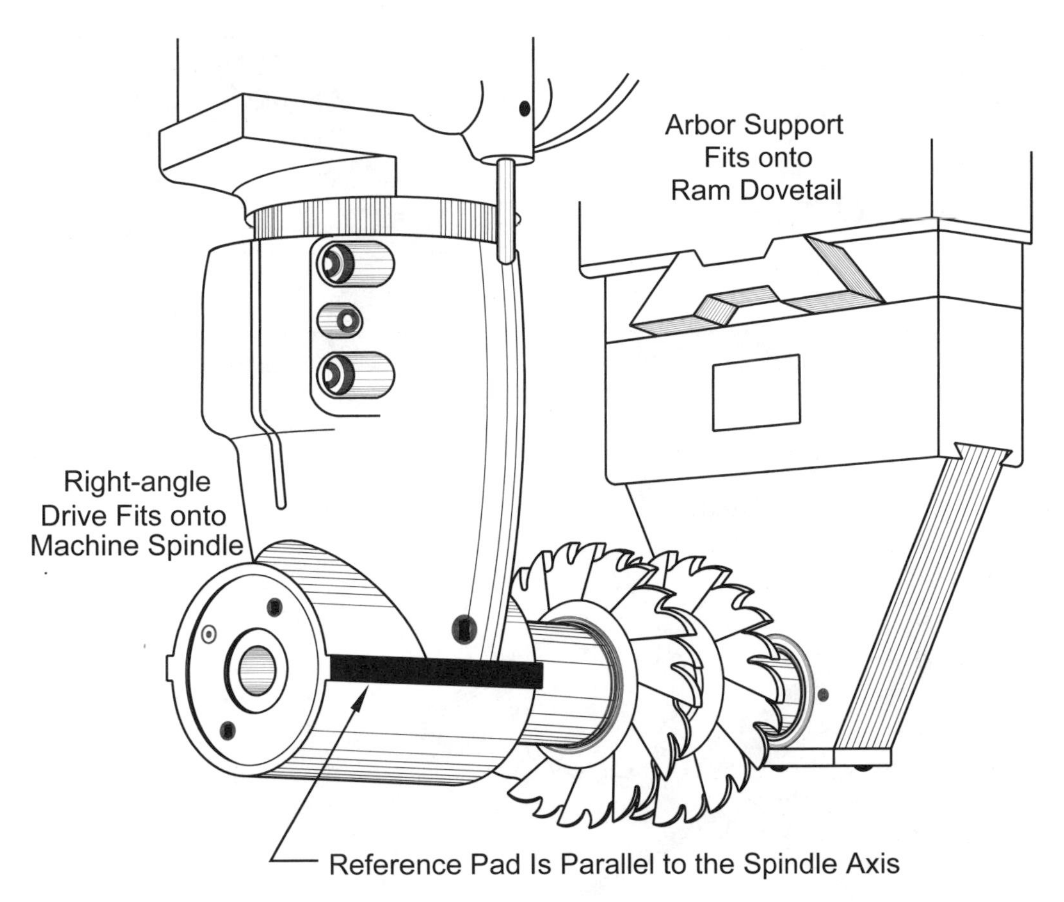

Figure 8–85. Horizontal milling attachment.

What is the *reference pad* on the right-angle drive spindle used for?
The reference pad, or surface, located on the side of the right-angle drive housing, is parallel to the spindle axis of the accessory, Figure 8–85. This reference pad is used to align the spindle axis to the Y-axis of the milling machine.

How is this alignment done?
Using a magnetic base, secure a dial indicator on the milling table and place its measuring leg against the reference pad. Rotate the right-angle drive on the milling machine spindle until the indicator does not move when the table is moved in and out on the Y-axis. The spindle on the right-angle drive is parallel to the Y-axis and cutters on the arbor will be parallel to the X-axis. Secure the bolts on the top of the right-angle drive so it will not rotate.

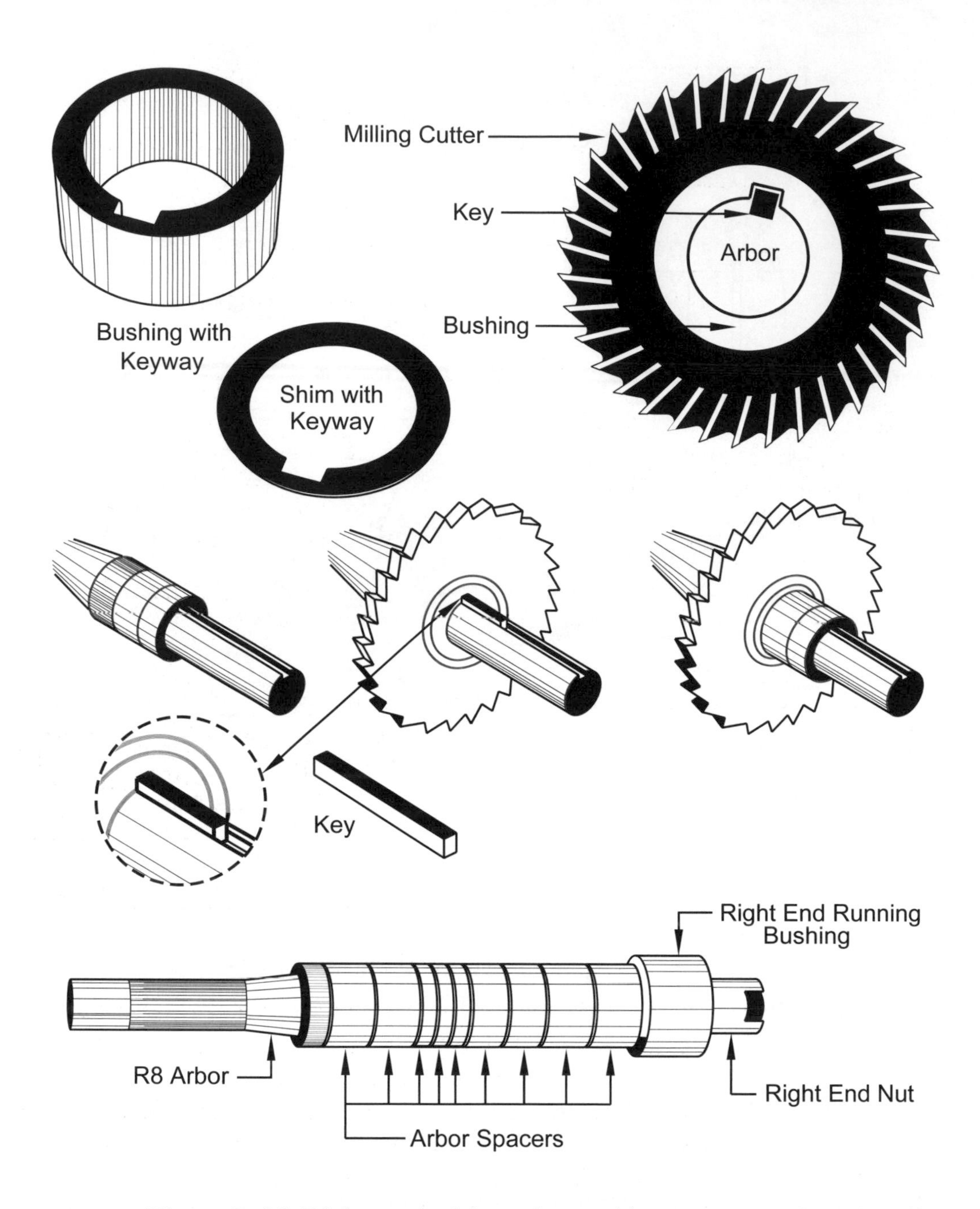

Figure 8–86. Right-angle drive arbor and its components for horizontal milling attachments.

What operations can the horizontal milling attachment perform?
Any operation that uses an arbor supported at both ends can be done with this attachment. Figure 8–87 shows typical operations.

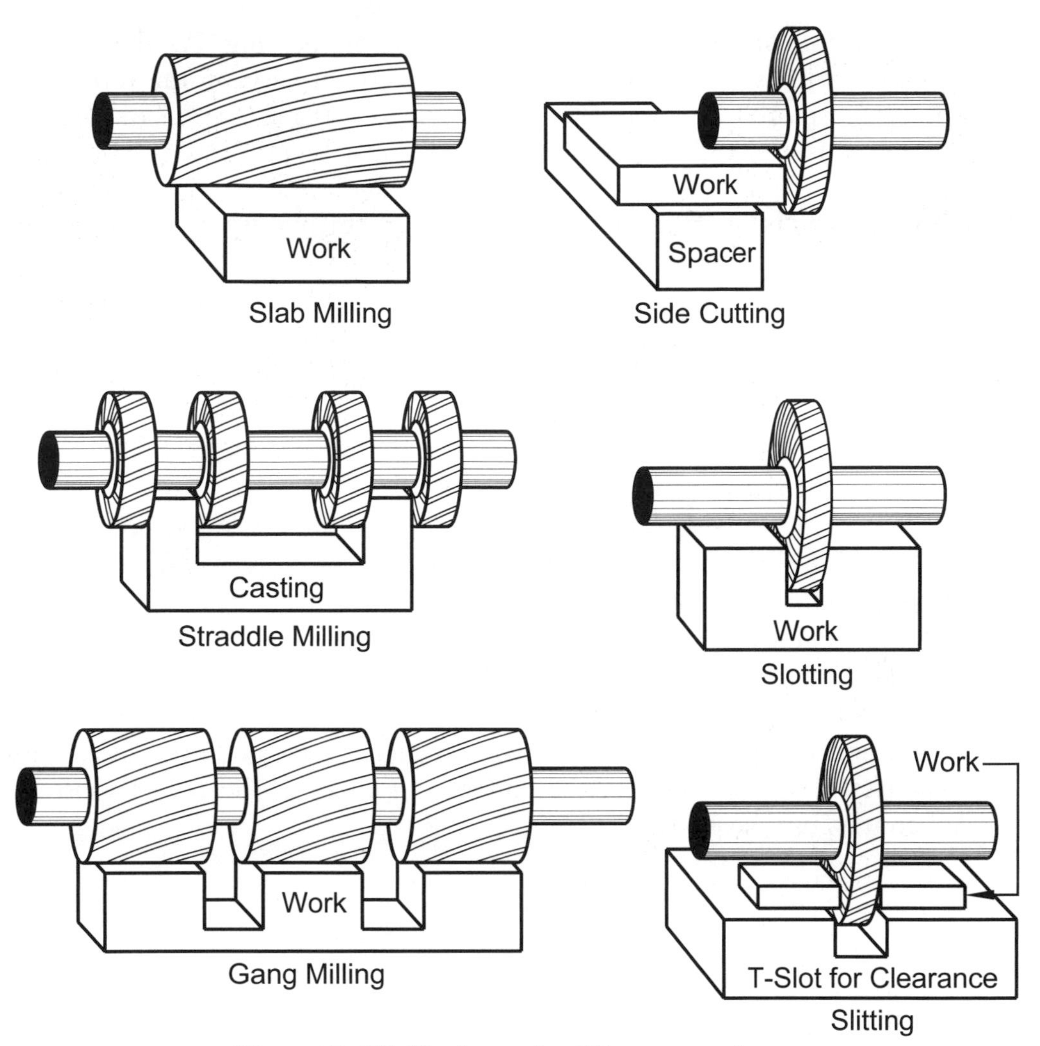

Figure 8–87. Horizontal milling operations.

How is the horizontal milling accessory set up in a vertical knee mill?

1. The right-angle drive is slipped onto the quill, pulled up and held in place by a special drawbar. The drive is also aligned to the spindle.
2. Using a DTI along the reference surface, verify that the arbor is parallel to the Y-axis. When the head is parallel with the Y-axis, the DTI will not move as the table is moved in and out along this axis.
3. Milling cutters are slipped onto the horizontal arbor. Each cutter put on the arbor replaces an arbor spacer of the same size. Assorted thin shims are used to make small position adjustments of the cutters. A key is installed to drive each cutter as the arbor turns, Figure 8–86. A right-end nut on the arbor holds the spacers and cutters in place.
4. The right-end arbor support is installed on the ram dovetail.
5. The outside, or left-end of the arbor, is inserted into the right-angle drive,

the right end of the arbor slips into its support bearing. As the milling machine head is moved back towards the vertical base on the turret, the arbor is captured between the right-angle drive and the arbor support bearing. It is now rigidly supported at both ends.

What precautions should be observed when setting up and using the horizontal milling attachment?

- Take care to prevent dirt and debris from getting between the arbor spacers and the cutters. They must seat together solidly.
- Determine and set the proper spindle speed. Remember that most horizontal milling attachments have some speed reduction in the right-angle head, so the indicated speed on the milling machine head will not show the true speed of the horizontal arbor.
- Make sure the work is firmly secured. Climb milling is a good choice for horizontal milling because it exerts large downward force on the work and tends to hold it in place. Very large cutting forces, seen in horizontal milling, cannot only ruin work, they can be dangerous. When slitting flat stock, clamp the stock to the milling table and let the cutter project below the work into the T-slot space.
- Determine the feed rate for the work material and cutter(s). Begin by manually feeding work into the cutters and check if the feed is correct. When it is, use the power feed.

Section X – Safety

In addition to normal machine shop safety precautions, what other steps are necessary when using a milling machine?

- Always wear safety glasses in the shop.
- Learn where your machine's emergency stop button is located.
- Keep your hands and arms away from the revolving cutters.
- Never make adjustments or take measurements on the machine while the cutter is in motion.
- Make sure the work-holding device and the work are securely fastened.
- Avoid talking while operating a machine tool and do not allow anyone else to turn your machine on for you.
- When installing milling cutters, hold them with a rag to avoid cutting your hands.
- When setting up work, install the cutter last to avoid being cut on their sharp edges.

- When using large twist drills, lock the quill and raise the table to drill the work. Large drills can draw the quill into the work and jam the drill.
- Learn what your milling machine sounds like when cutting properly and shut it down when the noise is different or there is excessive vibration.
- Make a plywood or wooden tray on which to keep tools to protect the milling table.
- Do not wear gloves when operating a milling machine as they may draw your hands into the machine.
- Get help or use a hoist to move heavy work on and off the milling table.
- Do not use compressed air to clean out T-slots. Make and use a chip rake, as in Figure 8–88, to clear the T-slots on your milling machine.

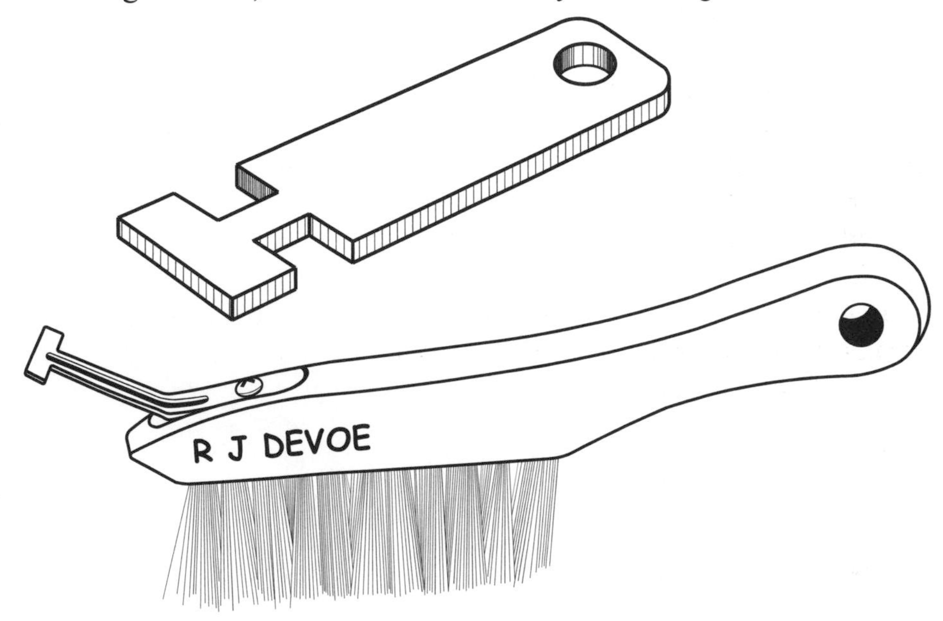

Figure 8–88. Use a T-slot cleaner or a chip brush like these to avoid hurting your fingers when removing chips from the milling table.

- When cutting Woodruff keyslots, direct a compressed air stream on the cutter teeth as they emerge from the slot. This will blow away the emerging chips and prevent them from jamming in the work on the next cutter revolution and snapping off the cutter.
- Do not place anything except the work on the milling machine table as it can become a dangerous distraction when it moves about from machine vibrations.
- Check that shaft-mounted milling cutters are mounted to turn forward—in their cutting direction. Running milling cutters backwards instantly destroys them.

- Never leave a running machine unattended without extensive precautions: the limit stops must be set, chips must be contained and others warned that the machine is running.
- When applying cutting oil manually, put it on the cutter teeth as they emerge from the work. Prevent the oil can spout from coming in contact with the cutter.
- Remember the peculiarities of Bridgport-type milling machines: Changing from the HIGH-SPEED range setting to the LOW-SPEED range setting *reverses spindle direction, and the machine's motor must be set to run in reverse to make the spindle run forward*—clockwise as viewed from above.

Chapter 9

Fastening Methods

The man who makes no mistakes does not usually make anything.
—Edward John Phelps

Introduction

Most machine shop time is spent turning raw materials into finished parts, but the next manufacturing step is often assembling and fastening these parts together. This chapter looks at both fastening *devices:* screws, bolts, pins and rivets, and fastening *processes:* adhesive bonding, soldering and welding. Cost, weight, strength, availability, reliability, corrosion resistance and simplicity are some of the many factors that influence the fastening method used by the prototype or non-production machinists. There are over one hundred different fastener designs and tens of thousands of variations when size, finish and material are counted. This chapter presents the most common fastening methods. Information on other methods can be obtained from *Machinery's Handbook*, industrial tool supply catalogs and manufacturers' product literature. These sources often include excellent application notes.

Section I – Threaded Fasteners

Parts of a Bolt

What are the parts of a bolt?

See Figure 9–1.

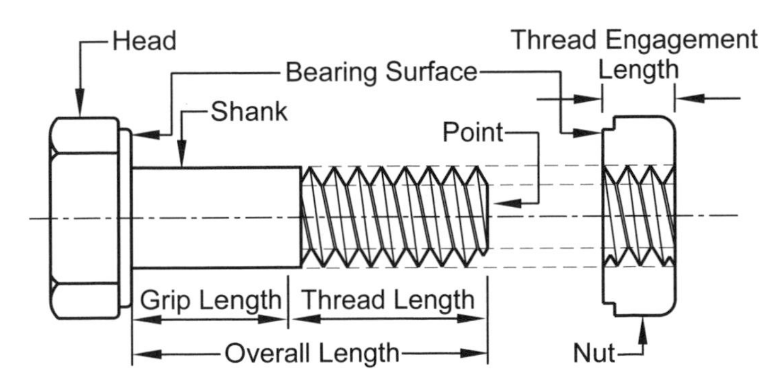

Figure 9–1. Bolt nomenclature.

Screws vs. Bolts

What are the differences between *screws* and *bolts*?

- *Screws* are externally threaded fasteners capable of being inserted into holes in assembled parts, of mating with a preformed internal thread, of forming their own thread and of being released by torquing their heads.
- *Bolt*s are externally threaded fasteners designed for insertion through holes in assembled parts and are normally intended to be tightened or released by torquing a nut.

These definitions are from ANSI-ASME B18.2.1 1981 and *Fastener Standards*, 6th Edition, International Fastener Institute, Independence, OH 44131 and are recognized by the U.S. Government.

What is the difference between *drive design* and *head design*?

Drive design is the external or internal shape on the fastener head that couples the screwdriver, socket wrench or other driving tool to turn the fastener. The head design is the overall shape of the head that makes the fastener better suited to a particular application such as a round, flat, oval or fillister head. However, there is some overlap: all external hex drives are hex heads, but internal hex drives, also called Allen heads, may take a variety of shapes. Even though they are commonly called hex or Allen heads, these two terms are imprecise.

Screw Drive Designs

What are the most common drive designs for screws?

The most common designs are shown in Figure 9–2:

- *Slot heads* have a single slot and are driven by a flat-bladed screwdriver. Their main drawback is that they do not work well with power screwdrivers since flat blades have a tendency to come out of the slot and damage surrounding work. Slot heads are used on screws and small bolts.
- *Phillips® heads*, sometimes called *cross-head screws*, have a "+"-shaped slot. They were originally designed for use with power screwdrivers in production. The rounded corners in the tool recess make the driver *cam out* when the fastener is tight and also makes unscrewing difficult. Phillips heads are used on screws and bolts.

Figure 9–2. Common screw and bolt drive designs.

- *Combo heads* accept either a Phillips screwdriver or slot-blade driver. The National Electric Code now requires this head on all wiring devices, switches, outlets and the like. Combo heads are only used on screws.
- *Pozidriv*® heads are similar to Phillips heads, but have more metal-to-metal contact to permit higher torque application without camming out. Phillips screwdrivers will usually work in Pozidriv screws, but Pozidriv screwdrivers are likely to slip or tear out the screw head when used in Phillips screws. Popular in Europe, Pozidriv is used on screws and bolts and may be identified by the four small lines at 45° to the main cross. It is a growing practice to put Phillips heads on inch-based fasteners and Pozidriv heads on metric ones.

Phillips, Pozidriv, Bureau de Normalisation de l'Aéronautique et de l'Espace (often marked B.N.A.E.) and Supadriv® drives appear nearly identical as viewed from above, but they require their own matching bits because their internal cavity shapes differ.

What are some other screw drive designs?

See Figure 9–3. These designs are usually chosen for their use with power drivers and for their tamper resistance.

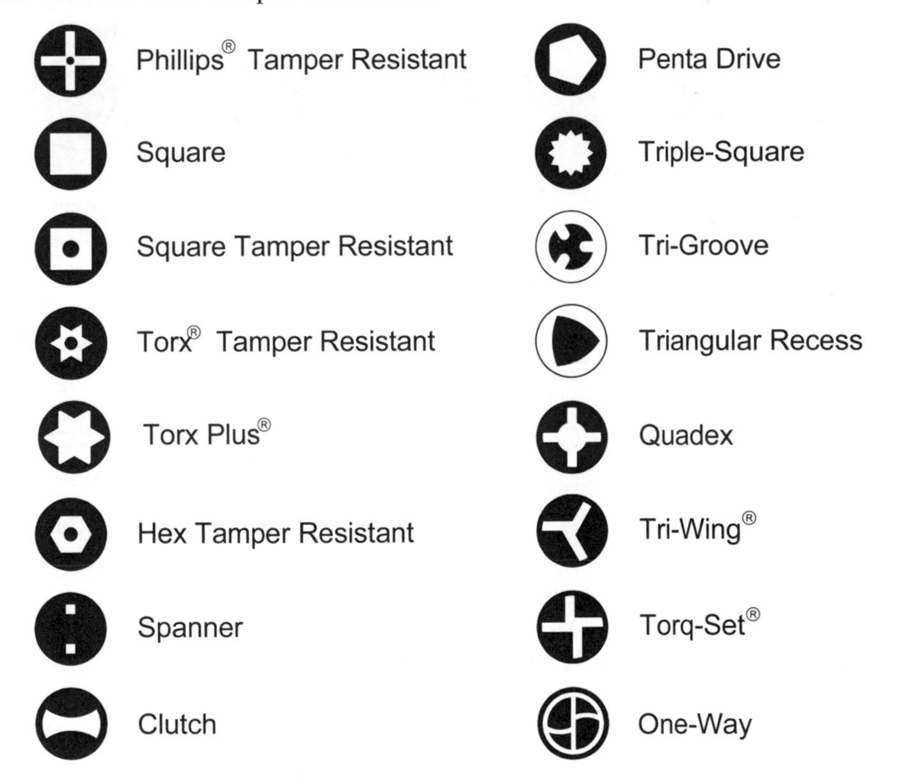

Figure 9–3. Other screw drive designs.

Bolt Drive Designs

Although some bolts are slotted or Phillips, heavy-duty, industrial and aerospace bolts often have one of the following drive designs:

- *External hex drive* is the most common design for bolts. They work with socket wrenches, box wrenches and open-end wrenches. These bolts are also called *hex head bolts.*
- *Internal hex* or *Allen drive* have a hexagonal hole and are driven by a hexagonal wrench, sometimes called an Allen key, or by a power tool with a hexagonal bit. Allen heads are used on screws and bolts.
- *Internal Torx® drive* is star-shaped with six rounded internal points. This design works well with power screwdrivers, permits the application of high torque without camming out and resists tampering. Torx heads are used on all sizes of screws and bolts.
- *External Torx drive* is the male version of the internal Torx fastener.
- *12-point external wrenching drive* for high-strength bolts.

All these designs work well with power drives and have much higher torque transmission than with a slot or Phillips head. See Figure 9–4.

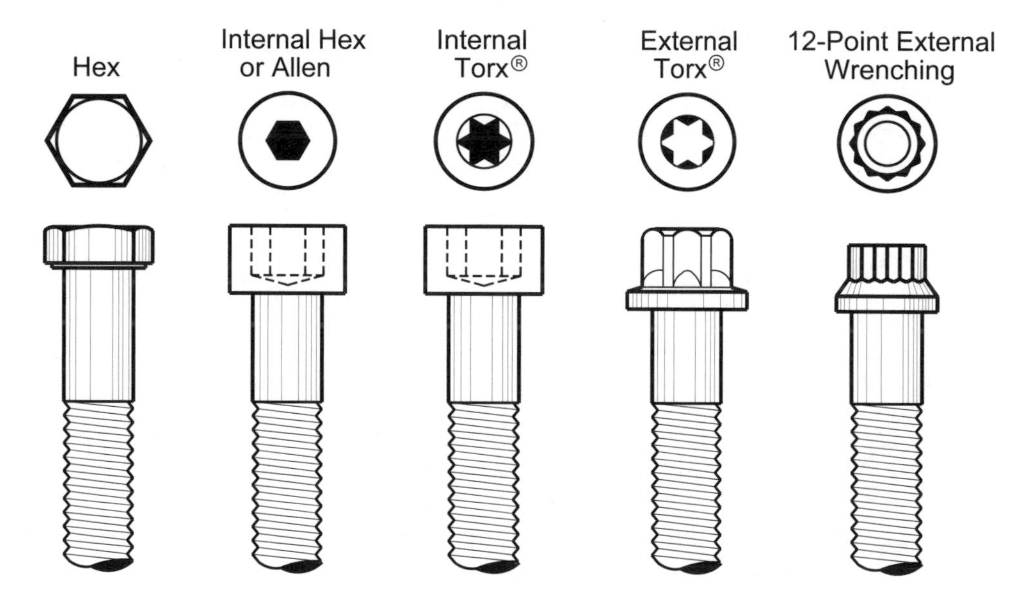

Figure 9–4. The most common bolt drive designs.

Important Threaded Fasteners

What are the main threaded fasteners?

- *Machine screws,* Figure 9–5, are made in a variety of head shapes, drive styles and materials. They are available as small as #0000 (0.021-inches diameter) to ½-inch diameter and in lengths from ⅛ to 3 inches.

Figure 9–5. Machine screw head shapes.

- *Machine bolts,* Figure 9–6, are most often available in hex or square head in diameters from ½ to 3 inches. Mating nuts are available for them. Dimensional control on machine bolts makes them suitable only for rough work.

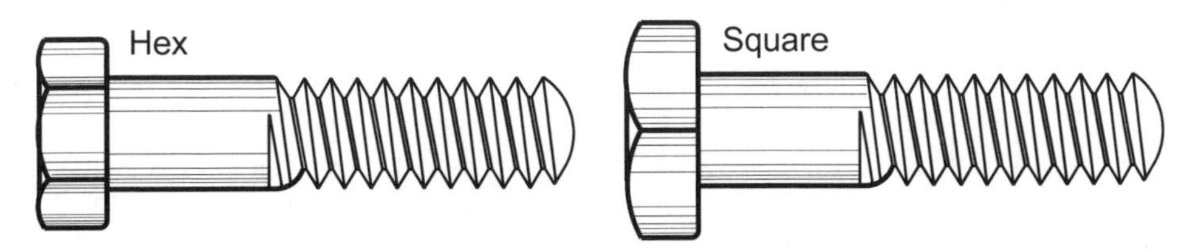

Figure 9–6. Machine bolts.

- *Socket head cap screws* and *bolts* are similar to machine screws and machine bolts, but of higher quality. They usually have a distinctive cylindrical head, a semi-finished bearing surface under their heads, and are made to higher dimensional standards. They are heat treated, making them substantially stronger than machine screws and bolts of the same diameter, and are available from #0 through 2-inches diameter and from ¼ to 10 inches in length. Although they come in several different lengths within a given diameter, their minimum and maximum lengths are proportional to their diameters. Both inch and metric sizes are available. Cap screws with drives other than hex and Phillips head shapes are available; some are shown in Figure 9–4. They are most commonly available with a black oxide finish, but stainless steel is available. Socket head cap screws are the most readily available quality fastener because most industrial tool supply companies and heavy equipment dealers stock them. Even higher quality fasteners can be obtained, but only through fastener specialty houses and at substantial additional cost.
- *Set screws*, Figure 9–7, are threaded fasteners that hold pulleys and collars on shafts. They also lock and hold mechanical settings or adjustments in place and come in more than a dozen basic shapes. They are usually heat-treated steel, but are also available in stainless, nylon and Delrin®. Inch-based sizes follow the pattern used for screws and bolts: numbered sizes, #0 through #12, and then fractional-inch sizes. See *Machinery's Handbook.*

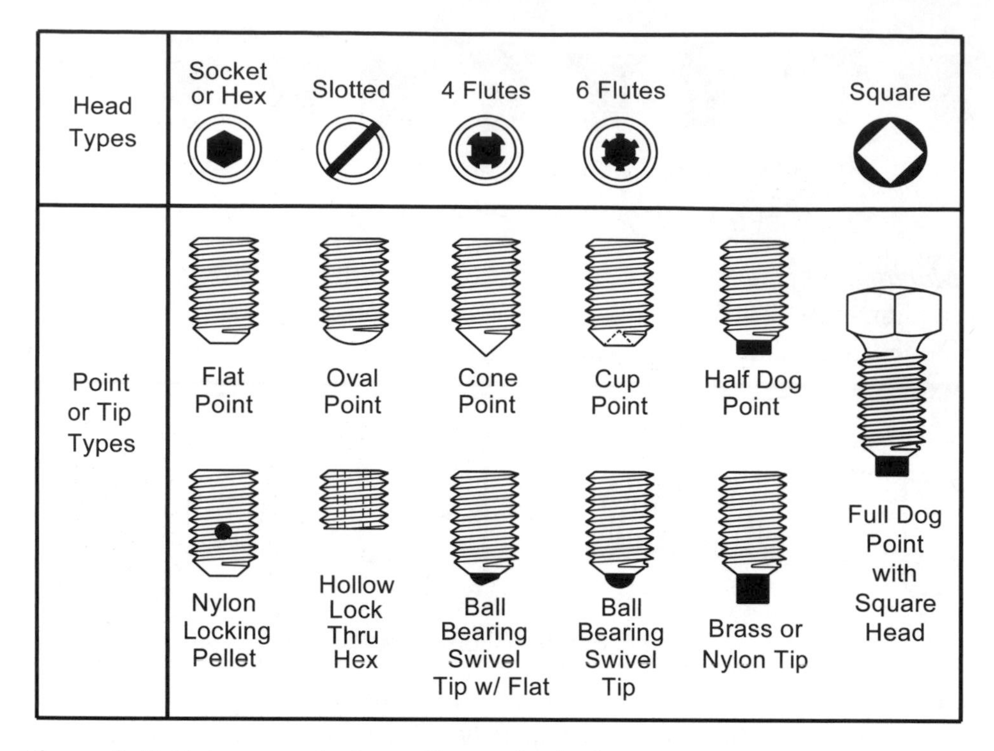

Figure 9–7. Set screw designs. Cup point is the most common set screw and often used to secure gears and pulleys onto shafts. The sharp edge of its rim digs into the shaft metal anchoring the gear or pulley to its shaft.

- *Carriage bolts*, Figure 9–8, are used to join heavy lumber. A square shank section just under the head that is larger than the shank diameter keeps the bolt from turning as its nut is tightened.

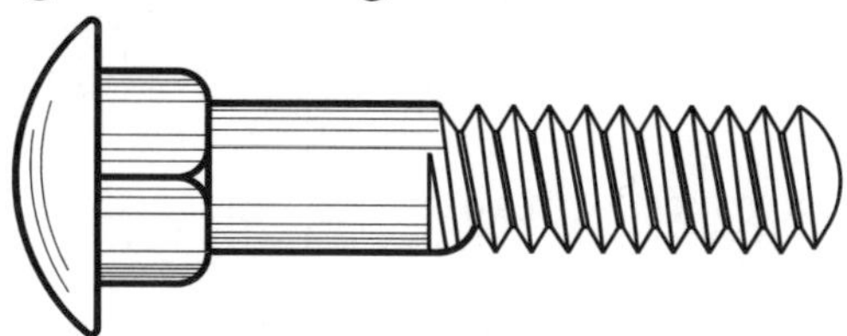

Figure 9–8. Carriage bolt.

- *Self-tapping screws,* Figure 9–9, require a pre-drilled hole. There are many combinations of size, length, finish and material. Many self-tapping screws have integral washers or under-the-head serrations to lock them in place. They can be tightened with a screwdriver, but power drivers work well on most designs. They work best in softer metals and plastics. Self-drilling screws have mostly replaced self-tapping screws in sheet metal. Figure 9–10 shows several self-tapping screw tip designs. Depending on their thread design, these screws either tap threads or form them.

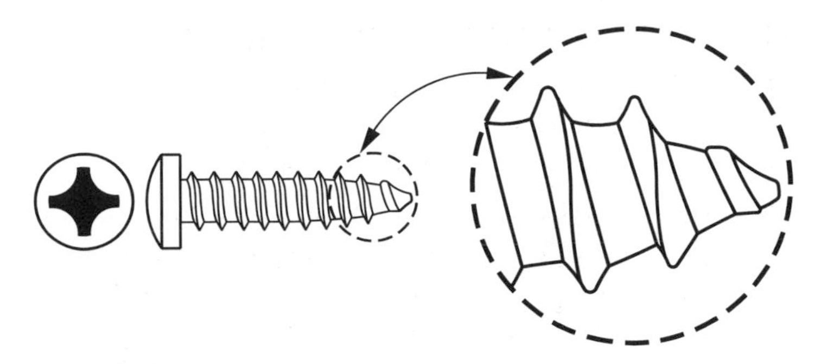

Figure 9–9. Typical self-tapping screw.

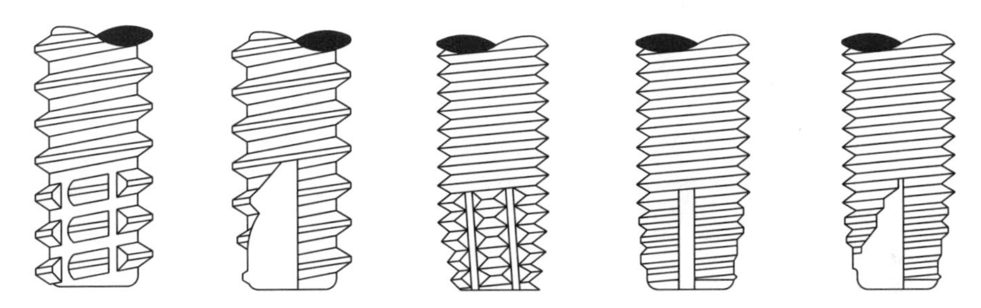

Figure 9–10. Self-tapping screw tip designs.

- *Self-drilling screws*, Figure 9–11, drill their own holes, tap threads and fasten materials together in a single step. They work well in sheet metal, stainless steel, brass, aluminum and plastics.

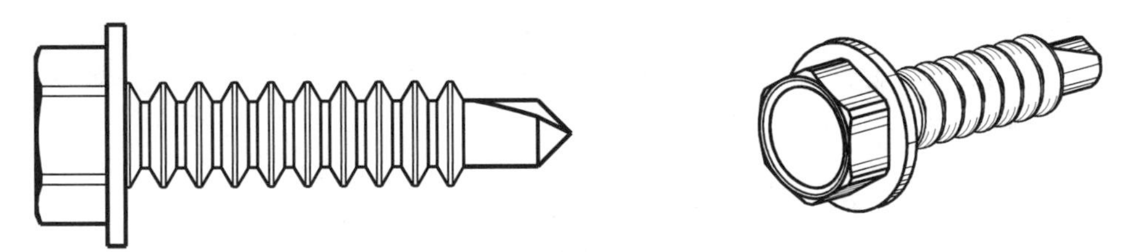

Figure 9–11. Self-drilling screw.

- *Drywall and deck screws*, Figure 9–12, are designed to fasten drywall, plywood, wood, or sheet metal to wood or steel studs. The shorter screws are called *drywall screws* and the longer ones are called *deck screws*. They have several valuable properties:
    - They are much stronger than traditional, wedge-shaped wood screws and less likely to split the wood they fasten.
    - On softer woods, like pine, and on thin sheet metal, they require no pilot holes and their bugle-shaped heads allow them to be sunk flush without countersinking. Harder woods require pilot holes and countersinking.

- They usually have Phillips head or square drives, so they work well with power drivers.
- They come as short as 1 inch and as long as 6 inches.
- They are usually steel or galvanized. Stainless steel screws are available too, but are much more expensive.

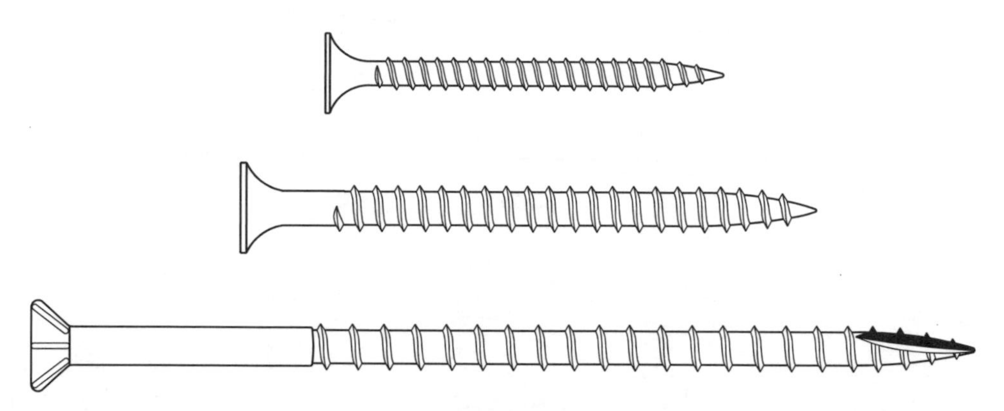

Figure 9–12. Drywall and deck screws have Phillips or square drives.

- *Shoulder* or *stripper screws* and *bolts,* Figure 9–13, are used for locating parts, retaining components, as guides, axle bolts and pivots. Their extended shank is usually ground to size. They are available in steel, alloy steels and stainless steel, and with nylon locking pellets or dry thread adhesive.

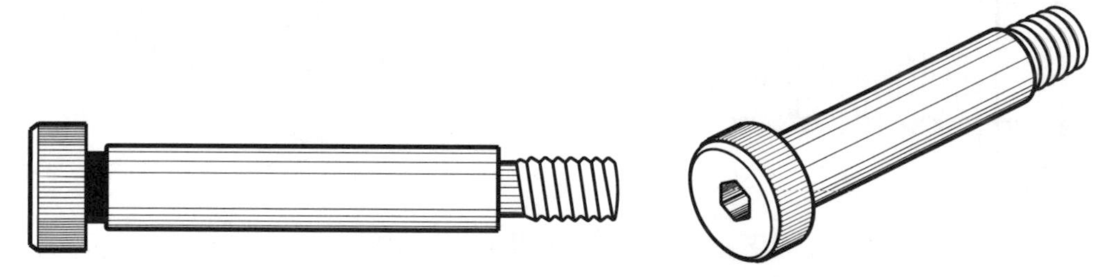

Figure 9–13. Shoulder screw or bolt.

- *Studs* are double-ended threaded fasteners, which are usually used to join a relatively thin cover plate, such as an engine-head casting, to a thicker component, like an engine block. Studs thread into a pre-threaded hole on one end and are pulled up with a hex nut on the other, Figure 9–14. For added gripping power in softer materials, sometimes the stud threads in the pre-tapped hole have coarser threads than the end holding the nut. Studs should be threaded in finger tight, not pulled up with a wrench. When the hex nut is tightened, the stud will lock in place. Thread locking adhesive may be used in the pre-threaded hole.

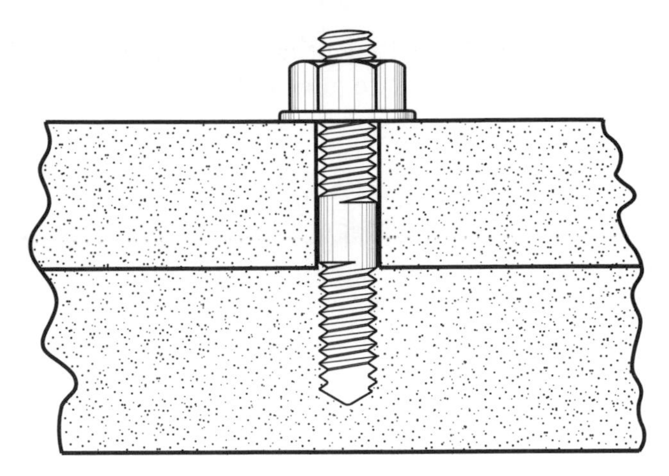

Figure 9–14. Studs thread into a pre-threaded hole on one end and pull up with a hex nut on the other.

Nuts

Figure 9–15 shows a few of the dozens of nut designs. Always use hardened nuts with cap screws and bolts. Also, hardened washers are often needed or the screw or bolt will embed in the base metal and loosen.

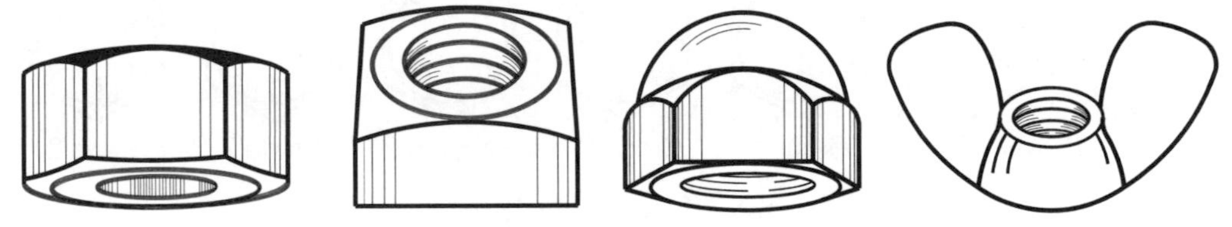

Figure 9–15. Nuts: finished, square, acorn and wing (left to right).

Thread Fit Classifications

Why are there different fit classifications for fasteners, and what are they?

Different applications require different degrees of fit, or tightness, between male and female thread pairs. Since there is a great difference in production costs, it makes sense to match cost and fit with application.

- *Classes 1A and 1B:* For work of rough commercial quality where loose fit for spin-on-assembly is desirable.
- *Classes 2A and 2B:* The recognized standard for normal production of the great bulk of commercial bolts, nuts and screws.
- *Classes 3A and 3B:* Used for high-quality work where a close fit between mating parts is required.
- *Class 4:* Obsolete, not used.

- *Class 5:* For a wrench fit. Used principally for studs and their mating tapped holes. This is a force fit requiring the application of high torque for semi-permanent assembly.

Threaded Fastener Quality

What are the principal quality categories of *threaded fasteners*, where are they used, and what organizations control their specifications?

There are five main threaded fastener groups:

- *Commercial fasteners* are sold by hardware stores and home improvement centers. They are the least expensive fasteners and are frequently made overseas with poor quality control. Their use should be limited to non-critical applications. These usually correspond to SAE Grade 1 or 2.
- *Automotive fasteners* are used in all types of motor vehicles. In the U.S. the Society of Automotive Engineers (SAE) is the specifying authority. The minimum tensile strength of automotive fasteners runs from 60,000 psi for Grade 1 to 150,000 psi for Grade 8. The number and pattern of lines forged into their heads indicates their Grade, Figure 9–16. These fasteners run from ¼- to 1½-inches diameter.

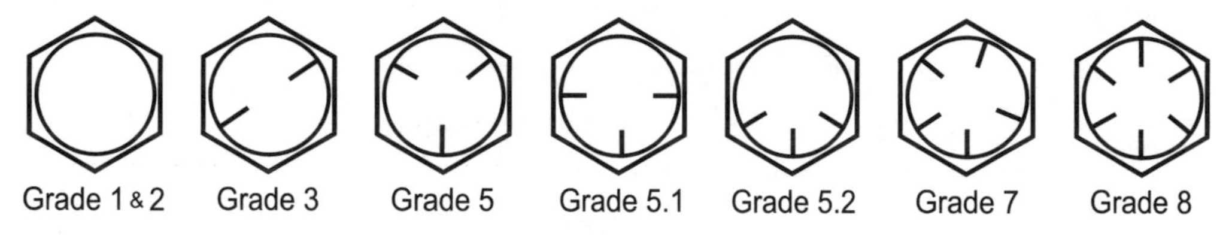

Figure 9–16. SAE fastener bolt-head markings.

- *Structural steel fasteners,* bolts, washers and nuts are specified by the American Society of Testing Materials (ASTM). Their bolt-head markings indicate the specification they meet.
- *Aircraft fasteners* are also used in missiles, space vehicles, satellites, racing cars and boats. MS (Military Standards) and NAS (National Aerospace Standards) specifications apply. These fasteners are more expensive than similar SAE fasteners. They are made of a wide variety of materials from steel and stainless steel to Inconel® and titanium.
- *Metric fasteners* are used in many U.S. vehicles and widely used in Europe for most applications. They are covered by specifications from the American Society of Mechanical Engineers (ASME), International Standards Organization (ISO), SAE and ASTM.

Bolt Pre-Tensioning

Why are bolts *pre-tensioned* or *torqued up*?

A tightened bolt stretches like a spring and clamps the workpieces together,

Figure 9–17. How much the bolt is tightened determines this clamping force. Bolts should be pre-tensioned to 65 to 70% of their proof load, Figure 9–18. Pre-tension is performed:

- To increase friction between the clamped layers to prevent them from moving with respect to each other, and so the bolt is never put into shear.
- To prevent the nut and bolt pair from loosening.
- Without proper pre-tensioning, bolts experience fatigue forces. If the fluctuating forces on bolts remain below the pre-tensioning force, the bolt does not *see* them and is not subject to fatigue forces that could eventually destroy the bolt.
- If the bolt clamps a gasket, adequate tension is required to make a seal.

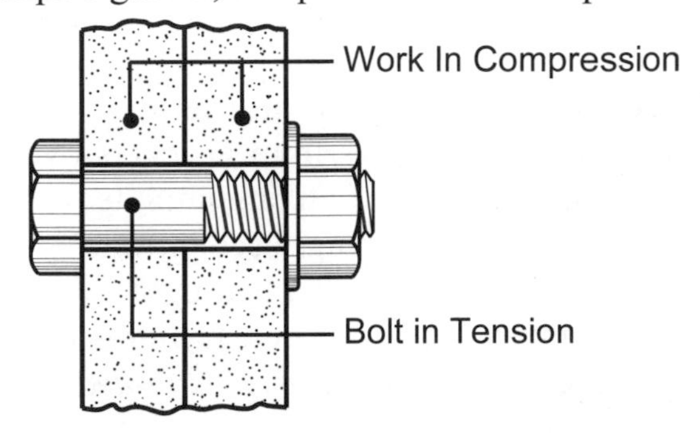

Figure 9–17. Bolt pre-tensioning.

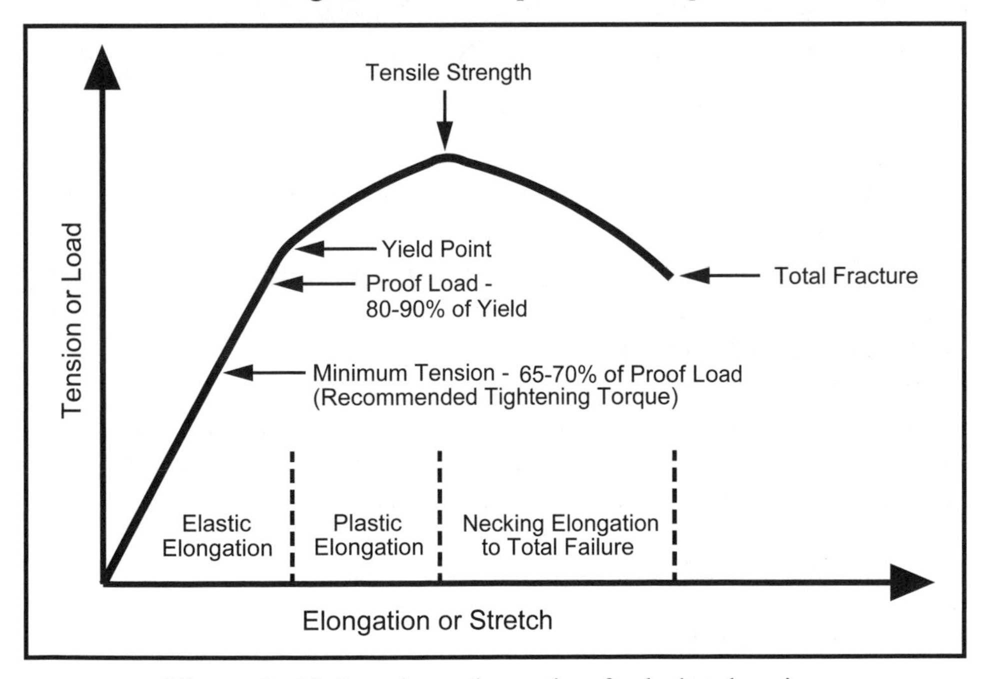

Figure 9–18. Load vs. elongation for bolts showing minimum tension for proper preload.

How is pre-tensioning done?

Using a torque wrench is the most common pre-tensioning method. Although this tool accurately measures *total* applied torque, the important torque needed to stretch the bolt is just 8 to 15% of the total torque. The remaining torque overcomes thread and nut-face friction, which varies depending on lubrication and thread conditions. The large variations in these two torques prevent accurate measurement of the stretching torque itself. No matter how accurate the torque wrench, bolt stretching torque is at best an estimate. Drilling a hole down through the center of the bolt nearly to its base and measuring its stretch though this hole is very accurate and is done on critical engine parts. There are also good ultrasonic methods as well as torque-indicating nuts.

Flat Washers

Why are washers used under nuts, bolts and screws?

Washers, Figure 9–19, distribute forces from fasteners over a greater surface area to prevent *embedding*, the fastener sinking into the work and allowing the fastener to lose its pre-load and loosening. Second, washers provide a smooth surface for the bolt or nut to turn against when tightened so sufficient torque can be applied without rupturing the fastener. If high-strength bolts are used, matching *hardened* washers must also be used. *Non-heat treated washers will yield.*

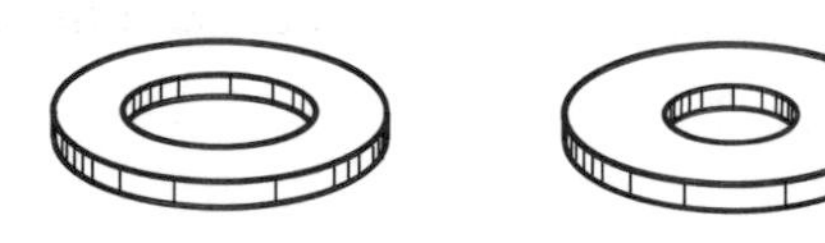
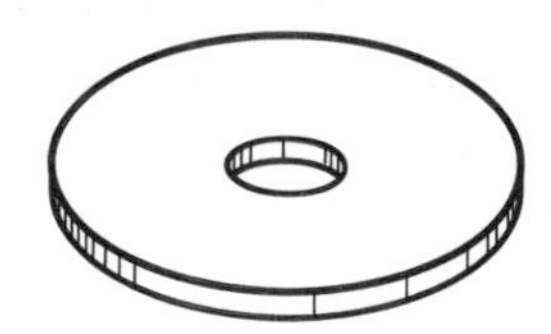

Figure 9–19. Washers: SAE-type, USS-type and fender (left to right).

Section II – Locking Threaded Fasteners

Fastener Loosening

Why do threaded fasteners loosen under vibration?

Although many machinists and engineers believe that vibration causes loosening, the most frequent cause is side sliding of the nut or bolt head relative to the joint. This results in relative motion between their threads.

This motion occurs from bending of the bolted parts at the friction surface, uneven heating of the top and bottom of the joint, or shifting of the joint surfaces from loading forces.

What can be used to prevent threaded fasteners from loosening?

- *Lock Washers* provide limited protection and should never be used in critical applications. They are more an expression of intention than a solution.
- *Nylocks*® are nuts with a nylon collar insert which the bolt thread cuts into. This increases the torque needed to rotate the nut and is effective in many applications. Do not use exposed to elevated temperatures.
- *Nylon pellets* in the threads of screws and bolts work on the same principle as Nylocks; they increase the static torque, sometimes called prevailing torque, on the fastener. They are reusable several times.
- *Distorted threads* on screws and bolts are very effective because they increase prevailing torque. Some nuts have their top threads designed to pinch mating screw threads.
- *Castellated nuts* are a traditional positive locking method for critical applications. A cotter pin fits through both the bolt and the castellated nut, preventing their relative rotation. Excellent at high temperatures.
- *Safety locking wires* on nuts and bolts are a positive, traditional, and effective way to prevent critical fasteners from turning by looping stainless steel wire through them. It is labor intensive. This method is frequently used on critical aircraft and racing car components.
- *Thread locking adhesives* work well and can be used alone or with other locking methods. They can be used on studs where there are no other locking alternatives. These adhesives can be released by raising their temperature to about 300°F.

See Figure 9–20.

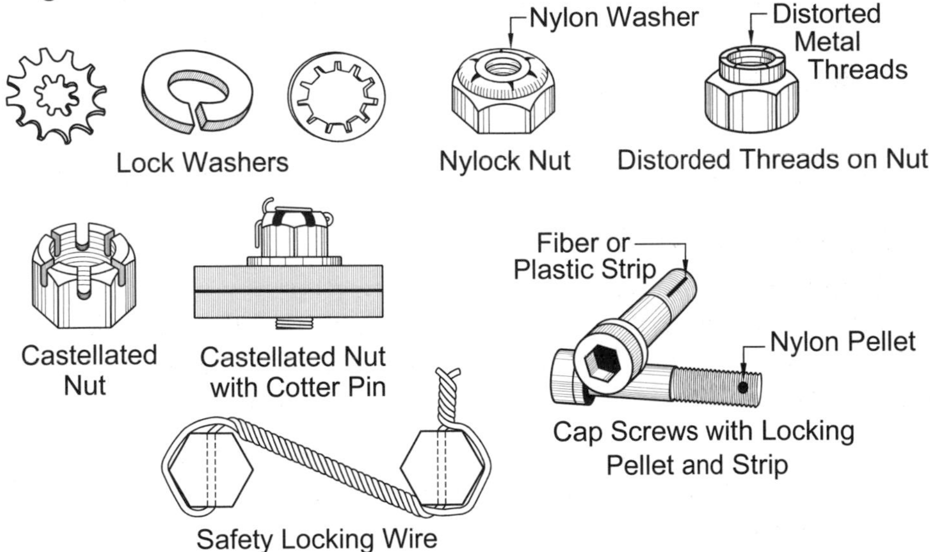

Figure 9–20. Methods to prevent fasteners from loosening.

Section III – Non-Threaded Fasteners

Rivets

What are the common non-threaded fasteners?

- *Solid Rivets,* Figure 9–21, are available in steel, copper, aluminum, brass and stainless steel in many sizes and shapes. They are inexpensive and easy to install with simple tools. Properly applied, they are strong, provide a finished appearance and resist vibration. In critical aerospace applications, rivet holes are drilled and reamed to size so stresses are evenly distributed.

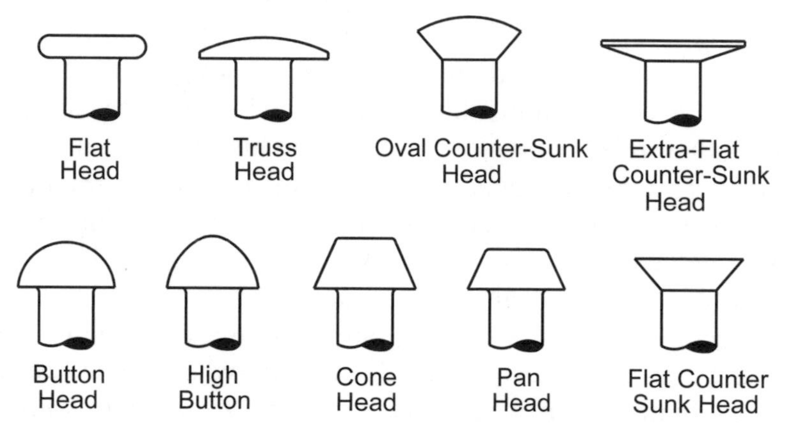

Figure 9–21. Solid rivet head designs.

- *Blind Rivets* are designed for applications where there is access to only one side of the part. There is a wide range of styles, sizes and materials including steel, aluminum and stainless steel. Liquid-tight designs are available. They are strong and vibration and tamper-resistant. They draw together the layers of materials they fasten as they are pulled up. A special tool is needed to install them, but these tools are not expensive. Both manual and air-actuated blind rivet pullers are available.

  A blind rivet consists of two parts: the rivet body and the setting mandrel, which is inside the rivet body, Figure 9–22. Here is how they work:

  1. The rivet body is inserted in a pre-drilled hole through the materials to be joined.
  2. The rivet insertion tool is actuated and the jaws of the manual or power-operated riveting tool grip the mandrel of the rivet.
  3. The rivet is set by pulling the mandrel head into the rivet body, which expands it, pulling the parts together, forming a strong, tight, reliable joint. At a predetermined setting force, the mandrel breaks apart.

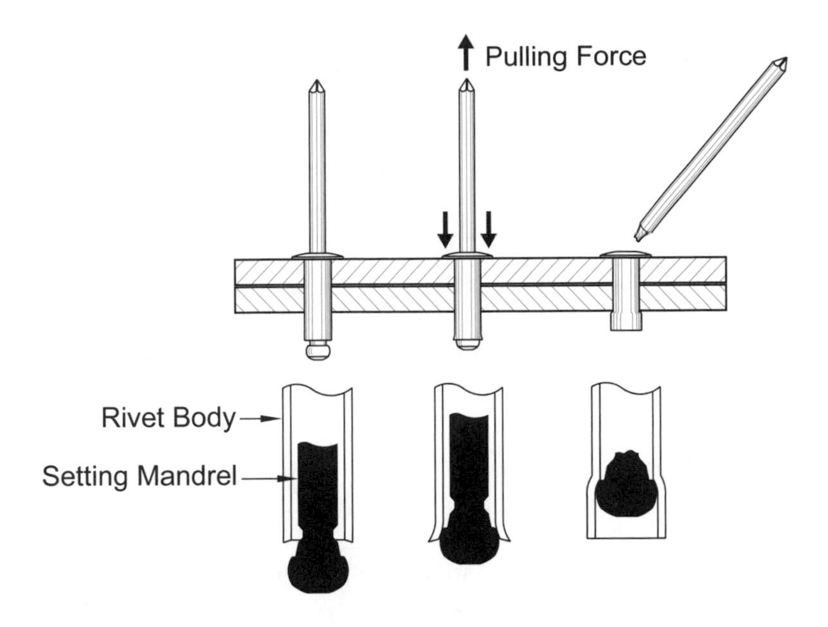

Figure 9–22. Blind rivet joining process.

Pins

- *Dowel pins,* Figure 9–23 (left), retain parts in a fixed position or preserve their alignment. The most common dowel pins are ⅛ to 1-inch diameter, but larger and smaller ones are available along with metric sizes. Usually only two pins are required. In applications where frequent pin removal would subject the pins to wear, taper pins should be used. Use an undersized hole on one end of the dowel to retain it and a tight slip fit on the other, Figure 9–24. If both dowel ends are a press-fit, separating the doweled parts will be difficult or impossible. A good rule of thumb is that dowel length should be 1½ to 2 times its diameter in each plate doweled. Dowels are typically supplied +0.002 to − 0.0000 inches of their nominal diameter. Dowel holes are usually reamed using over/under reamers. It is good practice to provide access to remove the dowel pin with a punch.

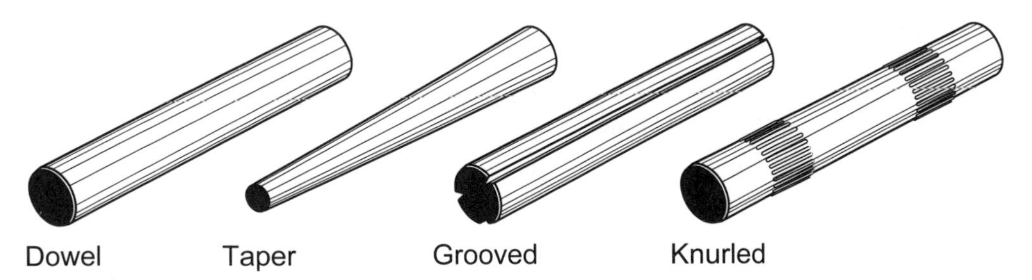

Figure 9–23. Common round-pin designs.

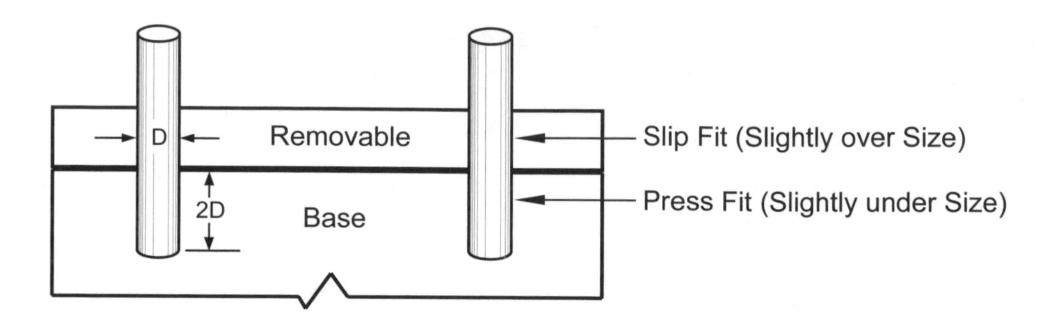

Figure 9–24. A dowel pin secures and aligns two plates.

- *Taper pins,* Figure 9–23 (second from left), require more installation time, but are not as subject to wear as dowel pins and are easily removed with a punch. They lock two components together more accurately than a dowel pin. Taper pins are hardened and ground, and have a ¼ inch/foot uniform taper. Their ends are slightly rounded. Both steel and stainless steel taper pins are available. Larger taper pins require a step-drilled hole that is then reamed to final size, Figure 9–25. Reamers are made specifically for this purpose. See *Machinery's Handbook* for dimensional details.

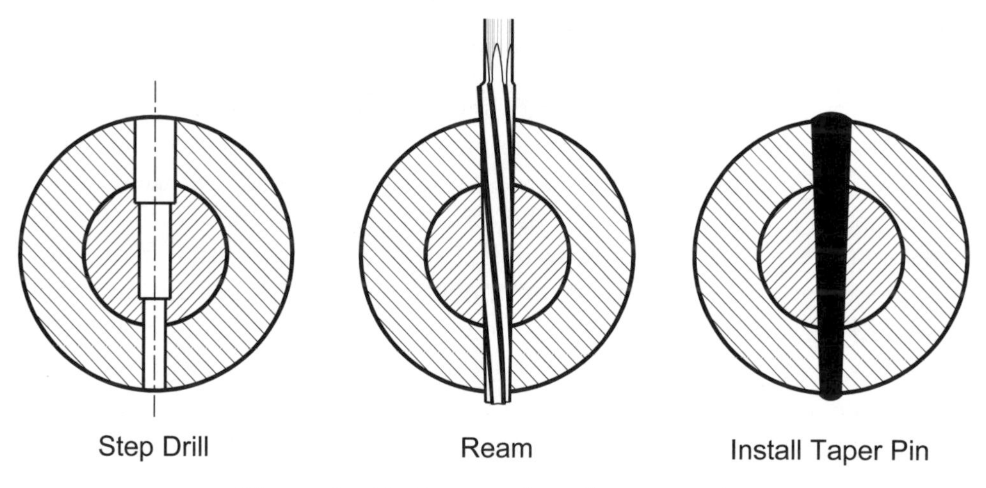

Figure 9–25. Steps to install a taper pin.

- *Grooved pins,* Figure 9–23 (second from right), are non-threaded press-fit fasteners, usually of carbon steel. The material displaced from the three parallel longitudinal grooves is displaced beyond the nominal diameter of the pin and secures the pin in place through an interference fit. The mounting hole must never be smaller than the nominal pin diameter. They are less expensive than either dowel or taper pins and are used in less critical applications where there is not a lot of end load. Taper pins are soft steel, unlike dowel pins which are hardened and ground to size.

- *Knurled pins,* Figure 9–23 (right), have one or more knurls parallel to their longitudinal axis. These knurls form an interference fit that locks the pin in place. They work well in both metals and plastics. They are inexpensive and not reusable.
- *Coiled spring pins* and *roll pins*, also called *slotted pins, spring pins* and *split pins*, Figure 9–26, are installed in holes slightly smaller than their nominal diameter; no reaming is needed for these pins. When pins are pressed into place, radial forces hold them in position resisting shock and vibration. Coiled spring pins are superior to roll pins as forces are more evenly distributed through the cross section of the fastener and better resist shearing forces. Roll pins compress by closing the gap. These stresses concentrate opposite the gap and compression is not uniform around the circumference of the pin. Roll pins are inexpensive and are used as dowels, spacers, stop pins, T-handles and hinge pins. When installed into a blind hole, they add tamper resistance to the part. Figure 9–26 (bottom) shows the tool for driving these pins, but transfer punches also work well. The center projection on the drive punch prevents it from damaging the hole or pin by keeping the punch aligned with the pin.

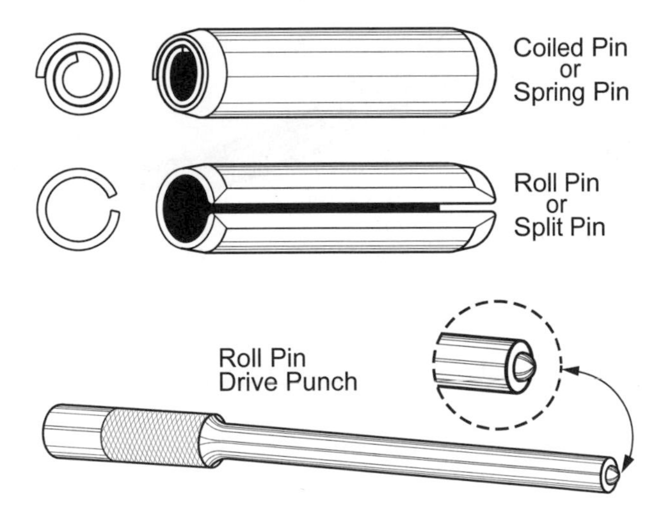

Figure 9–26. Coiled or spring pin (top), roll or split pin (middle), and drive punch for these fasteners (bottom).

- *Snap rings*, also called *retaining rings*, Figure 9–27, are available in internal and external designs. The internal design acts as a removable shoulder to retain an object *inside* the housing, while external snap rings, like a shoulder on a shaft, acts to retain it. Their materials include steel, stainless steel, brass and beryllium copper. Figure 9–28 shows the tool that compresses or expands snap rings for installation or removal.

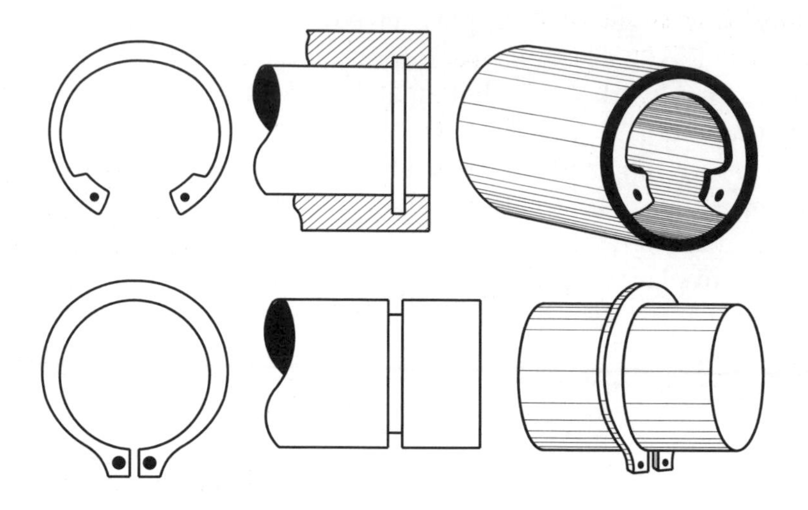

Figure 9–27. Internal (top) and external (bottom) snap rings.

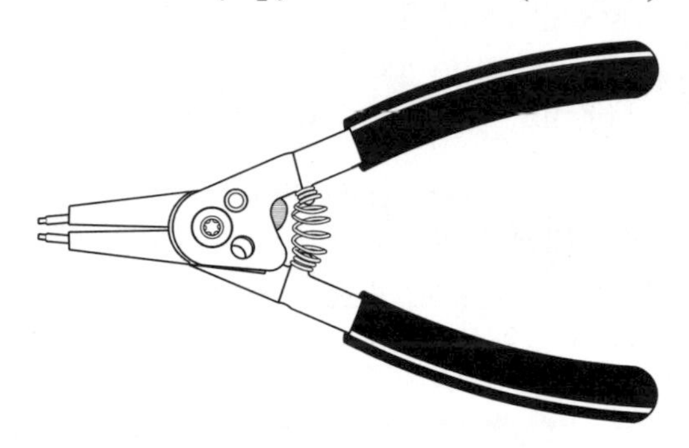

Figure 9–28. Pliers for installing or removing snap rings. Their jaws are reversible for internal and external snap rings.

Section IV – Assembly Processes

Force Fits

How are *force fits* used to assemble and secure parts?

Cylindrical components, Figure 9–29, are squeezed together in a vise, arbor press or hydraulic press, depending on the force needed. Hammering parts together is poor practice as the parts tend to get knocked out of coaxial alignment and jam. To reduce assembly forces, the inside part may be shrunk by chilling in a freezer, dry ice or liquid nitrogen. The external part may also be heated to expand it and reduce assembly forces. In general, the difference in part sizes runs between 0.001 to 0.0025 inches/inch of diameter, but parts measuring the same diameter will still be a press fit. *Appendix Table A–4* provides fit allowance data for various conditions.

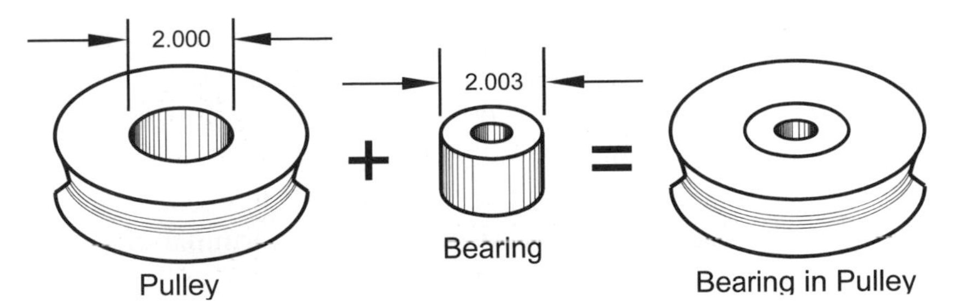

Figure 9–29. Assembling parts by force fit.

Adhesives

What are the most common adhesives needed in the prototype machine shop?

- *Rubber cement* is a low-strength adhesive suitable for paper, plastic and wood. It is excellent for temporary assembly and very light loads.
- *Contact bond cement* is a much stronger version of rubber cement. It works well on metals, laminates, wood, leather, cloth, fiberglass, rubber, glass and many plastics. It is applied with a brush or roller to both surfaces to be joined and allowed to dry for 15 to 20 minutes. Then the two surfaces are aligned and pressed together. You have only one chance to get their alignment right because once the two cement-covered surfaces touch, they bond together and cannot be readjusted. Many contact bond cements use a volatile and explosive component, so use it in a well-ventilated area without ignition sources. On plastics always test solvent-based adhesives on scrap to see if it is compatible and does not dissolve the work.
- *Silicone rubber cement* and *sealant* is available in clear, white and black. A one-part system, it cures to a tough, rubbery solid when exposed to moisture in the air. They have excellent adhesive strength, high elongation and outstanding electrical insulation and heat-resistance qualities. They join metal, glass, wood, silicone resin, vulcanized silicone rubber, ceramic, natural and synthetic fibers, most plastics and painted surfaces. They resist weathering, vibration and exposure to oil, moisture, ozone and temperatures from sub-zero to 450°F (230°C).
- *Epoxy adhesives* are available in one-part or two-part systems, and are cured either at room temperature or with heat. Most adhere well to metals, plastics and glass, but tend to be relatively rigid and inflexible. They are useful for both bonding and potting—filling up a volume with resin to secure components—to keep moisture away or prevent tampering. Many products are supplied in dual syringes for easy measuring. Bonding surfaces must be clean and dry.

- *Cyanoacrylate adhesives* spontaneously cure in the absence of air. Most develop a bond in under a minute, cure overnight, and have a working temperature range of − 65° to +200°F. A spray accelerator makes them set up in seconds. They are often used to bond metal surfaces or dowel pins that may be easily removed later with a heat gun. They work well on metals, rubbers, plastics, wood and glass. Some formulations are thin to get into cracks, others are thick to fill gaps.
- *Thread locking adhesives*, often called *thread lockers,* are cyanoacrylate adhesives formulated specifically for securing threaded fasteners. They fill the small voids between threads, exclude water to prevent rust, and prevent fastener loosening. Most are applied to screw or bolt threads before fastening, but there are also penetrating versions for preassembled fasteners. Different strength versions are identified by their colors. They can be removed by heating above 300°F.

Section V – Soldering, Brazing & Welding

What are the differences between soldering, brazing and welding?

- *Soldering* uses the adhesion of solder between two surfaces to join them. Solder is usually a tin-lead alloy and melts below 840°F (450°C). The advantages of soldering are that its relatively low melting temperatures leave the base metals un-melted, which minimizes shrinkage and distortion, and permits the entire assembly to be exposed to the soldering temperature. The surfaces are cleaned, heat is applied to melt the solder, and capillary attraction draws the solder into the joints and along the work surface. After the solder solidifies, the two surfaces are joined. Soldering is a relatively simple process, requires limited skill, and can be done with inexpensive tools like a soldering iron, propane or a MAPP® gas torch.
- *Hard soldering*, also called *silver soldering,* or *silver brazing*, uses filler metals that melt at temperatures above tin-lead alloys, but still below brazing temperatures. These joints have greater strength than tin-lead soldered joints.
- *Brazing* is similar to soldering except it is carried out at temperatures above 840°F (450°C). Usually copper or nickel-based alloys are used with an oxygen-fuel torch. Brazed joints are stronger than silver soldered ones. Like soldering, the equipment can be simple, and limited skill is needed.
- *Welding* melts the two base metals to be joined with or without adding filler metal. These joints are stronger than brazed joints and often stronger than the base metals themselves. Although oxyacetylene torches may be used, most welding is performed electrically with specialized electrical power from welding machines.

Chapter 10

Machine Shop Steel Metallurgy

Success is simply a matter of luck.
Ask any failure.
—Earl Wilson

Introduction

The hardening and softening of metals using heat is essential to making the tools and products of the modern world, from kitchen knives to jet engine blades. Because steel is the most common material heat-treated, as well as the most often used metal in the machine shop, this chapter focuses on steel.

Metallurgy and heat-treatment is a complex subject. Metallurgists, physicists and heat-treat specialists often spend a lifetime in its study, so it is not possible to completely cover the subject in this chapter. For critical or complex applications, professional guidance and equipment are essential. However, there are several important heat-treating tasks that can be done in the machine shop with limited equipment, and we will cover these tasks.

This chapter looks at the crystalline structure of metals and explains why steel can be hardened using heat. It reviews practical aspects of heating, quenching, tempering and annealing steel. The equipment needed to perform these processes, and their advantages and limitations are explained, as are common heat-treatment problems and their solutions. Several types of tool steels, which the machinist can readily heat-treat, are detailed. In addition, the uses and differences between drill rod, drill blanks and reamer stock are discussed, along with their common alloys.

There are other uses for heat in the shop too: annealing work-hardened copper and brass, softening metals to bend them, soldering and brazing and expanding parts prior to slipping them over other parts. These operations are easily done and offer practical solutions and alternatives to the machinist.

There is also an explanation of metal fatigue and the steps to avoid it.

Section I – Metals Made from Iron

Engineering Materials

What are the properties of iron-based metals?

The element iron makes up all but a few percent of the contents of iron-based metals. These metals have very different properties depending on their carbon content. Table 10–1 lists the iron-based metals, their carbon content, characteristics and applications.

Material	Percent Carbon	Characteristics & Applications
Wrought Iron	< 0.008	Very soft material.
Low-carbon Steel	< 0.30	Also called *mild steel* or *machine steel.* Does not contain enough carbon to be hardened, but can be case hardened. Used for parts, which do not need to be hardened, like nuts, bolts, washers, sheet steel and shafts. Also used in structural steel and machine tool frames.
Medium-carbon Steel	0.30–0.60	May be hardened. Used where greater tensile strength than low-carbon steel is needed. Used for tools like hammers, wrenches and screwdrivers, which are drop-forged and then hardened.
High-carbon Steel	0.60–1.7	May be hardened. Also known as *tool steel.* Used for cutting tools, punches, taps, dies, reamers and drills.
High-speed Steel (HSS M2)	0.85–1.5	Hardenable. Used for lathe and milling machine cutting tools and drill bits. Retains hardness and cutting edge at red heat. This is the least costly member of the HSS family. See Table 10-2 for details on the alloying elements of all the HSS alloys.
Cast Iron	> 2.1	Cannot be hardened, but available in several forms with differing properties.

Table 10–1. Carbon content of major iron-containing metals.

HSS Steel Alloy	Percent					
	Carbon	Chromium	Tungsten	Molybdenum	Vanadium	Cobalt
M2	0.85	4.00	6.00	5.00	2.00	—
M42	1.10	3.75	1.50	9.50	1.15	8.00
T1	0.75	4.00	18.00	—	1.00	—
T15	1.50	4.00	12.00	—	5.00	5.00

Table 10–2. Principal alloying elements of high-speed steels.

Section II – Why Steel Hardens

Structure of Metals

What is the difference between the structure of metals when molten and solid?

During the hot liquid state metals have no particular structure. There is no orderly, defined or regular organization among their atoms. However, at lower temperatures, the atoms have less energy, move less rapidly, and atomic forces tend to arrange them into particular structures or patterns called *crystals*. All metals and alloys are crystalline solids. Figure 10–1 shows the three most common crystal structures in metals and alloys.

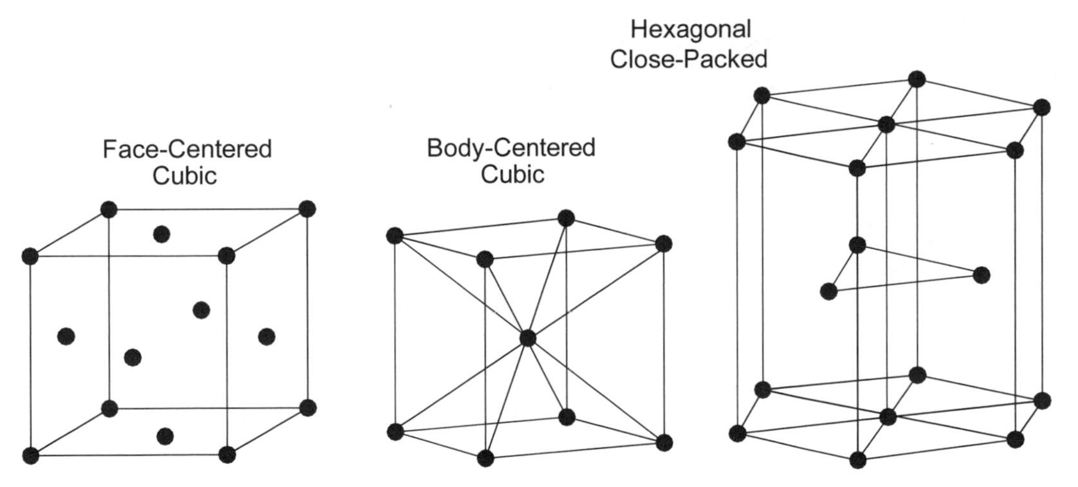

Figure 10–1. The three most common crystal structures in metals and alloys.

Behavior of Pure Iron

What behavior does pure iron exhibit as its temperature is gradually raised from room temperature and before it becomes a liquid?

Iron changes its crystalline structure as its temperature increases while remaining solid:

- From room temperature to 1670°F (910°C), it is body-centered cubic.
- From 1670°F (910°C) to 2535°F (1390°C), it is face-centered cubic.
- From 2535°F to its melting point of 2800°F (1538°C), it is again body-centered cubic.

What makes these changes important to us?
These changes between different crystalline structures for *pure iron* are not particularly important by themselves. However, if we add as little as one percent carbon by weight to iron, we make the iron-carbon alloy called steel. Because of the changes in iron's crystalline structure at different temperatures, an entirely new series of microstructures form. These microstructures not only have radically different physical properties from pure iron and from each other, but they can be modified by heating and cooling cycles.

Iron-Carbon Diagram

What are these new *iron-carbon structures* and when are they formed?
The best way to visualize the interaction of these two elements and the resulting microstructures they form is to study the *iron-carbon diagram*. This diagram shows the microstructures that exist for a small percentage of carbon in iron at various temperatures, Figure 10–2.

Looking at this diagram, we can see that:
- Steel exists for iron-carbon alloys with between 0.2 and 2 percent carbon.
- Iron-carbon mixtures with more than 2 percent carbon are called *cast iron.*
- Before the iron-carbon alloy becomes a liquid, there are six different microstructures represented by areas on the diagram:
  - Pearlite and ferrite
  - All pearlite
  - Pearlite and cementite
  - Austenite and ferrite
  - All austenite
  - Austenite and cementite
- In the areas showing two microstructures simultaneously, the structure changes from all one microstructure, to a mixture of the two, to all the other components as we move along the bottom axis, which represents the carbon content. For example, in the "pearlite and ferrite" area of the diagram (bottom left), the structure goes from all ferrite at 0.2 percent carbon, to a mixture of ferrite and pearlite, to all pearlite at 2 percent carbon. This means that in the areas of two microstructures, both the

microstructure and its properties are altered with changes in its carbon content.

- Austenite exists only at a temperature of 1333°F (723°C) or higher and can contain up to 2 percent carbon.

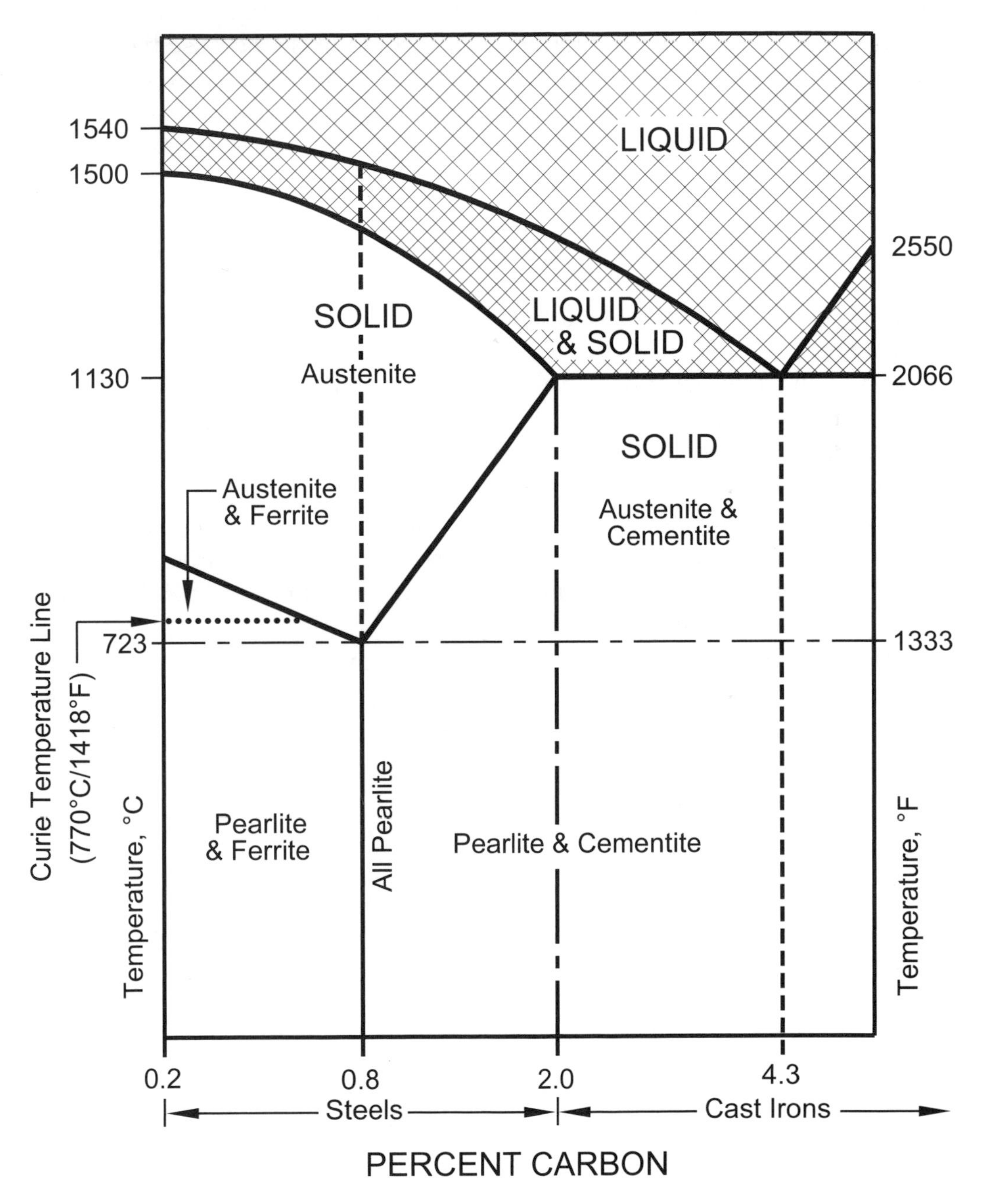

Figure 10–2. Iron-carbon diagram.

This iron-carbon diagram has a special qualification: *it represents only the microstructures that exist at equilibrium.* This means that the diagram shows the structure when the alloy has been allowed to remain at a particular temperature long enough to reach its final condition where no further microstructure changes can occur no matter how much longer the alloy is held at this temperature. This equilibrium qualification exists because the changes in crystalline structure do not occur instantly, but take time to complete. Rapid temperature changes can "freeze" a microstructure, so it continues to exist at a lower temperature than it would at equilibrium. This is the key to hardening steel.

What are the characteristics of the microstructures appearing in the iron-carbon diagram, Figure 10–2?

- *Ferrite*–A solid solution of small amounts of carbon (>0.02%) in iron. It is magnetic.
- *Cementite*–Contains about 6.69% carbon. Because our iron-carbon diagram cuts off at about 5% carbon, no region of pure cementite appears. Only structures containing cementite-pearlite and cementite-austenite mixtures appear.
- *Pearlite*–Alternating layers of ferrite and cementite make up pearlite. Pearlite *always* contains 0.77% carbon. Steels and cast irons with more than 0.77% carbon will form mixtures of cementite and pearlite. Steels with pearlite are usually ductile.
- *Austenite*–A solid solution of carbon in gamma iron, which can dissolve up to 2% carbon. Austenite is never stable below 727°F (386°C) in carbon steel. Above 1418°F (770°C) it will not be attracted to a magnet.

Hardening Steel

What important iron-carbon microstructure *does not* appear in the iron-carbon diagram and why is it missing?

Martensite is missing because it is an entirely new structure. It forms if austenite is cooled rapidly, or *quenched*, and the iron-carbon diagram shows only structures formed when iron and carbon are cooled slowly. Quenching causes a distorted crystal structure that has the appearance of fine needles. *This is the structure that hardens steel.*

For example, suppose we begin with a steel alloy of 0.8 percent carbon content. The vertical line in the iron-carbon diagram marked "all pearlite" represents this steel alloy. Here are the steps which form martensite:

1. Begin by raising the temperature of this alloy above 1333°F (723°C), called the alloy's *critical temperature.* Although the alloy remains solid,

its microstructure changes from pearlite to austenite. The conversion process from pearlite to austenite is not instantaneous. It takes from several seconds to several minutes to complete. In practice, steel is usually raised 100°F above the critical temperature to make sure the conversion to austenite is complete. The black V-shaped area in Figure 10–3 shows the temperature range used for forming austenite.

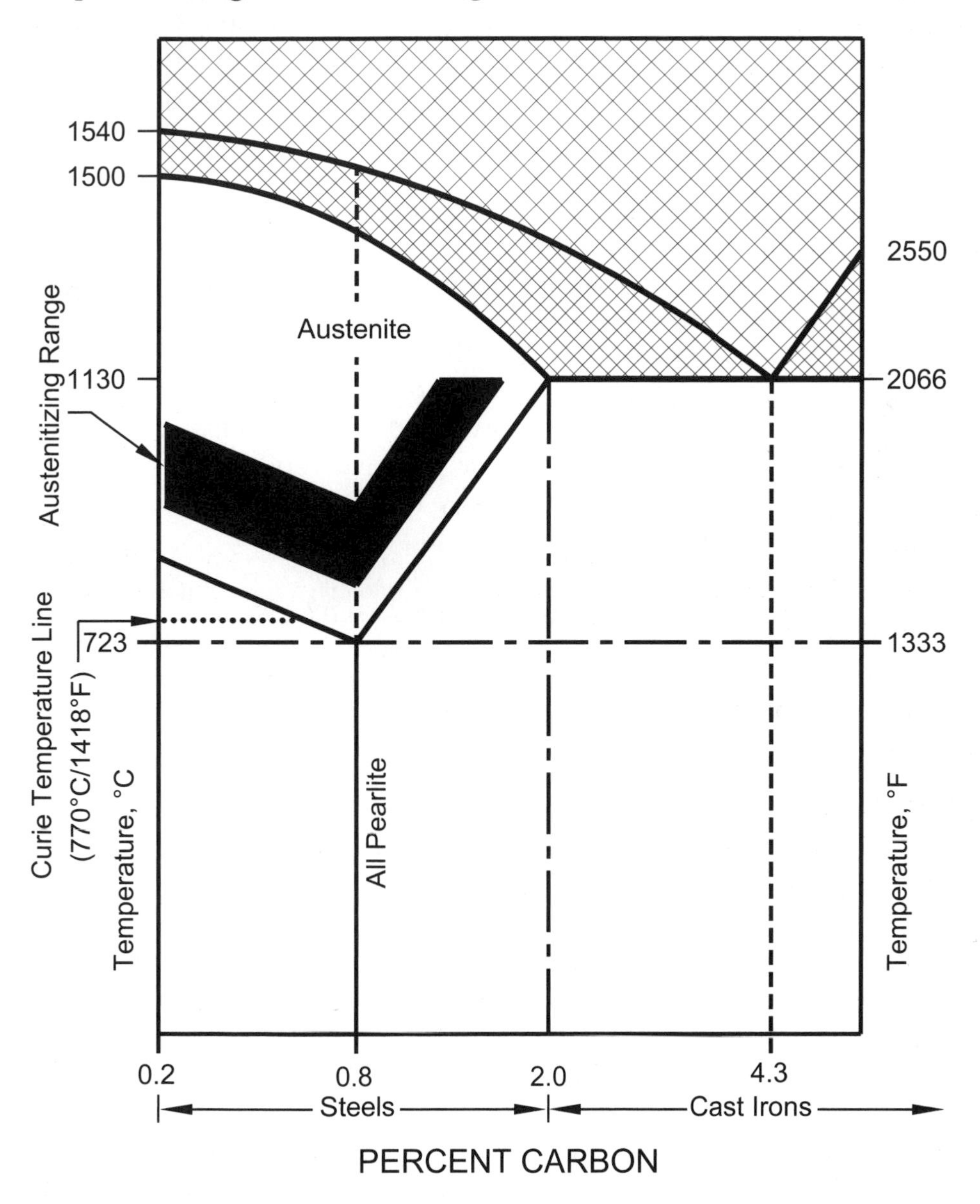

Figure 10–3. The black area in this iron-carbon diagram shows the temperature range used to form austenite.

2. If we slowly lower the temperature of austenite, its previous crystalline structure, pearlite, re-forms and no martensite forms. However, if we *quench* the austenite by lowering its temperature rapidly from above its critical temperature to below 752ºF (400ºC) by immersing it in water, brine or oil, an entirely new and different microstructure called martensite forms. This occurs because the rapid temperature drop associated with quenching does not give the carbon time to move out of the austenite crystal structure in an orderly fashion to form pearlite. This results in a new structure that is very hard—martensite.

Martensite is much harder and stronger than pearlite. It is also less ductile and is generally too hard and brittle for most engineering applications. But, by tempering martensite, we can restore some ductility without significantly reducing its strength. Martensite is very useful for metal cutting tools, punches, dies and many other applications. The formation of martensite is what makes the hardening of steel possible. Hardening steel also increases its abrasion resistance, an important factor for many parts subjected to wear. There is no partial transformation associated with martensite, it forms or it doesn't. However, only the parts of a section that cool fast enough will form martensite. In a thick section, it will only form to a certain depth, and if the shape is complex, it may only form in small pockets.

Besides carbon, are other alloying elements used?
Additional alloying elements are manganese, tungsten, cobalt, nickel, vanadium, molybdenum, chromium and silicon. Usually two or more alloying elements are added. Alloying elements further modify the steel's properties. They can affect the final hardening, quenching, tempering characteristics, hardness at high temperatures and nearly all other mechanical properties. Most high-performance alloys contain additional alloying elements. When we extend the iron-carbon diagram to include other alloying elements, the diagram becomes very complex, but the basic concepts of hardening, tempering and annealing are unchanged.

Section III – Practical Steel Hardening & Tempering

Equipment for Hardening Steel

What equipment is used to heat steel to above its critical temperature for hardening?
The most common equipment used is:

- Metallurgical furnace heated by gas or electricity and usually controlled electronically.
- Oxyacetylene torch or MAPP® gas-air torch. Torches are inexpensive, but have many limitations for heat treatment.

A metallurgical furnace is far superior to a torch. First, it provides excellent control of maximum temperature as well as cooling rate. Second, larger parts must soak in the furnace for considerable time at the maximum temperature before the entire part, especially the interior, reaches hardening temperature. Using a torch to heat a part with a varying cross section or sharp corners is difficult and the heating is uneven. Larger parts must sometimes be heated for one hour per inch of thickness to insure that the entire part reaches the same temperature. This is virtually impossible to do with a torch, and certainly not practical.

How large a part can be hardened using just torch heat?
Somewhere between a pack of cigarettes and a 1-2-3 block, depending on the skill of the operator. The objective is to bring the entire part up to the same temperature without over-heating, which can lead to cracking. *Larger parts must be sent to the heat treater.*

How can you determine when steel has been heated to its critical temperature so austenite is formed?
There are four methods:

- Electrical instruments, called *pyrometers*, accurately determine the temperature of the steel. Furnaces are usually equipped with them and hand held instruments are also available. Using a pyrometer is the most accurate temperature measuring method.
- At about $10 each, a much less expensive alternative is a *temperature-indicating crayon*, Figure 10–4. The two common brands are Tempilstiks® from Tempil, Inc. and Temp Sticks from LA-CO® Markal. These crayons are slightly less than ½-inch diameter and 5-inches long and are supplied in a holder/dispenser with a pocket clip. Each crayon melts within one percent of its marked temperature value. Over 100 different temperature crayons span temperatures from 100 to 2500°F (38 to 1371°C). They make measuring work surface temperatures quick, easy and accurate.

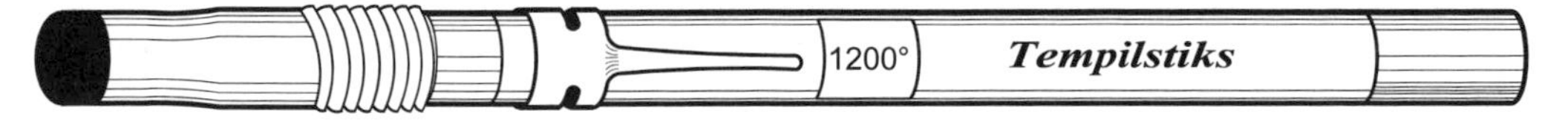

Figure 10–4. Temperature indicating crayon.

- Test the work with a magnet to determine when it is no longer magnetic. This will assure that the work is at or above the critical temperature. The dotted horizontal line in Figure 10–2 indicates this transition from a magnetic to a non-magnetic structure. This is called the *Curie temperature.* When the steel cools, it will become magnetic again.
- Visually observe that the work is cherry red in subdued light. This is the traditional blacksmiths' method and has been used for centuries, even before the hardening process was understood.

Quenching Steel

How is *quenching* performed?

After the work's microstructure is completely converted to austenite, or *austenized*, quenching is done by immersing the work in oil, water or brine. Slower cooling in still air is also used.

Why are different quenching media used?

Different quenching media, also called *quenchants*, have different cooling rates. For a particular steel alloy, too fast a cooling rate will crack or warp the work, too slow a rate, and martensite will not form. The quenching media must match the particular steel alloy and the size (heat capacity) of the part. Each type of steel has its own optimum quenchant. Steels are often designated by their quenchant such as O-1 (oil quench), W-1 (water quench) and A-2 (air quench). Cooling in still air is relatively slow, oil is faster, and water and brine are even faster, Figure 10–5.

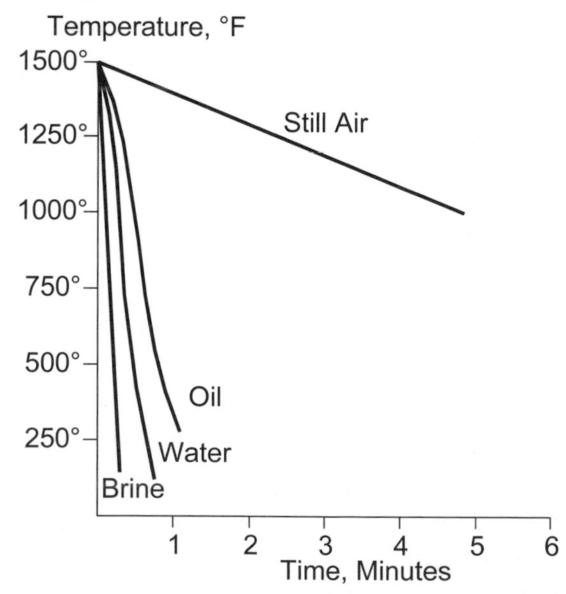

Figure 10–5. Work temperatures versus time using air, oil, water and brine.

Brine cools faster than water because no steam pockets form around the work. Steam pockets, which form in water, both retard heat transfer and disrupt uniform heat transfer over the part. This leads to uneven hardening and warping or cracking.

The important interaction between the work and the quenching media is how quickly and evenly heat is transferred from the work to the media. There is no chemical interaction.

What oils are commonly used for quenching?
Some excellent choices are:
- Mineral oil
- Automatic transmission fluid (ATF)
- Synthetic motor oil
- Commercial quenching oil

Johnson & Johnson's Baby Oil is perfumed mineral oil and it will also work fine if you don't mind the scent. Unscented mineral oil is sold over the counter at drug stores as a laxative and is commercially available in 55-gallon drums. Avoid vegetable oil as is can go rancid.

What precautions should be followed when doing an oil quench?
- Quench outside or in a well ventilated area so you are not exposed to the resulting fumes.
- Use a steel container, not plastic, which can melt through or catch fire.
- Have a metal cover for the container handy if the oil should catch fire. An even better solution is to have a *hinged* cover on the container to flip shut in case of fire. Keep a Class B fire extinguisher handy.
- Monitor the oil temperature and keep it well below its flash point.
- To avoid splashes, use an open mesh basket, rack or screen to lower the part into the oil and to retrieve it later. Such an arrangement also avoids having your face and arms over the oil during the quench.
- If the oil has been heated in a previous use, stir the oil to assure an even temperature distribution.

Quenching Problems & Solutions

The part did not harden evenly when oil quenched. What was wrong?
Most likely, there was not enough oil volume available to cool the part evenly. Cooling rate is determined by the *temperature difference* between the work and the oil. When the oil absorbed heat from the part, its temperature increased, slowing the cooling rate.

There are two rules of thumb to determine minimum oil volume:

- Provide a quenching oil volume equal to 20 times the volume of the part.
- Another approach is to provide a large enough oil volume and oil container to leave a space between the side of the part and the side of the oil vessel equal to 2 times the largest dimension of the part, Figure 10–6.

The most even quenching occurs when the work is symmetrical and is dunked up and down vertically in the quenchant, not swirled around in a circle. Swirling cools the front or leading face of the part more than the trailing edge causing warping or cracking. Even if the workpiece is symmetrical, if it has a shape that disturbs the even flow of quenchant around it, it will not harden evenly. An example of such a part would be a rod with stepped diameters or a cylinder with a larger disk-shaped projection. Regions of the part not cooled fast enough will fail to harden properly. Heat treaters have quenching tanks with special spray heads to reach these hard-to-cool spots.

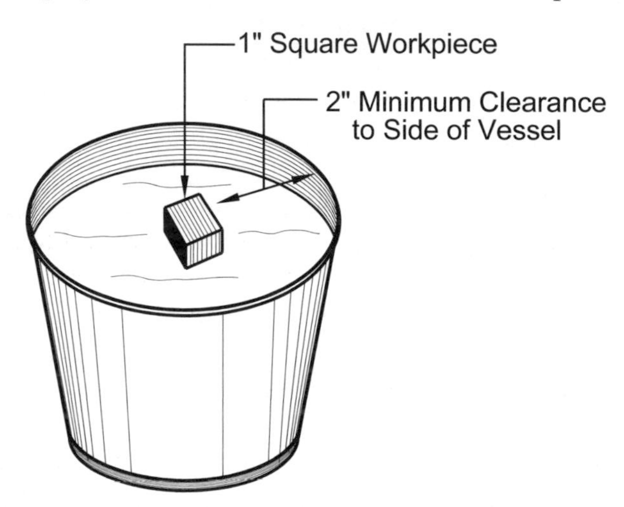

Figure 10–6. Quenching container showing adequate side clearance to assure proper cooling.

Several identical parts were hardened. The initial ones hardened properly, but the later ones did not. What was wrong?

The oil, or other quenching medium, was heated up by the initial parts and slowed the cooling rate in the later parts. To avoid this problem, use a larger volume of quenching medium. The best quenching results are usually obtained with quenching media between 70 and 100ºF (21 and 38ºC), although some materials require a higher starting temperature for quenching fluid. Too low a quenching media temperature can cause the metal to warp or crack.

When using an air quench, the work was set on a brick to cool. The upper part of the work hardened, but the bottom did not. What was wrong?
The brick prevented the bottom from cooling rapidly enough to form martensite, so it did not harden. The solution is to bend a piece of sheet metal into a Z-shaped cooling stand and use it set on edge to cool the work, Figure 10–7. This method allows the bottom of the work to cool more rapidly.

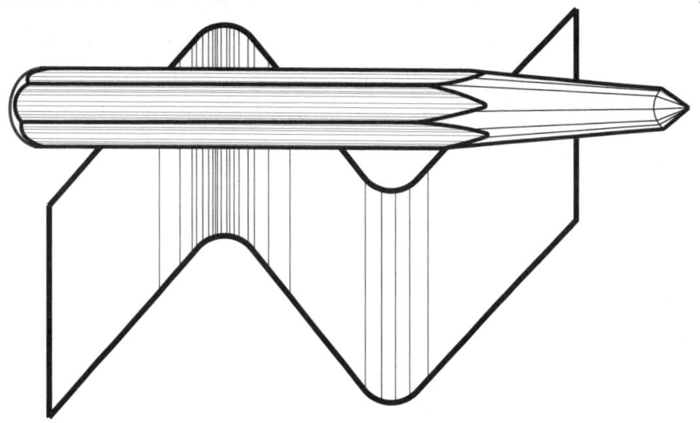

Figure 10–7. Shop-made Z-shaped sheet metal cooling stand.

Tempering Steel

What is the objective of *tempering*?
Tempering, also called *drawing*, reduces the brittleness of hardened steel and removes the internal strains caused by quenching. It is performed by reheating hardened steel to a temperature *below* its critical temperature, and then allowing it to cool to room temperature at a controlled rate. This allows some carbon to escape from inside the martensite structure, reducing its brittleness.

Nearly all hardened steel must be tempered or it will shatter in use. By using different tempering cycles, we can control the balance between hardness and brittleness in the final product.

What temperatures are used for tempering?
Temperatures from 350 to 1350°F (177 to 732°C) are used, but most steels are tempered at about 450°F (232°C). For a given steel, the higher the peak tempering temperature, the lower the hardness, Figure 10–8. Because controlling peak temperature is much more important in tempering than in heating prior to quenching, just using a torch for tempering is crude and not usually satisfactory.

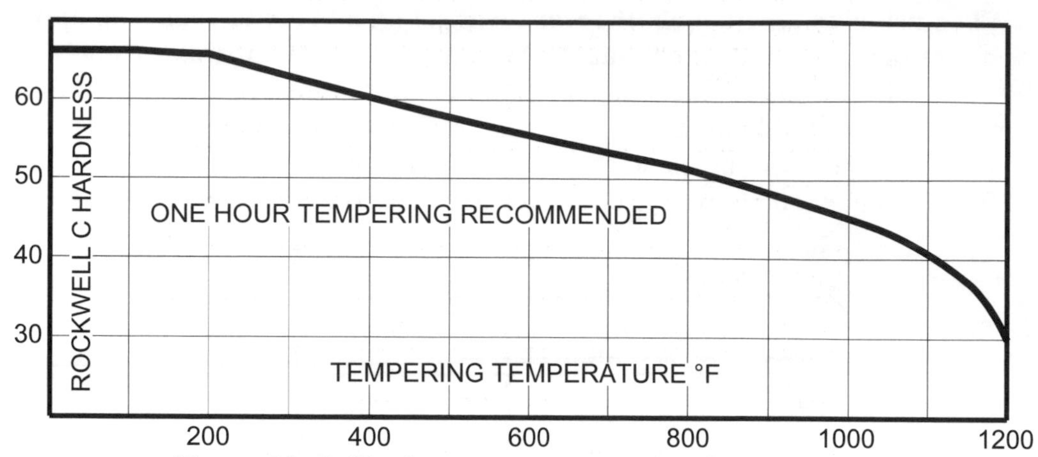

Figure 10–8. Hardness versus tempering for, Starrett No. 496, oil-hardening steel, AISI 0-1.

How is tempering actually performed?

There are several ways:

- A metallurgical furnace provides the best process control. The work is heated to the tempering temperature and allowed to cool slowly. In some instances, wrapping the work in 0.002-inch (0.05-mm) thick stainless steel foil and sealing the edges by creasing them, is used to prevent oxidation or scaling.
- In heat-treating operations, the work is heated in oil, molten salt or lead baths. Sometimes the work is warmed, or preheated, then placed in the tempering bath. Other times the work is placed in a cold bath and the bath is brought up to tempering temperature. A pyrometer is used to control the bath temperature.
- Temperature indicating crayons can be helpful.
- Visual observation of the tempering colors is also helpful.

Tempering Colors

What are *tempering colors*?

Oxides that form on the surface of steel above 430ºF (220ºC) are called tempering colors. Specific colors are related to the maximum temperature seen by the metal, Table 10–3. A skilled worker can use these oxide colors to determine when the work has reached a specific temperature. This is the same oxide that appears when a drill or lathe bit has been overheated or *burned*, and a blue oxide appears on its surface. Tempering colors cannot be observed unless the work surface has been cleaned and polished *before* heat is applied.

Tempering Color	°Fahrenheit	°Celsius
Pale yellow	420	216
Very pale yellow	430	221
Light yellow	440	227
Pale straw-yellow	450	232
Straw-yellow	460	238
Deep straw-yellow	470	243
Dark yellow	480	249
Yellow-brown	490	254
Brown-yellow	500	260
Spotted red-brown	510	266
Brown with purple spots	520	271
Light purple	530	277
Full purple	540	282
Dark purple	550	288
Full blue	560	293
Dark blue	570	299
Pale blue	590	310
Light blue	610	321
Greenish blue	630	332
Light blue	640	338
Steel gray	650	343

Table 10–3. Tempering colors and their temperatures for carbon steel.

When steel reaches even higher temperatures, the atoms in the steel become so excited they radiate light. For example, at 1652°F (900°C) steel will glow a cherry red. These colors are also used to judge work surface temperature during heating, forging and blacksmith work, but are *not* oxide colors.

Another instance colors are mentioned in relation to tempering steel is when steel is surface-hardened, called *case hardening*. In case hardening, carbon is absorbed into a thin skin (the case) of the work, resulting in beautiful oxide colors. These colors are frequently seen as very desirable on firearms components, but they are relatively fragile and can be destroyed by polishing and strong light. Air is sometimes bubbled through the quenchant to enhance oxide color formation.

Is there an effective method for tempering small tools and workpieces?
Yes, and it is a traditional one. Here is how it is done:

1. Clean and polish the end of the tool to be tempered so the oxides will be visible.
2. Heat a piece of scrap steel plate red hot with a volume of 30 to 50 times the volume of the work. The greater the volume the better.
3. Remove the steel plate from the heat and place it on a firebrick.
4. Place the tool to be tempered on the heated steel plate. Put the body of the tool on the plate and the end to be tempered so it hangs off the plate, Figure 10–9.
5. Heat will transfer from the steel plate to the work. The part of the work on the plate will heat first and the heat will gradually work its way out to the end overhanging the plate. Watch the polished portion on the end of the tool for the proper oxide color. This indicates the correct tempering temperature has been reached.
6. Drop the part in water to stop the tempering process.

The tempered end of the tool will be harder than the body of the tool on the steel plate. Do not grind a final sharp point on the tool before the hardening and tempering is complete. The tip will be blunted in these tempering processes.

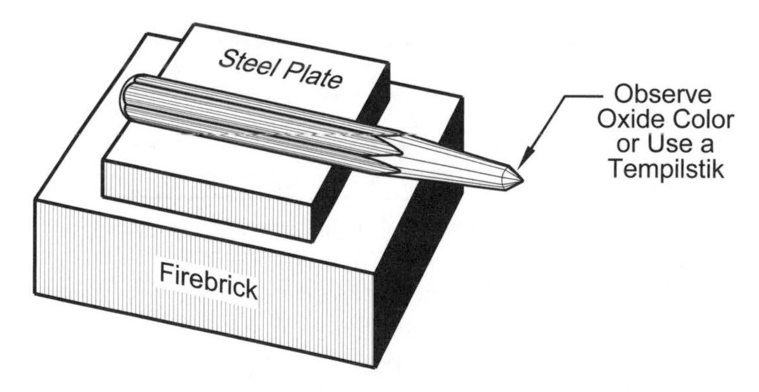

Figure 10–9. Tempering work on a heated steel plate by observing tempering colors.

What does the complete hardening, quenching and tempering temperature cycle look like?
See Figure 10–10.

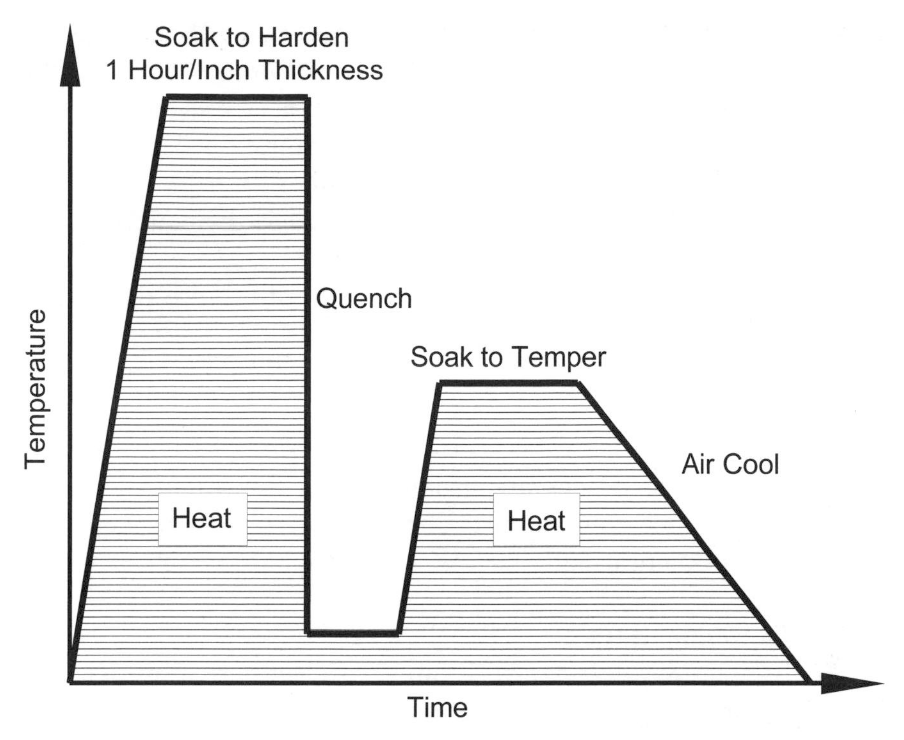

Figure 10–10. Hardening, quenching and tempering temperature cycle for carbon steel.

Where can hardening and tempering temperature data be obtained?
There are a couple of sources:

- The steel company or industrial supply house which sold the metal.
- *Machinery's Handbook.*

Case Hardening

What is *case hardening*, and what are its properties and advantages?
It is the three-step hardening process of (1) adding carbon to the outer skin, then hardening the portions with this additional carbon by (2) austenitizing and (3) quenching. The result is a hard outer skin around a softer and less brittle core. This hardened skin typically runs from 0.002- to 0.040-inches deep. Here are the advantages of case hardening:

- Low-carbon steel, which cannot be hardened by austenizing and quenching, can be case hardened. In fact, it is the only way to harden low-carbon steel.
- Medium-carbon steel can also be hardened, but high-carbon steels and alloys do not benefit from it.

- The depth of the hardened skin can be controlled by the carbon addition cycle time, or in the case of a product like *Kasenit*, more application cycles. More time (or cycles) adds more carbon. Case hardening results from carbon added on the surface gradually migrating deeper and deeper into the steel and increasing the carbon content.
- While the part is made of one piece of metal, it has the benefit of having a hard outer skin and a resilient interior.
- It is a relatively tolerant process, which does not require instrumentation and can be performed under primitive conditions.
- Case hardening is gentle on parts and works well on precision parts with sharp corners.
- Thin layers of colorful oxides, prized by firearms owners, form during case hardening. The oxides are usually just a few molecules thick, are relatively fragile, and cannot be polished without removing them.
- Stainless steels do not usually benefit from case hardening.

How is *case hardening* done?

Pack hardening and *Kasenit treatment* are two case hardening variations of interest to the machinist. However, there are many other methods available in a heat-treating operation.

Pack hardening is a traditional case hardening method used long before the actual chemistry of hardening was understood. First, the steel is carefully cleaned of dirt, scale and fingerprints. Then, the work is placed inside a vessel with a source of carbon. Ashes from bone charcoal, leather or walnut shells are favorite carbon sources. Commercial carbon mixes are also used. The work and carbon source are in contact with each other and the vessel is sealed to keep air away from the work. The vessel can be an iron box, ceramic or graphite crucible, or can be as simple as an unfired clay pot with a sealing cover. The vessel is brought to a red heat in a fire or furnace for several hours depending on the depth of hardening wanted. This time and temperature permits carbon atoms to migrate into the outer layers of the work and form austenite. The inner work material is not affected because the carbon does not diffuse far inside the work. After the time needed to deposit the carbon and austenize the outer skin, or *case depth* desired, the entire contents of the vessel are dumped into water for quenching. Sometimes the parts are cooled, reheated to austenize them, and then quenched. The work may be tempered if needed.

A widely available product, *Kasenit Surface Hardening Compound*, offers another approach to case hardening using sodium ferrocyanide. Here is how it is done:

1. The work is carefully cleaned of scale, dirt and fingerprints.
2. Heat the workpiece to 1652°F (900°C), a light red color. This may be done in a metallurgical furnace or with an oxyacetylene torch. When using a torch, try to support the part so the pliers or tongs holding it do not prevent even distribution of heat. Supporting the work with several loops of soft iron wire may be better than pliers.
3. With the part at temperature, dip and roll it in Kasenit powder, which will melt. Some will stick to the work. Make sure the part is completely coated. It is from this material that the carbon is drawn to harden the work.
4. Heat the part for 2 to 3 minutes, and then chill the part in water or oil to perform the actual hardening. Make sure your pliers are dry when picking up the part from the Kasenit. *Caution: the coating may flare up briefly when the flame is applied. Use this product in a well-ventilated area.*
5. Repeat Step 3 two or more times to increase the depth of hardening.

Section IV – Annealing Steel

Definitions

What is *annealing*?

It is the reverse of hardening. Heating hardened steel above its critical temperature converts the martensite to austenite, and then slow cooling reforms pearlite or other soft microstructures leaving the steel in the annealed or softened condition. We can reform either pearlite or martensite as many times as we wish by reheating and adjusting the cooling rate. Fully annealing steel removes all residual stress whether it was induced by cold working, grinding or by heat-treating operations.

What is the difference between *tempering* and *annealing* steel?

Tempering allows some of the carbon trapped in the martensite to escape the martensite structure reducing brittleness, but still leaves the steel hardened and predominantly of martensite. Annealing converts the microstructure to austenite, and by slow cooling, converts all martensite to other softer structures. No martensite remains.

Practical Annealing

What factors limit steel annealing in the metal shop?

Two major factors limit your ability to anneal steel. They are:

- To completely anneal a workpiece, the entire part must be raised *above* its critical temperature. While this may be done easily with a torch for a small part, not larger than a cigarette pack, larger parts must be put into an oven or fire to heat thoroughly. Most machine shops do not have an oven.

- Equally critical to successful annealing is the ability to cool the part slowly. Most carbon steel can be annealed either in still air, sand, vermiculite or lime. These last three materials have no chemical interaction with the part, but are merely insulators to retard cooling. High-alloy steels are easy to harden, but hard to anneal. Annealing these alloys requires a very slow cooling rate, approximately 10 to 40°F per hour. It is hard to achieve this rate without a metallurgical furnace as the part must be maintained at temperature for hours for complete transition—and annealing—to occur.

When annealing a previously hardened part in a metallurgical furnace, sealing the part in a stainless steel foil pouch will prevent the formation of surface scale. Some machinists coat parts with soap to retard surface deterioration.

You must anneal some ½-inch diameter ball bearings or similarly small high alloy parts. You have tried to anneal them, but your cooling rate was too high and they reharden when cooled. What can be done?
Increase the mass of heated material to decrease the cooling rate by either heating more ball bearings at the same time or by heating a larger piece of scrap steel with the bearings. Ideally, the scrap should be more cube- than rod-shaped and its volume should be 50 to 100 times the volume of the bearings.

Section V – Drill Rod & Drill Blanks

What is *drill rod* and what is it used for?
Drill rod, called *silver steel* in Britain, is a ground, polished and annealed high-carbon tool steel rod. Drill rod is used for punches, tools and dies, drills, taps, arbors, roller bearings, dowel pins, power drive shafts and a variety of machine parts where high strength and long life are required. It is easy to cut and machine as supplied in the annealed or softened state, and then readily hardened and tempered when the machining is complete.

In the U.S., drill rod is widely available in inch-fraction sizes from 1/16- to 3-inches diameter in 1/64-inch increments. Metric sizes from 1- to 40-mm diameter in 0.5-mm increments are also available. Inch-decimal sizes, number-drill sizes and letter-drill sizes are available from specialty suppliers. Not all suppliers carry all sizes and alloys. Diameter and concentricity tolerance typically runs from ±0.0003 inches in small diameters to ±0.00075 inches in the larger ones.

How is drill rod hardened?

In general, drill rod is heated to a cherry red color, somewhat above its transition temperature of 800°C, and quenched in air, oil or water depending on the alloy. Usually tempering is necessary and done by reheating to a lower temperature and quenching again. Determining when to quench is done by visually monitoring the metal's tempering colors.

How do *drill blanks* differ from drill rods?

Drill blanks are supplied as *hardened* HSS rods, or "rounds" in the same lengths as twist drills of the same diameter. They are physically identical to HSS twist drills, but have no flutes. Drill blanks are available in inch-fraction sizes from 1/16 to 1/2-inch in 1/64-inch increments, in number-drill sizes from 1 to 90, and in letter-drill sizes. Metric sizes are available, but hard to find. Drill blanks may be purchased individually, but are usually supplied in complete sets in a metal index.

Drill blanks are used to check hole and drill bushing sizes, as guide pins, punches and cutting tools. Because they are supplied in the hardened condition, the user shapes and forms them by grinding. This must be done under coolant or relatively slowly so that they are not accidentally annealed and their hardness destroyed.

How do *reamer blanks* differ from *drill blanks*?

The difference is in their tolerance limits. Drill blanks typically have tolerances of +0.0000 to −0.0003 inches while reamer blanks are +0.0002 to −0.0000 inches. Put another way, drill blanks may be undersized, but not over sized, while reamer blanks may be slightly oversized, but not undersized.

What types of drill rod are available?

There are three common types:

- Air hardened.
- Oil hardened.
- Water hardened.

Table 10-4 shows the common drill rod types available.

Quenching Medium (AISI Grade)	Properties
Air (A-1)	Excellent wear and abrasion resistance.
Air (A-2)	Maximum abrasion resistance. A-2 is recommended rather than O-1 when increased wear resistance, safer hardening and less distortion are required. Less likely to crack on cooling because of its slower cooling rate. Machinability: 65%.*
Air (D-2)	High abrasion resistance, but hard to machine. Good for tools, dies and long production runs. Non-free machining. Machinability: 50%.
Air (S-7)	Shock resistance and high impact strength with moderate wear resistance. Machinability: 95%.
Oil (O-1)	Most widely used alloy for general tool room use. Moderate resistance to abrasion. Non-shrinking. Temper immediately. Machinability: 90%.
Water (W-1)	Least expensive grade. For average work where the best grades of tool steel are not required. Offers a combination of high abrasion resistance and toughness when hardened. Often usable without heat treatment. Machinability: 100%.

* 1.0% Carbon Tool Steel = 100% machinability.

Table 10–4. Drill rod AISI grades and alloy properties.

Section VI – Ground Flat Stock

What is *precision ground flat stock* and why is it useful to the machinist? Precision ground flat stock, also called *gage plate*, consists of carbon or alloy steel of known composition and properties. It is ground flat and square with parallel sides and a fine surface finish. Several different grades are available. Typical Starrett offerings are:

- Oil-hardening steel, Starrett No. 496, AISI 0-1.
- Air-hardening chromium alloy, Starrett No. 497, AISI A-2.
- Case-hardenable, low-carbon free machining steel, Starrett No. 498, similar to AISI 1117.

It is supplied in 18-, 24- or 36-inch (457-, 610- or 914-mm) lengths, minimum thicknesses from 1/64 to 2 inches (0.4 to 51 mm) and widths from 3/16 to 4 inches (4.8 to 102 mm). Not all grades are available in every dimension.

It is useful to the machinist because:

- The user need not have a surface grinder to have a ready and convenient supply of high-quality steel stock in many sizes.
- Their composition and hardening properties are provided so the user does not risk problems or losses in heat-treating.
- This stock is dimensionally stable so the user does not risk warping it when machining removes portions of it before heat treatment.

What are the specific properties of the following ground flat stock steels? See Table 10–5.

Steel Designation	Applications and Characteristics
Oil hardening steel, Starrett No. 496, AISI 0-1	R_c 63–65 as hardened. Harden @ 1450–1500°F. Oil quench @ 120–140°F. Temper Cutting Tools @ 300–350°F. Temper Punches & Dies @ 400–450°F. Temper Springs @ 750–800°F. Temper for 1 hour. See tempering curve in Figure 10–8.
Air hardening, chromium alloy, Starrett No. 497, AISI A-2	R_c 63.5–65 as hardened. Harden @ 1700–1775°F; cool in still air. Temper Punches & Dies @ 400–425°F. Temper Heavy Punches & Dies @ 700°F. Anneal @ 1525–1575°F furnace: at no more than 50°F per hour to 800°F for maximum softness. Detailed tempering curve in the Starrett catalog.
Case hardenable, low-carbon, free-machining, Starrett No. 498, similar to AISI 1117	Best machining and lowest cost of the three steels. May be used in many applications without hardening. Case hardening @ 1700°F for 3 hours will produce a 1/32-inch hardening depth.

Table 10–5. Three Starrett ground flat stock steels.

These three products will probably meet most shop requirements that do not require the involvement of a metallurgist or heat-treating shop. Purchasing these alloys costs more than using steels of unknown properties, but their dimensional stability and consistent hardening/tempering properties justifies the additional cost. You will get predictable results. In almost every instance, the cost of machining the steel is many times greater than the cost of the steel itself, so using junk stock of unknown properties is usually a false economy. Similar alloys are available from many steel companies.

Section VII – Other Heat Applications

What are some other applications of heat in the machine shop?

- Annealing copper and brass. This may be done in a furnace or an oven by heating it red hot and quenching in water or air cooling. This process may be done repeatedly after each work-hardening cycle.
- Separating metal parts "frozen" together by corrosion can often be accomplished by heating the parts red hot and then quenching. The idea is to use heat to expand one portion of the assembly against another. Naturally, the heating and quenching is likely to destroy the parts' factory heat treatment. Both a torch or heat gun make good heat sources.
- Soldering and brazing are excellent ways to join many metals. These processes use much lower temperatures than welding, usually avoiding the warpage problems associated with the high temperatures of welding, but they still provide good strength.
- Expanding parts prior to assembly with a torch or heat gun. Sometimes it is necessary to *heat* the outer part and *cool* the inner one in a freezer to get them together. Do not apply torch heat to bearings or similar assemblies. To heat them there are commercial kits that use the bearing as the secondary winding of a transformer. The electrical current circulating *inside* the bearing metal heats it evenly and safely.
- Heating parts with a torch greatly reduces the force needed for bending, especially in steel.

Section VIII – Metal Fatigue

What is *metal fatigue*?

Metal fatigue is the development and gradual growth of a crack, which after repeated cyclic stress suddenly fractures, destroying the part. Metal fatigue is a phenomenon of crystalline materials.

At what stress levels do metal fatigue failures occur?

Fatigue-induced failures often occur at one-half to one-quarter of the maximum elastic load the metal can sustain under constant stress. In extreme fatigue cases, as little as one-eighth the maximum elastic load causes failure after enough load cycles.

Why is metal fatigue important?
Even if a metal part is properly sized to carry its anticipated maximum loads, fatigue failure can eventually occur, causing inconvenience, property damage and even loss of life.

Metal fatigue first became important in the eighteenth century with the advent of the steam engine. Although engineers had enough understanding of physics and materials to design steam engine components to handle the anticipated *static* loads, the cyclic loads developed by the steam engine introduced these early engineers to metal fatigue, sometimes with catastrophic results. Even at two hundred revolutions per minute, a machine will see over 100 million load cycles after a year. Before steam power, most metal structures were bridges or farm machines which were subject to few load cycles. Suddenly, metal fatigue was an issue.

Today 90% of all "in use" failures are fatigue failures, not stress overload failures. Proper design prevents fatigue failures. Also, replacing critical parts after a set number of load cycles or hours of service prevents in-service failures. Periodic inspections can spot signs of fatigue failure *before* it occurs.

How does *fatigue failure* begin?
Despite the fact that the result of fatigue failure is a sudden and complete breakage of the part, fatigue failure is a *progressive* failure. After beginning, it may take a very long time to reach complete failure.

Fatigue cracks begin where a geometric irregularity in the part, such as a hole, crack, scratch, sharp corner or metallic defect, creates a *stress concentration.* Often this increase in stress occurs in a very small area. Although the average unit stress across the entire part cross-section may be well below the metal's yield point, a non-uniform distribution of stresses causes stress to exceed the yield point in some minute area and causes plastic deformation. This eventually leads to a small crack. This crack aggravates the already non-uniform stress distribution and leads to further plastic deformation. With repeated load cycles, the crack continues to grow until the cross section of the part can no longer sustain the load. The origin of the crack is on or near the part's surface because that is where stresses are greatest.

Each load cycle works to break the bonds between metal atoms and enlarge the crack. In general, these cracks split the metal's grains, rather than preferentially running along grain boundaries. This *intragrain* fracturing creates a concentric ring pattern radiating from the failure's origin called *beach marks* after the pattern left by waves as they retreat from a sandy beach, Figure 10–11 (left). When the crack becomes large enough that the remaining metal cannot sustain the stress, catastrophic failure occurs. This failure is *intergranular*, or between the grains, and leaves a different kind of pattern between the failed surfaces. The beach mark pattern is caused by uneven crack growth and the rubbing together of the sides of the crack as it develops.

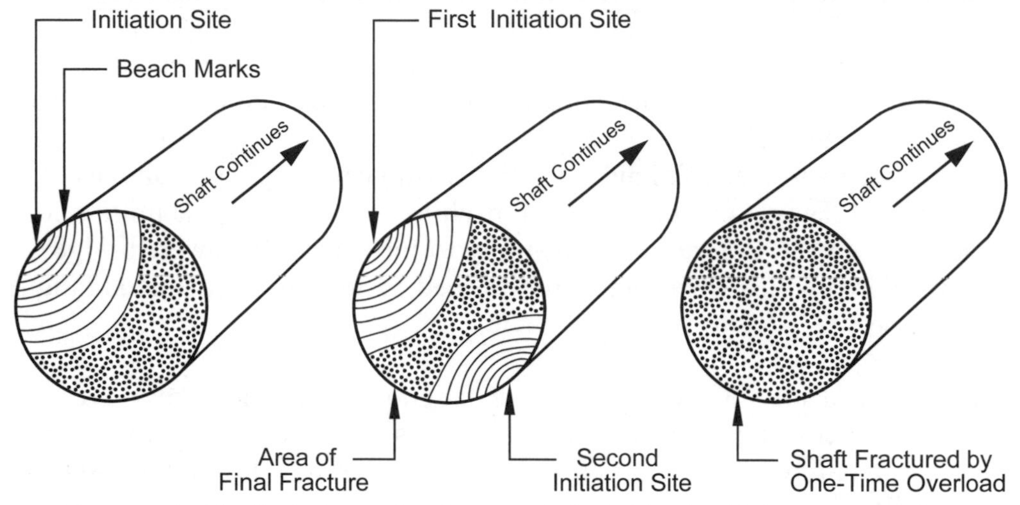

Figure 10–11. End view of a fractured shaft: Typical beach marks from fatigue crack growth (left), growth of a second fatigue crack with two sets of beach marks (center) and shaft failure by one-time overload (right).

What can examining the fracture interfaces tell us?

Because the semicircular-shaped beaching pattern inside the crack radiates from the crack's initiation point, we can determine its point of origin. Sometimes after one crack initiates, another crack develops on the opposite side of the part. This second initiation site will also be visible, Figure 10–11 (center). Another clue to understanding what happened to cause fatigue failure is that the plane of fatigue fracture is always perpendicular to the direction of maximum stress. By examining the ratio of the beaching pattern area versus the intergranular final fracture area, we can see how much margin of safety the part had initially. The greater the area of beaching, the more material the crack had to work through before failure. A failed part with a small beaching area had a smaller margin of safety than a similar part with a larger beaching area. A failed part without beaching indicates a one-time overload, not a fatigue failure, Figure 10–11 (right).

What causes stress concentrations?

Figure 10–12 (top) shows a metal bar under stress with no defects or irregularities. The stress trajectories shown by the dotted lines show stress divided evenly across the part. Average stress is the maximum stress. Figure 10–12 (bottom) shows a similar part with a notch or crack also under stress. The stress trajectories represented by the dotted lines "bunch up" around the tip of the notch. These are areas of stress concentration. Peak stress is much higher than average stress. Stress concentration in this area will cause a crack as the part is load cycled, and the crack will grow.

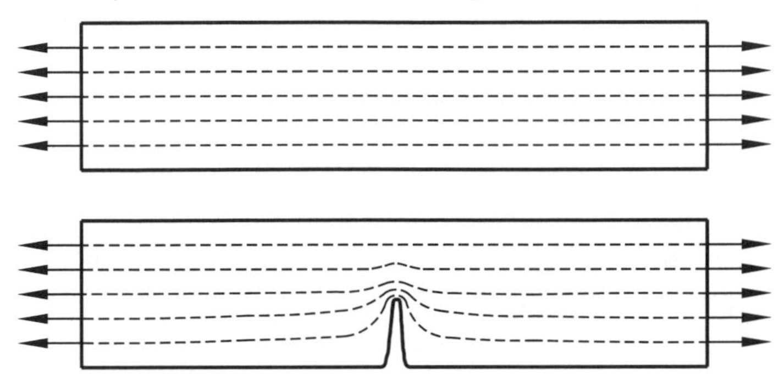

Figure 10–12. Stress trajectories in a flat bar under tension: Bar without defects (upper) and bar with crack entering its side creating stress concentration at crack tip (lower).

In the simple case of a round hole drilled through a metal bar, Figure 10–13, how much of a stress increase does adding the hole cause?

The peak stress level increases around a hole to 82.5 from the non-hole average of 25, or about 2.5 times. See Table 10–6.

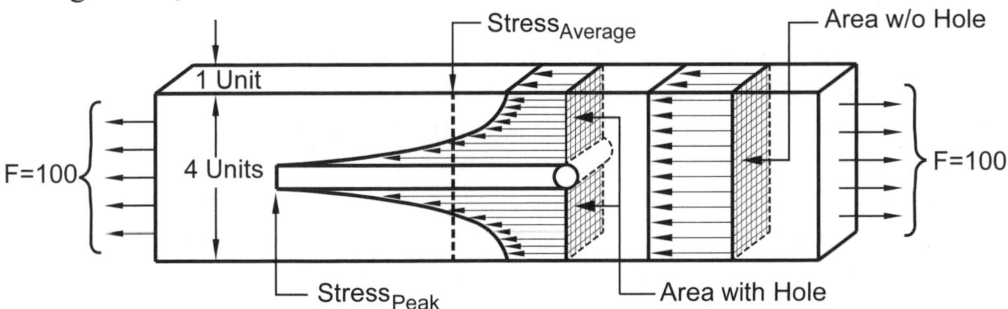

Figure 10–13. Stress increases in a rectangular metal bar under tension caused by the addition of a hole.

When a hole is added, two factors affect stress levels in the bar:

- The cross sectional area of the bar is reduced from 4 area units to 3 area units, so $Stress_{Average}$ increases from 25 to 33.3 units, Table 10–6.
- The hole disrupts the smooth flow of stress through the bar, and produces stress concentrations at the edges of the hole. Unlike sections of the bar without holes, the average stress is no longer peak stress. In fact, the stress around a hole, like the one in Figure 10–13, is about 2.5 times the average stress. This is called the *stress concentration factor* or *SCF* for short. Figure 10–13 and Table 10–6 shows how the SCF increases $Stress_{Peak}$ to 82.5 units.

In this example, $Stress_{Peak}$ increased from 25 to 82.5 units because of the addition of a hole. Stress concentrations caused by holes must be taken into account by designers for both static and fatigue strength. Mechanical engineering handbooks have charts to estimate stress concentration factors for various part shapes.

Bar Cross Section	**Area**	**Force**	**$Stress_{Average}$**	**$Stress_{Peak}$**
No Hole	4	100	$\frac{F}{A}=\frac{100}{4}=25$	$Stress_{Average}=$ $Stress_{Peak}=25$
With Hole	3	100	$\frac{F}{A}=\frac{100}{3}=33$	$Stress_{Peak}=$ $SCF \times Stress_{Average}=$ $2.5 \times 33.3 =$ 82.5

All units are relative.

Table 10–6. Average and peak stress for cross-sections of a bar under tension.

How much does a crack or other irregularity increase stress above the average stress?

Usually holes, fillets, steps and sharp corners create stress concentrations from 1.2 to 3 times the average stress at the cross-section. The increase of stress above average stress depends on the shape of the specific workpiece, and the size, shape and location of the stress raiser.

Stress concentrations at the corners of a square hole can be much larger, and small cracks and scratches that have sharp points have stress concentration factors between 100 and 1000. Shipbuilders learned about these factors the hard way, hence round port holes. Stress concentrations are more than enough to cause crack growth and fatigue failure. Figure 10–14 shows a metal bar under tension and a shaft under torsion with stress concentrators, and the same parts with modifications to reduce stress concentrations caused by these abrupt shape changes.

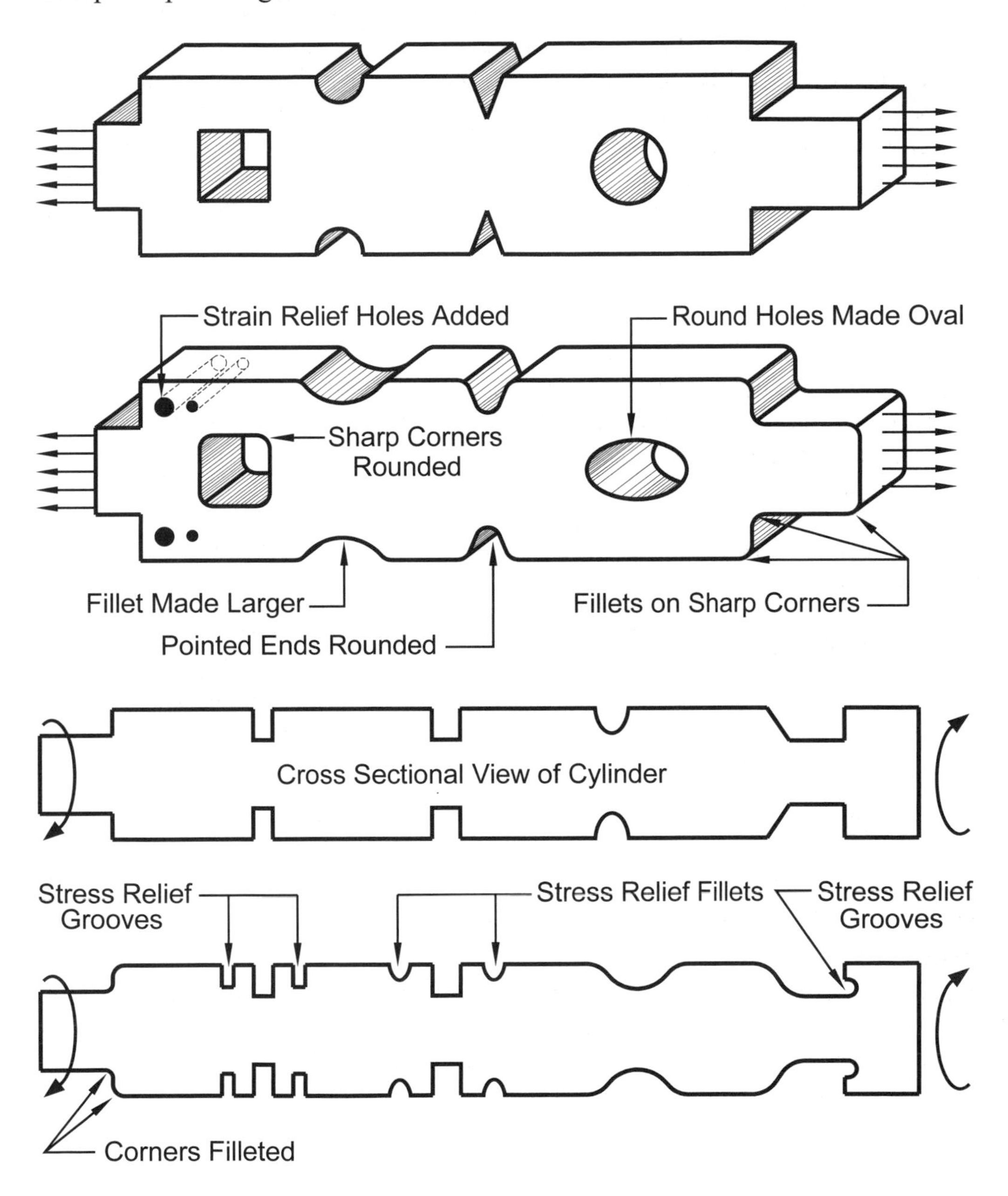

Figure 10–14. Typical stress concentrators and modifications to reduce stress concentrations in these parts. The upper two bars are in tension, and the lower two shafts are in torsion.

What are some of the important factors that influence fatigue life?

- Sharp corners and abrupt changes in shape create stress points.
- The larger the part, the greater the chance of fatigue cracks. This is because a greater volume of metal is under high stress and there is also a greater chance of residual stress increasing the stress. Residual stress is added to external stress in calculating fatigue stress.
- Type of stress loading—tension, bending or torsion—affects results.
- Surface finish, particularly scratches, scribe marks, machining tool marks and irregular points along a weld bead add points of stress concentration. Removing surface scratches by polishing may be helpful.
- Surface irregularities caused by rolling, forging or stamping processes create stress points on the part surface.
- Surface treatments like plating and grinding increase surface stress.
- Fine-grained materials, particularly high-strength steels, are more sensitive to surface finish than larger-grained, lower-strength materials.
- Corrosion attacks the metal surface and causes stress concentrations.
- Exposure to high temperatures usually reduces fatigue life.
- If possible, place welds in low stress and low flex locations.

Chapter 11

Safety & Good Shop Practices

The work of the individual still remains the spark that moves mankind ahead even more than teamwork.
—Igor Sikorsky

Introduction

Machine shops contain many potential hazards, but with proper training, knowledge, common sense and caution, you can avoid these dangers. You must remember that human flesh is no match for machines powered by several horsepower and designed to rapidly remove large volumes of metal. These machines take no prisoners and rarely offer second chances. In addition to machine hazards, there are chemical, electrical, fire and environmental hazards. Your best chance of avoiding injury comes when you:

- Understand how tools and machines work before attempting to use them.
- Know and follow basic shop safety rules.
- Learn the precautions necessary for safely handling hazardous materials by studying their MSDSs.
- Plan ahead and anticipate what is likely to happen next.
- Take your time and do not rush your work.

The material in this chapter cannot cover every potential hazard and circumstance, but it will get you started in the right direction. Additional safety precautions associated with specific machines appear in relevant chapters.

What are the principal hazards in a machine shop?

- Contact with cutting tools, particularly rotating cutters and saw blades.
- Pinches, cuts and punctures from hand and power tools.
- Capture of loose clothing, hair, jewelry or rags by rotating tools or machinery, which draws in the victim.
- Eye injury inflicted by flying metal chips, sparks, compressed-air-driven debris, compressed gases or splashing liquids.

- Eye injury from exposure to ultraviolet or infrared radiation from welding, torch or laser operations.
- Hearing damage from prolonged exposure to loud noises.
- Burns from contact with hot cutting tools, work, torches, soldering irons, furnaces, ovens and tempering baths.
- Respiratory damage from particulates and mists from machine shop cutting fluids, grinding dusts, toxic metals like beryllium and cadmium, welding smoke and spray painting operations.
- Chemical vapors from paints, solvents, adhesives, fluxes, fuels and chemicals.
- Injury from falling and tipping objects.
- Trips and falls.
- Electrical shock and secondary injuries resulting from being thrown by involuntary muscle contraction from the shock.
- Injury from lifting heavy loads.
- Accidental ingestion of non-food materials.

Personal Safety Practices

What are the essential personal steps you must take for machine shop safety?

- *In the shop, always wear safety glasses with side shields.*
- Add additional eye and face protection by using a full-face shield or over-glasses goggles when grinding, chipping, handling chemicals, blowing out chips with compressed air or using a rotary wire brush.
- Dress properly for the shop:
  - Remove ties.
  - Put up long hair in a hairnet or a baseball cap so it will not become caught in machinery. Tie up long beards because they too can be caught in machinery.
  - When working with machine tools, wear short sleeves or roll up long sleeves and fasten securely; do not wear loose clothing since it may be caught in rotating machinery and draw you into it.
  - Never wear open-toed shoes or sneakers in the shop: wear steel-toed shoes if there is a possibility of objects falling on your feet.
- Use appropriate eye, face, hand, arm and body protection when welding or brazing.
- Continue to wear safety glasses under welding helmets and full-face shields; they offer protection from shattered face shields and stray sparks.

- Remove all jewelry including rings, watches, bracelets, dangling earrings, bolo ties and necklaces.
- Use clear plastic chip and splash guards on machine tools, Figures 11–1 and 11–2. Never reach over or near a rotating cutter.

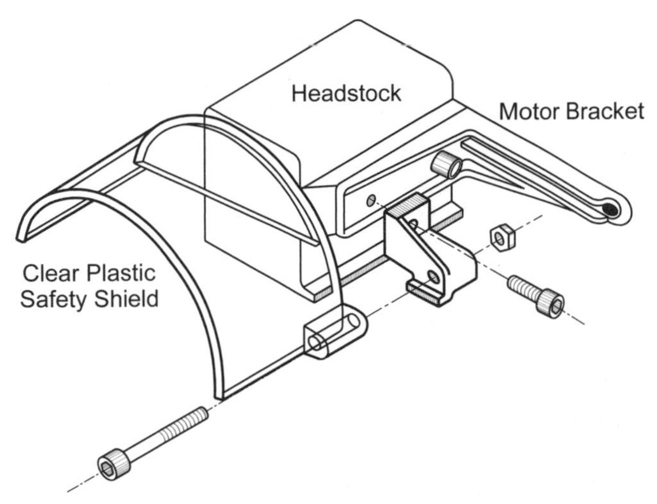

Figure 11–1. Clear plastic safety shield over Sherline lathe spindle shown from the back of the machine.

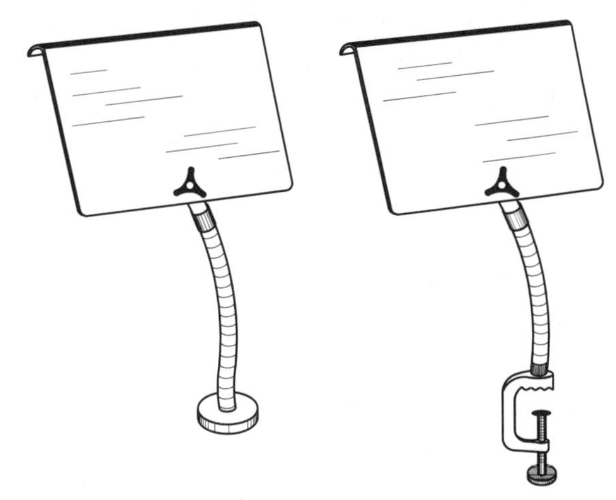

Figure 11–2. Clear plastic chip and splashguards for milling machines.

- Use hearing protection when exposed to loud noses such as when hammering, sawing or grinding.
- Do not operate machinery when fatigued, taking drugs of any kind or using alcohol.
- Wear dust and vapor protection when these hazards are present.
- Avoid horseplay in the shop; it is extremely dangerous.

- Never have more than one person operating a machine.
- Know the location of fire/emergency exits, first-aid kits, emergency eyewash stations, fire extinguishers and fire alarm stations.
- Locate the nearest phone to call for fire or rescue help and have emergency phone numbers posted by the phone.

Work Environment

What factors are helpful in performing quality work?

- A light illuminating the work in the drill press, lathe, grinder or milling machine is extremely helpful in accurately locating layout marks. These are not expensive or difficult to install and will improve the quality of your work.
- The work area should be maintained at a comfortable humidity and temperature, about 70ºF (20ºC).

Housekeeping

What are the main housekeeping steps?

- Keep the floor clear of tools, metal stock, metal chips and scrap to eliminate tripping hazards.
- Clean up liquid spills immediately.
- Do not place tools on the lathe ways; put them on a wooden tray on the ways or over the spindle.
- Do not place tools on the milling machine table; put them on a nearby worktable, protective tool tray, rubber mat or cart.
- Remove metal turnings and chips using a metal hook, pliers or a brush. Never attempt to remove them by hand. Wait until the machine has stopped before removing chips and scrap. Besides being sharp-edged, chips may catch on the rotating work or tool and draw in your hand.
- Keep all hand tools sharp and clean. A sharp tool is much safer than a dull one.
- Do not let chips accumulate in the lathe chip pan where they may get caught by the rotating chuck and be thrown at you.

Shop Safety Practices

What essential safety rules must be followed in the shop?

- Never attempt to operate a machine until you understand how to use it properly. An unusually high percentage of accidents occur during a machinist's first days on the job.
- Learn how to stop all machines quickly and the location of the emergency stop button, if any.
- Do not operate machinery until all guards are in place.

- Do not use a machine when it is broken.
- Stop your machine to make adjustments or measurements.
- Listen to your machine; if it sounds like something is wrong, it probably is. Stop the machine and find out what is wrong.
- Make sure the workpiece and tool are securely mounted or clamped *before* turning on the machine.
- Rotate lathe work manually before applying power to make sure the tool is not touching the work and the work rotates freely. On larger lathes, use the jog button to make sure the work turns freely before applying full motor power.
- Lock out machinery so someone else cannot turn on the machine while you are working on it; this is especially important if you are not in full view when working.
- Avoid eating and drinking in the work area. However, if you must, keep food in distinctive containers reserved for food and drink alone to avoid accidental ingestion of oils, solvents and chemicals.
- It is bad practice to talk with someone while operating machinery because your attention is distracted.
- Do not walk away and leave a machine running.
- Some tasks require wearing gloves and other tasks become hazardous when wearing them. Be sure to know the difference.
- Never use a rag near rotating machinery; it can be pulled into the machine and draw you in too.
- After a workpiece has been cut, remove all burrs and sharp edges with a file, belt sander or grinder.
- Store pointed and bladed tools in boxes or with protective covers.
- Lay tools out in an orderly fashion; do not heap them in a tangled pile.
- When using a chuck key, never take your hand off it until you have removed it from the chuck. Turning on a lathe with a key in the chuck turns the key into a flying projectile.
- Wear nitrile gloves, a special kind of rubber glove, when handling degreasing solvents.
- Secure compressed gas cylinders so they will not tip.
- Do not use compressed air or any other compressed gas to blow metal chips or debris from your work, machine or clothing. An exception to this rule may be made to remove material from blind holes with compressed air, but extreme caution—and goggles—must be used.
- Report and get immediate medical attention for cuts, burns, scrapes, pinches or any other injury no matter how minor.
- Know where the first aid kit is kept.

Fire-Related Safety

What are the precautions for fire safety?

- Dispose of oil-soaked rags daily to avoid spontaneous combustion, which is ignition by rapid oxidation of oils, and occurs without heat from an external source.
- Be aware that some metal chips, like magnesium, can catch fire during machining operations. Make sure you know the hazards associated with the materials you machine.
- *Never* use gasoline as a cleaning solvent; it is a fire and explosive hazard, and also a carcinogen. Remember that many solvents present explosive vapor hazards and must be used outside.
- Disk and belt sanding machines used for plastics often contain enough plastic dust in their housings and exhausts to be ignited by the sparks from later work on metal parts. Such ignition will create a sudden and spectacular jet of flame from the machine's exit port. To avoid this, clean plastic grindings promptly from the body of the machine.
- Many common solvents and adhesives are toxic, flammable or present an explosion hazard. Make sure you know the properties of the materials you use.

Electrical Safety

What steps must be followed for electrical safety?

- Use only properly grounded hand-held power tools unless they are double-insulated.
- Do not use hand-held power tools outside or in a damp location unless protected by a ground-fault interrupter (GFI).
- Use caution in handling long lengths of metal so they do not touch overhead lighting fixtures or crane power rails.
- Employ extension cords large enough for the load and distance.

MSDS

What is an *MSDS*?

MSDS is the acronym for *Material Safety Data Sheet.* By U.S. Federal law, suppliers of potentially hazardous products are required to supply MSDSs to their customers. The law also requires ready access to these data sheets by all users in their work area. Sometimes they are posted; sometimes they are in a binder or a computer file. MSDSs are designed to provide both workers and emergency personnel with the proper procedures for handling or working with a particular substance. MSDS requirements apply to products found in the workplace.

Do MSDSs apply only to chemicals and solvents?
An MSDS is required for nearly all hazardous materials used in industry whether they are chemicals, solvents, adhesives, elements, gases or metals. Even products which contain no hazardous materials or components must have MSDSs.

Where can you get MSDSs?
- Your workplace should have a collection of them that came with the product. OSHA requires that an MSDS be shipped with the initial order of a hazardous product and whenever the information is updated.
- On request, you can get them from the customer service department of the supplier.
- Many industrial suppliers have fax-on-demand systems and internet sites from which MSDSs for their products can be downloaded.
- Many state governments and universities offer extensive, free MSDS databases.

What kind of products require an MSDS?
Industrial products with hazards that include:
- *Acute (short-term acting)* and *chronic (long-term acting) health hazards.* This means products which are carcinogens (cancer causing), toxic, irritants, corrosives, sensitizers or agents which damage the lungs, skin, eyes, liver, kidneys, nervous system, reproductive system or mucous membranes.
- *Physical hazards,* such as products that are a combustible liquid, a compressed gas, an explosive, an oxidizer, chemically unstable (reactive) or water-reactive.

Note: Unfortunately, Federal law does not require MSDSs for hazardous consumer products.

What kind of information is found in MSDSs?
- Manufacturer's name and contact information.
- Hazardous components by chemical and their common names.
- Listing of permissible exposure limit and threshold level value.
- Physical characteristics of the product: boiling point, vapor pressure, vapor density, specific gravity, melting point, evaporation rate, solubility in water, physical appearance and odor.
- Fire and explosion hazard data: flash point, flammability limits, extinguishing media, special firefighting procedures, unusual fire and explosion hazards.

- Reactivity and stability: conditions to avoid, incompatibility (materials to avoid), hazardous decomposition or byproducts, hazardous polymerization and how to avoid it.
- Health hazard data: routes of entry (inhalation, skin, ingestion), health hazards, cancer signs and symptoms of exposure, medical conditions generally aggravated by exposure, emergency and first aid procedures.
- Precautions for safe handling and use: steps to be taken in case material is released or spilled, waste disposal method, and precautions to be taken in handling or storage.
- Control measures: respiratory protection, ventilation, protective gloves, eye protection and other protective clothing or equipment.

Note: It is a law of nature that the chemicals and solvents, which work quickly and effectively, are the most potentially dangerous. They are effective *because* they are reactive.

Are companies required to disclose the components in their products that are trade secrets?

Companies may withhold listing of *bona fide* trade secret ingredients from the MSDS, but they must disclose the hazards these materials present and the special handling they require. On written request by a health professional, the company must disclose the specific ingredients, even if they are trade secrets. Of course, with today's analytical laboratory instruments, especially liquid and gas chromatography, there are very few real trade secrets.

Are there metals seen in the machine shop which are particularly hazardous and should prompt reading its MSDS immediately?

Yes, they are:

- Aluminum
- Antimony
- Beryllium
- Cadmium
- Chromium
- Copper
- Lead
- Selenium
- Zinc

These metals may appear in forms other than solid pieces of metal. They may be chips, powders, grindings, machining dust, vapors, compounds with other chemicals or welding fumes. They may be safe when solid, but hazardous in other forms.

Chapter 12

Other Shop Know-how

Depend on the rabbit's foot if you will,
but remember it didn't work for the rabbit.
—Laurence J. Peter

Introduction

The title of Master Machinist is earned only after years of training and even more years of experience. This chapter presents a collection of expert advice and practical know-how from these seasoned veterans. Some of their solutions are simple and straightforward, while others are down right ingenious. Their direct answers to common questions are of benefit in two ways. First, they show how others have approached specific machine shop problems, and second, they lead the way to find your own solutions.

Section I – Turning-Related Methods

Friction Driving on the Lathe

You require one or more metal, rubber or plastic discs with a diameter tolerance of ±0.001 inches. The workpiece material is supplied flat and must remain flat after cutting. How can this part be made?

1. Cut two metal back-up cylinders the same diameter as needed for the finished discs.
2. Center drill one of the cylinders.
3. Set up the work in the lathe as in Figure 12–1. Using the tailstock and live center, compress the slightly oversized workpieces between the metal back-up cylinders. This is called *friction driving*. When cutting more than one disc, use double-stick tape between the parts to keep them in place.
4. Reduce the diameter of the workpieces so they are slightly larger than the back-up cylinders, then set the lathe tool bit so it just clears the right-hand cylinder, and "wipe" away the excess workpiece material. *Take light cuts.*

This method works well whether the workpiece material is just a few thousandths of an inch thick to ¼-inch thick. It does not distort thin stock, works well with foil, and many identical discs can be cut at once. Rubber can also be trimmed, but be sure to allow for changes in part diameter caused by compression. This method does the work of an expensive punch and die set.

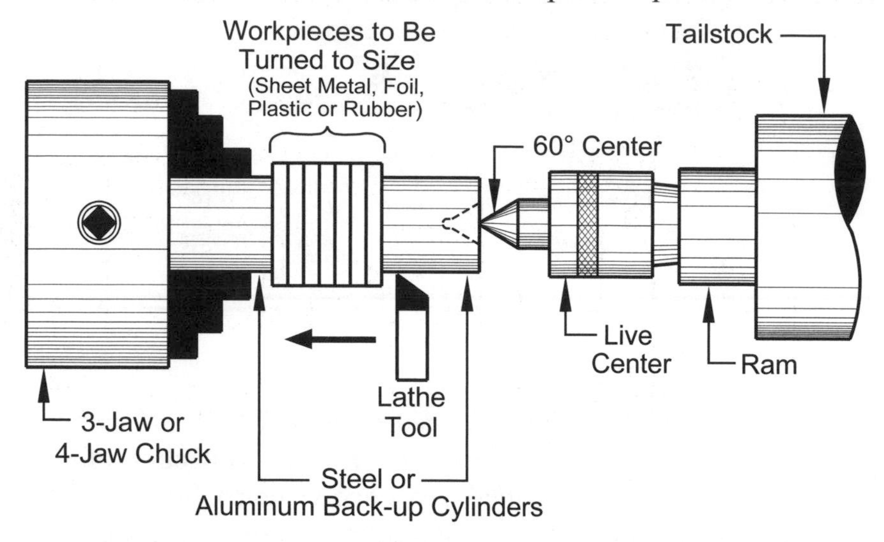

Figure 12–1. Cutting multiple discs to exact diameter using fiction driving.

A workpiece must be turned which may not be drilled for a mandrel and is too oversized to fit into the largest available chuck, how can this be done?

Set up the work as shown in Figure 12–2. Apply enough axial force with the tailstock through the live center to keep the workpiece in place. The chuck jaws may be adjusted to minimize dishing of the work from these forces.

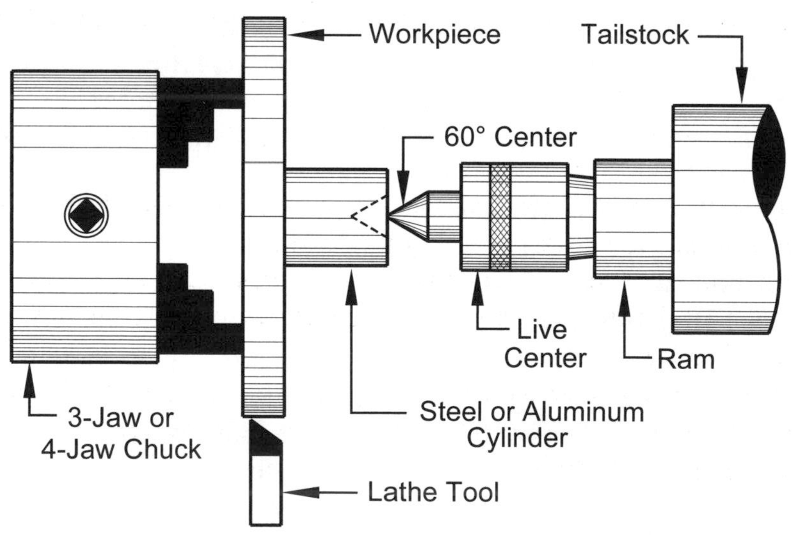

Figure 12–2. Turning oversized work.

A back-up cylinder may also be used as in Figure 12–3, but with the addition of abrasive cloth rubber-cemented to the cylinder face to increase friction. This prevents slippage and eliminates all dishing.

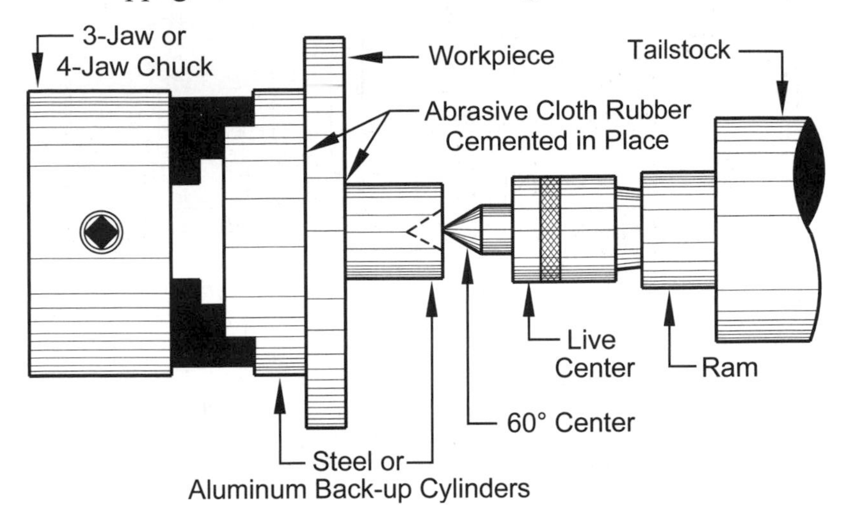

Figure 12–3. Turning oversized work with a large back-up cylinder to prevent dishing.

Grinding Accurate Radii on Lathe Tools

How can a specific radius be applied to HSS lathe tools?

Here are the steps:

1. Chuck a cylindrical aluminum oxide grinding stone in a drill press. This stone must have a radius equal to or larger than the desired lathe tool radius.
2. Mount a diamond wheel-dressing tool in a magnetic holder, and place the holder so the dressing tool can reduce the grinding stone to the desired radius. See Figure 12–4 (right).
3. Use the corner of a bench grinding wheel to rough out a U- or V-shape on the end of the HSS cutter as in Figure 12–4 (left).
4. Clamp a piece of steel or wood to the drill press table as a fence. This will guide the HSS cutter into the cylindrical stone to apply the desired radius, Figure 12–5 (left).
5. Apply a relief angle to the tool 0.030 to 0.050 inch below the cutter face. Either tilt the drill press table or use a wedge-shaped fixture to apply the relief, Figure 12–5 (right).

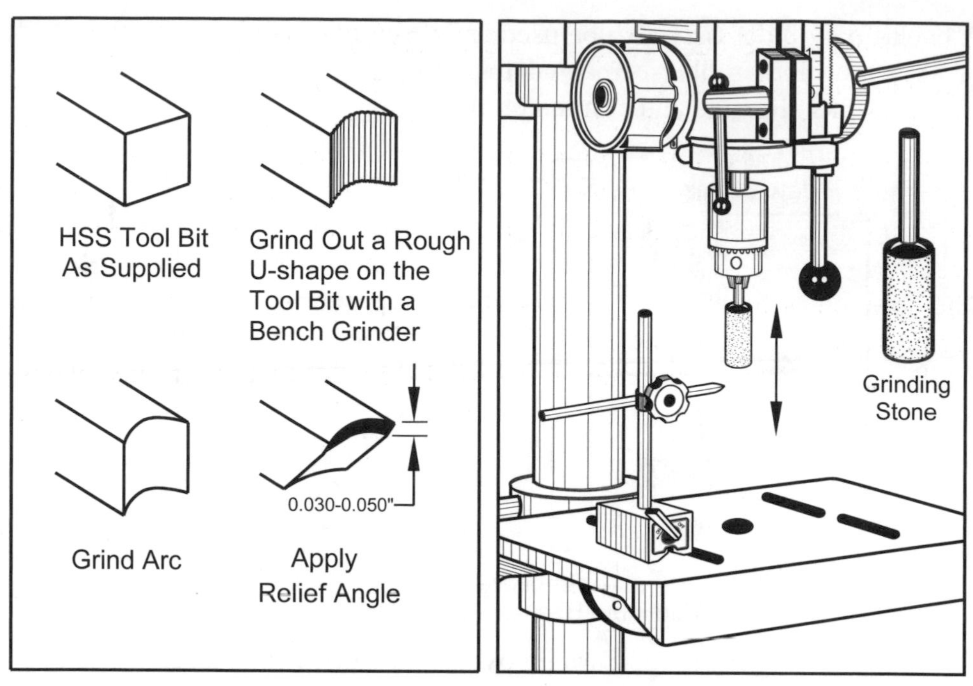

Figure 12–4. Steps to cut a lathe tool bit with a radius (left) and trimming the grinding stone to size with a diamond wheel dresser (right).

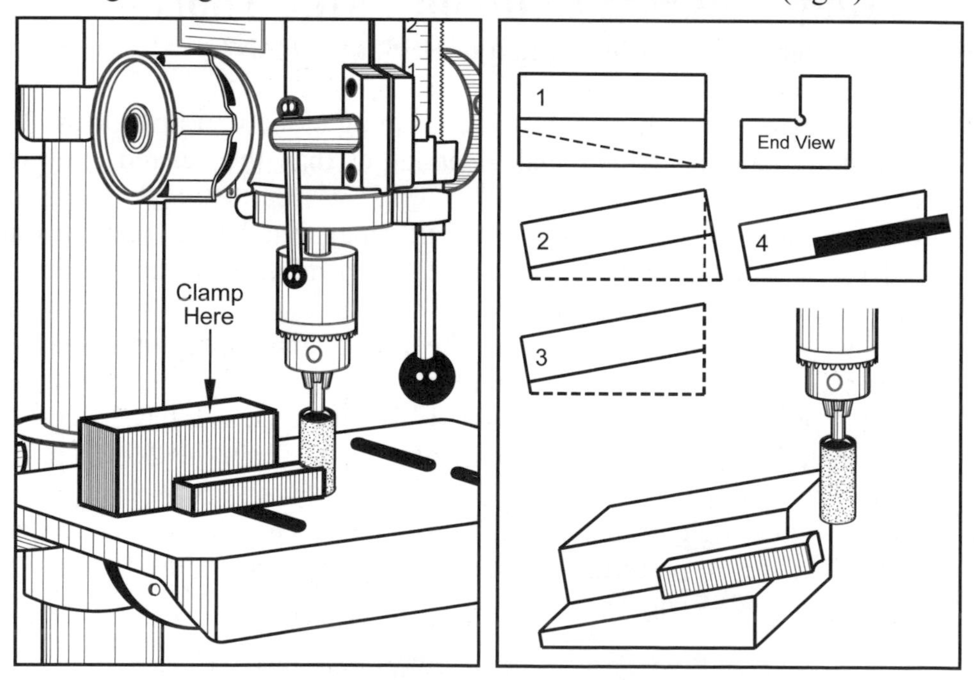

Figure 12–5. Grinding a radius on a lathe cutter (left) and applying the relief angle (right).

Turning a Bushing from Small Stock

You need a bushing made of Delrin® acetyl resin as shown in Figure 12–6 (A). There are two problems. First, Delrin has a slippery surface with a wax-like feel and is hard to hold and, second, the available raw material is only as thick as the finished bushing, making it impossible to hold the material in a chuck and cut off the finished part. How can we solve this problem?

The solution is to make a mandrel of brass, aluminum or steel as shown in Figure 12–6. Then drill the center hole to finished size in a drill press, cut the small part from the raw material, and remove its corners to reduce shock on the tool bit. Finally, mount the Delrin square on the mandrel and turn it down to finished size. This method works for most plastics.

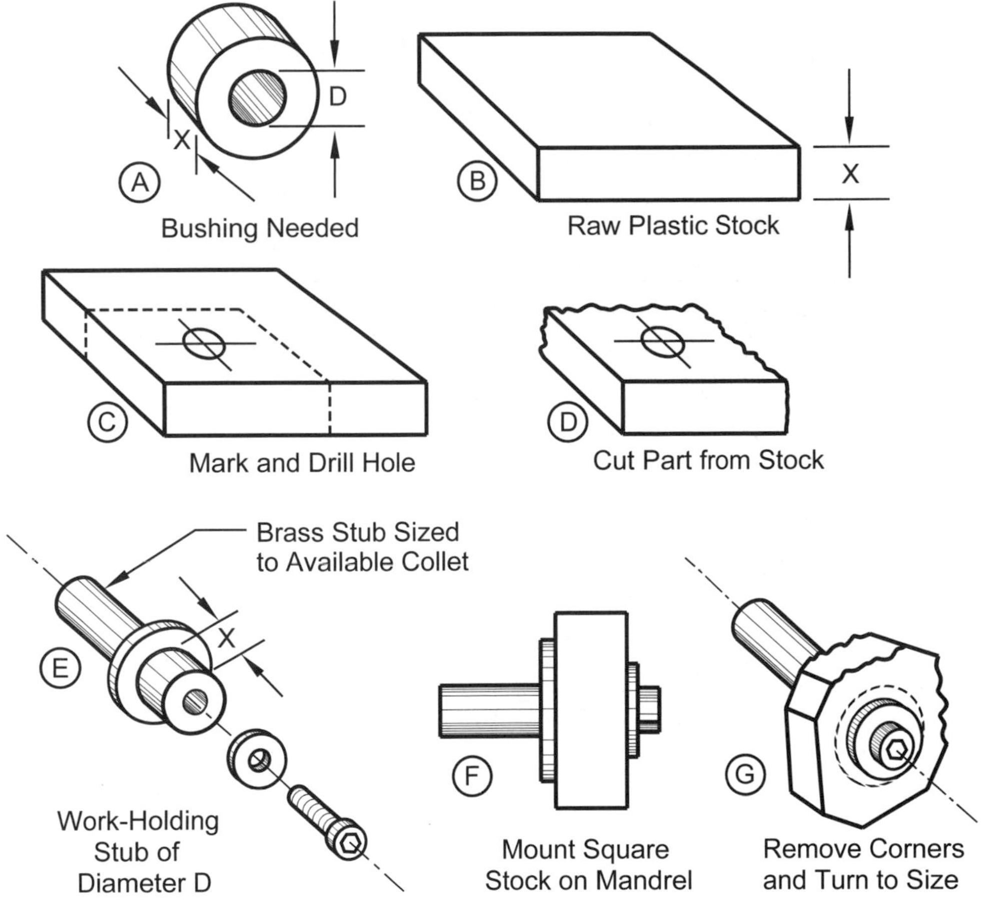

Figure 12–6. Making a bushing from a small piece of material.

Reconditioning Rubber Rollers

How can rubber rollers be turned smooth on a lathe?

There are two methods:

1. Use a tool-post grinder to apply a new surface. This method is often used for printing rollers.
2. Rollers may be frozen in dry ice and then turned at high speed. Take a series of light cuts. Refreeze rollers if they thaw and become soft during turning.

Fast Rotary Table Alignment

How can rotary tables be aligned to the spindle axis within 0.0005 inches? Make the centering tool in Figure 12–7. Its small diameter fits in an available collet, and its large diameter is 0.002-inch under the diameter of the rotary table's center hole. To center the table:

1. Fasten the rotary table to the milling table in its approximate location using T-slot bolts.
2. Mount a collet to hold the centering tool, insert the tool and tighten the collet.
3. Apply light downward pressure on the quill feed handle so the centering tool touches the edge of the hole in the rotary table's center.
4. Move the table back and forth while observing the centering tool.
5. When the spindle and rotary table are coaxial, the centering tool will drop into the center hole. Alignment is now complete. Set the DRO (or table collars) to zero.

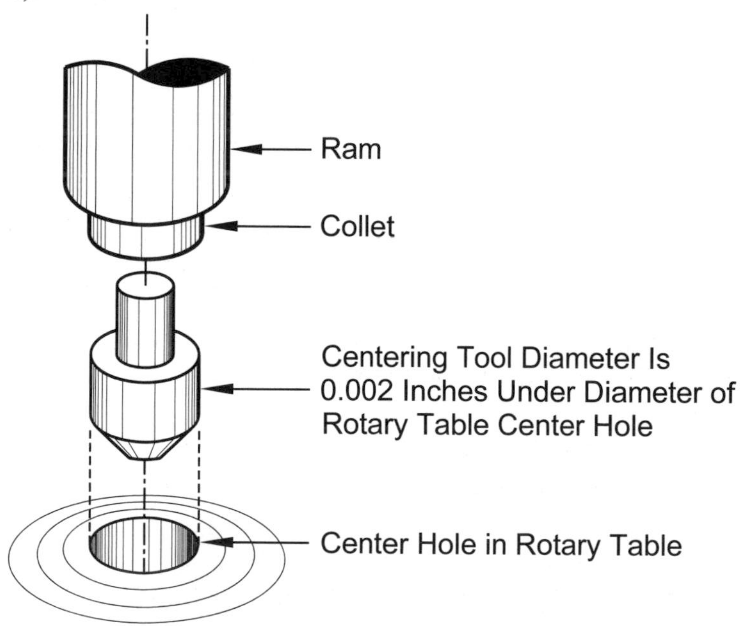

Figure 12–7. Method for quickly aligning a rotary table.

Cutting Hollow Hemispheres

A hollow hemisphere is required with a radius of up to 5 inches in a steel or aluminum block. How can this part be made?

Here are the steps:

1. Calculate the cutter diameter, which equals the length of chord AB, Figure 12–8 (A).
2. Set the cutter to the calculated diameter. Make several test cuts in scrap to verify the cutting diameter is correct, Figure 12–8 (B).
3. Lay out the center point and the finished diameter of the hemisphere on top of the workpiece. Also, scribe the X-axis and Y-axis on the top of the work. These marks will be needed to reposition the work if it is removed from the mill and to position the milling cutter. Center the spindle over the center of the hemisphere and zero the DRO.
4. Remove most of the hemisphere material. This is easy to do on a large lathe by drilling, boring and then cutting out six or seven steps, Figure 12–8 (C). However, the entire job can also be handled on a Bridgeport-style milling machine. Apply the table locks on the Y-axis (in-out movement) and do not remove them until the hemisphere is cut. Clamp the work to a rotary table, drill and bore starting holes, then use a milling cutter to remove annular rings (donuts) of material from inside the workpiece. Make up a fork to fit a ⅜-inch variable-speed electric drill and use it to turn the crank on the rotary table. This will save hand cranking and produce a smoother finish.
5. Using a DTI and the method shown in Figure 8–56, set the milling machine head to 45° off vertical, turning the head in the X-Z plane.
6. After making the adjustments in Step 2, install the cutter into the collet.
7. Apply layout dye to the scribed outline of the hemisphere. Position the cutter parallel to the X-axis. While manually swinging the cutter in a narrow arc across the X-axis, use the quill and X-axis table feed to position the tip of the cutting tool to just touch the scribed hemisphere outline where it crosses the X-axis. The tip of the cutter will make a mark in the layout dye along the hemisphere. Set the Z-axis DRO to zero, Figure 12–8 (A)
8. Apply the quill lock and X-axis locks.
9. Lower the table so the cutting tool clears the workpiece as it turns, Figure 12–8 (B).
10. Turn on the milling machine.
11. Raise the table until the tool begins cutting, then crank the rotary table through a full revolution.
12. Repeat Step 11 until the hemisphere has reached full depth, Figure 12–8 (C). The Z-axis DRO reading will indicate final cut depth.

Figure 12–9 shows the approximate tool path in the hemisphere. Cutting a 10-inch diameter steel hemisphere for molding basket balls on a manual Bridgeport-style mill is a three-day task. Figure 12–10 shows a top view.

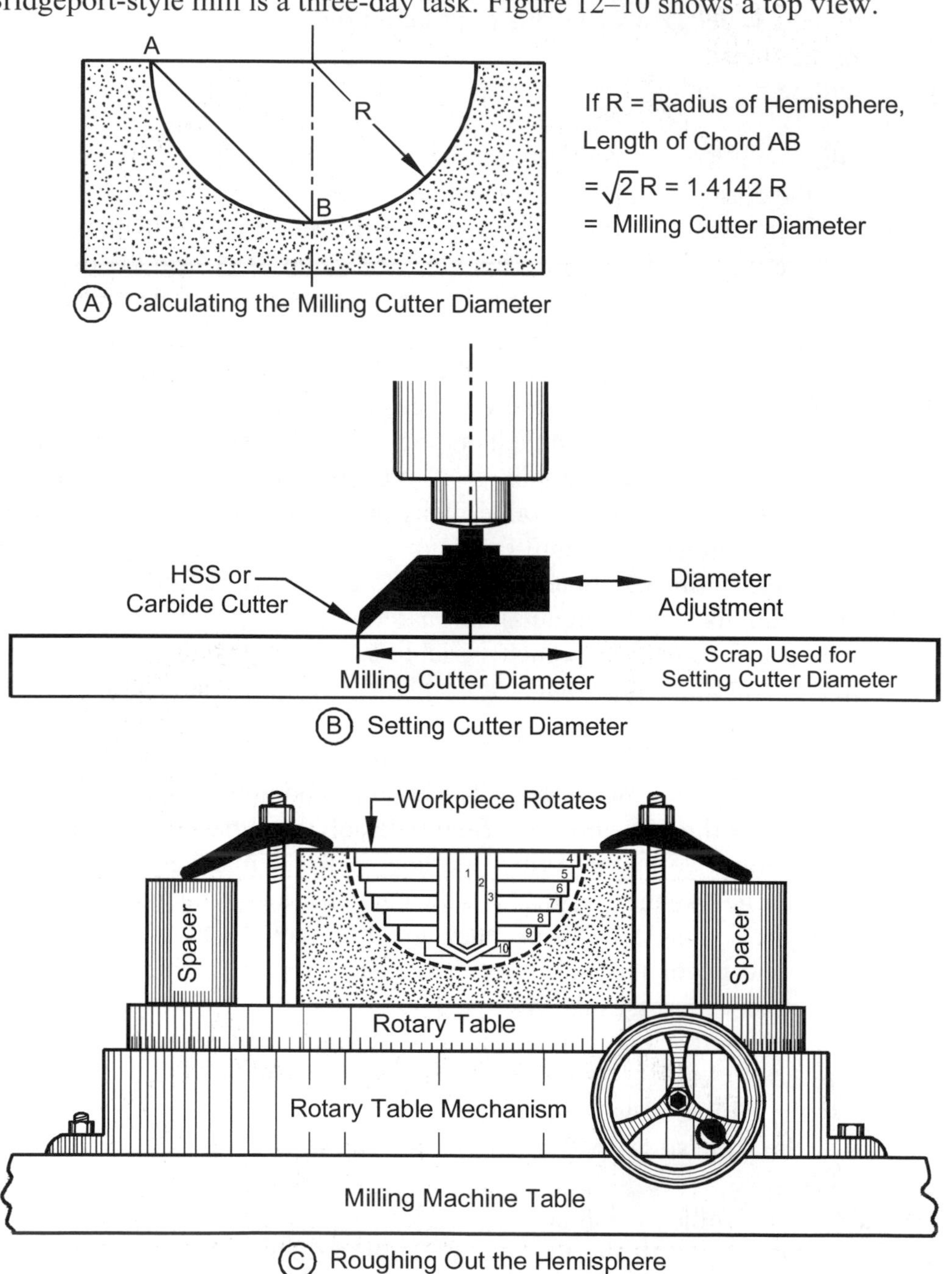

Figure 12–8. Preparing to cut a hollow hemisphere.

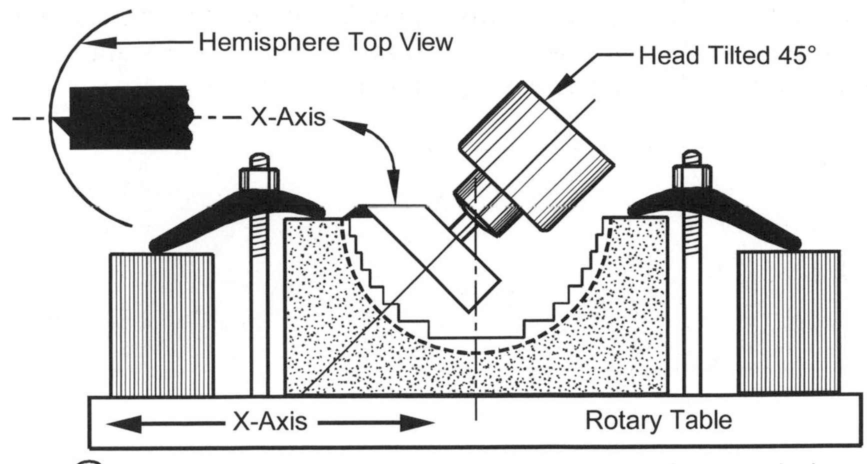

(A) Set Cutter Position on the X-Axis to the Edge of the Hemisphere

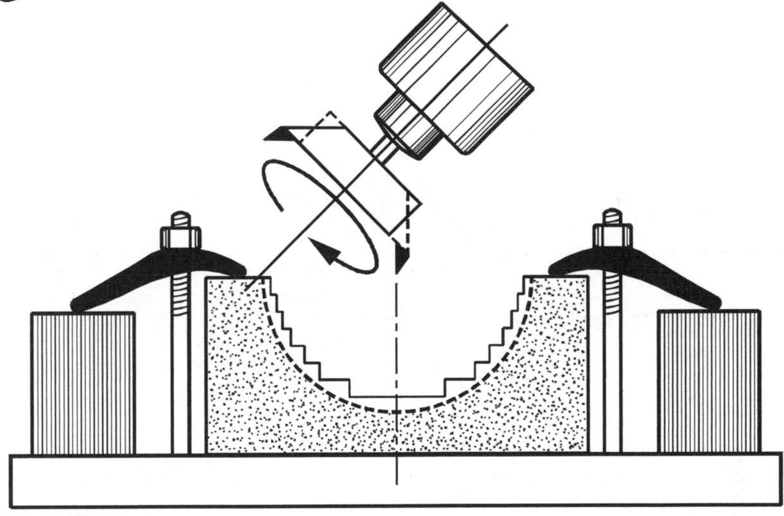

(B) Lower the Milling Table So the Cutter Clears the Work

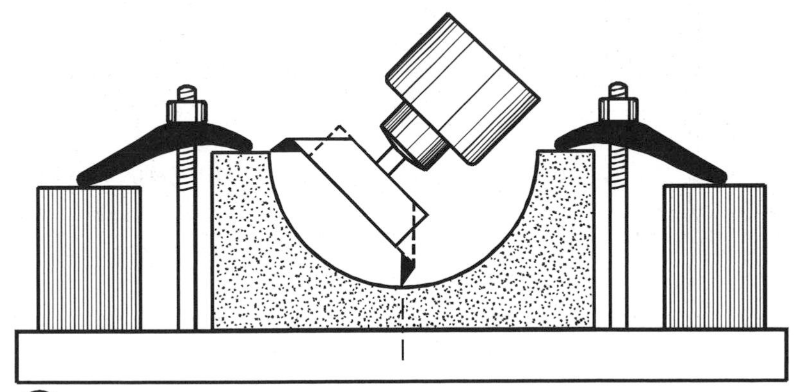

(C) Milling Table Fully Raised to Complete Cutting of Hemisphere

Figure 12–9. Cutting a hollow hemisphere.

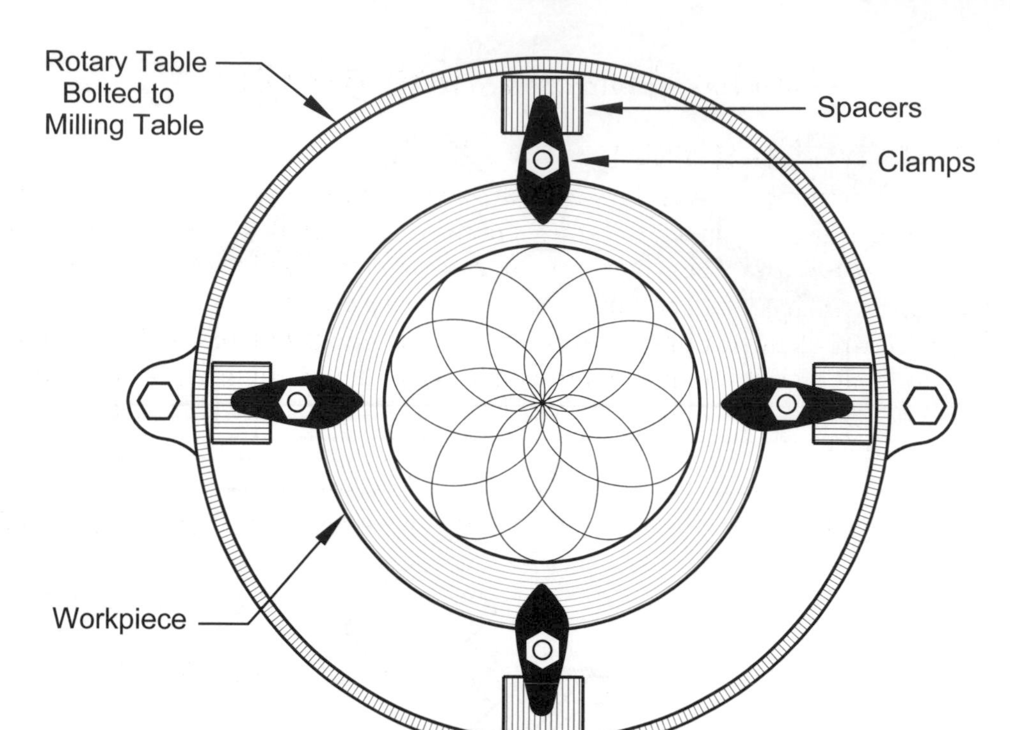

Figure 12–10. Top view of tool path when cutting a hemisphere.

Corner-Rounding Tools

You need to apply a radius to the edge of an aluminum workpiece, but do not have a profile milling cutter. How can you proceed?

Use a carbide corner-rounding cutter usually used in a router for wood. These work as well as expensive carbide milling machine cutters, are less expensive and are cheaper to resharpen. *Use them on aluminum only.*

Miniature Spacer Blocks

When milling a small part you require small spacers to use in place of the normal parallels. What common component can be used?

Key stock, normally used to lock gears and pulleys to shafts, is available from industrial supply houses in 12-inch lengths and from 1/16- to 1-inch square in 1/32-inch increments. Although available in many different metals, the best choice is annealed and heat-treatable high-carbon plain steel. Be sure to specify "Standard," which is within ±0.0015 inches of its nominal size. Note that there are both undersized and oversized key stock which will deviate from its nominal dimensions. Metric key stock is also available. Lathe tool cutter bits are usually accurately ground and also make good spacers.

Section III – Drilling Methods

Chain Drilling with Step Drills

You must remove an interior section from a plate. Typical chain drilling involves drilling a series of closely spaced holes around the interior edge of the cut-line, and then using a hacksaw or cold chisel to connect the drill holes and free the section. Is there a better way?

Yes, perform chain drilling with a step drill, which will bring the opening close to final size without sawing or chiseling. As in Figure 12–11, here is how to do it:

1. Mark the final opening cut line and inside the final cut line, scribe a line for chain drilling half the large diameter of the step drill.
2. Using a divider or compass, tick off increments half the large diameter of the step drill along the chain-drilling line.
3. Center punch these tick marks and drill each hole with the small step drill diameter.
4. Drill every other hole with the large step drill diameter. This will free the area to be removed. File off the remaining ridges.

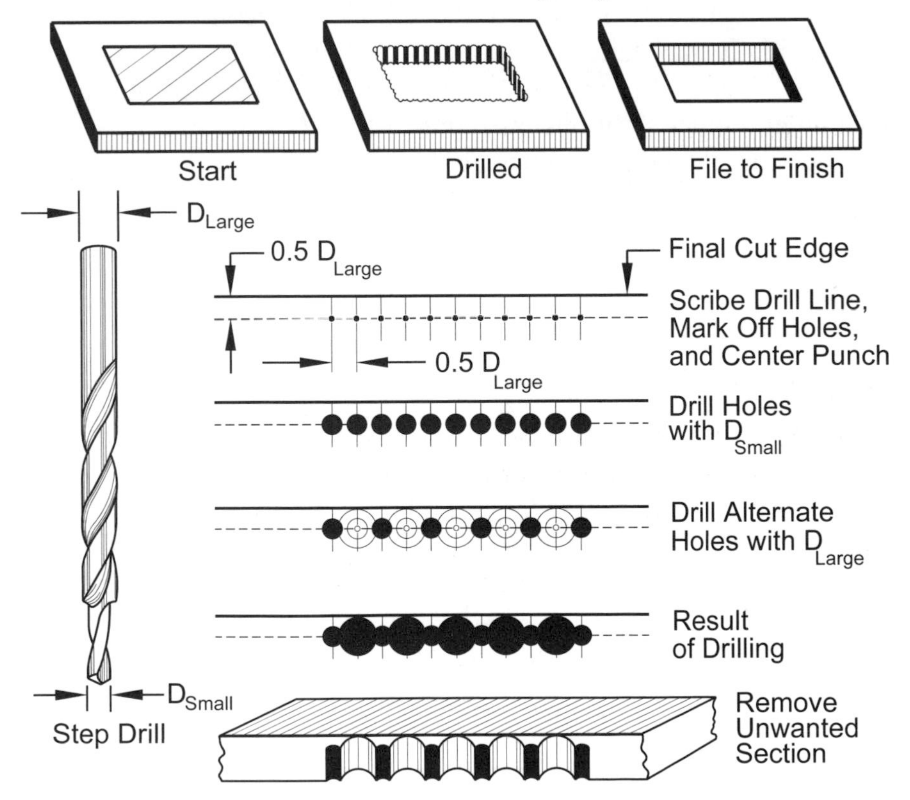

Figure 12–11. Using a step drill for chain drilling.

Drilling Spring Steel

How can an ⅛-inch diameter hole be drilled in spring steel or other material such as a hacksaw blade?

1. Set the drill press to 200 rpm.
2. Place an aluminum block under the stock to be drilled and clamp it if needed. Using wood will result in a funnel-shaped hole. The aluminum block also removes heat from the work.
3. Use a 135° drill bit with cutting oil. Moderate to heavy drill pressure will be needed to get drilling started.

Drilling Socket Head Cap Screws or Bolts

You need a small hole through the head of a socket head cap screw or bolt. How can this be done without annealing the head?

1. Grind a small flat on the side of the socket head cap screw. Apply this flat *parallel* to one of the hex flats, Figure 12–12 (left). Center punch a starting point for drilling.
2. Secure the screw or bolt so its head is well supported when drilled.
3. Set the drill press to 200 rpm.
4. Use a 135° carbide drill bit with cutting oil. Second choice is a HSS Cobalt drill. It may take several seconds before the drill starts cutting. Moderate to heavy drill pressure will be needed to get drilling started.

Drilling a hole in other locations, as Figure 12–12 (center and right), weakens the hex head more than drilling through the center of a flat.

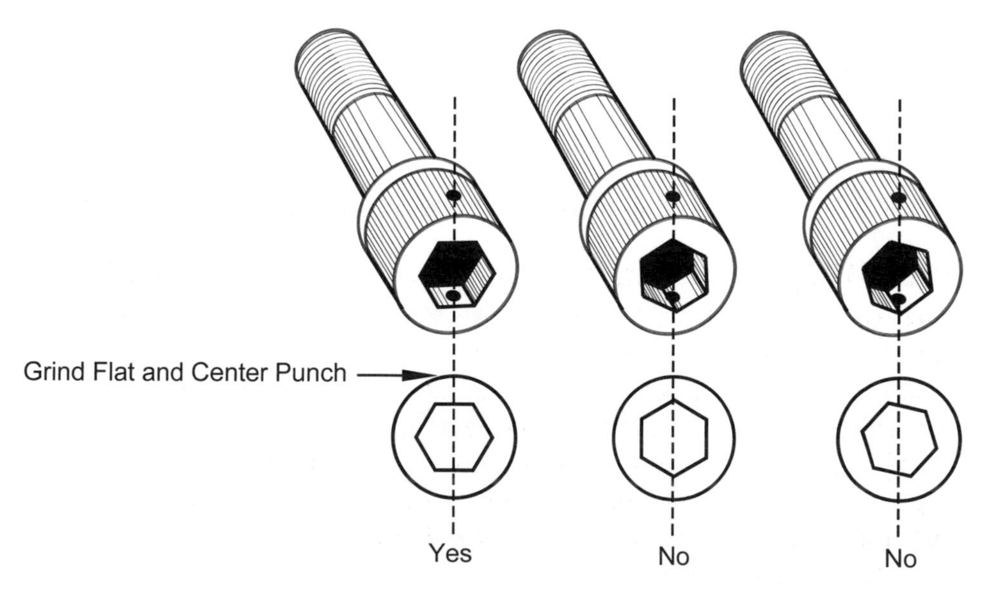

Figure 12–12. Drilling a socket head cap screw or bolt.

Section IV – Tricks of the Trade

Filing Round Corners

A metal or plastic base plate must have rounded corners with identical radii. How can this be done?

Here is a solution:

1. Select a piece of drill rod whose radius matches the radius desired on the base plate.
2. Chuck several inches of this drill rod in a lathe and cut off several ⅛-inch lengths using the cutoff tool.
3. Harden these coin-like pieces by using a torch. Bring them up to bright red, and quench them in air, oil or water, depending on the rod.
4. Position and clamp two of these drill rod coins to form a template for filing each corner. When using these templates, a disc or belt sander can easily apply the corners.
5. After removing most of the corner waste material, use light pressure on a file and let the hardened drill rod rounds guide the file to make the final radius cut. See Figure 12–13.

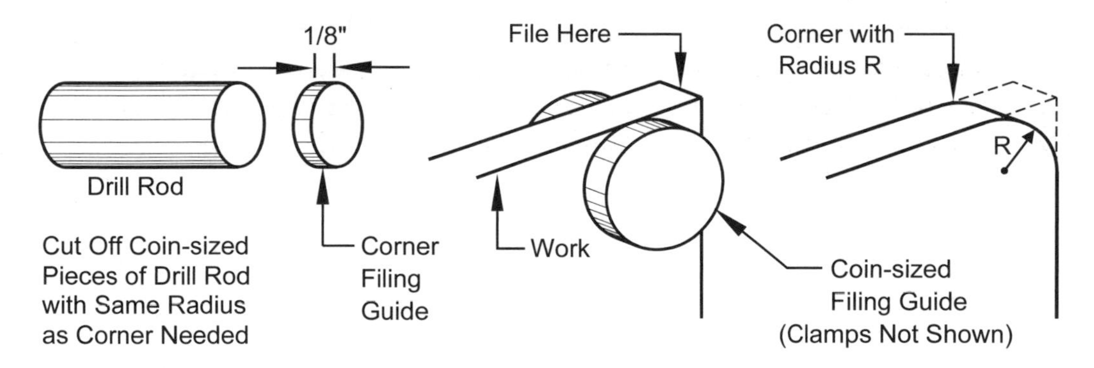

Figure 12–13. Using coin-shaped sections of drill rod to round corners.

Annealing, Drilling & Tapping Ball Bearings

Ball bearings are difficult to anneal without a furnace because, when heated, their alloy hardens easily and they cool too rapidly to remain annealed. How can annealing, in preparation for drilling and tapping, be done reliably without a heat-treating oven?

Figure 12–14 shows how this is done. The mass of the heated lead slows the cooling rate enough for the ball bearing to remain annealed. By keeping the bearing immersed in lead above its equator, the lead captures the bearing so it can be easily ground, drilled and tapped.

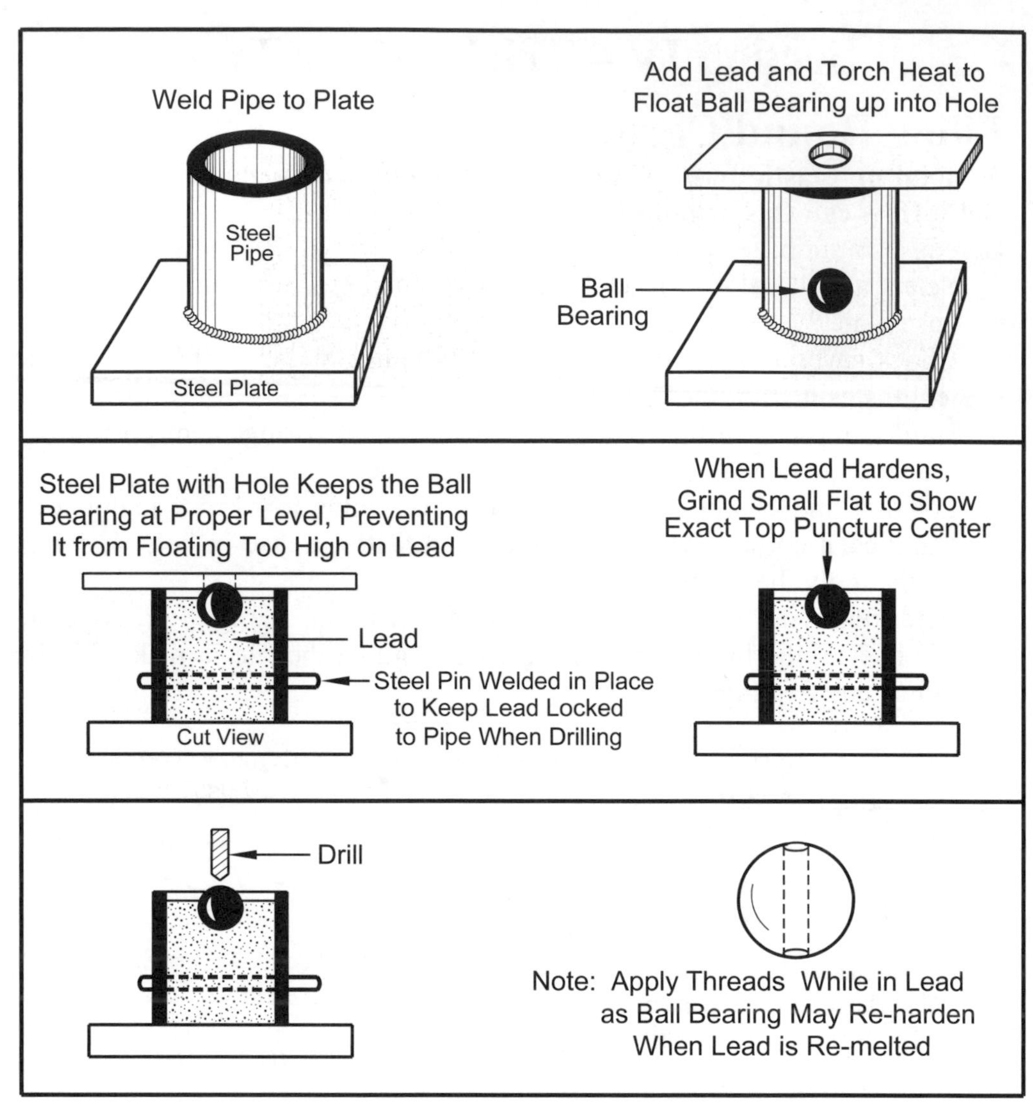

Figure 12–14. Annealing a ball bearing in a lead bath for drilling and tapping.

Holding Ball Bearings & Spheres

It is necessary to mill a flat and drill a hole in a sphere. What is the best way to hold it securely?

Figure 12–15 shows how two nuts can be used to capture the sphere. For even better results with less chances of marking the sphere, cut slight chamfers on the inside diameter of the nuts. Note how a bottom spacer provides stability and raises the sphere to allow better access to the milling cutter.

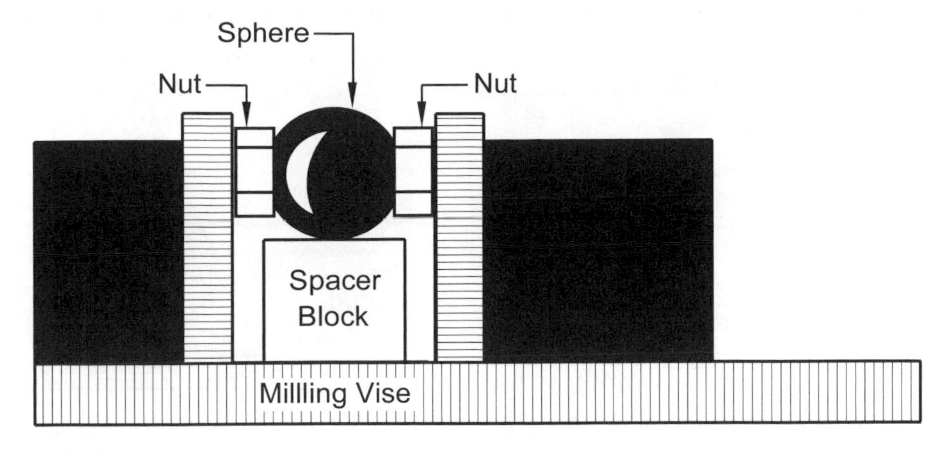

Figure 12–15. Method for securely holding a sphere in a vise.

Balls

What type balls are commercially available and what are their uses?

Balls or spheres are commercially available in more than a dozen different materials:

- Metals–aluminum, aluminum oxide, brass, bronze, chromed steel, stainless steel (Types 302 and 316) and tungsten carbide.
- Plastics–acrylic, Delrin®, Nylon 6/6, polypropylene and Teflon®.
- Semi-precious stones–ruby and sapphire.
- Elastomers–Buna-N and Viton®.

Metal balls are often available in both inch and metric sizes. Their sphericity, the maximum difference between diameters on a single ball, is measured in micro-inches, making them useful for precise measurement.

Some of their many uses are:

- Reducing hole size, Figure 12–16 (left).
- Measuring hole size. Balls are often secured on a wand to make handling them easier, Figure 12–16 (right).
- Stirring fluids using magnetic balls and an external magnetic field.
- Forming measuring instrument reference points and probe tips.
- *Ballizing,* the process of forcing balls through holes with an arbor press or hydraulic pressure, which slightly enlarges the hole and improves wall smoothness. Tungsten carbide or chrome steel balls are used in ballizing.
- Supporting precision mechanisms with bearings of metals or semi-precious stones, which both have very low coefficients of friction.
- In check valves and fluid flow indicators, plastic, metal or elastomer balls are used.

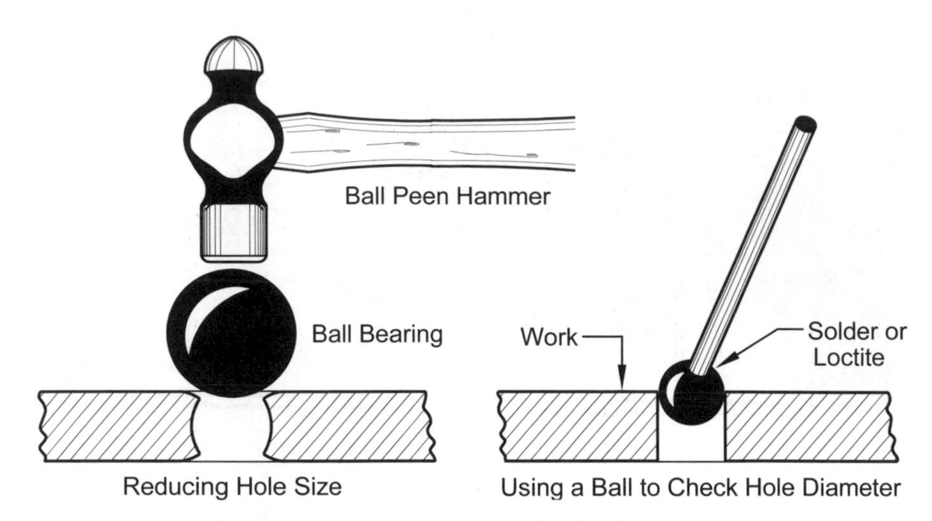

Figure 12–16. Some uses for ball bearings.

Cutting Acrylic Sheet Goods

What is a rapid method for cutting acrylic sheet along straight lines?

Acrylic sheet, up to 3/16 inches thick, may be cut by a method similar to that used to cut glass. Use a scribing knife, a metal scriber, an awl or a utility knife to score the sheet. Draw the scriber several times (7 or 8 times for a 3/16-inch sheet) along a straight edge held firmly in place. Then clamp the sheet or hold it rigidly under a straight edge with the scribe mark hanging just over the edge of a table. Apply a sharp downward pressure to break the sheet along the scribe line. Scrape the edges to smooth any sharp corners. With this method you can make accurate cuts rapidly, and without any scrap. It is not recommended for long breaks or thick material, and it will only cut straight lines.

Frozen Telescoping Aluminum Tubing

How can two pieces of aluminum tubing that are stuck together be separated without damaging them?

1. Begin by temperature cycling with a propane torch and water flooding to break the corrosion bond between the tubing. Try five or six cycles.
2. Turn a steel driving or hammer plug, which fits snugly into both tubes, Figure 12–17. This will prevent damage to both tubes and apply the hammering force to the inner tubing.
3. Hold the tubing in one hand and strike the hammer plug with a steel hammer bar, or strike the steel bar with a hammer. Do not rest the tubing against the floor or bench as its end will be deformed.

This method works on aluminum because its corrosive bonds are generally weak. It is not likely to work on steel as rust has a much stronger bond.

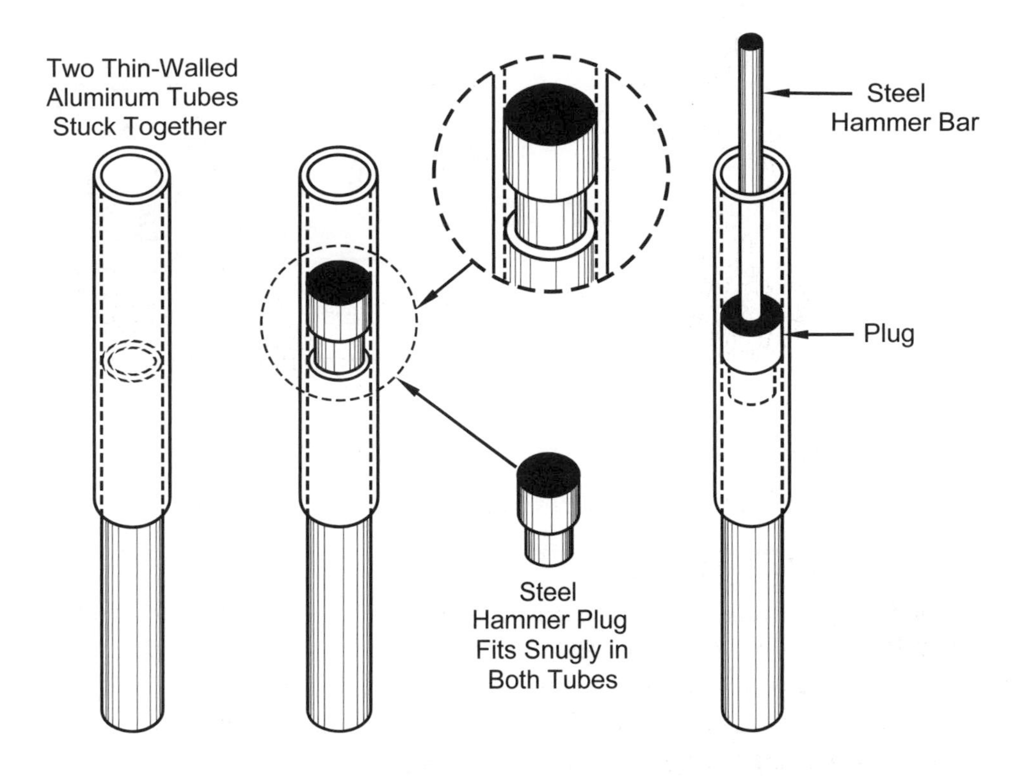

Figure 12–17. Removing stuck telescoping tubes.

Pipe Cleaning & Abrasive Burr Removal

What are two ways abrasive cloth may be used to clean the interior of pipe or tubing?

Figure 12–18 shows how aluminum oxide abrasive cloth can be wound around a saw-cut rod. The diameter of the pipe determines the length of the abrasive strip needed. When the outside section of abrasive cloth is worn, insert another strip under the first strip, which will hold it in place.

The arrangement in Figure 12–19 has the advantage of removing interior burrs from the pipe as well as enlarging its diameter. If the pipe is welded, the interior weld seam can be smoothed, but not completely removed. Use a drive punch on a lead block to make a center hole through the abrasive cloth to mount it on the arbor.

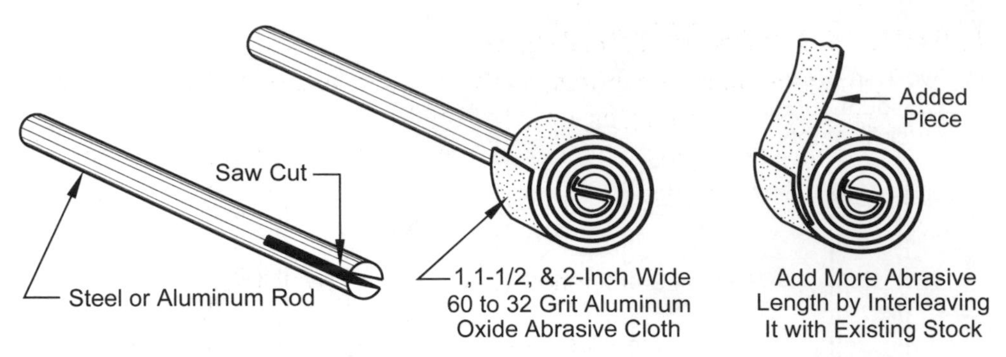

Figure 12–18. Holding abrasive cloth to clean the interior of a pipe.

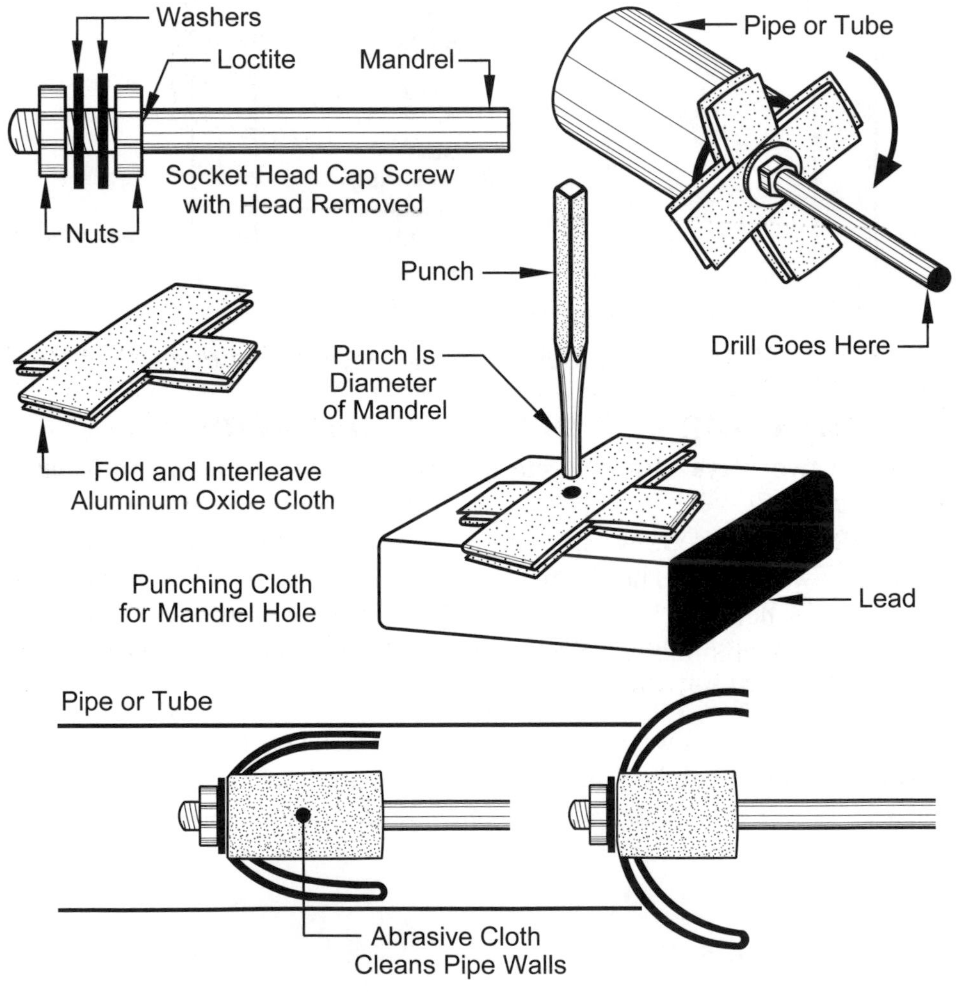

Figure 12–19. Holding abrasive cloth to clean pipe or tubing, reducing its inside diameter, removing interior weld seams and end-cutting burrs.

Camera Lens Wrenches

How can fragile threaded parts, such as filter lenses and lens mounts, be unscrewed without damaging or marking?

Use a friction, or pinch lens wrench. These wrenches are available commercially, but are easily shop-made from aluminum, brass or steel. See Figure 12–20.

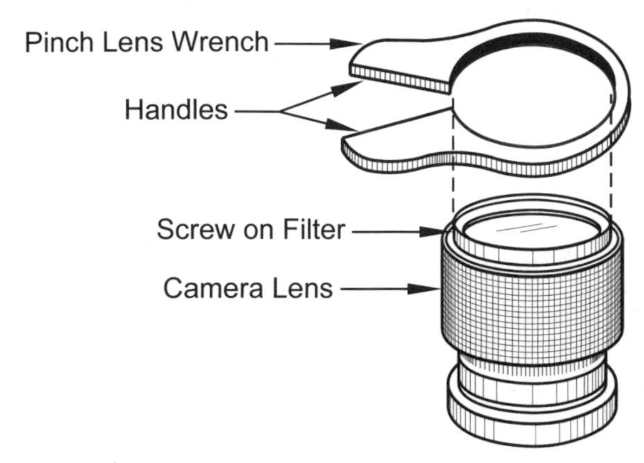

Figure 12–20. Friction wrench for camera lens.

Holding Thin-Walled Tubing

You have a two-piece part consisting of a hexagonal base threaded to a thin-walled tube, Figure 12–21 (A). The problem is that the threads are frozen together and the thin-walled tubing will collapse if clamped in a vise. How can the tubing be unscrewed?

The best route is to make a split-cylindrical sleeve as shown in Figure 12–21 (B through G). To work properly, the sleeve must just slip over the tubing. Place the part in the sleeve, put the sleeve in a vise, and gently snug up the vise. Apply a wrench to the hex-end of the part and continue to tighten the vise until the tubing stops turning in the sleeve. The parts will now unscrew if they are not rusted together.

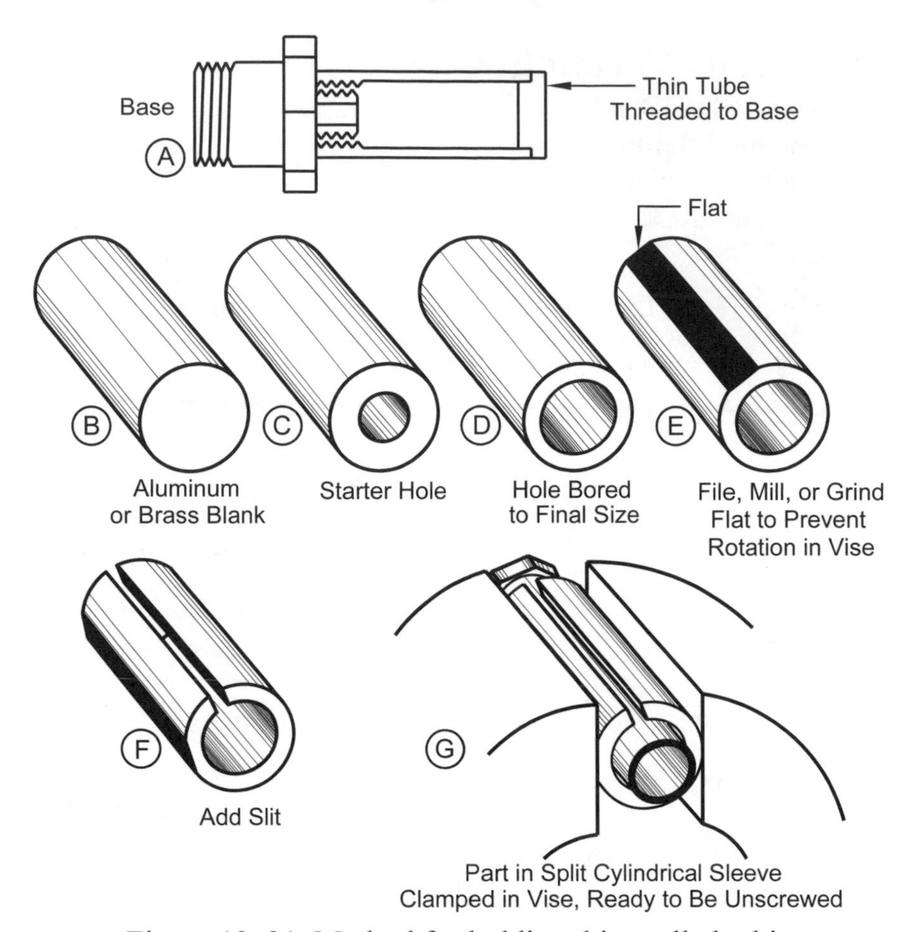

Figure 12–21. Method for holding thin-walled tubing in a vise without damaging the tubing.

Locking Dowel Pins in Place

When dowel holes are slightly oversized, dowel pins may work their way out from vibration or temperature variation. What are two methods that will insure a dowel pin remains in place?

Method 1 uses a wedge driven into a saw-slot in the dowel like those used to attach hammerheads to their handles, Figure 12–22 (left). This is a permanent solution and will be difficult to remove.

Method 2 uses two saw cuts at right angles to the dowel axis. Each must extend 15 to 20% of the radius beyond the full diameter, and they are made from opposite sides of the dowel, see Figure 12–22 (right). The saw cuts allow the built-up stresses in the dowel to become unbalanced, pulling the dowel sections around the cuts out of axial alignment. This wedges the dowel in place.

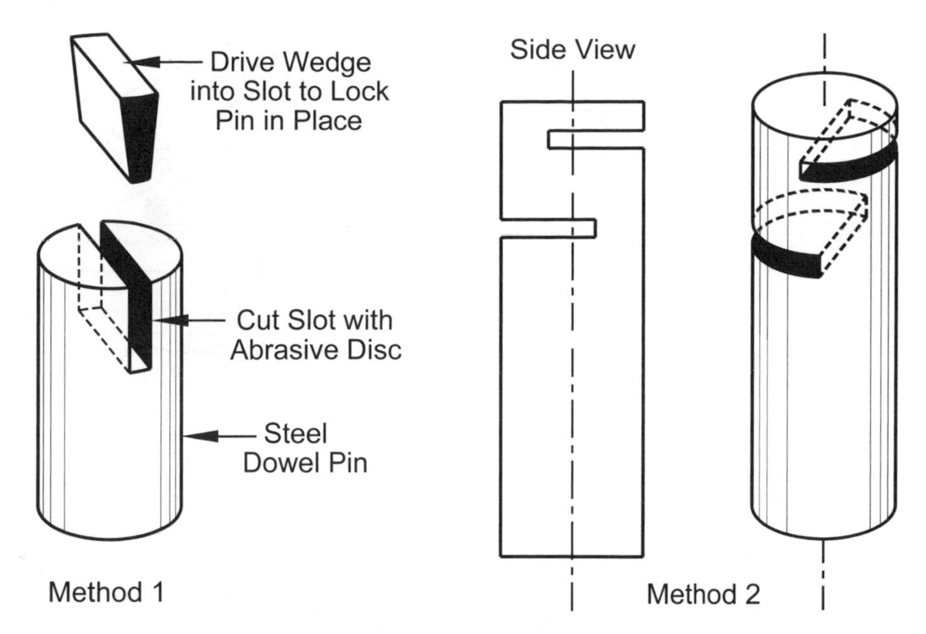

Figure 12–22. Two methods for locking a dowel in place.

Making Square or Hex Holes

You need a square or hex hole through a metal part, but do not have square broaches. How can this part be made?

Commercially available square, hexagonal and open-sided slotted sleeves in mild and hardened steel provide square or hex holes without the need for broaching, Figure 12–23. Installation is simple: Drill a hole through the metal part a few thousandths of an inch larger than the sleeve, slip the sleeve into the part and silver-solder the sleeve in place. The result is a square hole through the part.

These sleeves are used in shop-made boring bars, milling cutters, trepanning heads and other specialized tool holders. In fact, they are made in the most common cutting tool sizes from ⅛- to 1¼ -inches square.

The open-sided sleeves, Figure 12–23 (bottom), allow two or three set screws to bear directly against the side of the cutting tool. A single axial set screw in their ends makes precise tool adjustment easy.

Hex sleeves make excellent drive couplings for shafts with axial movement, specialized hex drivers and hex extensions.

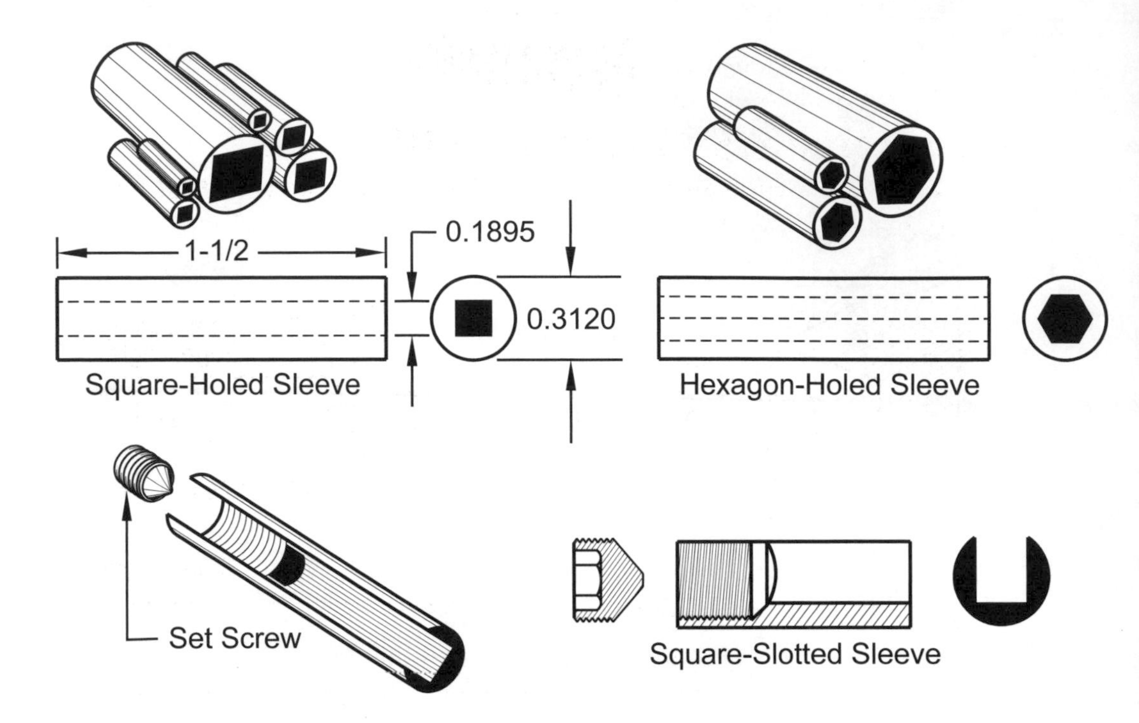

Figure 12–23. Broached sleeves are available in square-holed, hexagonal-holed or open-sided sleeves.

You are repairing a firearm and need a replacement screw. Other than ordering one from the factory, what other options do you have?

The screws in firearms may be Unified Extra Fine, Unified Fine, metric or Whitworth. The particular size, stem length and shape of many screws are specific to their function in that particular firearm. There are two options:

- Make a screw from the beginning.
- Begin with *unthreaded* screw stock (a slotted screw head on an unthreaded shank). Reduce the shaft diameter to the proper finished screw diameter, trim it to length, thread and blue it. Although the screw can be case hardened, most firearms frames are heat-treated, and most screws are not. Starting with unthreaded screw stock is a much better option than making a screw from scratch. Gunsmith supply companies, such as Brownells, offer selections of this unthreaded screw stock.

Appendix

Sharpening HSS Lathe Tool Bits

What are typical rake and clearance (relief) angles for HSS tool bits?
See Figure A–1 and Table A–1.

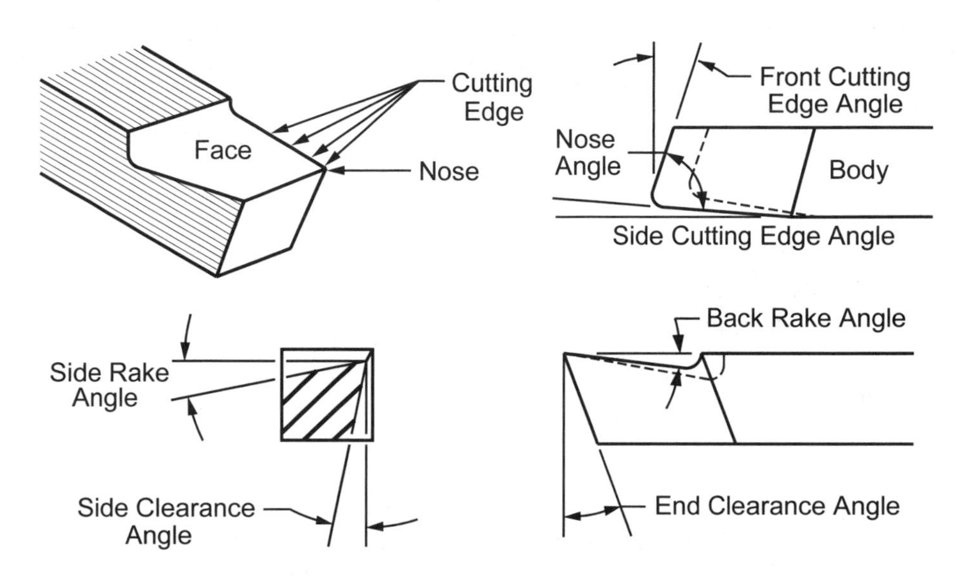

Figure A–1. Tool bit nomenclature.

Material	Side Clearance	Front Clearance	Side Rake	Back Rake
Aluminum	12	8	16	35
Brass	10	8	5 to −4	0
Bronze	10	8	5 to −4	0
Cast Iron	10	8	12	5
Copper	12	10	20	16
Machine Steel	10–12	8	12–18	8–15
Tool Steel (unhardened)	10	8	12	8
Stainless Steel	10	8	15–20	8

Table A–1. Clearance and rake angles in degrees for common metals.

What is the procedure for sharpening HSS general-purpose lathe tools? Begin by dressing the grinding wheel. Next, look up the typical angles for the workpiece material, and then follow the steps in Figure A–2. Dip the tool in coolant frequently to keep it from overheating and annealing. Any discoloration on the bit indicates it may no longer be hardened. If this happens, consider starting over again from the beginning. Tool bit angles are not critical and most tools will cut material satisfactorily, just less effectively.

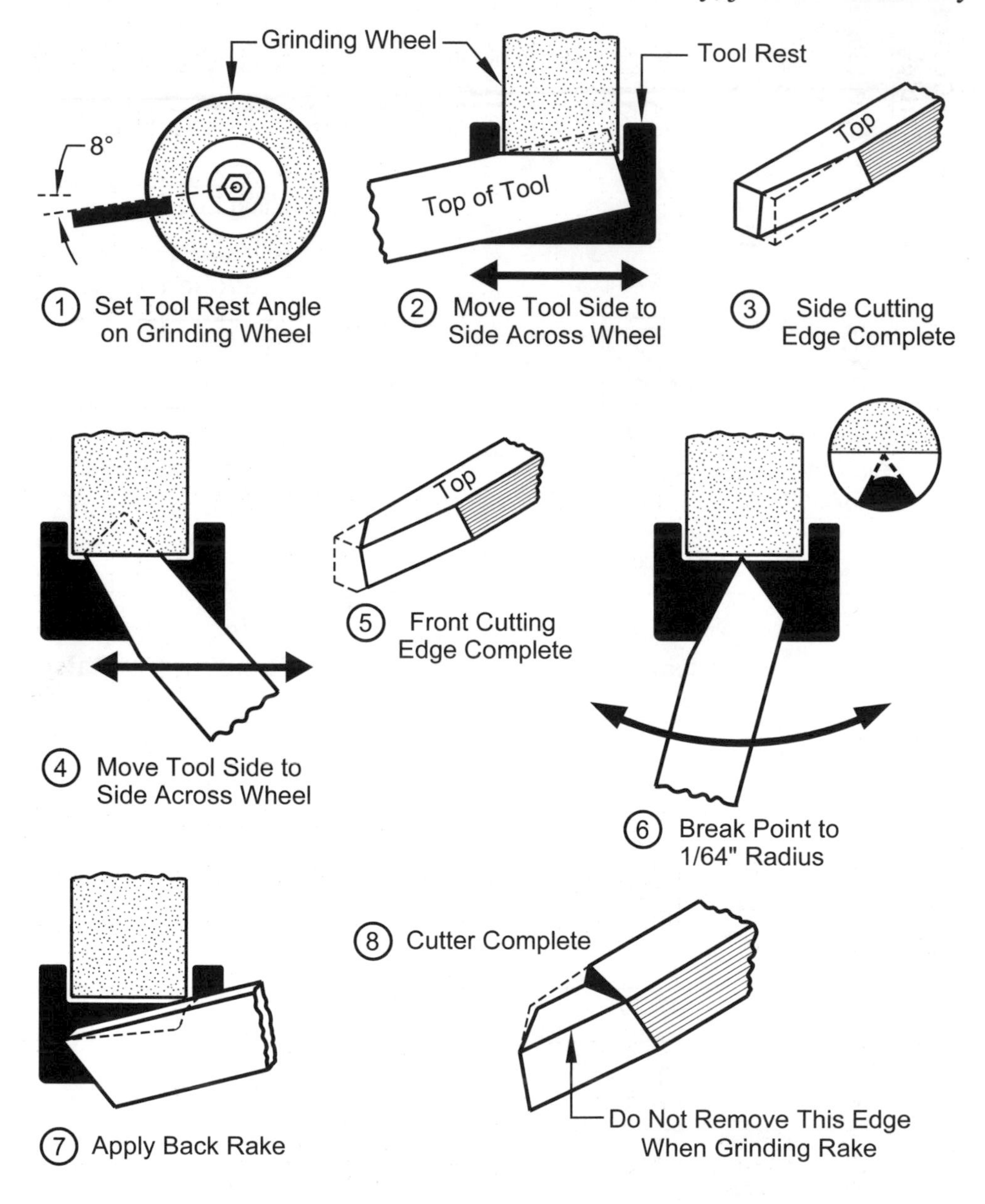

Figure A–2. Steps for sharpening HSS tool bits.

Table A–2. Surface Speed Table & Lathe Cutting Tool Selector Chart

SFM	High Speed Steel						Cobalt HSS						Uncoated Carbide			Single Coated Carbide			
Work Diameter Inches	**40**	**50**	**60**	**70**	**80**	**90**	**100**	**110**	**120**	**140**	**160**	**180**	**200**	**250**	**300**	**350**	**400**	**450**	**500**
	REVOLUTIONS PER MINUTE																		
¼	611	764	917	1070	1222	1375	1528	1681	1833	2140	2444	2750	3056	3820	4584	5348	6112	6882	7640
⅜	408	509	611	713	815	916	1018	1120	1222	1426	1630	1832	2037	2546	3056	3563	4074	4584	5092
½	306	382	458	535	611	688	764	840	916	1070	1222	1376	1528	1910	2292	2674	3056	3438	3820
⅝	244	306	367	428	489	550	611	672	733	856	978	1100	1222	1530	1833	2139	2445	2750	3056
¾	204	254	306	357	407	458	509	560	611	714	814	916	1018	1090	1528	1783	2037	2292	2546
⅞	175	218	262	306	349	393	436	480	523	612	698	786	872	1090	1310	1528	1746	1964	2183
1	153	191	229	267	306	344	382	420	456	534	612	688	764	955	1146	1337	1528	1719	1910
1⅛	136	170	204	238	272	305	339	374	407	476	544	610	678	850	1019	1188	1358	1528	1698
1¼	122	153	183	214	244	275	305	336	366	428	488	550	611	764	917	1070	1222	1375	1528
1⅜	111	139	166	194	222	249	277	305	332	388	444	498	554	695	833	972	1111	1250	1389
1½	102	127	153	178	204	229	254	280	305	356	408	458	509	637	764	891	1019	1146	1273
1¾	87	109	131	153	175	196	218	240	262	306	350	392	436	545	655	764	873	982	1091
2	76	95	114	133	153	172	191	210	229	266	306	344	382	477	573	668	764	859	955
2¼	68	85	102	119	136	153	170	187	204	238	272	306	340	425	509	594	679	764	849
2½	61	76	92	107	122	137	153	168	183	214	244	274	305	382	458	535	611	688	764
2¾	55	69	83	97	111	125	139	152	166	194	222	250	278	345	417	486	556	625	690
3	51	64	76	89	102	115	127	140	153	178	204	230	255	318	382	446	509	573	637
3¼	47	59	70	82	94	106	117	129	141	164	188	212	234	295	353	411	470	529	590
3½	44	54	65	76	87	98	109	120	131	152	174	196	218	270	327	382	437	491	540

Table A–2. Surface Speed Table (continued)

	High Speed Steel						Cobalt HSS						Uncoated Carbide			Single Coated Carbide				
SFM	40	50	60	70	80	90	100	110	120	140	160	180	200	250	300	350	400	450	500	SFM
Work Diameter Inches	REVOLUTIONS PER MINUTE																			Work Diameter Inches
3¾	41	51	61	71	81	92	102	112	122	142	162	184	204	255	306	357	407	458	509	3¾
4	38	48	57	67	76	86	95	105	114	134	152	172	191	239	286	334	382	430	477	4
4½	34	42	51	59	68	76	85	93	102	118	136	152	170	210	270	297	340	382	420	4½
5	30	38	46	53	61	69	76	84	93	106	122	138	153	191	229	267	306	344	382	5
5½	28	35	42	49	55	62	69	76	83	98	110	124	138	175	208	243	278	313	350	5½
6	25	32	38	44	51	57	64	70	76	88	102	114	128	160	191	223	255	286	318	6
6½	23	29	35	41	47	53	59	65	70	82	94	106	118	145	176	206	235	264	290	6½
7	22	27	33	38	44	49	54	60	65	76	88	98	109	136	164	191	218	246	273	7
7½	20	25	31	36	41	46	51	56	61	72	82	92	102	125	153	178	204	229	250	7½
8	19	24	29	33	38	43	48	52	57	66	76	86	96	120	143	167	191	215	239	8
8½	18	22	27	31	36	40	45	49	54	62	72	80	90	110	135	157	180	202	220	8½
9	17	21	25	30	34	38	42	47	51	60	68	76	85	106	127	149	170	191	212	9
9½	16	20	24	28	32	36	40	44	48	56	64	72	80	100	121	141	161	181	201	9½
10	15	19	23	27	31	34	38	42	46	54	62	68	76	95	115	130	153	172	191	10
11	14	17	21	24	28	31	35	38	41	48	56	62	70	87	104	122	139	156	174	11
12	13	16	19	22	25	29	32	35	38	44	50	58	64	79	95	111	127	143	159	12

To use this chart find the proper surface speed along the row second from the top, labeled SFM (surface feet per minute), and find the work diameter along the outside columns. The correct RPM is shown opposite the work diameter to be turned.

Table A–3. Decimal Equivalents of Fractional, Number, Letter & Metric Drills

Decimal	Inch	Wire	mm	Decimal	Inch	Wire	mm	Decimal	Inch	Wire	mm
.0135		80		.0430		57		.1285		30	
.0138			.35	.0465		56		.1360		29	
.0145		79		.0469	3/64			.1378			3.50
.0156	1/64			.0492			1.25	.1405		28	
.0158			.40	.0520		55		.1406	9/64		
.0160		78		.0550		54	1.25	.1440		27	
.0177			.45	.0591			1.50	.1470		26	
.0180		77		.0595		53		.1495		25	
.0197			.50	.0625	1/16			.1520		24	
.0200		76		.0635		52		.1540		23	
.0210		75		.0670		51		.1562	5/32		
.0217			.55	.0689			1.75	.1570		22	
.0225		74		.0700		50		.1575			4.00
.0236			.60	.0730		49		.1590		21	
.0240		73		.0760		48		.1610		20	
.0250		72		.0781	5/64			.1660		19	
.0256			.65	.0785		47		.1695		18	
.0260		71		.0787			2.00	.1719	11/64		
.0276			.70	.0810		46		.1730		17	
.0280		70		.0820		45		.1770		16	
.0292		69		0860		44		.1772			4.50
.0295			.75	.0890		43		.1800		15	
.0310		68		.0935		42		.1820		14	
.0312	1/32			.0938	3/32			.1850		13	
.0315			.80	.0960		41		.1875	3/16		
.0320		67		.0980		40		.1890		12	
.0330		66		.0994			2.50	.1910		11	
.0335			.85	.0995		39		.1935		10	
.0350		65		.1015		38		.1960		9	
.0354			.90	.1040		37		.1969			5.00
.0360		64		.1065		36		.1990		8	
.0370		63		.1094	7/64			.2010		7	
.0374			.95	.1100		35		.2031	13/64	5	
.0380		62		.1110		34		.2040		6	
.0390		61		.1130		33		.2055		5	
.0394			1.00	.1160		32		.2090		4	
.0400		60		.1181			3.00	.2165			5.50
.0410		59		.1200		31		.2188	7/32		
.0420		58		.1250	1/8			.221		2	

Table A–3. Decimal Equivalents (continued)

Decimal	Inch	Wire	mm	Decimal	Inch	Wire	mm	Decimal	Inch	mm
.2280		1		.3770		V		.6875	11/16	
.2340		A		.3860		W		.6890		17.50
.2344	15/64			.3906	25/64			.7031	45/64	
.2362			6.00	.3937			10.00	.7087		18.00
.2380		B		.3970		X		.7188	23/32	
.2420		C		.4040		Y		.7283		18.50
.2460		D		.4062	13/32			.7344	47/64	
.2500	1/4	E		.4130		Z		.7480		19.00
.2559			6.50	.4134			10.50	.7500	3/4	
.2570		F		.4219	27/64			.7656	49/64	
.2610		G		.4331			11.00	.7677		19.50
.2656	17/64			.4375	7/16			.7812	25/32	
.2660		H		.4528			11.50	.7874		20.00
.2677			6.8	.4531	29/64			.7969	51/64	
.2720		1		.4688	15/32			.8071		20.50
.2756			7.00	.4724			12.00	.8125	13/16	
.2770		J		.4844	31/64			.8268		21.00
.2810		K		.4921			12.50	.8281	53/64	
.2812	9/32			.5000	1/2			.8438	27/32	
.2900		L		.5118			13.00	.8465		21.50
.2950		M		.5156	33/64			.8594	55/64	
.2953			7.50	.5312	17/32			.8661		22.00
.2969	19/64			.5315			13.50	.8750	7/8	
.3020		N		.5469	35/64			.8858		22.50
.3125	5/16			.5512			14.00	.8906	57/64	
.3150			8.00	.5625	9/16			.9055		23.00
.3160		0		.5709			14.50	.9062	29/32	
.3230		P		.5781	37/64			.9219	59/64	
.3281	21/64			.5906			15.00	.9252		23.50
.3320		Q		.5938	19/32			.9375	15/16	
.3347			8.50	.6094	39/64			.9449		24.00
.3390		R		.6102			15.50	.9531	61/64	
.3438	11/32			.6250	5/8			.9646		24.50
.3480		S		.6299			16.00	.9688	31/32	
.3543			9.00	.6406	41/64			.9843		25.00
.3580		T		.6496			16.50	.9844	63/64	
.3594	23/64			.6562	21/32			1.0000	1	
.3680		U		.6693			17.00			
.3750	3/8			.6719	43/64					

Table A–4. Shaft Fit Allowances

Running Fits for Shafts – Speeds under 600 rpm – Light to Medium Duty

	Up to ½″	½″ to 1″	1″ to 2″	2″ to 3½″	3½″ to 6″
High Limit	–0.0005	–0.00075	–0.0015	–0.002	–0.0025
Low Limit	–0.001	–0.0015	–0.0025	–0.003	–0.004

Running Fits for Shafts – Speeds over 600 rpm – Heavy Duty

	Up to ½″	½″ to 1″	1″ to 2″	2″ to 3½″	3½″ to 6″
High Limit	–0.0005	–0.001	–0.002	–0.003	–0.004
Low Limit	–0.001	–0.002	–0.003	–0.004	–0.005

Sliding Fits for Shafts for Free Sliding of Gears & Clutches

	Up to ½″	½″ to 1″	1″ to 2″	2″ to 3½″	3½″ to 6″
High Limit	–0.0005	–0.00075	–0.0015	–0.002	–0.0025
Low Limit	–0.001	–0.0015	–0.0025	–0.003	–0.004

Standard Fits for Light Service – Parts Keyed to Shaft

	Up to ½″	½″ to 3½″	3½″ to 6″
High Limit	Standard	Standard	Standard
Low Limit	–0.00025	–0.0005	–0.00075

Standard Fits with Play Eliminated – Parts Assemble Readily

	Up to ½″	½″ to 3½″	3½″ to 6″
High Limit	Standard	+0.00025	+0.0005
Low Limit	+0.00025	+0.0005	+0.00075

Driving Fits for Permanent Assembly – Cramped Location for Driving

	Up to ½″	½″ to 1″	1″ to 2″	2″ to 6″
High Limit	Standard	+0.00025	+0.0005	+0.0005
Low Limit	+0.0025	+0.0005	+0.00075	+0.001

Driving Fits for Permanent Assembly & Severe Duty – Driving Possible

	Up to 2″	2″ to 3½″″	3½″ to 6″
High Limit	+0.0005	+0.00075	+0.001
Low Limit	+0.001	+0.00125	+0.0015

Forced Fits for Permanent Assembly Using Hydraulic Press – Severe Service

	Up to ½″	½″ to 1″	1″ to 2″	2″ to 3½″	3½″ to 6″
High Limit	+0.00075	+0.001	+0.002	+0.003	+0.004
Low Limit	+0.001	+0.002	+0.003	+0.004	+0.005

Figure A–3. R8 & 5C Collet Dimensions

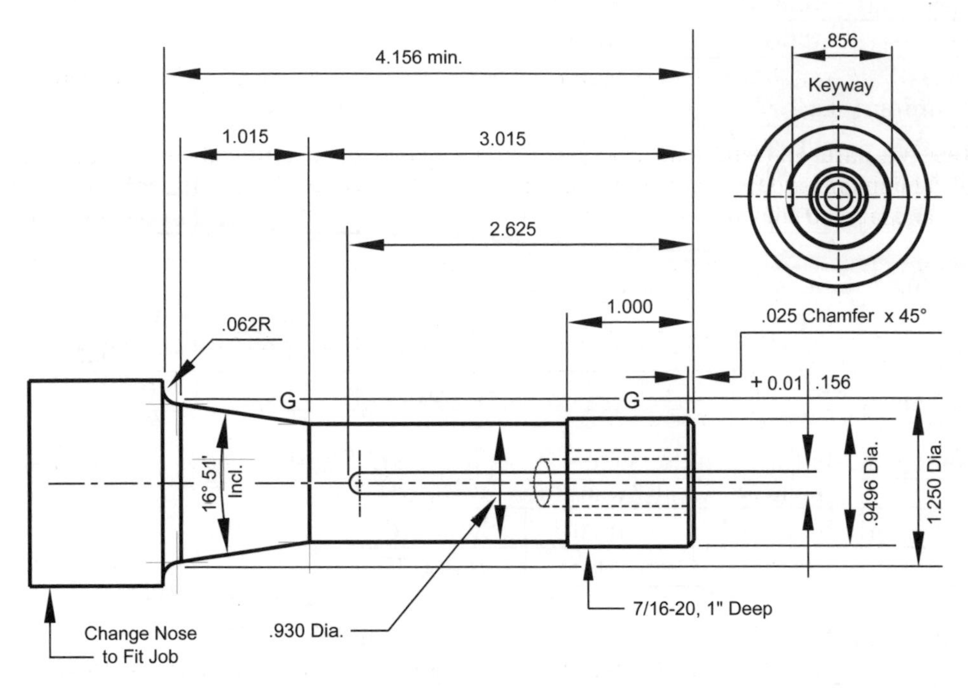

R8 Collet Used in Bridgeport-style Milling Machines

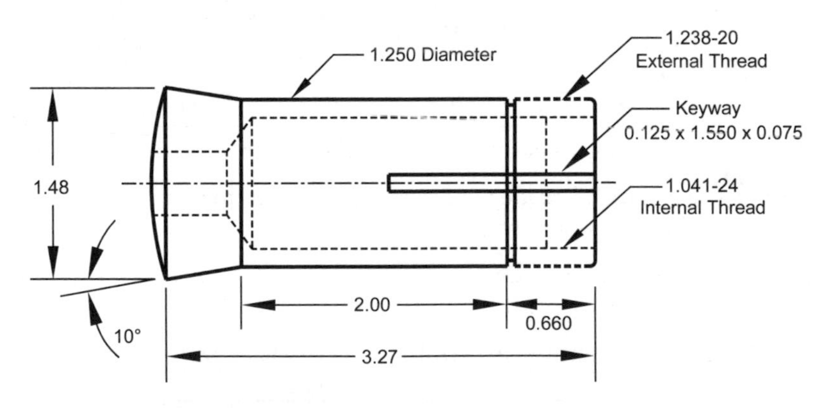

5C Collet Used in Many Lathes & Indexing Fixtures

All Dimensions in Inches

Glossary

A

abrasive wheels: Wheels of a hard abrasive, such as aluminum oxide or silicon carbide used for grinding.

abrasive, natural: (sandstone, emery, corundum, diamonds) or artificial (silicon carbide, aluminum oxide) material used for making grinding wheels, sandpaper, abrasive cloth and lapping compounds.

acetone: A colorless, flammable, volatile liquid used as a paint remover and as a solvent for oils and other organic compounds.

Acme thread: A screw thread having a 29° included angle and a flat top. Used largely for feed and adjusting screws on machine tools.

acute angle: An angle that is less than 90°.

adapter: A tool holding device for fitting together various types or sizes of cutting tools to make them interchangeable on different machines.

align: To adjust or set to a line or center.

alloy steel: A plain carbon steel to which another element, other than iron and carbon, has been added in a percentage large enough to alter its characteristics.

alloy: A substance with metallic properties, composed of two or more chemical elements of which at least one is a metal.

allthread: Steel or stainless steel rod threaded from end to end. It is available in a wide range of diameters and thread specifications.

aluminum: One of the chemical elements, a lightweight, ductile easily worked metal that resists corrosion and has a silvery to dull gray appearance depending on surface roughness.

angle plate: A precision holding fixture made of cast iron, steel or granite. The two principal faces are at right angles and may be slotted for holding the work or clamping to a table.

annealing: The controlled heating and cooling of a metal to remove stresses and to make it softer and easier to work with.

anvil: A heavy iron or steel block upon which metal is forged or hammered; also the fixed jaw on a micrometer against which parts are measured.

apron: That portion of a lathe carriage that contains the clutches, gears and levers for moving the carriage.

arbor press: A hand-operated or hydraulic machine tool designed for applying high pressure for the purpose of pressing together or removing parts.

arbor: A shaft or spindle for holding cutting tools; most usually on a milling machine or grinder.

ASTM: The American Society for Testing and Materials.

austenite: One of the basic steel microstructures wherein carbon is dissolved in iron. Austenite forms at 1330°F or higher.

automatic stop: A device which may be attached to any of several parts of a machine tool to stop the operation of the machine at any predetermined point.

AWS: The American Welding Society.

axis: The line, real or imaginary, passing through the center of an object about which it could rotate; a point of reference.

B

babbitt: An antifriction metal alloy used for bearing inserts; made of tin, antimony, lead and copper.

back gears: Gears fitted to a machine such as a lathe or milling machine to increase the number of spindle speeds obtainable with a cone or step pulley belt drive.

back rake: The angular surface ground back from the cutting edge of cutting tools. On lathe cutting tools the rake is positive if the face slopes down from the cutting edge toward the shank and negative if the face slopes upward toward the shank.

backlash: The lost motion or looseness (play) between the faces of meshing gears or threads.

bandsaw: A power saw the blade of which is a continuous, narrow, steel band having teeth on one edge and passing over two large pulley wheels.

bar stock: Metal bars of various lengths, made in flat, hexagon, octagon, round and square shapes from which parts are machined.

bastard: Threads, parts, tools and sizes that are not standard, such as bastard nuts, bastard plugs, bastard fittings, and so forth. The term also refers to a standard coarse-cut file.

bearing: The support and guide for a rotating, oscillating or sliding shaft, pivot or wheel.

bed: One of the principal parts of a machine tool, having accurately machined ways or bearing surfaces for supporting and aligning other parts of the machine.

bell mouth: The flaring or tapering of a machined hole, usually made at the entrance end because of misalignment or spring of the cutting tool.

bench grinder: A small grinding machine for hand shaping and sharpening the cutting edges of tools. Most grinders have two abrasive wheels 6 to 10 inches in diameter.

bevel: Any surface that is not at right angles to another surface. Also, the name given a tool used for measuring, laying out or checking the accuracy of work machined at an angle or bevel.

bit, tool (cutter): A hardened steel bar or plate that is shaped according to the operation to be performed and the material to be machined usually of HSS or Cobalt steel.

blind hole: A hole made in a workpiece that does not pass through it.

body-centered cubic (BCC): One of the common types of unit cells described as a cube with an atom at each of the eight corners and a single atom at the center of the cell.

bore: To enlarge and finish the surface of a cylindrical hole by the action of a rotating boring bar (cutting tool) or by the action of a stationary tool pressed (fed) against the surface as the part is rotated.

boring bar: A combination tool holder and shank for boring internal cylinders.

boring tool: A cutting tool in which the tool bit the boring bar and, in some cases, the tool holder are incorporated in one solid piece.

boss: A projection or an enlarged section of a casting through which a hole may be machined or a pad for alignment.

brass: A nonferrous alloy consisting essentially of copper and zinc.

brazing: A group of welding processes that produces a coalescence of materials by heating them to the brazing temperature in the presence of a filler metal having a liquidus above 850°F (450°C) and below the solidus of the base metal. The filler metal is distributed between the closely fitted faying surfaces of the joint by capillary action.

brine: A saltwater solution for quenching or cooling when heat treating steel.

Brinell hardness test: A common testing method using a ball penetrator. The diameter of the indentation is converted to units of Brinell hardness number (BHN).

Brinell hardness: A hardness scale for metals proposed by Swedish engineer Johan August Brinell in 1900. It was the first widely used and standardised hardness test. The large size of indentation and possible damage to test-piece limits its usefulness.

broach: A long, tapered cutting tool with tooth-like serrations which, when forced through a hole or across a surface, cuts a desired shape or size.

bronze: A nonferrous alloy consisting essentially of copper and tin.

buff: To polish to a smooth finish of high luster with a cloth or fabric wheel to which a polishing compound has been added.

burnishing: The process of finishing a metal surface by contact with another harder metal; to make smooth or glossy by or as if by rubbing; polish.

burr: The sharp edge left on metal after cutting or punching; also, a rotary cutting tool designed to be attached to a drill.

bushing: A sleeve or a lining for a bearing or a drill jig to guard against wear.

C

caliper: A device used to measure inside or outside dimensions.

cam: A device for converting regular rotary motion to irregular rotary or reciprocating motion. Sometimes the effect of producing off-center lathe operations.

capillary action: The force by which liquid in contact with a solid is distributed between closely fitted faying surfaces of the joint to be brazed or soldered.

carbide tool bits: Lathe and milling machine cutting tools to which carbide tip inserts have been brazed to provide cutting action on harder materials than HSS cutters are capable.

carbon steel: A broad term applied to hardenable tool steel other than high-speed or alloy steel.

carbon: A nonmetallic chemical element that occurs in many inorganic and all organic compounds. Carbon is found in diamond and graphite and is a major constituent of coal, petroleum, asphalt, limestone and other carbonates. In combination, it occurs as carbon dioxide and as a constituent of all living things.

carriage: A principal part of a lathe that carries the cutting tool and consists of the saddle, compound rest and apron.

case hardening: A heat treating process, basically carbonizing that makes the surface layer, or case, of steel substantially harder than the interior or core.

cast iron: A family of alloys, containing more than 2% carbon and between 1 and 3% silicon. Cast irons are not malleable when solid and most have low ductility and poor resistance to impact loading. There are four basic types of cast iron: gray, white, ductile and malleable. Machine tools are made of it because of its vibration damping properties.

casting: A part made by pouring molten metal into a mold.

castle nut: A nut with grooves cut entirely across the top face for the purpose of resisting loosening with vibration.

center drill: A combined countersink and drill used to prepare work for mounting between centers and to accurately locate a drilled hole.

center gage: A small, flat gage having 60° angles that is used for grinding and setting the thread cutting tools in a lathe. It may also be used to check the pitch of threads and the points of center.

center head: A part of a combination square set that is used to find the center of or to bisect a round or square workpiece.

center punch: A conically pointed hand tool made of hardened steel and shaped somewhat like a pencil. Used during layout work for starting drills precisely.

center, dead: A center that does not rotate; commonly found on the tailstock of a lathe. Also, an expression for the exact center of an object.

center, half: A dead center that has a portion of the 60° cone cut away.

center, live: Generally this is a *dead center* in the spindle which revolves with the work; however, the ball bearing type tailstock center is also called a *live center*.

center: A point or axis around which anything revolves or rotates. In the lathe one of the parts upon which the work to be turned is placed. The center in the headstock is referred to as the "live" center and the one mounted in the tailstock as the "dead" center.

chamfer: The bevel or angular surface cut on the edge or a corner of a machined part.

chasing threads: Cutting threads in a lathe or screw machine.

chatter: The vibrations caused between the work and the cutting tool which leave distinctive tool marks on the finished surface that are objectionable.

chip breaker: A small groove ground or formed back of the cutting edge on the top of a cutting tool to keep the chips short. Also, a separate adjustable part on the top of the cutting tool.

chipping: The process of cutting metal with a cold chisel and hammer or pneumatic hammer.

chisel: Any one of a variety of small hand or powered cutting tools, generally wedge-shaped.

chromium: A lustrous, hard, brittle, bluish steel-gray metallic element used to harden steel alloys in production of stainless steel, and has a hard, corrosion-resistant plating.

chuck, independent jaw: A chuck, each of whose jaws (usually four) is adjusted with a screw action independently of the other jaws.

chuck, universal (self-centering chuck, concentric scroll chuck): A chuck whose jaws are so arranged that they are all moved together at the same rate by a T-shaped wrench sometimes called a "key."

chuck: A device on a machine tool to hold the workpiece or a cutting tool.

clearance angle: The angle between the rear surface of a cutting tool and the surface of the work at the point of contact from the horizontal or machine axis.

clearance: The distance or angle by which one object or surface clears another.

climb milling: A method of milling in which the work table moves in the same direction as the direction of rotation of the milling center. Sometimes called down cutting or down milling.

coefficient of thermal expansion: The increase in length per unit length for each degree a metal is heated.

cohesion: Cohesion is the result of a perfect fusion and penetration when the molecules of the parent material and the added filler materials thoroughly integrate as in a weld.

cold work: Cold working refers to forming, bending or hammering a metal well below the softening or melting point. Cold working of metals causes work hardening, making them stronger but less ductile.

cold-rolled steel: Steel that has been rolled to accurate size and smooth finish after pickeling to remove scale. In contrast, hot-rolled steel may have a rough, pitted surface and slag inclusions.

collet: A precision work-holding chuck which centers finished round stock automatically when tightened. Specialized collets are also available in shapes for other than round stock such as square or hexagonal.

combination square: A checking and adjustable layout tool combining a rule with 90° and 45° head, a level, a protractor and a center head.

compound (rest): The part of a lathe set on the carriage that carries the tool post and tool holder. It is designed to swing in any direction and to provide feed for turning short angles or tapers.

concave: A curved depression in the surface of an object.

concentric: Accurately centered or having a common center.

contour: The outline of an object.

convex: The curved surface of a cylinder, as a sphere when viewed from without.

coolant: A common term given to the numerous cutting fluids or compounds used with cutting tools to increase the tool life and to improve surface finish on the material.

corrosion: Oxidation (rusting in the case of iron or steel) or similar chemical change in other metals.

counterbore: To enlarge the top part of a hole to a specific size, concentric with the original hole, as for the head of a socket-head or cap-screw. Also, the tool that is used normally with a pilot to insure concentricity.

countersink: To enlarge the top part of a hole at an 82° cone angle for a flat-head screw or 90° for a rivet. Also, the tool that is used for the operation.

cross feed: The feed that operates across the axis of the workpiece or at right angles to the main or principal feed on a machine.

cross section: A view showing an internal structure as it would be revealed by cutting through the piece in any plane.

CS: Cutting speed, usually expressed in feet/minute.

cutting fluid: A liquid used to cool and lubricate the cutting tool, to improve the work surface finish and remove chips.

cutting speed: The surface speed of the workpiece in a lathe or a rotating cutter, commonly expressed in feet per minute (FPM) and converted to revolutions per minute (rpm) for proper setting on the machine.

cutting tool: A hardened piece of metal (tool steel, usually HSS or tungsten carbide tipped) that is machined and ground so that it has the shape and cutting edges appropriate for the operation for which it is to be used.

D

dead center: See center, dead. Usually a 60° solid steel center on a lathe.

dead smooth: The term applied to the finest cut of a file or a very flat surface.

deburr: To remove sharp edges.

dial indicator: A precision instrument which shows variations of thousandths of an inch or less when testing the trueness or alignment of a workpiece, fixture or machine.

die stock: The frame and two handles (bars) which hold the dies (chasers) used for cutting (chasing) external screw threads.

die: A tool used to form or stamp out metal parts, also, a tool used to cut external threads.

dividers, spring: Dividers whose legs are held together at the hinged end by the pressure of a C-shaped spring.

dividing head (indexing head): A machine tool holding fixture which positions the work for accurately spacing holes, slots, flutes and gear teeth, and for making geometric shapes. When geared to the table leadscrew, it can be used for helical milling operations.

dog: A clamping device (lathe dog) used to drive work being machined between centers. Also, a part projecting on the side of a machine worktable to trip the automatic feed mechanism off or to reverse the travel.

dovetail: A two-part slide bearing assembly used in machine tool construction for the precise alignment and smooth operation of the movable components of the machine for linear movement.

dowel: A cylindrical pin fitted or keyed in two adjacent parts to accurately align the parts when assembling them.

down feed (climb cutting, climb milling): A seldom used method of feeding work into milling cutters. The work is fed in the same direction as the portion of the cutter which comes in contact with it.

draw: See tempering.

dressing: The act of removing the glaze and dulled abrasives from the face of a grinding wheel to make it clean and sharp and to make it cylindrical again. See truing.

drift: A tapered, flat steel wedge used to remove drills and other tapered shank tools from spindles, sockets or sleeves. Also a round, tapered punch used to align or enlarge holes.

drill bushing: A hardened steel guide inserted in jigs, fixtures or templates for the purpose of providing a guide for the drill in drilling holes in their proper or exact location.

drill chuck: A device used to grip drills and attach them to a rotating spindle.

drill jig: A jig which holds parts or units of a structure and, by means of bushings, guides the drill so that the holes are properly located.

drill press: A drilling machine with a counterbalanced spindle which makes it possible for the operator to accurately control the rate at which the drill is fed into the work. The sensitive drill press usually uses drills that are less than ½-inch diameter and which rotate at high speeds.

drill rod: A term given to an annealed and polished high carbon tool steel rod usually round and centerless ground. The sizes range in round stock from .013 to 1½ inches diameter. Commercial qualities include air, water and oil hardening grades. Drill rod is used principally by machinists and tool and die makers for punches, drills, taps, dowel pins, screw machine parts and small tools.

drill sleeve: An adapter with an internal and external taper (such as a Morse Taper) which fits tapered shank tools such as drills or reamers to adapt them to a larger size machine spindle.

drill socket: An adapter similar to a sleeve except that it is made to adapt a larger or smaller tapered-shank tool to a smaller size spindle. Also used for length extension of the tool.

drill, center: A combination drill and countersink, usually 60°.

drill, twist: A commonly used metal-cutting drill, usually made with two flutes running around the body.

drive fit: One of several classes of fits in which parts are assembled by pressing or forcing one part into another.

ductility: The property of a metal that permits it to be drawn, rolled or hammered without fracturing or breaking.

E

eccentric: A circle not rotating on its geometric center. Also, a device such as a crankshaft or a cam for converting rotary motion to reciprocating motion.

elastic limit: The greatest stress the material can withstand without permanent elongation when all load has been removed from the specimen. Any strains developed up to the elastic limit are both small and reversible.

element: Matter which cannot be broken up into simpler substances by chemical action, that is, whose molecules are all composed of only one kind of the most basic atom.

emery: A natural abrasive used for grinding or polishing. It has been largely replaced by artificial abrasives.

emulsion: A coolant usually formed by mixing soluble oils or compounds with water.

end mill: A cutting tool with helical flutes/teeth on the circumference and radial teeth on the end of the tool.

F

face milling: Milling a large flat surface with a milling cutter that operates in a plane that is at right angles to its axis.

face: To machine a flat surface, as in the end of a shaft in the lathe. The operation is known as facing.

face-centered cubic (FCC): One of the common types of crystallizing structure unit cells in which atoms are located on each corner and at the center of each face of a cube. Among the common FCC metals are aluminum, copper, nickel and austenitic stainless steel. This arrangement is typical of the austenitic form of iron.

faceplate: A large circular plate with slots and holes for mounting the workpiece to be machined. It is attached to the headstock of a lathe.

facing: The process of making a flat or smooth surface (usually the end) on a piece of stock or material.

fatigue failure: The sudden and complete breakage of a part as a result of the repeated application of a load. Fatigue failure is progressive and may not occur until after millions of load cycles.

fatigue strength: Ability of a material to withstand repeated loading cycles.

fatigue: The effect on certain materials, especially metals, undergoing repeated stresses.

feed mechanism: The mechanism, often automatic, which controls the advancing movement (feed) of the cutting tools used in machines such as lathes, mills and drills.

feed: The rate of travel of a cutting tool across or into the work, expressed in inches per minute, in inches per revolution, mm per minute or mm per revolution.

female part: A concave piece of equipment which receives a mating male (convex) part.

ferrous: A metal alloy in which iron is the major ingredient.

filler material: The material, metal or alloy to be added in making a welded, brazed or soldered joint.

fishtail: A steel gage used when grinding the 60° angle on threading tool bits and for positioning the tool perpendicular to the work on the lathe.

fit: The relation between mating or matching parts, that is, the amount of, or lack of, play between them.

fixture: A production work-holding device used for machining duplicate workpieces. Although the term is used interchangeably with a jig, a fixture is not designed to guide the cutting tools as the jig does. Also, a device designed to hold and maintain parts in proper relation to each other.

flange: A relatively thin rim around a part.

flute: The groove in a cutting tool which provides a cutting edge and a space for the chips to escape and permits the cutting fluids to reach the cutting edges.

flycutter: A single-point cutter mounted on a bar in a flycutter holder or a flycutter arbor. It is used for special applications for which a milling cutter is not available.

follower rest: A support for long, slender work turned in the lathe. It is mounted on the carriage, travels close to and with the cutting tool and keeps the work from springing away.

footstock (tailstock): Part of an indexing attachment for an indexing head which has a center and serves the same purpose as the tailstock of a lathe.

force fit: A fit in which one part is forced or pressed into another to form a single unit. There are different classes of force fits depending on standard limits between mating parts.

forge: To form or shape heated metal by hammering. Also, the name of the unit used for heating metal, as the blacksmiths' forge.

formed cutters: Milling cutters which will produce shaped surfaces with a single cut. They may be sharpened without changing their outline or shape.

forming tool: Tool ground to a desired shape to reproduce this shape on the workpiece.

FPM: Feet per minute at the workpiece or tool, usually expressed as SFPM.

free-cutting steel: Bar stock containing a higher percentage of sulfur making it very easy to machine. Sulfur combines with manganese to make the chips break.

free fit: A class of fit intended for use where accuracy is not essential or where large temperature variations are likely to be encountered, or both conditions.

fuel gas: A gas such as acetylene, natural gas, hydrogen, propane, stabilized methylacetylene propadiene and other fuels normally used with oxygen in one of the oxyfuel processes and for heating.

fulcrum: The point or support on which a lever pivots.

fusion: The joining of base material, with or without filler material, by melting them together.

G

gage, feeler (thickness gage): A gage consisting of a group of very thin blades, each of which is accurately rolled and finished to a specific thickness.

gage, indicating (dial indicator): A gage consisting of a dial, commonly graduated (marked) in thousandths of an inch, to which is fastened an adjustable arm.

gage, radius (fillet gage): Any one of a number of small, flat, standard-shaped metal leafs or blades used for checking the accuracy of regular concave and convex surfaces.

gage, screw pitch: A gage consisting of a group of thin blades used for checking the number of screw threads per unit of distance, usually per inch, on a screw, bolt, nut, pipe or fitting.

gage, telescoping: A T-shaped gage used in measuring the diameter or width of holes.

gage: Any one of a large variety of devices for measuring or checking the dimensions and shapes of objects.

gang milling: A milling setup where a number of cutters are arranged on an arbor so that several surfaces can be machined at one time. It is commonly used for production purposes.

gib: A tapered strip of metal placed between the bearing surface of two machine parts to ensure a precision fit and provide an adjustment for wear.

GTAW: The Gas Tungsten Arc Welding process; also called Heliarc™ and TIG (tungsten inert gas).

H

hacksaw frame: A U-shaped frame holding a metal blade of hardened steel having small, close teeth on one edge.

half-nut: A lever-operated mechanism that resembles a split nut that can be closed on the leadscrew of a lathe when threads are being cut.

handwheel: Any adjusting or feeding mechanism shaped like a wheel and operated by hand.

hardening: A heat-treating process for steel and aluminum which increases its hardness and tensile strength and reduces its ductility.

hardness tests: Tests to measure the hardness of metals, usually Rockwell C Scale (R_C) or Brinell Scale, and the Shore Scale for rubber.

headstock: The fixed or stationary end of a lathe or similar machine tool which is powered, usually to rotate.

heat treatment: The process of heating and cooling a solid metal or alloy to obtain certain desired metallurgical properties or characteristics.

helical gear: A gear with teeth cut at some angular curve other than at a right angle across the face of the gear, thus permitting more than one tooth to be engaged at all times and providing a smoother and quieter operation than a spur gear.

helix angle: The angle between the direction of the threads around a screw and a line running at a right angle to the shank.

helix: A path formed as a point advances uniformly around a cylinder, as the thread on a screw or the flutes on a drill.

hex: A term used for anything shaped like a hexagon, a six-sided polygon.

hexagonal close packed (HCP): A unit cell in which two hexagons (six-sided shapes) form the top and bottom of the prism. An atom is located at the center and at each point of the hexagon. Three atoms, one at each point of a triangle, are located between the top and bottom hexagons. Among the common HCP metals are zinc, cadmium and magnesium.

high-carbon steel: See carbon steel, usually contains more than 0.9% carbon by making it hardenable from a red heat.

high-speed steel (HSS): An alloy steel commonly used for cutting tools because of its ability to remove metal at a much faster rate than carbon steel tools. The two common alloy series are the Molybdenum-based one (M-2 through M-42) and the Tungsten-based one (T-1 shrough T-15).

hob: A cylindrical cutting tool shaped like a worm thread and used in industry to cut gears.

hobbing: The operation of cutting gears with a hob.

hog: To remove excess material in a hurry without regard to finish, sometimes causing tool breakage; also, to rough out.

hole saw: A cutting tool used to cut a circular groove into or a hole through solid material.

honing: The process of finishing ground surfaces to a high degree of accuracy and smoothness with abrasive blocks applied to the surface under a light, controlled pressure and with a combination of rotary and reciprocating motions.

hot-rolled steel: Steel which is rolled to finished size while hot. Identified by a dark oxide scale left on the surface.

hydrogen: The lightest chemical element, colorless, odorless and tasteless. It is found in combination with other elements in most organic compounds and many inorganic compounds. Hydrogen combines readily with oxygen in the presence of heat and forms water. It can cause embrittlement in steel during welding and heat treating.

I

ID: Inside diameter of an internal cylinder.

idler: A gear or gears placed between two other gears to transfer motion from one gear to the other gear without changing their speed or ratio; it may change the direction of rotation.

impact strength: The ability of a material to resist shock, dependent on both strength and ductility of the material.

independent jaw chuck: A chuck in which each jaw may be moved independently of the others.

index plate: A metal disk or plate machined with many holes arranged in a series of rings, one outside the other each ring containing a different number of holes.

indexing fixture: A complete indexing unit composed of a dividing head and tailstock. See dividing head.

indexing: The process of positioning a workpiece for machining it into equal spaces, positioning dimensions or angles using an index or dividing head.

inserted-tooth cutter: A milling cutter designed with replaceable HSS or carbide cutting tooth inserts to save the expense of a new cutter when the teeth become damaged or worn.

intermediate gear: See idler.

IPM: Feed rate in inches per minute.

iron carbide: A binary compound of carbon and iron; it becomes the strengthening constituent in steel.

iron-carbon phase diagram: A graphical means of identifying different structures of steel and percentages of carbon occurring in steel at various temperatures.

J

jack, leveling: Small jacks (usually screw jacks) for leveling and holding work on milling machine beds and similar places.

Jarno: A standard taper having 0.600-inch taper per foot used on some machine tools.

jig: A device that holds the workpiece in place and guides the cutting tool during the cutting operation.

Johannson blocks (Jo blocks): Common term for the precision rectangular gage blocks used and accepted as dimensional standards by machinists, toolmakers and inspectors. They are usually sold in sets.

joint clearance: The distance between the faying surfaces of a joint in brazing or soldering.

K

kerf: The width of a cut produced during a cutting process.

keyseat: A recessed groove (slot) machined into a shaft or a part going on the shaft (usually a wheel or gear).

key: One of the several types of small metal objects designed to fit mating slots in a shaft and the hub of a gear or pulley to provide a positive drive between them. The two most common are the rectangular and Woodruff keys. Also, the name of the T-handle wrench used on lathe or milling machine chucks.

knee: That part of a column of a knee-type milling machine which carries the saddle and the table and provides the machine with vertical feed adjustments. Also, the name of a precision angle plate called a "toolmakers' knee".

knurl: A decorative gripping surface of straight-line or diagonal design made by uniformly serrated rolls called knurls.

knurling: The process of finishing a part by scoring (pressing) patterns on the surface of the work. Also, using a knurling tool.

L

land: That surface on the periphery of a rotary cutting tool, such as a milling cutter, drill tap or reamer, which joins the face of the flute or tooth to make up the basic cutting edge.

lap: A tool made of soft metal and charged with fine abrasives for precision finishing of metal surfaces. Also, to perform the operation using a lap.

lard oil: A cutting oil made from animal fats usually mixed with mineral oils to reduce its cost and improve its qualities.

layout: To locate and scribe on blank stock the shape and size dimensions required to machine or form the part.

lead: The distance a thread will advance along its axis in one complete revolution. Also, a heavy, soft, malleable metal having a low melting point. It has a bright, silvery color when freshly cut or poured and turns to a dull gray over time with oxidation.

lead hole: See pilot hole.

leadscrew: The long, precision screw located in front of the lathe bed geared to the spindle and used for cutting threads. Also, the table screw on the universal milling machine when geared to the indexing head for helical milling.

LH: Left hand.

limits: The smallest and largest dimension which are tolerable (allowed).

lip of a drill: The sharp cutting edge on the end of a twist drill.

liquidus: The lowest temperature at which a metal or an alloy is completely liquid.

live center: See center, live.

loading: A condition caused by grinding the wrong material with a grinding wheel or using too heavy a grinding action. The spaces between the abrasive grit become filled with the material being ground and abrasive action is greatly reduced.

M

machinability: The degree of difficulty with which a metal may be machined; may be found in appropriate handbooks.

machine tool: A power-driven machine designed to bore, cut, drill or grind metal or other materials.

machining, finish: Machining a surface to give it the desired finish.

machining, rough (rough finishing): Removing excess stock (material) with a machine tool thus shaping it in preparation for finish machining.

machinist: A person who is skilled in the operation of machine tools. He must be able to plan his own procedures and have a knowledge of heat-treating principles.

magnesium: A lightweight, ductile metal similar to but lighter than aluminum.

magnetic chuck: A flat smooth-surfaced work holding device, or V-shaped block, which operates by magnetism to hold ferrous metal workpieces for grinding.

malleable: Capable of being extended or shaped by hammering, bending or rolling.

mandrel (for lathe or milling machine): A precision-made tapered shaft to support work for machining between centers on a lathe, grinder or milling machine.

manganese: A gray-white nonmagnetic metallic element resembling iron, except it is harder and more brittle. Manganese can be alloyed with iron, copper and nickel, for commercial alloys. In steel it increases hardness, strength, wear resistance and other properties. Manganese is also added to magnesium-aluminum alloys to improve corrosion resistance.

MAPP® gas: A trade name for a fuel gas methacetylene-propadiene, which burns hotter than propane.

martensite: A very hard, brittle microstructure of steel produced when steel is rapidly quenched after being transformed into austenite.

medium steel: Steel with a carbon content between 0.30 to 0.70%.

mesh: To engage, as the teeth between two gears.

metal: A class of chemical elements that are good conductors of heat and electricity, usually malleable, ductile, lustrous and more dense than other elemental substances.

metallic bond: The principal atomic bond that holds metals together.

metallurgy: The science explaining the properties, behavior and internal microscopic structure of metals.

methylacetylene propadiene: A family of alternative fuel gases that are mixtures of two or more gases (propane, butane, butadiene, methylacetylene and propadiene). Methylacetylene propadiene is used for oxyfuel cutting, heating, brazing and soldering.

mike: A term used for micrometer, or to measure with a micrometer.

micrometer, depth: A micrometer in which the spindle projects through a flat, accurately machined base used to measure the depth of holes or recesses.

micrometer, thread: A micrometer in which the spindle is ground to a point having a conical angle of 60º. The anvil, instead of being flat, has a 60-degree V-Shaped groove, which fits the thread.

microstructure: A term use to describe the structure of metals. Visual examination of etched metal surfaces and fractures reveal some configurations in etched patterns that relate to structure, but magnification of minute details yields considerably more information. Microstructures are examined with low-power magnifying glasses, optical microscopes or electron microscopes.

mild steel: Refer to carbon steel; also called low carbon steel.

mill: A milling machine; also, the act of performing an operation on the milling machine.

milling cutter: A cutting tool, generally cylindrical in shape used on a milling machine and operated essentially like a circular saw.

milling, climb: See climb milling; see face milling.

minor diameter: The smallest diameter of a screw thread. Also known as the root diameter.

modulus of elasticity: The ratio of stress to strain in material; also referred to as Young's modulus.

molybdenum: A hard, silver-white metal, a significant alloying element in producing engineering steels, corrosion resistant steels, tool steels and cast irons. Small amounts alloyed in steel promote uniform hardness and strength.

Monel®: A high tensile strength Ni-Cu alloy that exhibits high fatigue resistance in salt water, corrosive atmospheres and various acid and alkaline solutions. It is non-magnetic and spark-resistant.

Morse taper (MT): A self-holding standard taper largely used on small cutting tools such as drills, end mills and reamers and, on some machines, spindles in which these tools are used.

MSDS: Material Safety Data Sheet.

multiple-lead thread screw: A screw made of two or more threads to provide an increased lead with a specified pitch.

music wire: A high tensile strength steel wire used for making springs. Also called piano wire.

N

NC: National Coarse, a thread designation.

necking: Machining a groove or undercut in a shaft to permit mating parts to be screwed tightly against a shoulder or to provide clearance for the edge of a grinding wheel.

NF: National Fine, thread designation.

nickel: An alloying element which increases the strength, toughness and wear and corrosion resistance of steels.

nitrogen: A gaseous element that occurs freely in nature and constitutes about 78% of earth's atmosphere. Colorless, odorless and relatively inert, although it combines directly with magnesium, lithium and calcium when heated with them. Produced either by liquefaction and fractional distillation of air, or by heating a water solution of ammonium nitrate.

nominal pipe size (NPS): The size of pipe is identified by its *nominal pipe size.* For pipes between ⅛- and 12-inches nominal size, the outside diameter (OD) was originally selected so that the inside diameter was equal to the nominal size for pipes of standard wall thickness of the times. Today this is no longer true with the changes in metals and manufacturing processes, but the nominal size and standard OD continue in use in the trade.

noncorrosive flux: A soldering flux that in either its original or residual form does not chemically attack the base metal. It usually is composed of rosin-based materials.

nonferrous: Metal containing no iron, such as brass and aluminum.

normalizing: The process of heating a metal above a critical temperature and allowing it to cool slowly under room temperature conditions to obtain its softest state or to reduce internal stress and reduce distortion.

NPS: National Pipe (Thread) Straight, thread designation; it is non-tapered and often used on lighting fixtures.

NPT: National Pipe (Thread) Tapered, thread designation; forms a liquid-tight seal.

O

OD: Outside diameter.

off center: Not centered; offset, eccentric or inaccurate.

oil hardening: The process of quenching in oil when heat treating alloy steel to bring out certain qualities, such as hardness.

oilstones: Molded abrasives in various shapes used to hand-sharpen cutting tools. These stones can be used with a light oil to improve their cutting.

oxidizing flame: An oxyfuel flame in which there is an excess of oxygen, resulting in an oxygen-rich zone extending around and beyond the cone.

oxygen: A colorless, odorless, tasteless, gaseous chemical element, the most abundant of all elements. Oxygen occurs free in the atmosphere, forming 1/5 of its volume and in combination in water, sandstone, limestone, etc.; it is very active and able to combine with nearly all other elements. It is essential to most life forms.

P

pack hardening: A heat-treating process in which the workpiece is packed into a metal box together with charcoal, charred leather or other carbonaceous material to case-harden the part.

parallels: Hardened steel bars accurately ground to size and ordinarily made in pairs in many different sizes to support work in precision setups such as on a milling machine.

parting: The operation of cutting off a piece from a part held in the chuck of a lathe.

pattern: Wood, metal, paper or plastic sheet material that replicates the shape of a part. Patterns are used to transfer this shape to the work and may contain other information such as hole location, alignment marks and bend lines.

pd: pitch diameter of a thread or gear.

peen: To draw, bend or flatten, also, the formed side of a hammer opposite the face.

phase transitions: When metals or metal alloys go from solid to liquid or the reverse, this is a phase transition. Iron phase transitions are: at room temperature to 1,670°F (910°C) iron is body-center cubic, 1670°F (910°C) to 2535°F (1388°C) iron is face-center cubic and 2535°F (1390°C) the melting point of iron to 2800°F (1538°C) iron is again body-center cubic. These changes are also called allotropic transformations.

phosphoric acid: The acid, H_3PO_4, widely used in industrial metal cleaning.

phosphorous: A highly reactive, toxic, nonmetallic element used in steel, glass and pyrotechnics. It is almost always found in combination with other elements such as minerals or metal ores. Found in steel and cast iron as an impurity. In steel it is reduced to 0.05% or less otherwise phosphorous causes embrittlement and loss of toughness. However, small amounts in low-carbon steel produce a slight increase in strength and corrosion resistance.

pilot hole: A starting hole for large drills to serve as a guide, reduce the drilling force and aid in maintaining the accuracy of the larger hole. Also called a lead hole.

pilot: A guide at the end of a counterbore which keeps it aligned with the hole.

pin fixture: A tool for bending wire, bar or rod into a curve or series of curves.

pinning: A term used to describe the condition of a file clogged with metal filings causing it to scratch the work.

pitch diameter: The diameter of a thread at an imaginary point where the width of the groove and the width of the thread are equal.

pitch line: An imaginary line which passes through threads at such points that the length of the part of the line between adjacent threads is equal to the length of the line within a thread.

pitch: The distance from any point on a thread to the corresponding point on the adjacent thread, measured parallel to the axis.

plain cutter: A milling cutter with cutting teeth on the periphery (circumference) only.

play: The looseness of fit (slack) between two pieces.

plug weld: A weld made in a circular hole in one member of a joint fusing that member to another member.

powder coating: A durable, weather-proof, polymer coating for metals applied in a spray and then cured in an oven.

proportional limit: The greatest stress a material can withstand without deviation from the straight-line proportionality between stress and strain.

protective atmosphere: A gas or vacuum envelope surrounding the workpieces, used to prevent or reduce the formation of oxides and other detrimental surface substances and to facilitate their removal.

PTFE seal: A polyterafluroethylene polymer seal as a tape or paste; TEFLON® is the DuPont trademark for PTFE.

punch, center: A solid punch with a conical point used to make an indentation in a workpiece to get a drill started in the proper location. It is often used after the location is initially marked with a prick punch. A center punch has a more blunt point than a prick punch and is usually used after a prick punch.

punch, prick: A solid punch with a sharp conical point, used to mark centers or other locations on metal.

pyrometer: A device for measuring the high temperatures in a heat-treating furnace.

Q

quenching: The sudden cooling of heated metal by immersion in water, oil or other liquid. The purpose of quenching is to produce desired strength properties in hardenable steel.

quick return: A mechanism on some machine tools that provides rapid movement of the ram or table on the return or anointing stroke of the machine.

R

rack: An array of gear teeth in a straight line; a circular gear of infinite radius.

radial: In a direction directly outward from the center of a circle or sphere or from the axis of a cylinder. The spokes of a wheel, for example, are radial.

radius: The distance from the center of a circle to its circumference (outside); half the diameter of the circle.

rake: That surface of a cutting tool against which the chips bear while being severed. If this surface is less than 90° from the surface being cut, the rake is positive, if more, the rake is negative.

ram: The moving part of an arbor press or hydraulic press.

rapid traverse: A lever-controlled, power-operated feature of some lathes and mills that permits the rapid movement of the carriage or worktable from one position to another.

reaming, line: The process of reaming two or more holes to bring them into very accurate alignment.

recess: An internal groove. See undercut.

relief: A term for clearance or clearance angle.

residual stress: Stress present in a joint member or material that is free of external forces or thermal gradients.

RH: Right hand.

Rockwell hardness test: The most common hardness testing method. This procedure uses a minor load to prevent surface irregularities from affecting results. There are nine different Rockwell hardness tests corresponding to combinations of three penetrators and three loads.

root diameter: See minor diameter.

roughing: The fast removal of stock to reduce a workpiece to approximate dimensions, leaving only enough material to finish the part to specifications.

running fit: A class of fit intended for use on machinery with moderate speeds, where accurate location and minimum play are desired; a fit not too tight or too loose.

S

SAE: The Society of Automotive Engineers.

scale: The rough surface on hot finished steel and castings. Also, a shop term for steel rules and for the markings on a machine tool dial.

scraper: A hardened steel hand tool used to scrape surfaces very smooth by removing minute amounts of metal.

scribe (scratch awl): A steel rod 8- to 12-inches long and about 3/16 inches in diameter. It has a long, slender, hardened steel point on one or both ends.

sector: A device that has two radial, beveled arms which can be set to include any number of holes on the indexing plate of a dividing head to eliminate recounting the holes for each setting.

set screw: A plain screw used principally for locking adjustable parts in position; usually headless.

set: The bend or offset of a saw tooth to provide a clearance for the blade while cutting. Also, the permanent change in the form of metal as the result of repeated or excessive strain.

setup: The preparation of a machine tool to complete a specific operation. It includes mounting the workpiece and necessary tools and fixtures and selecting the proper speeds, feeds, depth of cut and coolants.

SF: Standard form.

shank: That part of a tool or similar object which connects the principal operating part to the handle, socket or chuck by which it is held or moved.

shear strength: The characteristic of a material to resist shear forces.

shims: Very thin sheets of metal made in precise thickness and used between parts to obtain desired spacing. Sometimes they are laminated, so layers can be pulled off to the desired depth or thickness.

shoulder: A term for the step made between two machined surfaces.

shrink fit: A class of fit made when the outer member is expanded by heating to fit over a shaft and then contracts or shrinks tightly to the shaft when cooled. Sometimes the inside part is cooled with refrigeration, dry ice or liquid nitrogen as well.

side cutter: A milling cutter that has cutting teeth on the side as well as on the periphery or circumference.

side rake: That surface which slopes to the side of the cutting edge. It may be positive or negative and is combined with the back rake. See rake.

sine bar: A precision instrument for laying out, setting, testing and otherwise dealing with angular work.

slabbing cutter: A wide, plain milling cutter having helical teeth. Used for producing large, flat surfaces.

slitting saw: A narrow circular milling cutter designed for cutoff operations or for cutting narrow slots.

slotter: An attachment which operates with a reciprocating motion. Used for machining internal slots and surfaces on a milling machine.

soft hammer: A hammer made of brass, copper, lead, leather or plastic so they will not mar finished surfaces on machines or workpieces.

soft jaws: Plastic, leather, lead or aluminum covers on the jaws of a vise or pliers used to prevent marking and damage to the work.

solder: The metal or alloy used as a filler metal in soldering, which has a liquidus not exceeding 840°F (450°C) and below the solidus of the base metal; usually lead- and tin-based.

spheroidizing: A stress relieving method of long-term heating of high-carbon steel at or near the lower transformation temperature, followed by slow cooling to room temperature which changes the crystalline structure.

spindle speed: The rpm at which a machine is set.

spindle: A rotating device widely used in machine tools such as lathes, milling machines, drill presses and so forth, to hold the cutting tools or the work and to give them their rotation.

spot facing: Finishing a bearing surface around the top of a hole.

spur gear: A gear with teeth parallel to the axis of the shaft on which it is mounted.

square surface: A surface at a right angle with another surface.

square threads: A thread having a depth, width and space between threads that are equal. It is used on heavy jack screws, vise screws and other similar items.

square, solid (toolmakers' try-square): A very accurate try-square in which a steel blade is set firmly into a solid, rectangular-shaped handle so that each edge of the blade makes an angle of exactly 90° with the inner face (side) of the handle.

steady rest: A support that keeps the work from flexing when machining a long workpiece in the lathe. The steady is clamped to the lathe bed and is sometimes called a center rest.

steel: A material composed primarily of iron, less than 2% carbon and (in an alloy steel) small percentages of other alloying elements.

step block: A fixture designed like a series of steps to provide support at various heights required for setups, usually on a milling machine.

stock: A term for the materials used to make parts in a machine tool. Also, the die stock used for threading dies.

stop: A device attached to a machine tool to limit the travel of the worktable and sometimes the work head.

straddle milling: A milling setup where two side-milling cutters are spaced on an arbor to machine two parallel surfaces with a single cut.

surface grinding: The process of grinding flat surfaces on a surface grinding machine. With special setups, angular and formed surfaces may also be ground.

surface plate: An accurately flat surface made of hard granite or cast iron used to check the flatness of surfaces and when performing layout operations.

swing: The dimension of a lathe determined by the maximum diameter of the work that can be rotated over the ways of the bed.

T

tailstock: That part of a machine tool, such as a lathe or cylindrical grinder, which supports the end of a workpiece with a center. It may be positioned at any point along the ways of the bed and may be offset from center to machine tapers. Also used with a dividing head.

tang: The flat on the shank of a cutting tool, such as a drill, reamer or end mill, that fits a slot in the spindle of a machine to keep the tool from slipping. Also, the part of a file that fits into a handle.

tap: A tool used to cut or form threads on the inside of a round hole.

taper: A uniform increase or decrease in the size or diameter of a workpiece.

tapping: The process of cutting or forming screw threads in a round hole with a tap.

T-bolt: Term for the bolts inserted in the T-slots of a worktable to fasten the workpiece or work-holding device to the table as on a milling machine.

tempering: A heat-treating process to relieve the stresses produced when hardening and to impart certain qualities, such as toughness; sometimes called "drawing."

template: A pattern or guide for laying out or machining to a specific shape or form.

tensile strength: The resistance at the breaking point exhibited by a material when subjected to a pulling stress. Measured in lb/in^2 or kPa.

thermal expansion: The expansion of materials caused by heat input.

thermal stress relieving: A process of relieving stresses by uniform heating of a structure or a portion of a structure, followed by uniform slow cooling.

thread axis: A line running lengthwise through the center of the screw.

thread crest: The top surface joining the two sides of a thread.

thread depth: The distance between the crest and the root of a thread.

thread pitch diameter: The diameter of a screw thread measured from the thread pitch line on one side to the thread pitch line on the opposite side.

thread pitch: The distance from a point on one screw thread to a corresponding point on the next thread.

thread root: The bottom surface joining the sides of two adjacent threads.

thread: A helical projection of uniform section on the internal or external surface of a cylinder or cone. Also, the operation of cutting a screw thread.

tool steel: A general classification for high-carbon steel that can be heat treated to a hardness required for metal cutting tools such as punches, dies, drills, taps, reamers, and so forth.

torsion: The stress produced in a body, such as a rod or wire, by turning or twisting one end while the other is held firm or twisting in the opposite direction.

tpf: Taper per foot.

tpi: Taper per inch.

traverse: One movement across the surface of the work being machined.

truing: The act of centering or aligning a workpiece or cutting tool so that an operation may be performed accurately. Also, correcting the eccentricity or out of round condition when dressing a grinding wheel.

T-slot: The slots made in the tables of machine tools for the square-head bolts used to clamp the workpiece, attachments, or work-holding fixtures in position for performing the machining operations.

tumbler gears: A pair of small lever-mounted gears on a lathe used to engage or to change the direction of the leadscrew.

two-lip end mill: An end-milling cutter designed with teeth that cut to the center so that it may be used to feed into the work like a drill.

U

ultimate tensile strength (UTS): The maximum tensile stress a material placed in tension can bear without breaking.

UNC: Unified National Coarse, a thread pitch designation.

UNF: Unified National Fine, a thread pitch designation.

universal milling machine: A milling machine with a worktable that can be swiveled for milling helical work.

universal vise: A vise designed for holding work at a double or compound angle. Also called a toolmakers' vise.

V

volt: A unit of electrical force or potential.

W

ways: The flat or V-shaped bearing surfaces on a machining tool that guide and align the parts which they support.

wheel dresser: A tool or device for dressing or truing a grinding wheel.

work hardening: Also called cold working; the process of forming, bending or hammering a metal well below the melting point to improve strength and hardness. Also, the increase in the hardness of most metals when they are deformed at room temperature.

Y

yield strength: The stress at the uppermost point on the straight-line portion of the stress-strain curve. Stress imposed on the sample below this level produces no permanent lengthening and stress can vary from zero up to the yield strength. Stress above yield strength causes permanent deformation lengthening.

Index

Credits

Albrecht Inc., Figure 5–29 (right drawing only).
Cameron Micro Drill Presses, Figure 5–1.
Clausing Industrial, Inc., Figures 5–2, 7–5, 7–6, 7–7, 7–8, 7–13, 7–14, 7–42, 7–87, 7–88, 7–136, 7–137, 7–153, 7–164, 8–8.
Cooper Hand Tools manufacturer of Nicholson® files, rasps, saws and rotary burrs, Figures 3–1, 3–2, 3–3, 3–5, 3–7, 3–10, 3–11, 3–12.
Robert J. DeVoe, Tables A–2, A–3, A–4.
DRIV-LOK, Inc., Figure 10–23.
Emhart Industries Inc., Figure 9–22.
Hardinge Inc. (Bridgeport Milling Machines), Figures 8–2, 8–3, 8–4, 8–5, 8–12, 8–13, 8–14.
Helical Lap & Mfg. Company, Figure 4–17.
Holdridge Mfg. Co. Inc., Figure 7–84 (right drawing only), 7–85.
Louis Levin & Sons, Inc., Figures 7–2, 7–70, 7–71.
Machinery's Handbook, Figure 8–65, Table 8–8.
Mil-Spec Fasteners Corporation, Figure 9–5.
Myford, Ltd., Figure 7–4.
Phase II Machine & Tool, Inc., Figure 8–77.
David Randal, Figures 5–14, 5–43, 12–1, 12–2, 12–3, 12–4, 12–5, 12–6, 12–7, 12–8, 12–9, 12–10, 12–11, 12–14, 12–18, 12–19, 12–22, 12–23, A–1, Table A–1.
Sherline Products, Inc., Figures 7–3, 7–15 (right), 7–22 (left), 7–69, 7–72, 7–82 (top), 7–84 (top), 7–86, 8–1, 8–21, 8–22, 8–33, 8–35 (bottom), 8–37, Table 8–5, 8–46, 8–52, 8–55, 8–69, 8–79, 8–73 8–74, 8–80 (Craig Libuse), 8–81, 8–83 (Craig Libuse).
The L.S. Starrett Company, Figures 1–1, 1–4, 1–5, 1–6, 1–7, 1–8, 1–13, 1–15, 1–16, 1–17, 1–18, 1–19, 1–20, 1–21, 2–1, 6–7, 7–170, 8–66, 10–8, Table 10–4.
Vise-Grip® – American Tool Companies, Inc., Figure 2–13, page 33 (two upper drawings).
Welding Essentials: Questions & Answers, Figures 10–1, 10–2, 10–3, 10–5, 10–10.